Goutam Brahmachari

**Handbook of
Pharmaceutical Natural
Products**
Volume 2

Related Titles

G. Eisenbrand, W. Tang

Handbook of Chinese Medicinal Plants

Chemistry, Pharmacology, Toxicology

2010

ISBN: 978-3-527-32226-8

D.G. Barceloux

Medical Toxicology of Natural Substances

Foods, Fungi, Medicinal Herbs, Plants, and Venomous Animals

2008

ISBN: 978-0-471-72761-3

M. Negwer, H.-G. Scharnow

Organic-Chemical Drugs and Their Synonyms

7 Volume Set

2007

ISBN: 978-3-527-31939-8

O. Kayser, W.J. Quax (Eds.)

Medicinal Plant Biotechnology

From Basic Research to Industrial Applications

2007

ISBN: 978-3-527-31443-0

A. Ahmad, F. Aqil, M. Owais (Eds.)

Modern Phytomedicine

Turning Medicinal Plants into Drugs

2006

ISBN: 978-3-527-31530-7

X.-T. Liang, W.-S. Fang (Eds.)

Medicinal Chemistry of Bioactive Natural Products

2006

ISBN: 978-0-471-66007-1

Goutam Brahmachari

Handbook of Pharmaceutical Natural Products

Volume 2

WILEY-VCH

WILEY-VCH Verlag GmbH & Co. KGaA

The Author

Dr. Goutam Brahmachari
Visva-Bharati University
Department of Chemistry
Santiniketan 731 235
India

All books published by Wiley-VCH are carefully produced. Nevertheless, authors, editors, and publisher do not warrant the information contained in these books, including this book, to be free of errors. Readers are advised to keep in mind that statements, data, illustrations, procedural details or other items may inadvertently be inaccurate.

Library of Congress Card No.: applied for

British Library Cataloguing-in-Publication Data
A catalogue record for this book is available from the British Library.

Bibliographic information published by the Deutsche Nationalbibliothek
The Deutsche Nationalbibliothek lists this publication in the Deutsche Nationalbibliografie; detailed bibliographic data are available on the Internet at http://dnb.d-nb.de.

© 2010 WILEY-VCH Verlag GmbH & Co. KGaA, Weinheim

All rights reserved (including those of translation into other languages). No part of this book may be reproduced in any form – by photoprinting, microfilm, or any other means – nor transmitted or translated into a machine language without written permission from the publishers. Registered names, trademarks, etc. used in this book, even when not specifically marked as such, are not to be considered unprotected by law.

Cover Design Formgeber, Eppelheim
Typesetting Thomson Digital, Noida, India
Printing and Binding T.J. International Ltd. Padstow

Printed in Great Britain
Printed on acid-free paper

ISBN: 978-3-527-32148-3

Dedicated To My Little Asanjan

Preface

Nature stands as an inexhaustible source of novel chemotypes and pharmacophores. There has been a history of success in developing drugs from natural sources, particularly in tropical countries such as India, China, Japan, Nepal, Mexico, and South Africa. Nature has been a source of medicinal agents for thousands of years, and an impressive number of modern drugs find their origin in nature. Although the use of bioactive natural products or herbal drug preparations dates back a long time ago, their application as isolated and characterized compounds to modern drug discovery and development started only in the nineteenth century, the dawn of the chemotherapy era. Natural product chemistry has now experienced an explosive and diversified growth, making natural products the subject of much interest and promise in the present-day research directed toward drug design and drug discovery. Natural products and their derivatives from plant, microbial, and marine sources are at various advanced stages of clinical development.

Owing to multidirectional promising aspects, the interest in natural products continues to this very day. The last decade has seen a greater use of botanical products among members of the general public through self-medication than ever before. The use of herbal drugs is once more escalating in the form of complementary and alternative medicine (CAM). This phenomenon has been mirrored by an increasing attention to phytomedicines as a form of alterative therapy by the health professionals; in many developing countries, there is still a major reliance on crude drug preparation of plants used in traditional medicines for their primary health care. The World Health Organization (WHO) estimates that approximately 80% of the world's population relies mainly on traditional medicine, predominantly originated from plants, for their primary health care. The worldwide economic impact of herbal remedies is noteworthy; in the United States alone, in 1997 it was estimated that 12.1% of the population spent $5.1 billion on herbal remedies. In the United Kingdom, sales of herbal remedies were worth of £75 million in 2002, an increase of 57% over the previous 5 years. Studies carried out in other countries, such as Australia and Italy, also suggest an increasing prevalence of the use of herbal medicines among the adult population. In India and China, respectively, the Ayurvedic and Chinese traditional medicine systems are particularly well developed, and both have provided potentials for the development of Western medicine. Pharmacognosists employed in different institutions are aware of the changing trends of herbal medications and a number of useful texts on the analysis, uses, and potential toxicities of herbal remedies have appeared recently, which serve as useful guides in pharmacy practice.

Medicinal chemistry of bioactive natural products spans a wide range of fields, including isolation and characterization of bioactive compounds from natural sources, structure modification for optimization of their activity and other physical properties, and also total and semisynthesis for a thorough scrutiny of structure–activity relationship (SAR). It has been well documented that natural products played crucial roles in modern drug development, especially for antibacterial and antitumor agents; however, their use in the treatment of other epidemics such as AIDS, cardiovascular, cancerous, neurodegradative, infective, and metabolic diseases has also been extensively explored. The need for leads to solve such health problems threatening the world population makes all natural sources important for the search of novel molecules. The development of separation techniques and spectroscopic methods allows the isolation of complex mixtures and the characterization of a diversity of complex structures, contributing to the importance of the investigation of terrestrial and marine sources in order to obtain novel bioactive organic compounds coming from nature. Such diversified structural architectures of the isolated molecules presented scientists with unique chemical structures, which are beyond human imagination most of the time, inspired scientists to pursue new chemical entities with completely different structures from known drugs.

The most striking feature of natural products in connection to their long-lasting importance in drug discovery is, thus, their structural diversity that is still largely untapped. Most natural products not only are sterically more complex than synthetic compounds but also differ in regard to the statistical distribution of functionalities. They occupy a much larger volume of the chemical space and display a broader dispersion of structural and physicochemical properties than compounds issued from combinatorial synthesis. It needs to be mentioned that in spite of massive endeavors adopted in recent times for synthesizing complex structures following "diversity-oriented synthesis" (DOS) strategy, about 40% of the chemical scaffolds found in natural products are still absent in today's medicinal chemistry. The chemical diversity and unique biological activities of a wide variety of natural products have propelled many discoveries in chemical and biological sciences, and provided therapeutic agents to treat various diseases as well as offered leads for the development of valuable medicines. Analysis of the properties of synthetic and natural compounds compared to drugs revealed the distinctiveness of natural compounds, especially concerning the diversity of scaffolds and the large number of chiral centers. This may be one reason why approximately 50% of the drugs introduced to the market during the past 20 years are directly or indirectly derived from natural compounds.

The reason for the lack of lead compounds from synthetic libraries in some therapeutic areas such as anti-infectives, immunosuppression, oncology, and metabolic diseases may, thus, be attributed to the different chemical space occupied by natural products and synthetic compounds. This difference in chemical space makes natural products an attractive alternative to synthetic libraries, especially in therapeutic areas that have a dearth of lead compounds. Natural products have also been used as starting templates in the synthesis of combinatorial libraries. Natural product pharmacophores are well represented in lists of "privileged structures," which makes them ideal candidates for building blocks for biologically relevant chemical libraries. Natural products still constitute a prolific source of novel lead compounds or pharmacophores for medicinal chemistry, and hence, they should be incorporated into a well-balanced drug discovery program. Besides their potential as lead structures in drug discovery, natural products also provide attractive scaffolds for combinatorial synthesis and act as indispensable tools for validation of new drug targets. The diversity of three-dimensional shapes of natural molecules still surpasses that of synthetic compounds, and this ensures that natural products will continue to be important for drug discovery.

The wide range of nature virtually remains unexplored; it is estimated that only 5–15% of the approximately 2 50 000 species of higher plants (terrestrial flora) have been investigated chemically and pharmacologically so far. The marine environment has become an important source of new structures with new activities; hence, marine kingdom stands as an enormous resource for the discovery of potential chemotherapeutics and is waiting to be explored. Another vast untapped area is the microbial world – less than 1% of bacterial species and less than 5% of fungal species are known – and recent evidence indicates that millions of microbial species remain undiscovered. Microbial sources are making an increasingly important contribution to bioactive natural products, and the complex structures of such microbial natural products have fascinated chemists for decades. The future of natural products in drug development thus appears to be a tale of justifiable hope. Faithful drives are needed in more intensified fashion to explore nature as a source of novel and active agents that may serve as the leads and scaffolds for elaboration into urgently needed efficacious drugs for a multitude of disease indications.

The *Handbook of Pharmaceutical Natural Products* provides a much needed and comprehensive survey of bioactive natural products and their potentials as "drug candidates" for prospective use of such significant molecules in the pharmaceutical world; more than 1500 such individual molecules have been selected and discussed under a total of 950 entries distributed in two volumes of this book. Systematic and trivial names, physical data, source(s), structure, natural derivative(s), and pharmaceutical potentials with an emphasis on the structure–activity relationship of each bioactive molecule are presented in this book; hence, the book would serve as a key reference for recent developments in the frontier research on natural products and would also find much utility to the scientists working in this area. The book serves as a valuable resource for researchers in their own fields to predict promising leads for developing pharmaceuticals to treat various ailments and disease manifestations.

I would like to express my deep sense of appreciation to all of the editorial and publishing staff members of Wiley-VCH, Weinheim, Germany, for their all-round help so as to ensure that the highest standards of publication are maintained in bringing out this book. My effort will be successful only when it is found helpful to the readers at large. Every step has been taken to make the manuscript error-free; in spite of that, some errors might have crept in. Any remaining error is, of course, of my own. Constructive comments on the contents and approach of the book from the readers will be highly appreciated.

Finally, I should thank my wife and my son for their well understanding and allowing me enough time throughout the entire period of writing; without their support, this work would not have been possible.

Santiniketan, September 2009　　　　　　　　　　　　　　　　*Goutam Brahmachari*

Further Reading

Balandrin, M.F., Kinghorn, A.D., and Farnsworth, N.R. (1993) Plant-derived natural products in drug discovery and development: an overview, in *Human Medicinal Agents from Plants, American Chemical Society Symposium Series, No. 534* (eds A.D. Kinghorn and M.F. Balandrin), American Chemical Society, Washington, DC, pp. 2–12.

Barnes, J. (2003) Pharmacovigilance: a UK perspective. *Drug Safety*, **26**, 829.

Brahmachari, G. (2006) Prospects of natural products research in the 21st century: a sketch, in *Chemistry of Natural Products: Recent Trends & Developments* (ed. G. Brahmachari), Research Signpost, Kerala, India, pp. 1–22.

Brahmachari, G. (2009) Mother Nature: an inexhaustible source of drugs and lead molecules, in *Natural Products: Chemistry, Biochemistry and Pharmacology* (ed. G. Brahmachari), Alpha Science International, Oxford, UK, pp. 1–20.

Burke, M.D., Berger, E.M., and Schreiber, S.L. (2003) Generating diverse skeletons of small molecules combinatorially. *Science*, **302**, 613.

Buss, A.D. and Waigh, R.D. (1995) Natural products as leads for new pharmaceuticals, in *Burger's Medicinal Chemistry and Drug Discovery*, 5th edn, vol. 1 (ed. M.E. Wolff), John Wiley & Sons, Inc., New York, pp. 983–1033.

Butler, M.S. (2004) The role of natural product chemistry in drug discovery. *J. Nat. Prod.*, **67**, 2141.

Carte, B.K. (1996) Biomedical potential of marine natural products. *Bioscience*, **46**, 271.

Cragg, G.M. and Newman, D.J. (2001) Medicinals for the millennia: the historical record. *Ann. N.Y. Acad. Sci.*, **953**, 3.

Cragg, G.M., Newman, D.J., and Snader, K.M. (1997) Natural products in drug discovery and development. *J. Nat. Prod.*, **60**, 52.

Donia, M. and Hamann, M.T. (2003) Marine natural products and their potential applications as anti-infective agents. *Lancet Infect. Dis.*, **3**, 338.

Haefner, B. (2003) Drugs from the deep: marine natural products as drug candidates. *Drug Discov. Today*, **8**, 536.

Horton, D.A., Bourne, G.T., and Smythe, M.L. (2003) The combinatorial synthesis of bicyclic privileged structures or privileged substructures. *Chem. Rev.*, **103**, 893.

Kaul, P.N. and Joshi, B.S. (2001) Alternative medicine: herbal drugs and their critical appraisal – part II. *Prog. Drug Res.*, **57**, 1.

Kingston, D.G.I. and Newman, D.J. (2002) Mother nature's combinatorial libraries: their influence on the synthesis of drugs. *Curr. Opin. Drug Discov. Dev.*, **5**, 304.

Lee, M.-L. and Schneider, G.J. (2001) Scaffold architecture and pharmacophoric properties of natural products and trade drugs: application in the design of natural product-based combinatorial libraries. *Combin. Chem.*, **3**, 284.

Newman, D.J., Cragg, G.M., and Snader, K.M. (2000) The influence of natural products upon drug discovery. *Nat. Prod. Rep.*, **17**, 215.

Newman, D.J. and Cragg, G.M. (2004) Advanced preclinical and clinical trials of natural products and related compounds from marine sources. *Curr. Med. Chem.*, **11**, 1693.

Newman, D.J., Cragg, G.M., and Snader, K.M. (1997) Natural products in drug discovery and development. *J. Nat. Prod.*, **60**, 52.

Newman, D.J., Cragg, G.M., and Snader, K.M. (2000) The influence of natural products on drug discovery. *Nat. Prod. Rep.*, **17**, 215.

Newman, D.J., Cragg, G.M., and Snader, K.M. (2003) Natural products as sources of new drugs over the period 1981–2002. *J. Nat. Prod.*, **66**, 1022.

Shu, Y.-Z. (1998) Recent natural products based drug development: a pharmaceutical industry perspective. *J. Nat. Prod.*, **61** 1053.

Vuorelaa, P., Leinonenb, M., Saikkuc, P., Tammelaa, P., Rauhad, J.P., Wennberge, T., and Vuorela, H. (2004) Natural products in the process of finding new drug candidates. *Curr. Med. Chem.*, **11**, 1375.

Abbreviations

A-549	human lung carcinoma
AA	arachidonic acid
ABTS$^{•+}$	2,2′-azinobis(3-ethylbenzothiozoline-6-sulfonate) radical cation
ACAT	acetyl-CoA:cholesterol acyltransferase
AChE	acetylcholinesterase
ACV	acyclovir
AD	Alzheimer's disease
AIDS	acquired immune deficiency syndrome
ALT	alanine aminotransferase
AP-1	activator protein
Apr	acyclovir/phosphonoacetic acid-resistant
AST	serum aspartate aminotransferase
ATP	adenosine triphosphate
AZT	3′-azido-3′-deoxythymidine
BChE	butyrylcholinesterase
BCS	bovine calf serum
BHA	2,6-di-*tert*-butyl-4-hydroxyanisol
BHT	2,6-di-*tert*-butyl-4-methoxyphenol
BSO	buthionine sulfoximine
Caco-2	human colon carcinoma
Caspase	cysteine proteases
CC_{50}	50% cytotoxic concentration
Cdk	cyclin-dependent kinase
CL	chemoluminescence
Col-2	human colon carcinoma
COX	cyclooxygenase
COX-1	cyclooxygenase-1
COX-2	cyclooxygenase-2
CPE	cytopathic effect
DHF	dihydroxyfumaric acid
DNA	deoxyribose nucleic acid
DOPA	2-amino-3-(3′, 4′-dihydroxyphenyl)propionic acid
DPPH	1,1-diphenyl-2-picrylhydrazyl radical

Handbook of Pharmaceutical Natural Products, Volume 2.
Goutam Brahmachari
Copyright © 2010 WILEY-VCH Verlag GmbH & Co. KGaA, Weinheim
ISBN: 978-3-527-32148-3

EC_{50}	effective concentration (50%)
ED_{50}	effective dose (50%)
EGP-R PTK	epidermal growth factor receptor protein tyrosine kinase
Egr-1	early growth response gene-1
EMSA	electrophoretic mobility shift
FAS	fungal fatty acid synthase
fMLP	formyl-methionyl-leucyl-phenylalanine
5-FU	5-fluorouracil
GalN	D-galactosamine
GGTase I	geranylgeranyltransferase type I
GI_{50}	concentration inhibiting cell growth by 50%
GSK-3	glycogen synthase kinase-3
GST	glutathione S-transferase
HBeAg	hepatitis B virus e antigen
HBsAg	hepatitis B virus surface antigen
HBV	hepatitis B virus
hCMV	human cytomegalovirus
HCT-8	human ileocecal carcinoma
HCT-116	human colon tumor cells
HCV	hepatitis C virus
HDAC	histone deacetylase
HDM	house dust mite
HEMn	human epidermal melanocytes
Hep-G2	human hepatocellular carcinoma
β-HEX	β-hexosaminidase
12-HHTrE	12-hydroxyheptadecatrienoic acid
HIF-1	hypoxia-inducible factor-1
HIV	human immunodeficiency virus
HIV-1/2-RT	human immunodeficiency virus type-1/2 reverse transcriptase
HLE	human leucocyte elastase
HNE	human neutrophil elastase
HPC	human prostate cancer
HPLF	human periodontal ligament fibroblasts
HPV	human papilloma virus
HSV	herpes simplex virus
HSV-1	herpes simplex virus type 1
HSV-2	herpes simplex virus type 2
HUVEC	human umbilical vein endothelial cell
HUVEC	human umbilical venous endothelial cell
IBMX	3-isobutyl-1-methylxanthine
IC_{50}	inhibitory concentration (50%)
ICE	interleukin-1β converting enzyme

ICM-1	intercellular adhesion molecule-1
IFN-γ	interferon-γ
IKK	IκB kinase
IL-6	interleukin-6
iNOS	inducible nitric oxide synthase
IκB	inhibitory subunit of NF-κB
JNK	c-Jun NH_2-terminal kinase
K-562	human chronic myelogenous leukaemia cell
KB	human oral epidermoid carcinoma cell
L1210	lymphocytic murine leukaemia cell
L5178	mouse lymphoma
LD_{50}	lethal dose (50%) concentration
L-NAMA	N^{ω}-monomethyl-L-arginine
LoVo/Dx	human colon adenocarcinoma
LOX	lipoxygenase
LPS	lipopolysaccharide
LRSA	linezolid-resistant methicillin-resistant *Staphylococcus aureus*
LTB_4	leukotriene B_4
Lu-1	human lung carcinoma
MAPK	mitogen-activated protein kinase
MCF-7	human breast adenocarcinoma
MDA	malondialdehyde
MDR	multiple-drug resistant
MIC	minimum inhibitory concentration
MLC	minimum lethal concentration
MLCR	mixed lymphocyte culture reaction
MMP	matrix metalloproteinase
MRSA	methicillin-resistant *Staphylococcus aureus*
MRSE	methicillin-resistant *Staphylococcus epidermidis*
MSSA	methicillin-susceptible *Staphylococcus aureus*
MSSE	methicillin-susceptible *Staphylococcus epidermidis*
MTT	3-(4,5-dimethylthiazol-2-yl)-2,5-diphenyl-2H-tetrazolium
NAC	N-acetyl-L-cysteine
NADPH	nicotinamide adenine dinucleotide phosphate (reduced)
NBT	nitroblue tetrazolium
NCI	National Cancer Institute
NCI-H187	human lung cancer cells
NFAT	nuclear factor of activated T-cells
NF-κB	nuclear factor kappaB
NGF	nerve growth factor
NIDDM	noninsulin-dependent diabetes mellitus
NIK	NF-κB-inducing kinase
NO	nitric oxide
NQO1	NAD(P)H:quinine oxidoreductase

NSCLC	non-small-cell lung cancer
6-OHDA	6-hydroxydopamine
OVCAR-8	ovarian cell line
P-388	lymphoid murine leukaemia cell
PA	phosphatidic acid
PAA	phosphonoacetic acid
PAF	platelet-activity factor
PARP-1	poly(ADP-ribose)polymerase-1
PC-3	human prostrate cancer cell
PDF	peptide deformylase
PEP	prolyl endopeptidase
PGE_2	prostaglandin E_2
PKC	protein kinase C
PLC	phospholipase C
PMA	phorbol 12-myristate 13-acetate
PP-1	protein phosphatase 1
PP-2A	protein phosphatase 2A
PRSP	penicillin-resistant *Streptococcus pneumoniae*
PSSP	penicillin-susceptible *Streptococcus pneumoniae*
PTK	protein tyrosin kinase
PTPase	protein tyrosine phosphatase
QRSA	quinolone-resistant *Staphylococcus aureus*
Quin-R	quinolone-resistant *Streptococcus pneumoniae*
REV	regulation of virion expression
RocB	didesmethyl-rocaglamide B
ROS	reactive oxygen species
RSV	respiratory syncytial virus
SAR	structural activity relationships
SI	selectivity index
SK-MEL-5	human malignant melanoma cell line
SPA	scintillation proximity assay
$sPLA_2$	secretory phospholipase A_2
SSAR	succinic semialdehyde reductase
STZ	streptozotocin
SV40	transformed fibroblast cells
TEAC	trolox equivalent antioxidant capacity
TI	therapeutic index
TMV	tobacco mosaic virus
TNF-α	tumor necrosis factor-alpha
TPA	12-O-tetradecanoylphorbol-13-acetate
TRAF	tumor necrosis factor (TNF) receptor-associated factor
TRAIL	TNF-related apoptosis-inducing ligand
Trolox	6-hydroxy-2,5,7,8-tetramethylchroman-2-carboxylic acid
TRP-2	tyrosinase-related protein-2
TSP-1	throbospondin-1

TXB$_2$	thromboxane B$_2$
U46619	9,11-dideoxy-11α,9α-epoxymethanoprostagrandin
VCAM-1	vascular cell adhesion molecule-1
VEGF	vascular endothelial growth factor
VISA	vancomycin-intermediate *Staphylococcus aureus*
VRE	vancomycin-resistant *Enterococcus*

Contents

Preface *VII*

Abbreviations *XIII*

Chapters	Entries	Pages
A	Abacopterins A–D — Azoxybacilin	*1*
B	Ballotenic acid and ballodiolic acid — 7-*n*-Butyl-6,8-dihydroxy-3(*R*)-pent-11-enylisochroman-1-one	*67*
C	Cadiyenol — Cytotrienin	*107*
D	Daedalin A — Dzununcanone	*207*
E	Eckol — Exiguaflavanones A and B	*271*
F	F390 — Fuzanin D	*315*
G	Gaboroquinone A — Gypsosaponins A–C	*341*
H	*Halidrys* monoditerpene — *Hyrtios* sesterterpenes 1–3	*373*
I	IC202A, B, and C — Ixerochinolide	*417*
J	Jacarelhyperols A and B — Juglanins A and B	*439*
K	Kadlongilactones A and B — Kweichowenol B	*449*
L	Laccaridiones A and B — Lyratols C and D	*465*
M	Maackiaflavanones A and B — Myrsinoic acids A, B, C, and F	*507*
N	Nalanthalide — Nymphaeols A–C and isonymphaeol B	*589*
O	Oblonganoside A — Oxypeucedanin hydrate acetonide	*611*
P	Pacificins C and H — 6- and 8-(2-Pyrrolidinone-5-yl)-(−)-epicatechin	*631*
Q	Quassimarin — Quillaic acid glycosidic ester	*689*
R	Ravenic acid — Rubrisandrin A	*697*
S	Salaspermic acid — Syncarpamide	*715*

Handbook of Pharmaceutical Natural Products, Volume 2.
Goutam Brahmachari
Copyright © 2010 WILEY-VCH Verlag GmbH & Co. KGaA, Weinheim
ISBN: 978-3-527-32148-3

Chapters	Entries	Pages
T	Tabucapsanolide A — Tungtungmadic acid	787
U	UCS1025 A — Uvaretin	837
V	Vaccihein A — *Vitex* norditerpenoids 1 and 2	843
W	Weigelic acid — Wrightiamine A	857
X	Xanthohumols — Xylocarpin J	863
Y	Yahyaxanthone — Yuccalan	867
Z	Zaluzanin D — Zinolol	873

Tables — 879

A.1	Alkaloids	880
A.2	Antibiotics	884
A.3	Flavonoids	890
A.4	Lignoids	896
A.5	Terpenoids	897
A.6	Xanthonoids	911
A.7	Miscellaneous	914

L

Laccaridiones A and B

Physical data:
- Laccaridione A: $C_{22}H_{24}O_6$, deep reddish solid, mp 164–165 °C
- Laccaridione B: $C_{23}H_{26}O_6$, deep reddish solid, mp 178–179 °C

Structures:

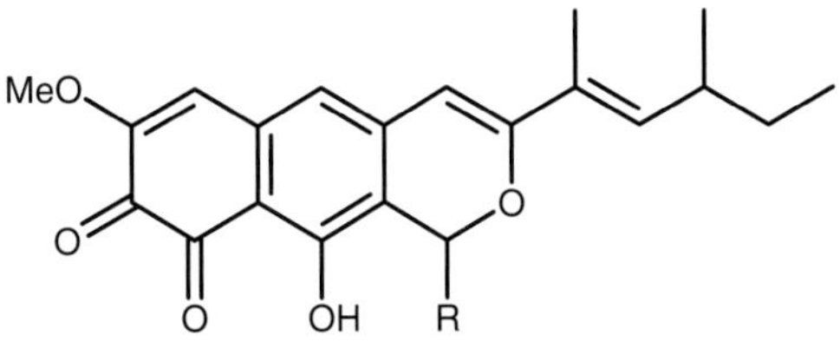

Laccaridione A: R = OCH$_3$
Laccaridione B: R = OCH$_2$CH$_3$

Compound class: *ortho*-Naphthoquinone derivatives

Source: *Laccaria amethystea*

Pharmaceutical potentials: *Protease inhibitors; antiproliferative*
Both of the metabolites were evaluated as potent inhibitors of a series of proteases such as trypsin, papain, thermolysin, collagenase, and zinc protease with respective IC$_{50}$ values of 14.7, 2.5, 18.8, 7.2, and 18.2 µg/ml (for laccaridione A) and 10.9, 5.1, 8.4, 5.7, and 3.0 µg/ml (for laccaridione B). Furthermore, laccaridione B showed strong antiproliferative effect on the murine fibroblast cell line L-929 (IC$_{50}$ = 2.4 µg/ml) and also on the human leukemia cell line K-562 (IC$_{50}$ = 1.8 µg/ml); however, the cytotoxic effect on HeLa was moderated with an IC$_{50}$ value of 13.9 µg/ml.

Reference

Berg, A., Reiber, K., Dorfelt, H., Walther, G., Schlegel, B., and Grafe, U. (2000) *J. Antibiot.*, **53**, 1313.

Lactonamycin

Physical data: $C_{28}H_{27}NO_{12}$, pale yellow powder, mp 168–171 °C, $[\alpha]_D^{25}$ +34° (MeOH, c 0.27)

Lactonamycin

Structure:

Lactonamycin

Compound class: Antibiotic

Source: *Streptomyces rishiriensis* MJ773-88K4 [1–3]

Pharmaceutical potentials: *Antimicrobial; cytotoxic*

Lactonamycin exhibited potent antimicrobial activity against various Gram-positive bacteria such as *Staphylococcus aureus* FDA209P (MIC = 0.39 μg/ml), *S. aureus* Smith (MIC = 0.39 μg/ml), *S. aureus* MS9610 (MIC = 0.78 μg/ml), *Micrococcus luteus* FDA16 (MIC = 0.78 μg/ml), *Bacillus anthracis* (MIC = 0.39 μg/ml), *B. subtilis* NRRL B-558 (MIC = 0.39 μg/ml), *B. cereus* ATCC10702 (MIC = 0.20 μg/ml), and *Corynebacterium bovis* 1810 (MIC = 0.78 μg/ml), while the antibiotic did not show antimicrobial activity against Gram-negative bacteria [1, 2]. The antimicrobial activities against *Escherichia coli, Shigella dysenteriae, Salmonella typhi, Proteus vulgaris, Providencia rettgeri, Serratia marcescens, Pseudomonas aeruginosa, Klebsiella pneumoniae, Mycobacterium smegmatis*, and *Candida albicans* were found to be weak (MIC > 100 μg/ml).

The antibiotic demonstrated effective potency against various *Streptococcus* species including streptomycin-resistant strains, tetracycline-resistant strains, and clinically isolated strains – the MIC values against *Streptococcus faecalis* 37787, *S. pyogenes* Cook, *S. pyogenes* group A St-92TC, *S. pyogenes* group A St-107TC, *S. pyogenes* group A St-108TC, *S. pyogenes* group A St-56-188SM, *S. pyogenes* TY-5727, *S. pyogenes* TY-5740, *S. pyogenes* TY-5914, *S. pyogenes* TY-5708, *S. pyogenes* TY-5745, *S. pyogenes* TY-5834, and *S. pyogenes* TY-5840 were determined to be 0.39, 0.39, 0.78, 0.78, 0.39, 0.39, 0.20, 0.78, 0.39, >100, 0.78, 0.78, and 0.39 μg/ml, respectively [2].

The present investigators [2] also evaluated that the test compound displayed potent antimicrobial activities against clinically isolated methicillin-resistant *Staphylococcus aureus* (MRSA) – the MIC values against the strains, *Staphylococcus aureus* TY-00930, *S. aureus* TY-00932, *S. aureus* TY-00933, *S. aureus* TY-00934, *S. aureus* TY-00936, *S. aureus* TY-01022, *S. aureus* TY-01033, *S. aureus* TY-01058, *S. aureus* TY-01759, *S. aureus* TY-01760, *S. aureus* TY-01796, *S. aureus* TY-01798, *S. aureus* TY-01800, *S. aureus* TY-01806, *S. aureus* TY-01809, *S. aureus* TY-Omi, *S. aureus* TY-01852, *S. aureus* TY-01856, *S. aureus* TY-01857, *S. aureus* TY-01859, *S. aureus* TY-03450, *S. aureus* TY-03454, *S. aureus* TY-03456, *S. aureus* TY-03460, *S. aureus* TY-03463, *S. aureus* TY-03466, *S. aureus* TY-03467, *S. aureus* TY-03468, and *S. aureus* TY-03470 were found to be 0.78, 0.78, 0.78, 0.78, 0.78, 0.78, 0.78, 0.78, 1.56, 0.78, 0.78, 0.78, 0.78, 0.39, 0.78, 0.78, 0.78, 0.78, 0.78, 0.78, 0.39, 0.78, 0.78, 0.78, 0.78, 0.78, 0.78, and 0.39 μg/ml, respectively. The compound also showed potent growth inhibitory activity against vancomycin-resistant *Enterococcus* (VRE) species such as

Enterococcus faecalis NCTC 12201 VCM R, *E. faecium* NCTC 12202 VCMR, *E. faecalis* NCTC 12203 VCM R, *E. faecium* NCTC 12204 VCMR, *E. faecalis* 5803, and *E. faecium* 5804 with respective MIC values of 0.20, 0.39, 0.78, 0.20, 0.39, and 0.78 µg/ml.

Lactonamycin was also reported to possess cytotoxic activity against various tumor cell lines such as L1210, P388, EL-4, Ehrlich, S180 IMC carcinoma, FS-3, Meth A, and B16-BL6 with IC_{50} values of 0.087, 0.123, 0.064, 1.290, 3.300, 1.970, 2.220, 0.150, and 0.860 µg/ml, respectively [2].

References

[1] Matsumoto, N., Tsuchida, T., Maruyama, M., Sawa, R., Kinoshita, N., Homma, Y., Takahashi, Y., Iinuma, H., Naganawa, H., Sawa, T., Hamada, M., and Takeuchi, T. (1996) *J. Antibiot.*, **49**, 953.

[2] Matsumoto, N., Tsuchida, T., Maruyama, M., Kinoshita, N., Homma, Y., Iinuma, H., Sawa, T., Hamada, M., Takeuchi, T., Heida, N., and Yoshioka, T. (1999) *J. Antibiot.*, **52**, 269.

[3] Matsumoto, N., Tsuchida, T., Nakamura, H., Sawa, R., Takahashi, Y., Naganawa, H., Iinuma, H., Sawa, T., Takeuchi, T., and Shiro, M. (1999) *J. Antibiot.*, **52**, 276.

Lamellarin α 20-sulfate

Physical data: $C_{29}H_{22}NO_{11}SNa$, white solid, mp 145–148 °C

Structure:

Lamellarin α 20-sulfate

Compound class: Lamellarin-type alkaloid

Source: An unidentified ascidian (IIC-197) collected from the Arabian Sea near Trivandrum, India

Pharmaceutical potential: *Anti-HIV*
Lamellarin α 20-sulfate was found to be an inhibitor of HIV-1 integrase, and was also found to show activity against the HIV-1 virus in cell culture; while sulfated molecules have been identified as integrase inhibitors, this compound is of particular interest since it is active against both preintegration complexes *in vitro* and HIV-1 virus in cell culture. The marine isolate inhibited the integrase terminal cleavage activity with an IC_{50} value of 16 µM, and the strand transfer activity

with an IC_{50} value of 22 µM. The antiviral IC_{50} values for the compound, using a variety of HIV assays, had a range of 8–62 µM. More interestingly, the test compound displayed low toxicity (LD_{50} 274 µM). Hence, lamellarin α 20-sulfate, having the ability to inhibit integrase protein *in vitro* and viral replication in cultured cells, might provide a new class of compounds for potential development of clinically useful integrase inhibitors.

Reference

Venkata Rami Reddy, M., Rama Rao, M., Rhodes, D., Hansen, M.S.T., Rubins, K., Bushman, F.D., Venkateswarlu, Y., and Faulkner, D.J. (1999) *J. Med. Chem.*, **42**, 1901.

Lancifodilactones F and G

Physical data:
- Lancifodilactone F: $C_{25}H_{40}O_6$, white crystals, mp 189–190 °C, $[\alpha]_D^{25.7}$ +28.07° (MeOH, *c* 0.095)
- Lancifodilactone G: $C_{29}H_{36}O_{10}$, white prisms, mp 171–172 °C, $[\alpha]_D^{25.9}$ +75° (MeOH, *c* 0.24)

Structures:

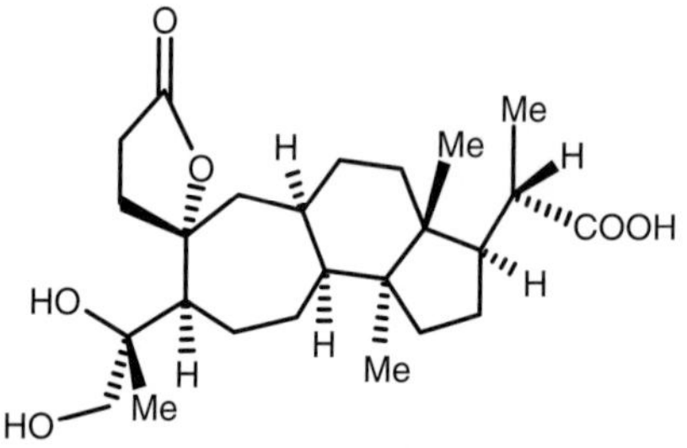

Lancifodilactone F

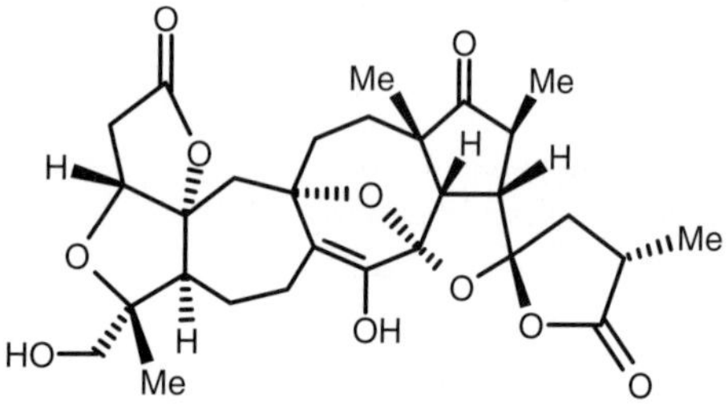

Lancifodilactone G

Compound class: Nortriterpenoids

Source: *Schisandra lancifolia* (Rehd. et Wils) A. C. Smith (leaves and stems) [1, 2]

Pharmaceutical potential: *Anti-HIV*

Lancifodilactones F and G displayed *in vitro* anti-HIV activity with respective EC_{50} values of 20.69 ± 3.3 and 95.47 ± 14.19 µg/ml; the selective index (SI) for each of the compounds were calculated to be >6.62 and in the range of 1.82–2.46, respectively [1, 2].

References

[1] Xiao, W.-L., Li, R.-T., Li, S.-H., Li, X.-L., Sun, H.-D., Zheng, Y.-T., Wang, R.-R., Lu, Y., Wang, C., and Zheng, Q.-T. (2005) *Org. Lett.*, **7**, 1263.
[2] Xiao, W.-L., Zhu, H.-J., Shen, Y.-H., Li, R.-T., Li, S.-H., Sun, H.-D., Zheng, Y.-T., Wang, R.-R., Lu, Y., Wang, C., and Zheng, Q.-T. (2005) *Org. Lett.*, **7**, 2145.

Lancilactone C

Physical data: $C_{30}H_{40}O_4$, obtained as colorless granules ($CHCl_3$–petroleum ether); mp 217–219 °C, $[\alpha]_D$ +43.37° ($CHCl_3$, c 3.60)

Structure:

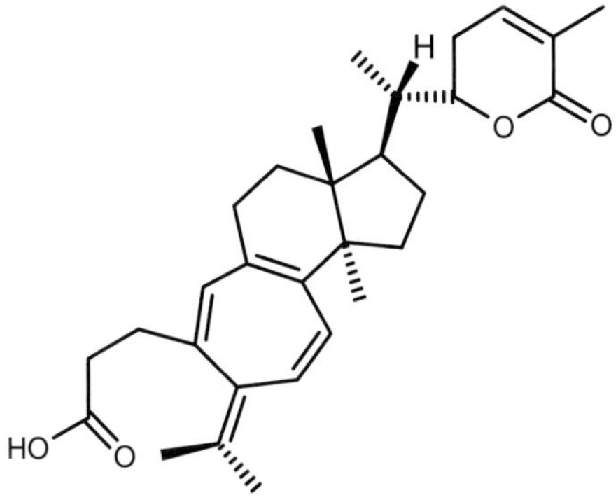

Lancilactone C

Compound class: Triterpene lactone

Source: *Kadsura lancilimba* How. (stems and roots; family: Schisandraceae)

Pharmaceutical potential: *Anti-HIV*
Lancilactone C was reported to exhibit inhibition toward HIV replication in H9 lymphocyte cells with an EC_{50} value of 1.4 µg/ml, and a therapeutic index of greater than 71.4.

Reference

Chen, D.-F., Zhang, S.-X., Wang, H.-K., Shi-Yuan Zhang, S.-Y., Sun, Q.-Z., Cosentino, L.M., and Lee, K.-H. (1999) *J. Nat. Prod.*, **62**, 94.

Lanigerol

Systematic name: 10,12-Dihydroxy-9(10→20)-abeo-8,11,13-abietatriene

Physical data: $C_{20}H_{30}O_2$, pale yellow oil, $[\alpha]_D^{25}$ +65.7° (MeOH, c 0.7)

Structure:

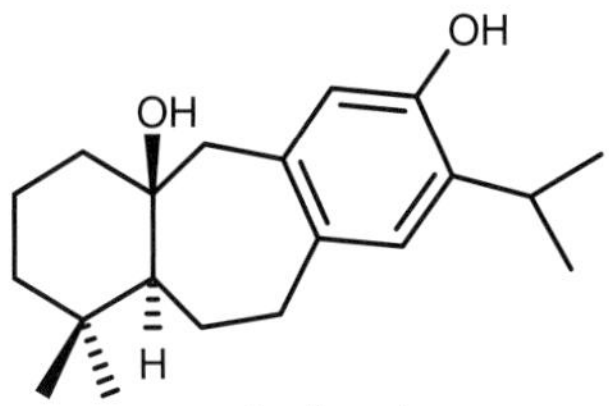

Lanigerol

Compound class: Icetaxane diterpene

Source: *Salvia lanigera* Poir. (roots; family: Labiatae)

Pharmaceutical potential: *Antimicrobial*
Lanigerol was reported to exhibit antimicrobial activity against Gram-positive bacterial strains *Bacillus subtilis*, *Mycobacterium luteus*, and *Staphylococcus aureus* with an MIC value of 1 mg/ml; however, the compound was found to be inactive (MIC > 10 mg/ml) against Gram-negative bacteria and yeasts.

Reference

El-Lakany, A.M., Abdel-Kader, M.S., Sabri, N.N., and Stermitz, F.R. (1995) *Planta Med.*, **61**, 559.

Lansioside A

Physical data: $C_{38}H_{61}NO_8$, colorless needles, mp 174–175 °C, $[\alpha]_D^{15}$ +26.5° (EtOH, *c* 1.06)

Structure:

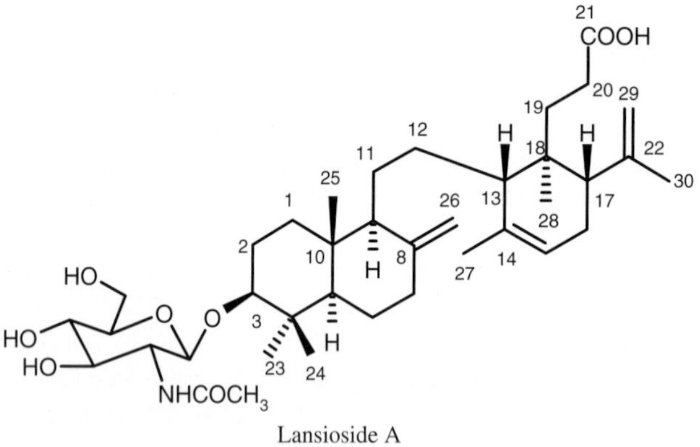

Lansioside A

Compound class: Triterpene glycoside (secoonocerane-type amino sugar glycoside)

Source: *Lansium domesticum* Jack var. Duku (fruit peels; family: Meliaceae)

Pharmaceutical potential: *Inhibitor of leukotriene D_4-induced contraction of ileum*
Lansioside A was evaluated to possess significant *in vitro* inhibitory activity against leukotriene D_4-induced contraction of guinea pig ileum in a dose-dependent manner with an IC_{50} value of 2.4 µg/ml.

Reference

Nishizawa, M., Nishide, H., Kosela, S., and Hayashi, Y. (1983) *J. Org. Chem.*, **48**, 4462.

β-Lapachone

Systematic name: 3,4-Dihydro-2,2-dimethyl-2*H*-naphtho[1,2-*b*] pyran-5,6-dione

Physical data: $C_{15}H_{14}O_3$

Structure:

β-Lapachone

Compound class: Quinone (*ortho*-naphphoquinonic compound)

Source: *Tabebuia avellanedae* (the lapacho tree; barks; family: Bignoniaceae) [1]

Pharmaceutical potentials: *Anticancer; antitumor; antiproliferative; antiviral; antibacterial; antifungal; antiparasitic; anti-inflammatory*

β-Lapachone, a natural *ortho*-naphphoquinonic plant product extracted from *Tabebuia avellanedae* (commonly known as the lapacho tree), has been found to possess a broad spectrum of pharmacological effects, including antiviral, antibacterial, antifungal, and antiparasitic activities [2–4]; the compound has also been established as a promising anticancer agent [5–7]. Preclinical studies on anticancer drugs are now in a developmental phase [8–10].

Frydman *et al.* [11] reported the *in vitro* cytotoxicity of β-lapachone against a variety of drug-sensitive and drug-resistant tumor cell lines, including MDR1-overexpressing cell lines, camptothecin-resistant (CPT-K5 and U93/CR), and also the multidrug-resistant CEM/V-1 cell line. In addition, β-lapachone was found to induce apoptosis in human promyelocytic leukemia cells (HL-60), human prostate cancer cells (DU-145, PC-3, and LNCaP) [5, 12], and MCF-7:WS8 breast cancer cells [13]. β-Lapachone treatment (4 h, 1–5 µM) resulted in a block at G_0/G_1 with decreases in S and G_2/M phases and increases in apoptotic cell populations over time in HL-60 and three separate human prostate cancer (DU-145, PC-3, and LNCaP) cells [12] – interestingly, β-lapachone treatment of p53 wild type containing prostate cancer cells (i.e., LNCaP) did not result in the induction of nuclear levels of p53 protein, as did camptothecin-treated cells. Like other Topo I inhibitors, β-lapachone may induce apoptosis by locking Topo I onto DNA, blocking replication fork movement, and inducing apoptosis in a p53-independent fashion. Furthermore, the β-lapachone-mediated cell death of human leukemic cells involves the upregulation of intracellular ROS generation and the subsequent initiation of c-Jun amino terminal kinase and caspase-3 activation [7, 14, 15]. Currently, it is still unclear whether β-lapachone mediates apoptosis in cancer cell lines by acting as a DNA topoisomerase inhibitor or a free radical generator [16]. From their experimental observations, Don *et al.* suggested that unlike β-lapachone-induced apoptosis of HL-60 cells, the β-lapachone-induced apoptosis of human prostate cancer (HPC) cells might be independent of oxidative stress [16]; in contrast to the 10-fold β-lapachone-induced increase in

H_2O_2 production seen in HL-60 cells, the investigators observed only a 2- to 4-fold increase in HPC cells. Furthermore, β-lapachone induced cell death is not prevented by antioxidant treatment – N-acetyl-L-cysteine (NAC) along with other antioxidants failed to inhibit the apoptosis in HPC cells subjected to β-lapachone treatment for 24 h; hence, the β-lapachone-induced cell death of HPC cells may not be mediated by ROS production, which has also been evident from the results of experiments using DCFH-DA (2′,7′-dichlorofluorescein diacetate) and HE (hydroethidine bromide) probes. From their detailed studies, Don et al. [16] suggested that damage to genomic DNA is the trigger for the apoptosis of HPC cells induced by β-lapachone; β-lapachone induces DNA-PK expression (DNA-dependent kinase) and PARP (poly(ADP-ribose) polymerase) cleavage in advance of any morphological changes. In addition, β-lapachone promotes the expression of cyclin-dependent kinase (Cdk) inhibitors ($p21^{WAF1}$ and $p27^{Kip1}$), induces bak expression, and subsequently stimulates the activation of caspase-7 but not of caspase-3 or caspase-8 during the apoptosis of HPC cells [16].

Choi et al. [17] also demonstrated a combined mechanism that involves the inhibition of pRB phosphorylation and induction of p21 as targets for β-lapachone exerting its antiproliferative effect on cultured human prostate cancer (HPC) cells; the compound arrested the cell cycle progression at the G1 phase. The effects were associated with the downregulation of the phosphorylation of the retinoblastoma protein (pRB) as well as the enhanced binding of pRB and the transcription factor E2F-1. Also, β-lapachone suppressed the cyclin-dependent kinases (Cdks) and cyclin E-associated kinase activity without changing their expressions. Furthermore, this compound induced the levels of the Cdk inhibitor p21(WAF1/CIP1) expression in a p53-independent manner, and the p21 proteins that were induced by β-lapachone were associated with Cdk2. β-Lapachone also activated the reporter construct of a p21 promoter; a combined mechanism for β-lapachone, involving the inhibition of pRB phosphorylation and induction of p21 as targets were proposed by the investigators [17].

Li et al. [18] found a synergism between β-lapachone and paclitaxel that potently inhibits tumor survival; ablation of tumor colonies was noticed in a wide spectrum of human carcinoma cells in culture, including ovarian, breast, prostate, melanoma, lung, colon, and pancreatic cancer cells, after treatment with the combination of these two low molecular mass compounds [18]. However, this synergism is schedule dependent – taxol must be added either simultaneously or after β-lapachone. This combination therapy displayed an unusually potent antitumor activity against human ovarian and prostate tumor prexenografted in mice, with a little host toxicity. Combining drugs imposes different artificial checkpoints; cells can undergo apoptosis at cell-cycle checkpoints, a mechanism that eliminates defective cells to ensure the integrity of the genome. β-Lapachone induces cell-cycle delays in late G_1 and S phase, and taxol arrests cells at G_2/M. Cells treated with both drugs were, thus, delayed at multiple checkpoints before committing to apoptosis. The investigators assumed that when cells are treated simultaneously with drugs activating more than one different cell-cycle checkpoint, the production of conflicting regulatory signaling molecules induces apoptosis in cancer cells [18].

Like camptothecin and topotecan, β-lapachone was also found to be a novel DNA topoisomerase I inhibitor; topoisomerase inhibitors seem to be effective against several types of cancer, including lung, breast, colon, and prostate cancers and malignant melanoma. As we know cancer cells grow and reproduce at a much faster rate than normal cells, they are more vulnerable to topoisomerase inhibition than are normal cells. The enzyme, topoisomerase I, is responsible to cleave DNA required to form chromosomes that must be unwound in order for the cell to use the genetic information to synthesize proteins; β-lapachone retains the chromosomes wound tight so that the cell cannot make proteins, as a result of which the cell stops growing. Li et al. [19] found that the quinone inhibits the catalytic activity of topoisomerase I from calf thymus and human cells, but, unlike camptothecin, it does not stabilize the cleavable complex, thereby indicating a different mechanism of action. From the results of their experiments, the investigators observed a direct interaction of β-lapachone with topoisomerase I rather than DNA substrate – incubation of topoisomerase I with β-lapachone before adding DNA substrate enhanced the inhibition most effectively, while incubation of the enzyme with DNA prior to β-lapachone made the enzyme refractory, and addition of DNA with β-lapachone before the topoisomerase yielded no effect. Thus, the quinone does not inhibit binding of enzyme to DNA substrate; in cells, β-lapachone itself did not induce a SDS-K(+)- precipitable complex, but inhibited complex formation with camptothecin [19] – hence, it is reflected that the direct interaction of β-lapachone with topoisomerase I does not affect the assembly of the enzyme-DNA complex, but does inhibit the formation of cleavable complex. Furthermore, it increases the sensitivity of tumor cells to DNA damaging agents, such as methylmethanesulfonate or ionizing radiation [20], and interrupts the cell cycle [21]. It is also suggested to be a DNA topoisomerase IIα inhibitor, because it suppresses the growth of yeast and blocks the DNA-unwinding enzyme topoisomerase IIα *in vitro* [22, 23]. The pharmacokinetics of β-lapachone shows to be biexponential with a rapid distribution phase after 40 mg/kg IP administration in nude mice; maximum liver concentration for β-lapachone was obtained at 5 min postinjection [24].

Ca^{2+} is recognized as an important regulator of apoptosis [25–30]; increase in intracellular Ca^{2+} levels appears to be an important cell-death signal in human cancer cells that might be exploited for antitumor therapy. The ion may act as a signal for apoptosis by directly activating key proapoptotic enzymes (e.g., calpain); the role of this ion in cell-death processes involving caspase activation has been examined in detail [31–34]. β-Lapachone triggers apoptosis in a number of human breast and prostate cancer cell lines through a unique apoptotic pathway that is dependent upon NAD(P)H:quinone oxidoreductase (NQO1); the apoptosis was found to be significantly enhanced by NAD(P)H:quinone oxidoreductase (NQO1) enzymatic activity [35] as evident from the fact that β-lapachone cytotoxicity was prevented by cotreatment with dicumarol (an NQO1 inhibitor) in NQO1-expressing breast and prostate cancer cells [35]. Pink et al. [36] characterized the activation of a novel cysteine protease in various breast cancer cell lines with properties similar to the Ca^{2+}-dependent cysteine protease, calpain, after exposure to β-lapachone; using NQO1-expressing breast cancer cells, they showed that β-lapachone elicits a rise in intracellular Ca^{2+} levels shortly after drug administration that eventually leads to apoptosis. On the basis of detailed studies, Tagliarino et al. [37] assumed a critical role of Ca^{2+} in the NQO1-dependent cell-death pathway initiated by β-lapachone, and suggested the use of β-lapachone to trigger an apparently novel, calpain-like-mediated apoptotic cell death could be useful for breast and prostate cancer therapy. Furthermore, an NAD(P)H:quinone oxidoreductase-1(NQO1)- and poly(ADP-ribose) polymerase-1 (PARP-1)-mediated cell-death pathway induced in non-small-cell lung cancer cells by β-lapachone was studied in detail by Bey et al. [10] on the basis of which they concluded that β-lapachone induces PARP-1-mediated cell death in an NQO1-dependent manner – an exploitable pathway for selective therapy of NSCLCs.

Li et al. [38] found potent induction of apoptosis by β-lapachone in human multiple myeloma (MM) cell lines and patient cells; the drug, at a dose of less than 4 µM, inhibited cell survival and proliferation by triggering cell death with characteristics of apoptosis in ARH-77, HS Sultan, and MM.1S cell lines, in freshly derived patient MM cells (MM.As), MM cell lines resistant to dexamethasone (MM.1R), doxorubicin (DOX.40), mitoxantrone (MR.20), and melphalan (LR5).

The results of the detailed studies made by Li et al. [38] suggest potential therapeutic application of β-lapachone against MM, particularly to overcome drug resistance in relapsed patients. Gupta et al. [39] also studied in detail on the efficacy of β-lapachone in multiple myeloma (MM) cell lines (U266, RPMI8226, and MM.1S); MM cell lines resistant to dexamethasone (MM.1R), melphalan (RPMI8226/LR5), doxorubicin (RPMI8226/DOX40), and mitoxantrone (RPMI8226/MR20); and MM cells from patients (MM1-MM4), and promising outcomes of the experiments may provide a framework for clinical evaluation of β-lapachone to improve the outcome for patients with MM [39].

Boothman and Pardee [40] showed that β-lapachone could act as a potent inhibitor of DNA repair in mammalian cells and activate topoisomerase I as a result of which radiation-induced neoplastic transformation is prevented; the investigators observed that the compound can prevent the oncogenic transformation of CHEF/18A cells by ionizing radiation. From their detailed study they concluded that the activation of topoisomerase I by β-lapachone may convert repairable single-strand DNA breaks into the more repair-resistant double-strand breaks, thereby preventing potentially lethal DNA damage repair (PLDR) and neoplastic transformation.

β-Lapachone has also been found to be a potent chemopreventive agent that may selectively target tumor cells in retinoblastoma (RB) cell line as reported by Shah et al. [41]; the compound inhibited proliferation and induced apoptosis in RB cell lines. β-Lapachone induced significant dose-dependent growth inhibitory effects in all three retinoblastoma cell lines (i.e., Y79, WERI-RB1, and RBM human retinoblastoma cell lines) – the respective IC_{50} values were 1.9, 1.3, and 0.9 μM. β-Lapachone also induced proapoptotic effects in RB cells. Thus, such potent cytotoxic effects of β-lapachone in RB cell lines at low micromolar concentrations suggest that this agent could be useful in the clinical management of RB [41].

β-Lapachone was found to reduce endotoxin-induced macrophage activation and lung edema and mortality; Tzeng et al. [42] carried out a detailed study (using primary cell culture of rat alveolar macrophages and mouse macrophage cell line RAW264.7 for *in vitro* studies, and a LPS-treated BALB/c mouse model for *in vivo* studies) to evaluate the intracellular mechanism of β-lapachone on lipopolysaccharide (LPS)-induced iNOS (inducible nitric oxide synthase) and TNF-α (tumor necrosis factor-alpha) production in alveolar macrophages, and *in vivo* LPS-induced lung edema, ERK1/2 phosphorylation, NF-κB (nuclear factor kappaB) activation, lethal toxicity and increased NO and TNF-α. It was observed that the drug was capable of suppressing LPS-induced iNOS and TNF-α production by inhibition of protein tyrosine phosphorylation and NF-κB activation. This overall result is of significance in an animal model of sepsis.

Kim et al. [43] reported on the novel functions of β-lapachone on antimetastasis and anti-invasion using human hepatocarcinoma cell lines, HepG2 and HepG3; the drug inhibited cell viability and migration of both HepG2 and Hep3B cells, dose dependently, as studied in MTT and wound healing assays. On the basis of detailed studies, the investigators suggested that β-lapachone may be expected to inhibit progression and metastasis of hepatoma cells, at least in part, by inhibiting the invasive ability of the cells via upregulation of the expression of the early growth response gene-1 (Egr-1), throbospondin-1 (TSP-1), and E-cadherin [43].

A notable limitation of administrating β-Lapachone is that the drug is poorly soluble in most pharmaceutically acceptable solvents; but this limitation has been succesfully resolved by Jiang and Reddy [44] – they have developed an effective method for a pharmaceutical composition comprising a therapeutically effective amount of β-lapachone (or a derivative or analog thereof) and a pharmaceutically acceptable solubilizing carrier molecule, which may be at water-solubilizing carrier molecule such as hydroxypropyl-β-cyclodextrin or an oil-based solubilizing carrier molecule, for enhancing the solubility of β-lapachone in aqueous solution.

References

[1] Woo, H.J., Park, K.Y., Rhu, C.H., Lee, W.H., Choi, B.T., Kim, G.Y., Park, Y.M., and Choi, Y.H. (2006) *J. Med. Food*, **9**, 161.
[2] Docampo, R., Cruz, F.S., Boveris, A., Muniz, R.P., and Esquivel, D.M. (1979) *Biochem. Pharmacol.*, **28**, 723.
[3] Goncalves, A.M., Vasconcellos, M.E., Docampo, R., Cruz, F.S., De Souza, W., and Leon, W. (1980) *Mol. Biochem. Parasitol.*, **1**, 167.
[4] Schaffner-Sabba, K., Schmidt-Ruppin, K.H., Wehrli, W., Schuerch, A.R., and Wasley, J.W. (1984) *J. Med. Chem.*, **27**, 990.
[5] Li, C.J., Wang, C., and Pardee, A.B. (1995) *Cancer Res.*, **55**, 3712.
[6] Lai, C.C., Liu, T.J., Ho, L.K., Don, M.J., and Chau, Y.P. (1998) *Histol. Histopathol.*, **13**, 89.
[7] Chau, Y.P., Shiah, S.G., Don, M.J., and Kuo, M.L. (1998) *Free Radical. Biol. Med.*, **24**, 660.
[8] Ravelo, A.G., Estévez-Braun, A., Chávez-Orellana, H., Pérez-Sacau, E., and Mesa-Siverio, D. (2004) *Curr. Top. Med. Chem.*, **4**, 241.
[9] Pardee, A.B., Li, Y.Z., and Li, C.J. (2002) *Curr. Cancer Drug Targets*, **2**, 227.
[10] Bey, E.A., Bentle, M.S., Reinicke, K.E., Dong, Y., Yang, C.-R., Girard, L., Minna, J.D., Bornmann, W.G., Gao, J., and Boothman, D.A. (2007) *Proc. Natl. Acad. Sci.*, **104**, 11832.
[11] Frydman, B., Marton, L.J., and Sun, J.S. (1995) *Cancer Res.*, **57**, 620.
[12] Planchon, S.M., Wuerzberger, S.M., and Frydman, B. (1995) *Cancer Res.*, **55**, 3706.
[13] Wuerzberger, S.M., Pink, J.J., and Planchon, S.M. (1998) *Cancer Res.*, **58**, 1876.
[14] Planchon, S.M., Wuerzberger-Davis, S.M., Pink, J.J., Robertson, K.A., Bornmann, W.G., and Boothman, D.A. (1999) *Oncol. Rep.*, **6**, 485.
[15] Shiah, S.G., Chuang, S.E., Chau, Y.P., Shen, S.C., and Kuo, M.L. (1999) *Cancer Res.*, **59**, 391.
[16] Don, M.-J., Chang, Y.-H., Chen, K.-K., Ho, L.-K., and Chau, Y.-P. (2001) *Mol. Pharm.*, **59**, 784.
[17] Choi, Y.H., Kang, H.S., and Yoo, M.A. (2003) *J. Biochem. Mol. Biol.*, **36**, 223.
[18] Li, C.J., Li, Y.-Z., Pinto, A.V., and Pardee, A.B. (1999) *Proc. Natl. Acad. Sci. USA*, **96**, 13369.
[19] Li, C.J., Averboukh, L., and Pardee, A.B. (1993) *J. Biol. Chem.*, **268**, 22463.
[20] Boorstein, R.J. and Pardee, A.B. (1983) *Biochem. Biophys. Res. Commun.*, **117**, 30.
[21] Boothman, D.A. and Pardee, A.B. (1989) *Proc. Natl. Acad. Sci. USA*, **86**, 4963.
[22] Frydman, B., Marton, L.J., Sun, J.S., Neder, K., Witiak, D.T., Liu, A.A., Wang, H.M., Mao, Y., Wu, H.Y., Sanders, M.M., and Liu, L.F. (1997) *Cancer Res.*, **57**, 620.
[23] Neder, K., Marton, L.J., Liu, L.F., and Frydman, B. (1998) *Cell. Mol. Biol.*, **44**, 465.
[24] Glen, V.G., Hutson, P.R., and Boothman, D.A. *et al.* (1996) *Int. Pharm. Abstr.*, **33**, 2316.
[25] Distelhorst, C.W. and Dubyak, G. (1998) *Blood*, **91**, 731.
[26] Fang, M., Zhang, H., Xue, S., Li, N., and Wang, L. (1998) *Cancer Lett.*, **127**, 113.
[27] Marks, A.R. (1997) *Am. J. Physiol.*, **272**, H597.
[28] McConkey, D.J., Hartzell, P., Amador-Perez, J.F., Orrenius, S., and Jondal, M. (1989) *J. Immunol.*, **143**, 1801.
[29] McConkey, D.J. and Orrenius, S. (1997) *Biochem. Biophys. Res. Commun.*, **239**, 357.
[30] McConkey, D.J. (1996) *Scanning Microsc.*, **10**, 777.
[31] McColl, K.S., He, H., Zhong, H., Whitacre, C.M., Berger, N.A., and Distelhorst, C.W. (1998) *Mol. Cell. Endocrinol.*, **139**, 229.

[32] Wertz, I.E. and Dixit, V.M. (2000) *J. Biol. Chem.*, **275**, 11470.
[33] Lotem, J. and Sachs, L. (1998) *Proc. Natl. Acad. Sci. USA*, **95**, 4601.
[34] Petersen, A., Castilho, R.F., Hansson, O., Wieloch, T., and Brundin, P. (2000) *Brain Res.*, **857**, 20.
[35] Pink, J.J., Planchon, S.M., Tagliarino, C., Varnes, M.E., Siegel, D., and Boothman, D.A. (2000) *J. Biol. Chem.*, **275**, 5416.
[36] Pink, J.J., Wuerzberger-Davis, S., Tagliarino, C., Planchon, S.M., Yang, X., Froelich, C.J., and Boothman, D.A. (2000) *Exp. Cell Res.*, **255**, 144.
[37] Tagliarino, C., Pink, J.J., Dubyak, G.R., Niemineni, A.-L., and Boothman, D.A. (2001) *J. Biol. Chem.*, **276**, 19150.
[38] Li, Y., Li, C.J., Yu, D., and Pardee, A.B. (2000) *Mol. Med.*, **6**, 1008.
[39] Gupta, D., Podar, K., Tai, Y.T., Lin, B., Hideshima, T., Akiyama, M., LeBlanc, R., Catley, L., Mitsiades, N., Mitsiades, C., Chauhan, D., Munshi, N.C., and Anderson, K.C. (2002) *Exp. Hematol.*, **30**, 711.
[40] Boothman, D.A. and Pardee, A.B. (1989) *Proc. Natl. Acad. Sci.*, **86**, 4963.
[41] Shah, H.R., Conway, R.M., Van Quill, K.R., Madigan, M.C., Howard, S.A., Qi, J., Weinberg, V., and O'Brien, J.M. (2008) *Eye*, **22**, 454.
[42] Tzeng, H.-P., Ho, F.-M., Chao, K.-F., Kuo, M.L., Lin-Shiau, S.-Y., and Liu, S.-H. (2003) *Am. J. Respir. Crit. Care Med.*, **168**, 85.
[43] Kim, S.O., Kwon, J.I., Kim, D.-E., Nam, S.W., Choe, W.-K., and Choi, Y.H. (2007) *FASEB J.*, **21**, 1b45.
[44] Jiang, Z. and Reddy, D.G. (2006) US Patent 7,074,824 (www.patentstorm.us/patents/7074824.html)

Few more references for further reading

[45] Schaffner-Sabba, K., Schmidt-Ruppin, K.H., Wehril, W., Schuerch, A.R., and Wasley, J.W.F. (1984) *J. Med. Chem.*, **27**, 990.
[46] Guiraud, P., Steiman, R., Campos-Takaki, G.M., Seigle-Murandi, F., and De Buochberg, M.S. (1984) *Planta Med.*, **60**, 373.
[47] Li, Y.Z., Li, C.J., Pinto, A.V., and Pardee, A.B. (1999) *Mol. Med.*, **5**, 232.
[48] Chau, Y.P., Shiah, S.G., Don, M.J., and Kuo, M.L. (1998) *Free Radical Biol. Med.*, **24**, 660.
[49] Krishnan, P. and Bastow, K.F. (2001) *Cancer Chemother. Pharmacol.*, **47**, 187.
[50] Liu, S.H., Tzeng, H.P., Kuo, M.L., and Lin-Shiau, S.Y. (1999) *Br. J. Pharmacol.*, **126**, 746.
[51] Manna, S.K., Gad, Y.P., Mukhopadhyay, A., and Aggarwal, B.B. (1999) *Biochem. Pharmacol.*, **57**, 763.
[52] Pardee, A.B., Li, Y.Z., and Li, C.J. (2002) *Curr. Cancer Drug Targets*, **2**, 227.

Largamides D–G

Physical data:
- Largamide D: $C_{56}H_{82}N_9O_{17}Br$, colorless amorphous powder, $[\alpha]_D^{20}$ −43.5° (MeOH, *c* 0.26)
- Largamide E: $C_{56}H_{82}N_9O_{17}Cl$, colorless amorphous powder, $[\alpha]_D^{20}$ −42.7° (MeOH, *c* 0.15)
- Largamide F: $C_{59}H_{80}N_9O_{18}Br$, colorless amorphous powder, $[\alpha]_D^{20}$ −55.0° (MeOH, *c* 0.04)
- Largamide G: $C_{60}H_{82}N_9O_{18}Br$, colorless amorphous powder, $[\alpha]_D^{20}$ −70.0° (MeOH, *c* 0.04)

Structures:

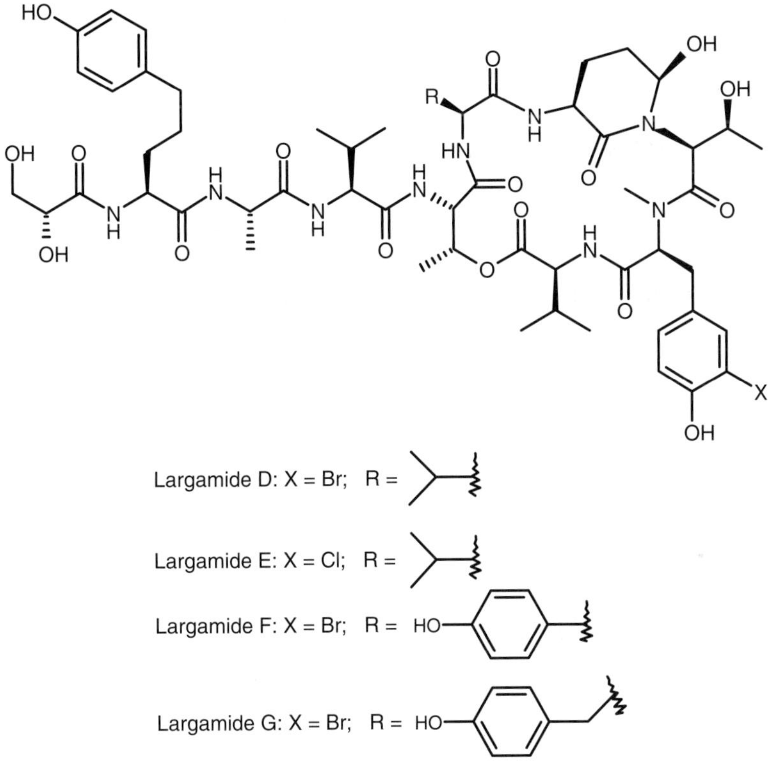

Largamide D: X = Br; R = isopropyl (isobutyl group)

Largamide E: X = Cl; R = isopropyl (isobutyl group)

Largamide F: X = Br; R = HO–C₆H₄–CH₂–

Largamide G: X = Br; R = HO–C₆H₄–CH₂–CH₂–

Compound class: Cyclic peptides

Source: *Oscillatoria* sp. (marine cyanobacterium)

Pharmaceutical potential: *Chymotrypsin inhibitors*
All the four cyclic peptides were evaluated as significant inhibitors of chymotrypsin; the present investigators reported that largamides D and E, which contain the sequence Ahp-Leu-Thr2, inhibited chymotrypsin with an IC$_{50}$ value of 10 μM, while largamides F and G, which contain the respective sequences Ahp-Tyr-Thr2 and Ahp-hTyr-Thr2, inhibited chymotrypsin activity with IC$_{50}$ values of 4.0 and 25.0 μM, respectively. However, none of them was found to inhibit trypsin.

Reference

Plaza, A. and Bewley, C.A. (2006) *J. Org. Chem.*, **71**, 6898.

Laurebiphenyl

Physical data: $C_{30}H_{38}O_2$, crystalline solid, mp 232–232.5 °C, $[\alpha]_D^{25}$ +15.2° (CHCl$_3$, *c* 0.092)

Structure:

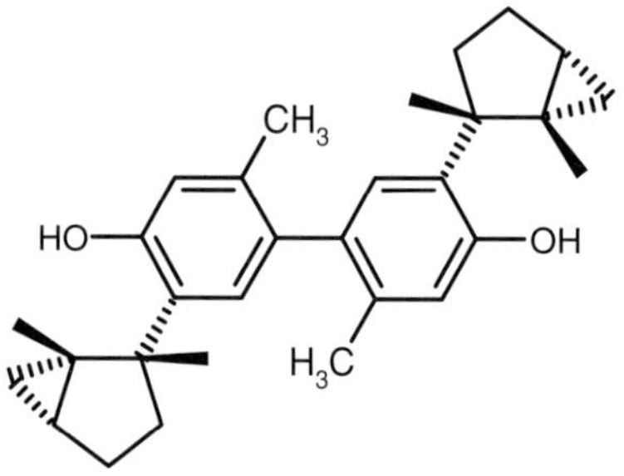

Laurebiphenyl

Compound class:: Dimeric sesquiterpene (cyclolaurane type)

Sources: *Laurencia nidifica* (marine red algae, Rhodomelaceae) [1]; *Laurencia tristicha* (marine red algae Ceramiales, Rhodomelaceae) [2]

Pharmaceutical potential: *Cytotoxic (anticancerous)*
Sun *et al.* [2] reported that laurebiphenyl possesses moderate cytotoxic activity against a number of human cancer cell lines including lung adenocarcinoma (A549), stomach cancer (BGC-823), hepatoma (Bel 7402), colon cancer (HCT-8), and HeLa cell lines with IC_{50} values of 1.68, 1.22, 1.91, 1.77, and 1.61 µg/ml, respectively, using MTT assay [2].

References

[1] Shizuri, Y. and Yamada, K. (1985) *Phytochemistry*, **24**, 1385.
[2] Sun, J., Shi, D., Ma, M., Li, S., Wang, S., Han, L., Yang, Y., Fan, X., Shi, J., and He, L. (2005) *J. Nat. Prod.*, **68**, 915.

Laurinterol and debromolaurinterol

Physical data:
- Laurinterol: $C_{15}H_{19}BrO$, colorless oil
- Debromolaurinterol: $C_{15}H_{20}O$, colorless oil

Structure:

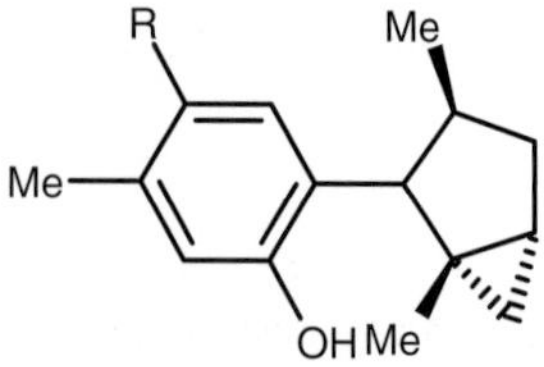

Laurinterol: R = Br
Debromolaurinterol: R = H

Compound class: Sesquiterpenoids

Sources: *Laurencia intermedia* Yamada (marine plant; family: Rhodomelaceae) [1]; *Aplysia kurodai* (sea hare; class: Gastropoda, order: Anaspidea, family: Aplysiidae) [2]

Pharmaceutical potential: *Na,K-ATPase inhibitors*
Okamoto *et al.* [2] evaluated that both laurinterol and debromolaurinterol possess inhibitory activities against the Na,K-ATPase enzyme with respective IC_{50} values of 0.04 and 0.4 mM (ouabain, a cardiac glycoside, was used as a positive control; $IC_{50} = 0.1$ µM). Natural and less toxic Na,K-ATPase inhibitors are demanding for their use in clinical purposes in treating heart diseases [3].

References

[1] Irie, T., Suzuki, M., Kurosawa, E., and Masamune, T. (1970) *Tetrahedron*, **26**, 3271.
[2] Okamoto, Y., Nitanda, N., Ojika, M., and Sakagami, Y. (2001) *Biosci. Biotechnol. Biochem.*, **65**, 474.
[3] Rose, A.M. and Valdes, R., Jr. (1994) *Clin. Chem.*, **40**, 1674.

Laurolistine (norboldine)

Physical data: $C_{18}H_{19}NO_4$

Structure:

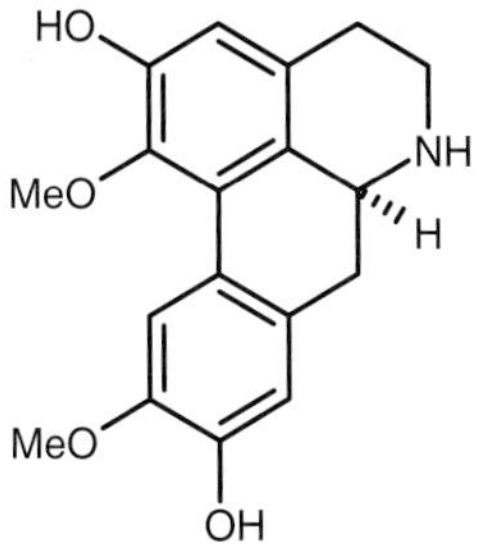

Laurolistine (norboldine)

Compound class: Alkaloid

Sources: *Litsea glutinosa* var. *glabraria* Hook. (leaves and stems; family: Lauraceae) [1]; *Lindera chunii* Merr. (roots; family: Lauraceae) [2]

Pharmaceutical potential: *Anti-HIV*
The alkaloid was evaluated to possess significant antihuman immunodeficiency virus type-1 (HIV-1) integrase activity with an IC_{50} value of 7.7 µM (suramin was used as a positive control; $IC_{50} = 2.4$ µM) [2].

References

[1] Tewari, S., Bhakuni, D.S., and Dhar, H.M. (1972) *Phytochemistry*, **11**, 1149.
[2] Zhang, C., Nakamura, N., Tewtrakul, S., Hattori, M., Sun, Q., Wang, Z., and Fujiwara, T. (2002) *Chem. Pharm. Bull.*, **50**, 1195.

8-Lavandulylkaempferol

Physical data: $C_{25}H_{26}O_6$, amorphous pale yellow powder, $[\alpha]_D^{20}$ +1.96° (MeOH, *c* 0.0158)

Structure:

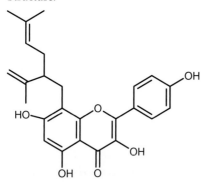

8-Lavandulylkaempferol

Compound class: Flavonoid

Source: *Sophora flavescens* Aiton (roots; family: Leguminosae)

Pharmaceutical potential: *Antioxidant*
The lavandulyl flavonoid derivative was evaluated as a strong scavenger of both DPPH radical and ONOO⁻ anion, with IC_{50} values of 12.91 ± 0.21 μM (IC_{50} of L-ascorbic acid (positive control) = 12.7 ± 0.10 μM), and 4.06 ± 0.41 μM (IC_{50} of DL-penicillamine (positive control) = 3.04 ± 0.74 μM), respectively.

Reference

Jung, H.J., Kang, S.S., Woo, J.J., and Choi, J.S. (2005) *Arch. Pharm. Res.*, **28**, 1333.

Leachianone A

Systematic name: (2*S*)-5,7,4′-Trihydroxy-8-lavandulyl-2′-methoxyflavanone

Physical data: $C_{26}H_{30}O_6$

Structure:

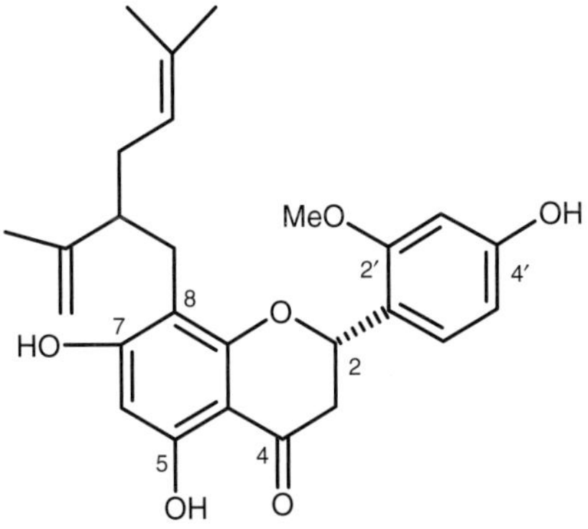

Leachianone A

Compound class: Flavonoid (flavanone)

Sources: *Sophora flavescens* (roots; family: Leguminoase) [1]; *S. leachiano* (roots) [2]

Pharmaceutical potentials: *Cytotoxic; antitumor; antimalarial*

Leachianone A showed moderate cytotoxic activity against human myeloid leukemia HL-60 cells with an IC_{50} value of 11.3 µM (IC_{50} of cisplatin, the positive control = 2.3 µM) [1]. However, it was found to possess profound *in vitro* cytotoxic activity against human hepatoma cell line HepG2 with an IC_{50} value of 3.4 µg/ml post-48-h treatment; the investigators suggested that it induces apoptosis through both extrinsic and intrinsic pathways [3]. Its antitumor effect was further demonstrated *in vivo* by 17–54% reduction of tumor size in HepG2-bearing nude mice, in which no toxicity to heart and liver tissues was observed [3]. In addition, the flavanone also showed moderate *in vitro* antimalarial activity against *Plasmodium falciparum* with an EC_{50} value of 2.6 µM [4].

References

[1] Kang, T.-H., Jeong, S.-J., Ko, W.-G., Kim, N.-Y., Lee, B.-H., Inagak, M., Miyamoto, T., Ryuichi, R., and Kim, Y.-C. (2000) *J. Nat. Prod.*, **63**, 680.
[2] Iinuma, M., Tanaka, T., Mizuno, M., Shirataki, Y., Yokoe, I., Komatsu, M., and Lang, F.A. (1990) *Phytochemistry*, **29**, 2667.
[3] Cheung, C., Chung, K., Lui, J., Lau, C., Hon, P., Chan, J., Fung, K., and Au, S. (2007) *Cancer Lett.*, **253**, 224.
[4] Kim, Y.C., Kim, H.-S., Wataya, Y., Sohn, D.H., Kang, T.H., Kim, M.S., Kim, Y.M., Lee, G.-M., Chang, J.-D., and Park, H. (2004) *Biol. Pharm. Bull.*, **27**, 748.

Lespeflorins A₃, B₂₋₄, C₃, D₁, G₂₋₃, G₅, G₈, G₁₀, and H₂

Physical data:
- Lespeflorin A$_3$ (1): $C_{30}H_{34}O_4$, colorless amorphous solid, $[\alpha]_D^{23}$ −35.6° (MeOH, c 0.65)
- Lespeflorin B$_2$ (2): $C_{25}H_{28}O_5$, colorless amorphous solid, $[\alpha]_D^{23}$ −2.5° (MeOH, c 1.14)
- Lespeflorin B$_3$ (3): $C_{30}H_{36}O_5$, colorless amorphous solid, $[\alpha]_D^{23}$ −10.2° (MeOH, c 4.06)
- Lespeflorin B$_4$ (4): $C_{30}H_{36}O_6$, colorless amorphous solid, $[\alpha]_D^{23}$ −27.9° (MeOH, c 0.29)
- Lespeflorin C$_3$ (5): $C_{21}H_{24}O_5$, colorless amorphous solid, $[\alpha]_D^{23}$ −2.9° (MeOH, c 1.0)
- Lespeflorin D$_1$ (6): $C_{26}H_{30}O_5$, colorless amorphous solid, $[\alpha]_D^{23}$ +25.1° (MeOH, c 0.88)
- Lespeflorin G$_2$ (7): $C_{26}H_{30}O_5$, colorless amorphous solid, $[\alpha]_D^{23}$ −50.2° (MeOH, c 1.35)
- Lespeflorin G$_3$ (8): $C_{27}H_{32}O_6$, colorless amorphous solid, $[\alpha]_D^{23}$ −178.1° (MeOH, c 1.55)
- Lespeflorin G$_5$ (9): $C_{26}H_{30}O$, colorless amorphous solid, $[\alpha]_D^{23}$ −55.6° (MeOH, c 1.7)
- Lespeflorin G$_8$ (10): $C_{22}H_{24}O_5$, colorless amorphous solid, $[\alpha]_D^{23}$ −121.8° (MeOH, c 0.74)
- Lespeflorin G$_{10}$ (11): $C_{21}H_{22}O_4$, colorless amorphous solid, $[\alpha]_D^{23}$ −91.7° (MeOH, c 0.42)
- Lespeflorin H$_2$ (12): $C_{21}H_{18}O_6$, colorless amorphous solid

Structures:

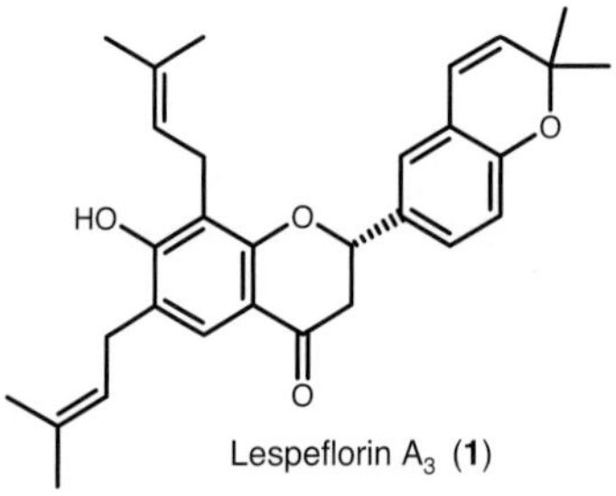

Lespeflorin A$_3$ (1)

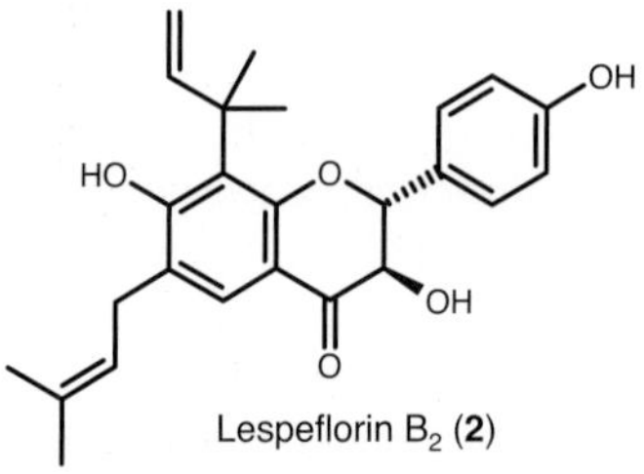

Lespeflorin B$_2$ (2)

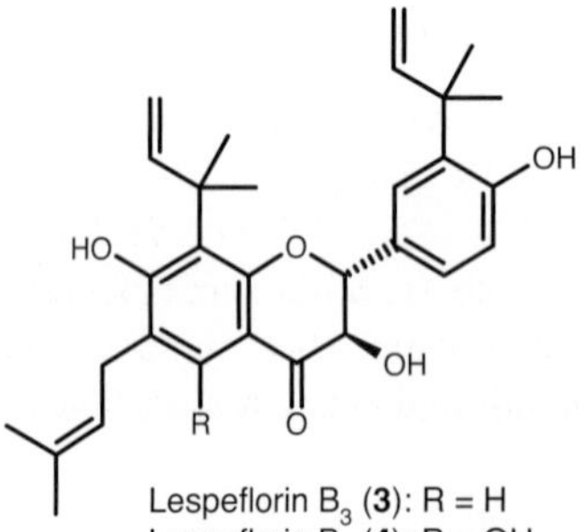

Lespeflorin B$_3$ (3): R = H
Lespeflorin B$_4$ (4): R = OH

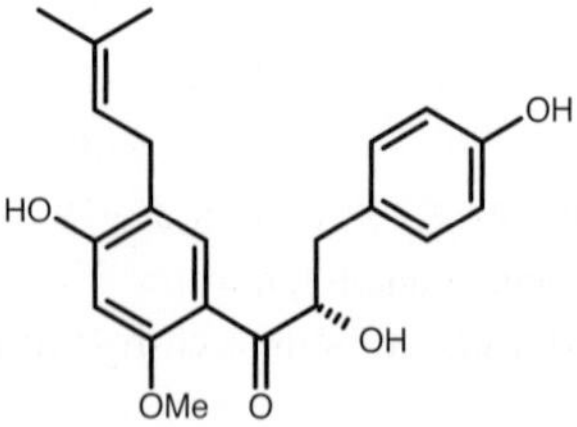

Lespeflorin C$_3$ (5)

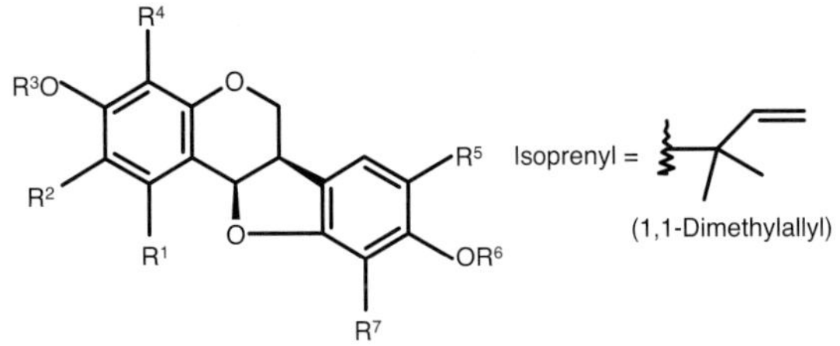

Lespeflorin G$_2$ (**7**): R^1 = R^3 = R^4 = H; R^2 = R^7 = isoprenyl; R^5 = OH; R^6 = CH$_3$

Lespeflorin G$_3$ (**8**): R^1 = OCH$_3$; R^3 = R^4 = H; R^2 = R^7 = isoprenyl; R^5 = OH; R^6 = CH$_3$

Lespeflorin G$_5$ (**9**): R^1 = R^3 = R^4 = R^6 = H; R^4 = R^7 = isoprenyl; R^5 = CH$_3$

Lespeflorin G$_8$ (**10**): R^1 = R^2 = R^4 = H; R^5 = OH; R^3 = R^6 = CH$_3$; R^7 = isoprenyl

Lespeflorin G$_{10}$ (**11**): R^1 = R^2 = R^3 = R^4 = R^6 = H; R^5 = CH$_3$; R^7 = isoprenyl

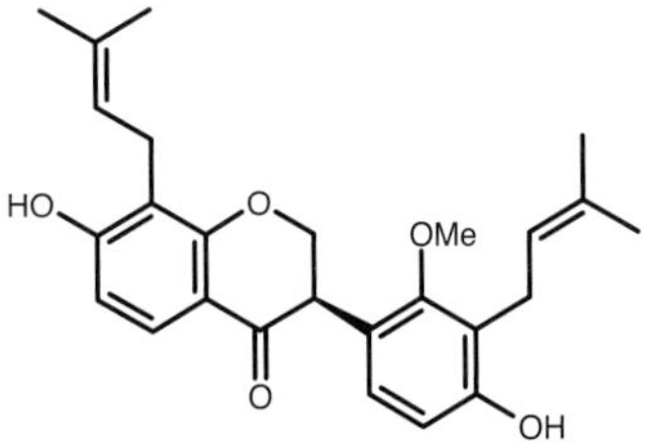

Isoprenyl = (1,1-Dimethylallyl)

Lespeflorin D$_1$ (**6**)

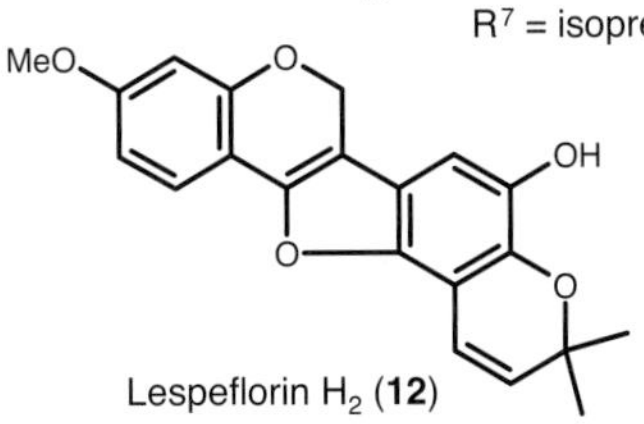

Lespeflorin H$_2$ (**12**)

Compound class: Flavonoids

Source: *Lespedeza floribunda* Miq. (roots; family: Leguminosae)

Pharmaceutical potential: *Melanin synthesis inhibitors*

The present investigators evaluated all the flavonoid isolates as promising melanin synthesis inhibitors as studied in normal human epidermal melanocytes (NHEM); the IC$_{50}$ values for the compounds 1–12 were determined to be 1.5, 2.19, 0.68, 1.23, 1.68, 1.47, 2.07, 2.03, 1.52, 1.63, 1.44, and 1.48 µM, respectively. It was observed that the IC$_{50}$ value of each of these compounds was lower than that of hydroquinone, a positive control (IC$_{50}$ = 2.2 µM); thus, these compounds were found to be more potent than hydroquinone, a compound widely used as a skin-lightening agent, at least in cell cultures.

Reference

Mori-Hongo, M., Takimoto, H., Katagiri, T., Kimura, M., Ikeda, Y., and Miyase, T. (2009) *J. Nat. Prod.*, **72**, 194.

Leucisterol

Physical data: $C_{29}H_{48}O_2$, colorless amorphous solid, mp 258–260 °C, $[\alpha]_D^{20}$ −26.0° (MeOH, c 0.02)

Structure:

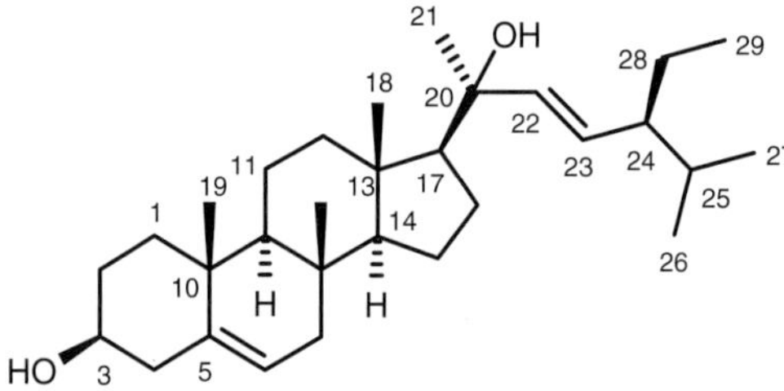

Leucisterol

Compound class: Steroid

Source: *Leucas urticifolia* Vahl (whole plants; family: Labiatae) [1]

Pharmaceutical potential: *Inhibitor of butyrylcholinesterase enzyme (BChE)*
Leucisterol exhibited significant inhibitory activity against butyrylcholinesterase enzyme (BChE) with an IC_{50} value of 3.2 ± 0.85 µM (eserine was used as a positive control; IC_{50} = 0.93 ± 0.3 µM) [1]. Butyrylcholinesterase (BChE) inhibition has been reported as an effective tool for the treatment of Alzheimer's disease and related dementia [2].

References

[1] Fatima, I., Ahmad, I., Anis, I., Malik, A., Afza, N., Iqbal, L., and Latif, M. (2008) *Arch. Pharm. Res.*, **31**, 999.
[2] Yu, S.Q., Holloway, H.W., Utsuki, T., Brossi, A., and Greig, N.H. (1999) *J. Med. Chem.*, **42**, 1855.

Leucosesterterpenone and leucosesterlactone

Physical data:
- Leucosesterterpenone: $C_{25}H_{36}O_7$, colorless crystals, mp 154–157 °C, $[\alpha]_D^{25}$ +240° ($CHCl_3$, c 0.04)
- Leucosesterlactone: $C_{25}H_{36}O_6$, colorless needles, mp 155–160 °C, $[\alpha]_D^{25}$ +80° ($CHCl_3$, c 0.04)

Structures:

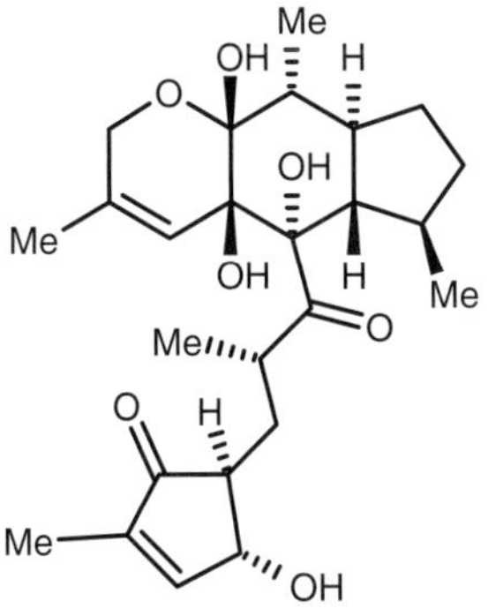

Leucosesterterpenone

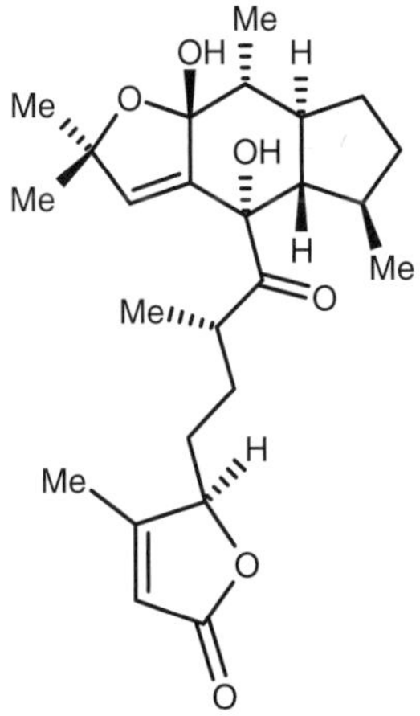

Leucosesterlactone

Compound class: Sesterterpenoids

Source: *Leucosceptrum canum* Sm. (aerial parts; family: Lamiaceae) [1]

Pharmaceutical potential: *Prolylendopeptidase (PEP) inhibitors*
Prolylendopeptidase (PEP) enzyme is reported to catalyze the degradation of proline-containing neuropeptides such as vasopressin, substance P, and thyrotropin-releasing hormones that are involved in the processes of learning and memory [2, 3]. The sesterterpenoid isolates, leucosesterterpenone, and leucosesterlactone, were found to exhibit moderate inhibitory activity against the PEP enzyme with IC_{50} values of 322.21 ± 5.23 and 296.91 ± 3.28 µM, respectively (bacitracin was used as a standard reference; $IC_{50} = 129.26 \pm 3.28$ µM) [1].

References

[1] Choudhary, M.I., Ranjit, R., Atta-ur-Rahman, Hussain, S., Devkota, K.P., Shrestha, T.M., and Parvez, M. (2004) *Org. Lett.*, **6**, 4139.
[2] Kobayashi, W., Miyase, T., Sano, M., Umehara, K., Warashina, T., and Noguchi, H. (2002) *Biol. Pharm. Bull.*, **25**, 1049.
[3] Fan, W., Tezuka, Y., Ni, K.M., and Kadota, S. (2001) *Chem. Pharm. Bull.*, **49**, 396.

Leucospiroside A

Systematic name: (25R)-5α-Spirostane-2α,3β,6β-triol 3-O-β-glucopyranosyl-(1→3)-β-glucopyranosyl-(1→2)-[β-glucopyranosyl-(1→3)]-β-glucopyranosyl-(1→4)-β-galactopyranoside

Physical data: $C_{57}H_{94}O_{30}$, white amorphous solid, $[\alpha]_D^{25}$ −51.9° (CHCl$_3$–MeOH–H$_2$O 26: 14: 3, *c* 0.3)

Leufolins A and B

Structure:

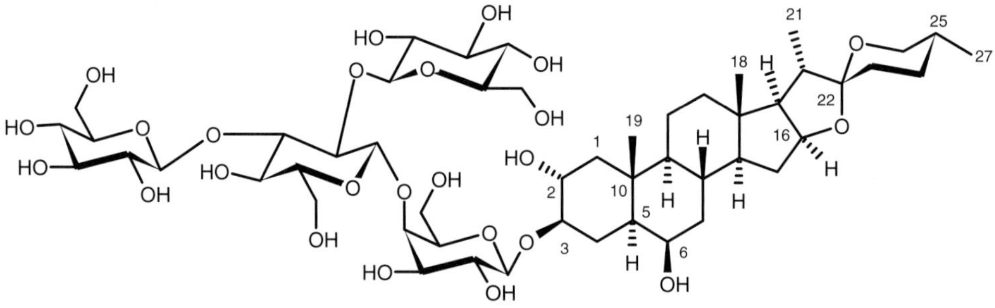

Leucospiroside A

Compound class: Steroidal saponin

Source: *Allium leucanthum* C. Koch (flowers; family: Alliaceae)

Pharmaceutical potential: *Cytotoxic*

Leucospiroside A exhibited *in vitro* cytotoxicity against human lung adenocarcinoma (A-549) and human colorectal adenocarcinoma (DLD-1) with IC_{50} values of 5.0 ± 0.1 and 7.2 ± 0.1 µM, respectively; 5-fluorouracil was used as a positive control (respective IC_{50} values 48 ± 18 and 11 ± 2 µM).

Reference

Mskhiladze, L., Legault, J., Lavoie, S., Mshvildadze, V., Kuchukhidze, J., Elias, R., and Pichette, A. (2008) *Molecules*, **13**, 2925.

Leufolins A and B

Systematic names:
- Leufolin A: {6-[4-(5,7-dihydroxy-4-oxo-3,4-dihydro-2*H*-chromen-2-yl)phenoxy]-3,4,5-trihydroxytetrahydro-2*H*-pyran-2-yl}methyl-(*E*)-3-(4-hydroxyphenyl)-2-propenoate
- Leufolin B: {6-[4-(6,8-dihydroxy-4-oxo-4*H*-chromen-2-yl)phenoxy]-3,5-dihydroxytetrahydro-2*H*-pyran-2-yl}methyl-(*E*)-3(4-hydroxyphenyl)-2-propenoate

Physical data:
- Leufolin A: $C_{30}H_{27}O_{12}$, light yellow amorphous powder (EtOH), mp 110–112 °C, $[\alpha]_D$ +43° (MeOH, *c* 0.12)
- Leufolin B: $C_{30}H_{27}O_{12}$, light yellow amorphous powder (EtOH), mp 140–142 °C, $[\alpha]_D$ +16° (MeOH, *c* 0.12)

Structures:

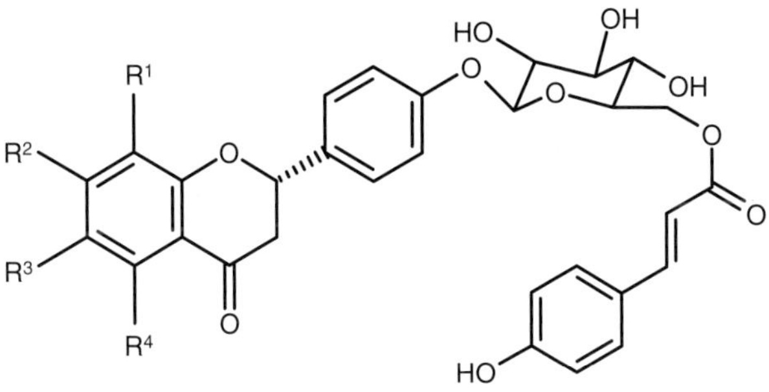

Leufolin A: $R^1 = R^3 = H$; $R^2 = R^4 = OH$
Leufolin B: $R^1 = R^3 = OH$; $R^2 = R^4 = H$

Compound class: Flavonoids (flavanones)

Source: *Leucas urticifolia* (whole plants; family: Lamiaceae) [1]

Pharmaceutical potential: *Butyrylcholinesterase (BChE) inhibiting*
Both leufolins A and B displayed potent butyrylcholinesterase (BChE)-inhibiting potential with respective IC$_{50}$ values of 1.6 ± 0.98 and 3.6 ± 1.7 µM (IC$_{50}$ of eserine (standard BChE inhibitor) = 0.93 ± 0.3 µM) [1]. BChE inhibition is considered to be an effective tool for the treatment of Alzheimer's disease and related dementias [2, 3].

References

[1] Atia-tun-Noor, Fatima, I., Ahmad, I., Malik, A., Afza, N., Iqbal, L., Latif, M., and Khan, S.B. (2007) *Molecules*, **12**, 1447.
[2] Yu, S.Q., Holloway, H.W., Utsuki, T., Brossi, A., and Greig, N.H. (1999) *J. Med. Chem.*, **42**, 1855.
[3] Schwarz, M., Glick, D., Loewensten, Y., and Soreq, H. (1995) *Pharmacol. Ther.*, **67**, 283.

Lindechunine A

Physical data: $C_{19}H_{13}NO_6$, yellow powder, $[\alpha]_D^{26.5}$ +0° (MeOH, *c* 0.1)

Structure:

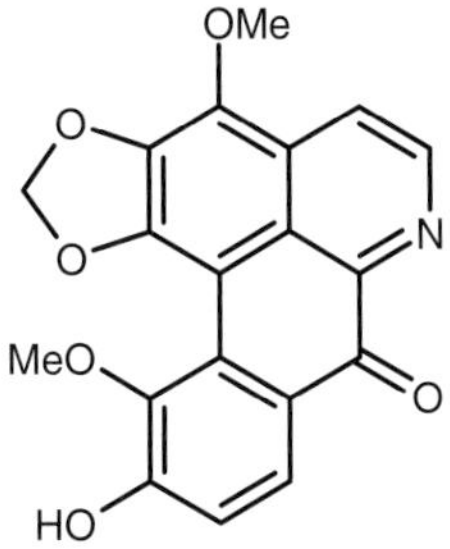

Lindechunine A

Compound class: Alkaloid

Source: *Lindera chunii* Merr. (roots; family: Lauraceae)

Pharmaceutical potential: *Anti-HIV*
The alkaloid was evaluated to possess significant antihuman immunodeficiency virus type-1 (HIV-1) integrase activity with an IC$_{50}$ value of 21.1 µM (suramin was used as a positive control; IC$_{50}$ = 2.4 µM).

Reference

Zhang, C., Nakamura, N., Tewtrakul, S., Hattori, M., Sun, Q., Wang, Z., and Fujiwara, T. (2002) *Chem. Pharm. Bull.*, **50**, 1195.

Liphagal

Physical data: C$_{22}$H$_{28}$O$_4$, amorphous yellow solid, $[\alpha]_D^{25}$ +12.0° (MeOH, *c* 3.7)

Structure:

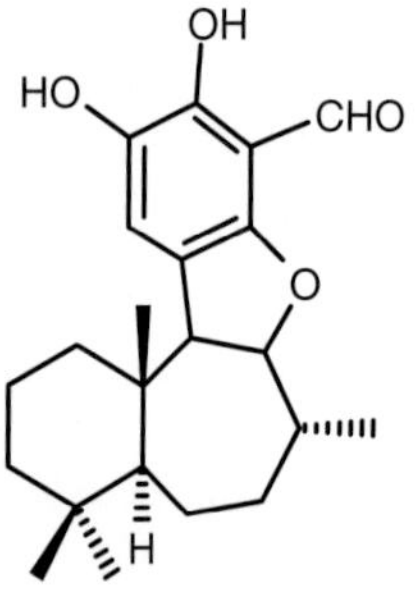

Liphagal

Compound class: Meroterpenoid

Source: *Aka coralliphaga* (marine sponge) [1]

Pharmaceutical potentials: *PI3 kinase-α inhibitor; cytotoxic (antitumor)*
Liphagal was evaluated to possess selective phosphatidylinositol-3-kinase-α (PI3K-α) inhibitory activity with an IC$_{50}$ value of 100 nM in a primary fluorescent polarization enzyme assay [1]. It has been reported that PI3K signaling pathway plays a crucial role in various biological factors, such as in regulating cell proliferation, cell survival, membrane trafficking, glucose transport, neurite outgrowth, and superoxide production and many others, and hence, selective drugs capable of attenuating PI3K signaling should have significant therapeutic potential for treatment of inflammatory and autoimmune disorders as well as cancer and cardiovascular diseases [2–5]. In addition, the isolate was found to be cytotoxic *in vitro* to human tumor cell lines such as LoVo (human colon), CaCo (human colon), and MDA-468 (human breast) with IC$_{50}$ values of 0.58, 0.67, and 1.58 µM, respectively [1].

References

[1] Marion, F., Williams, D.E., Patrick, B.O., Hollander, I., Mallon, R., Kim, S.C., Roll, D.M., Feldberg, L., Van Soest, R., and Andersen, R.J. (2006) *Org. Lett.*, **8**, 321.
[2] Ward, S.G., Sotsoios, Y., Dowden, J., Bruce, I., and Finan, P. (2003) *Chem. Biol.*, **10**, 207.
[3] Wymann, M.P., Zvelebil, M., and Laffargue, M. (2003) *Trends Pharmacol. Sci.*, **24**, 366.
[4] Ward, S.G. and Finan, P. (2003) *Curr. Opin. Pharmacol.*, **3**, 426.
[5] Yang, L., Williams, D.E., Mui, A., Ong, C., Krystal, G., van Soest, R., and Andersen, R.J. (2005) *Org. Lett.*, **7**, 1073.

Lissoclibadins 1–3

Physical data:
- Lissoclibadin 1: $C_{39}H_{57}N_3O_6S_7$, isolated as *tris*-TFA salt
- Lissoclibadin 2: $C_{26}H_{38}N_2O_4S_5$, isolated as *bis*-TFA salt
- Lissoclibadin 3: $C_{26}H_{38}N_2O_4S_4$, isolated as *bis*-TFA salt

Structures:

Lissoclibadin 1   Lissoclibadin 2   Lissoclibadin 3

Compound class: Polysulfur alkaloids

Source: The ascidian *Lissoclinum* sp. (cf. *L. badium* Monniot, F. and Monniot, C., 1996) [1–3]

Pharmaceutical potentials: *Cytotoxic; antitumor; anticancerous; antibacterial; antifungal*
Oda et al. [4] studied the effects of the isolates on IL-8 production in PMA-stimulated human promyelocytic leukemia (HL-60) cells; lissoclibadin 2 was evaluated to induce IL-8 production in a dose-dependent manner. Liu et al. [2] also reported that all of them displayed cytotoxicity against the human leukemia cell line HL-60 with IC_{50} values of 0.37 (0.33), 0.21 (0.13), and 5.5 (3.16) µM (µg/ml), respectively. The same group of investigators [3] studied antitumor effects of the polysulfur alkaloids on a number of human solid tumor-derived cell lines *in vitro*. All of them showed strong inhibitory potentials; however, lissoclibadin 2 was found to be the most interesting one possessing potent inhibitory activity against colon (DLD-1 and HCT116), breast (MDA-MB-231), renal (ACHN), and non-small cell lung (NCI-H460) cancer cell lines and showing no toxicity following a 50 mg/kg single treatment to mice and preferable stability in rat plasma.

Lissoclibadins 1 and 2 were found to inhibit growth of the marine bacterium *Ruegeria atlantica* (15.2 mm at 20 mg/disk and 12.2 mm at 5 mg/disk, respectively); lissoclibadin 2 also showed antifungal activity to *Mucor hiemalis* (13.8 mm at 50 mg/disk) [2].

References

[1] Liu, H., Pratasik, S.B., Nishikawa, T., Shida, T., Tachibana, K., Fujiwara, T., Nagai, H., Kobayashi, H., and Namikoshi, M. (2004) *Tetrahedron Lett.*, **45**, 7015.
[2] Liu, H., Fujiwara, T., Nishikawa, T., Mishima, Y., Nagai, H., Shida, T., Tachibana, K., Kobayashi, H., Mangindaan, R.E.P., and Namikoshia, M. (2005) *Tetrahedron*, **61**, 8611.
[3] Oda, T., Kamoshita, K., Maruyama, S., Masuda, K., Nishimoto, M., Xu, J., Ukai, K., Mangindaan, R.E.P., and Namikoshia, M. (2007) *Biol. Pharm. Bull.*, **30**, 385.
[4] Oda, T., Fujiwara, T., Liu H., Ukai, K., Mangindaan, R.E.P., Mochizuki, M., and Namikoshi, M. (2006) *Mar. Drugs*, **4**, 15.

Litseaefoloside C

Systematic name: 3-Hydroxy-4-*O*-β-D-(6-*O*-caffeoylglucopyranosyl)benzyl vanilloate

Physical data: $C_{30}H_{30}O_{14}$, yellow needles (MeOH), mp 122.4–124.5 °C, $[\alpha]_D^{20}$ −43.3°, $[\alpha]_{436}^{20}$ −100.7°, $[\alpha]_{546}^{20}$ −52.2°, $[\alpha]_{578}^{20}$ −45.6° (MeOH, *c* 0.50)

Structure:

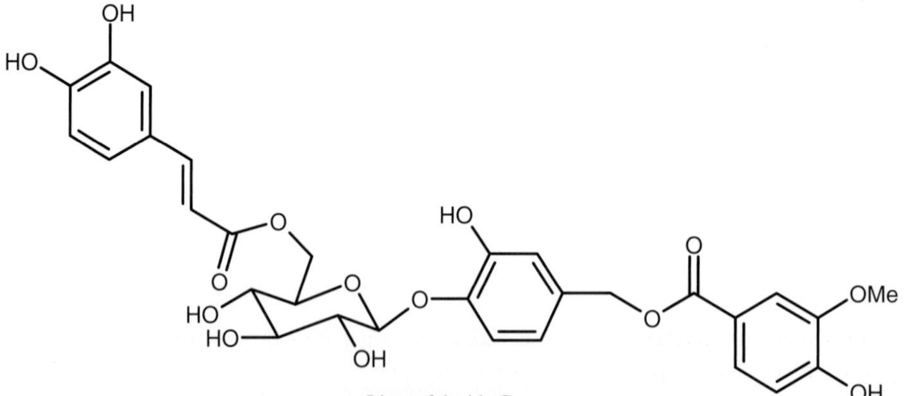

Litseaefoloside C

Compound class: Phenolic glycoside

Source: *Ilex latifolia* (stems; family: Aquifoliaceae)

Pharmaceutical potential: *Enzyme (α-glucosidase and lipase) inhibitory*
Litseaefoloside C was evaluated to possess *in vitro* inhibitory activities against α-glucosidase and lipase with IC_{50} values of 34.0 and 0.31 µg/ml, respectively.

Reference

Zhang, A.-L., Ye, Q., Li, B.-G., Qi, H.-Y., and Zhang, G.-L. (2005) *J. Nat. Prod.*, **68**, 1531.

Littorachalcone

Physical data: $C_{30}H_{26}O_8$, yellow powder (MeOH), mp 178–180 °C

Structure:

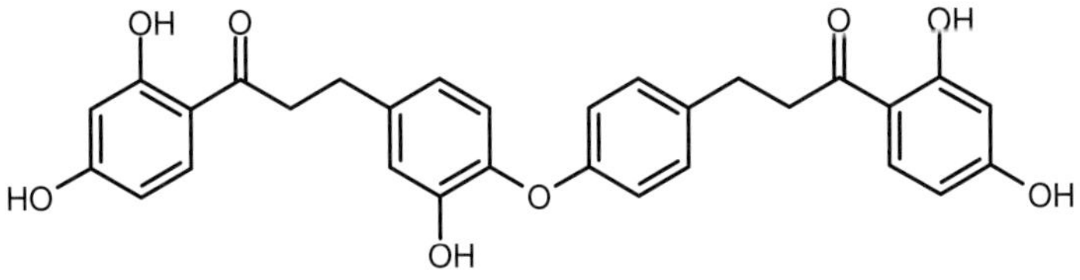

Littorachalcone

Compound class: Flavonoid (dihydrochalcone dimer)

Source: *Verbena litoralis* H.B.K. (aerial parts; family: Verbenaceae)

Pharmaceutical potential: *Nerve growth factor (NGF) stimulator*
The isolate was evaluated to have the potentiality in stimulating the effect of nerve growth factor (NGF) as assessed in NGF-mediated neurite outgrowth in PC12D cells; littorachalcone (1–30 µM) was found not to affect morphologically PC12D cells in the absence of NGF, but the test compound markedly enhanced the NGF (2 ng/ml)-induced increase in the proportion of neurite-bearing cells. The percentages of neurite-bearing cells presented by littorachalcone (3–30 µM) plus NGF (2 ng/ml) were observed to be approximately equal to or greater than 30 ng/ml NGF alone.

Reference

Li, Y., Ishibashi, M., Chen, X., and Ohizumi, Y. (2003) *Chem. Pharm. Bull.*, **51**, 872.

LMG-4

Systematic name: 1-*O*-[(*N*-Acetyl-α-D-neuraminosyl)-(2→8)-(*N*-acetyl-α-D-neuraminosyl)-(2→3)-β-D-galactopyranosyl-(1→4)-β-D-glucopyranosyl]-ceramide

Physical data: $C_{73}H_{133}N_3O_{31}$, amorphous powder

Lobohedleolides

Structure:

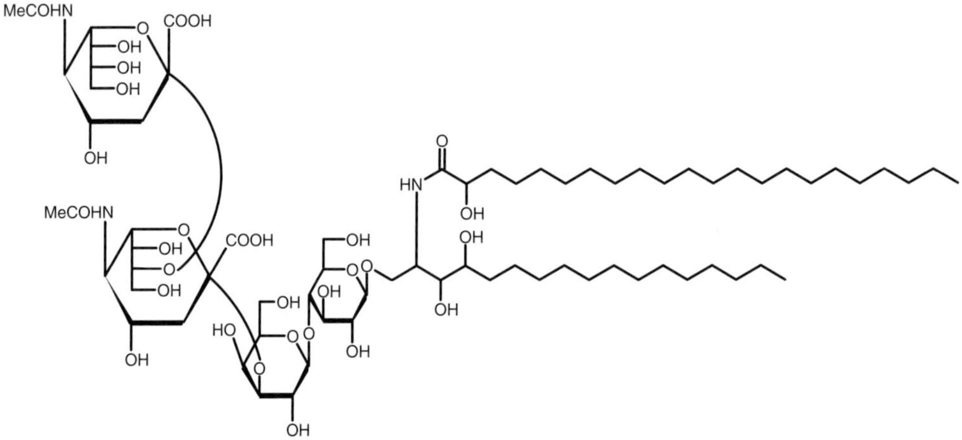

LMG-4

Compound class: GD_3-type ganglioside molecular species

Source: *Luidia maculata* (starfish)

Pharmaceutical potential: *Neuritogenic activity*
The investigators evaluated the effect of the ganglioside molecular species on the neuritogenesis of the rat pheochromocytoma cell line (PC12 cells); it exhibited neuritogenic activity in the presence of nerve growth factor (NGF). The proportion of the neurite-bearing cells of the compound at a concentration of 10 µM was determined to be 47.7% when compared with the control (NGF, 5 ng/ml; 20.6%). Interestingly, such effect was found to be almost similar to that of the mammalian ganglioside GM_1 (47.0%).

Reference

Kawatake, S., Inagaki, M., Isobe, R., Miyamoto, T., and Higuchi, R. (2004) *Chem. Pharm. Bull.*, 52, 1002.

Lobohedleolides

Physical data:
- Lobohedleolide: $C_{20}H_{26}O_4$, white amorphous powder, $[\alpha]_D$ +97.3° ($CHCl_3$, *c* 0.35) [2], $[\alpha]_D$ +104.2° [1]
- (7Z)-Lobohedleolide: $C_{20}H_{26}O_4$, white gum, $[\alpha]_D$ +35.0° ($CHCl_3$, *c* 0.07) [2], $[\alpha]_D$ +61.4° [1]
- 17-Dimethylaminolobohedleolide: $C_{22}H_{33}NO_4$, white gum, $[\alpha]_D$ +13.1° ($CHCl_3$, *c* 0.25)

Structures:

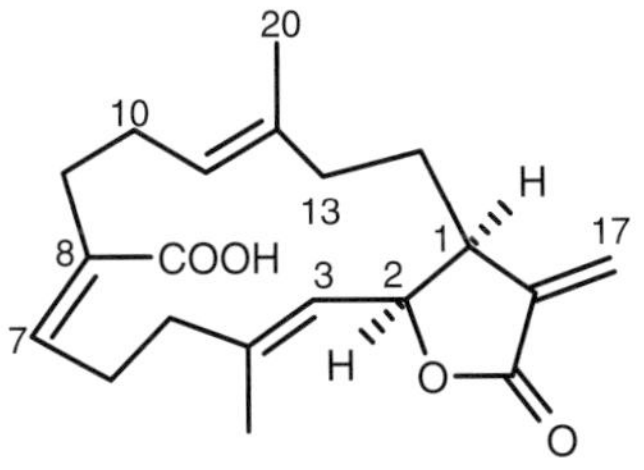

Lobohedleolide

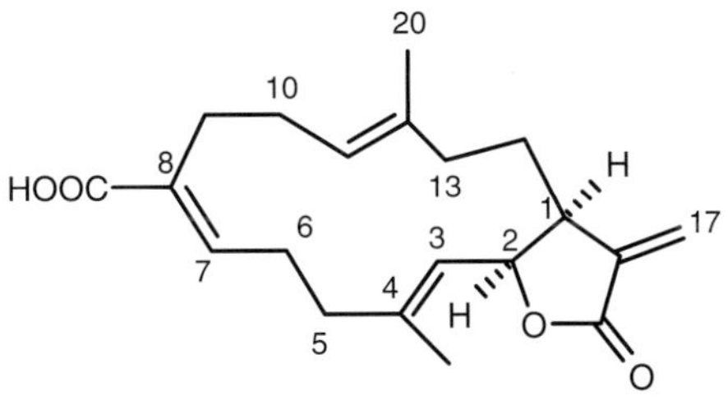

(7Z)-Lobohedleolide

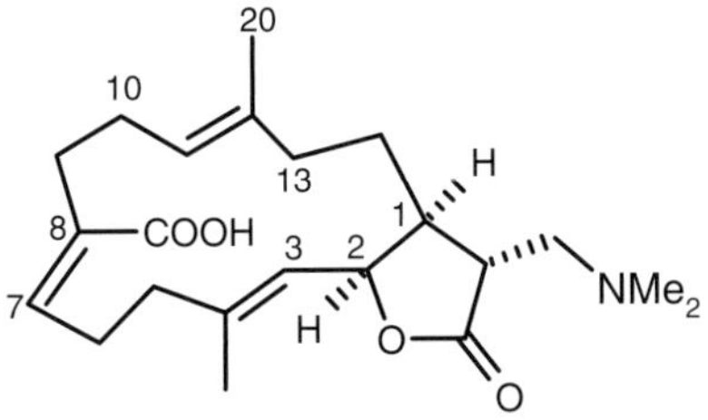

17-Dimethylaminolobohedleolide

Compound class: Cembranoid diterpenes

Sources: Lobohedleolide and (7Z)-lobohedleolide: *Lobophytum hedleyi* Whitelegge (Japanese soft coral) [1]; all the diterpenes from an aqueous extract of *Lobophytum* sp. (soft coral) [2]

Pharmaceutical potentials: *Anti-HIV; antitumor*
Rashid *et al.* evaluated all the isolates in an *in vitro* HIV assay, and all of them were found to inhibit the cytopathic effect of HIV-1 infection in a cell-based assay [2]; however, the compounds were evaluated to be cytoprotective only over a modest concentration range. The respective EC_{50} and IC_{50} values (µg/ml) for the test compounds were determined to be 3.6 and 9.0, 4.6 and 7.6, and 3.3 and 10.2 µg/ml with a maximum cellular protection of approximately 55–70% [2]. Lobohedleolide also exhibited *in vitro* growth inhibitory activity against HeLa cells [1].

References

[1] Uchio, Y., Toyota, J., Nozaki, H., Nakayama, M., Nishizono, Y., and Hase, T. (1981) *Tetrahedron Lett.*, **22**, 4089.
[2] Rashid, M.A., Gustafson, K.R., and Boyd, M.R. (2000) *J. Nat. Prod.*, **63**, 531–533.

Locoracemosides B and C

Physical data:
- Locoracemoside B: $C_{27}H_{33}O_{14}$, amorphous solid, $[\alpha]_D^{23}$ +17.6° (MeOH, *c* 0.01)
- Locoracemoside C: $C_{29}H_{38}O_{18}$, amorphous solid, $[\alpha]_D^{23}$ +19.4° (MeOH, *c* 0.01)

Longicalycinin A

Structures:

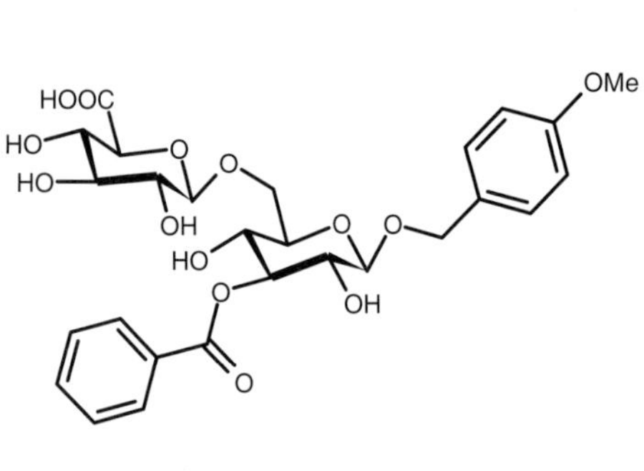

Locoracemoside B

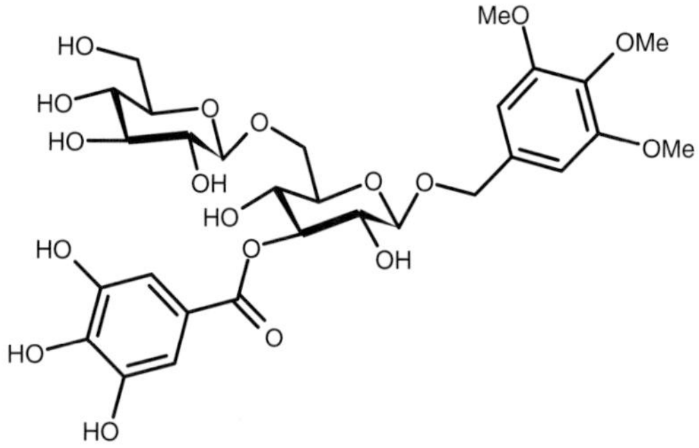

Locoracemoside C

Compound class: Benzylated glycosides

Source: *Symplocos racemosa* Roxb. (stem barks; family: Symplocaceae)

Pharmaceutical potential: α-Chymotrypsin inhibitors

Locoracemosides B and C were reported to display strong inhibitory activity *in vitro* against α-chymotrypsin enzyme with respective IC$_{50}$ values of 11.95 ± 1.85 and 6.04 ± 0.31 µM, which are comparable to that of chymostatin used as a positive control (IC$_{50}$ = 7.21 ± 2.31 µM).

Reference

Rasis, M.A., Ahmad, V.U., Abbasi, M.A., Ali, Z., Rasool, N., Zubair, M., Lodhi, M.A., Choudhary, M.I., and Khan, I.A. (2008) *Phytochem. Lett.*, **1**, 54.

Longicalycinin A

Physical data: $C_{34}H_{37}N_5O_6$, pale yellow powder, $[\alpha]_D^{25}$ −12° (MeCN, *c* 0.01)

Structure:

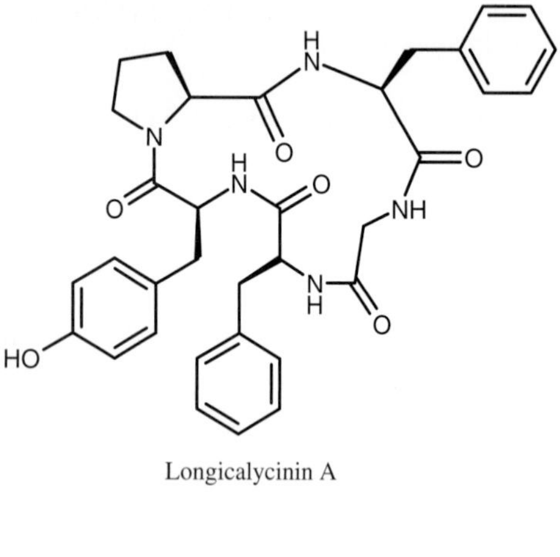

Longicalycinin A

Compound class: Cyclic peptide

Source: *Dianthus superbus* var. *longicalycinus* (Maxim.) Will. (whole plants; family: Caryophyllaceae)

Pharmaceutical potential: *Cytotoxic*
The test compound showed cytotoxicity against human hepatocellular carcinoma HepG2 cells with an IC_{50} value of 13.52 µg/ml.

Reference

Hsieh, P.-W., Chang, F.-R., Wu, C.-C., Li, C.-M., Wu, K.-Y., Chen, S.-L., Yen, H.-F., and Wu, Y.-C. (2005) *Chem. Pharm. Bull.*, **53**, 336.

Longipedumin A

Physical data: $C_{31}H_{32}O_8$, colorless prisms (MeOH), mp 176–178 °C, $[\alpha]_D$ +52.2° (MeOH, *c* 0.2)

Structure:

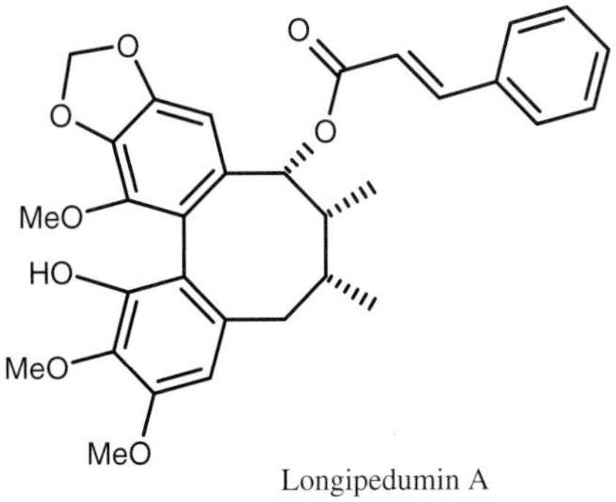

Longipedumin A

Compound class: Lignan

Source: *Kadsura longipedunculata* Finet. et Gagnep. (roots and stems; family: Schisandraceae)

Pharmaceutical potential: *HIV-1 protease inhibitor*
The lignan was evaluated for its HIV-1 protease inhibitory potential and the test compound exerted appreciable inhibition to the target by 77.8 ± 3.3% at 100 µg/ml concentration, thus having an IC_{50} value of 50 µg/ml.

Reference

Sun, Q.-Z., Chen, D.-F., Diang, P.-L., Ma, C.-M., Kakuda, H., Nakamura, N., and Hattori, M. (2006) *Chem. Pharm. Bull.*, **54**, 129.

Lucensimycin E

Physical data: $C_{34}H_{41}NO_{12}S$, colorless amorphous powder, $[\alpha]_D^{23}$ +58.8° (MeOH, c 2.5)

Structure:

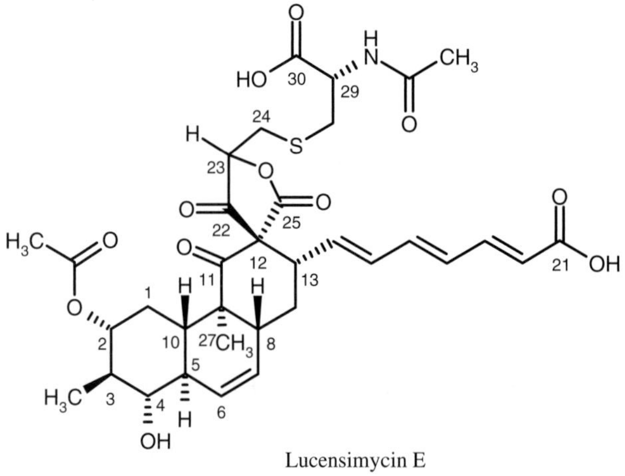

Lucensimycin E

Compound class: Antibiotic

Source: *Streptomyces lucensis* MA7349 (bacterial strain)

Pharmaceutical potential: *Antibacterial*
Lucensimycin E was evaluated to possess significant antibacterial activities against *Staphylococcus aureus* and *Streptococcus pneumoniae* with MIC values of 32 and 8 μg/ml, respectively.

Reference

Singh, S.B., Zink, D.L., Dorso, K., Motyl, M., Salazar, O., Basilio, A., Vicente, F., Byrne, K.M., Ha, S., and Gunilloud, O. (2009) *J. Nat. Prod.*, **72**, 345.

Lucialdehydes B and C

Systematic names:
- Lucialdehyde B: 24(*E*)-3,7-dioxo-5α-lanosta-8,24-diene-26-al
- Lucialdehyde C: 24(*E*)-3β-hydroxy-7-oxo-5α-lanosta-8,24-diene-26-al

Physical data:
- Lucialdehyde B: $C_{30}H_{44}O_3$, amorphous powder (MeOH–H_2O), $[\alpha]_D$ +31° (CHCl$_3$, c 0.105)
- Lucialdehyde C: $C_{30}H_{46}O_3$, amorphous powder (MeOH–H_2O), $[\alpha]_D$ +18° (CHCl$_3$, c 0.092)

Structures:

Lucialdehyde B: R = O
Lucialdehyde C: R= ⋯OH, H

Compound class: Lanostane-type triterpenes

Sources: Lucialdehyde B and C: *Ganoderma lucidum* Karst (fungus; family: Ganodermataceae) [1], *G. pfeifferi* [2]

Pharmaceutical potentials: *Cytotoxic; antiviral*
Gao et al. [1] evaluated cytotoxic efficacies of the isolate against murine and human tumor cell lines; lucialdehyde C showed appreciable cytotoxicity against LLC (mouse lung carcinoma), T-47D (human carcinoma), S-180 (mouse sarcoma), and Meth-A (mouse sarcoma) cell lines with ED_{50} values of 10.7, 4.7, 7.1, and 3.8 µg/ml, respectively. Lucialdehyde B was found to be relatively less effective – the respective ED_{50} values were determined to be 14.3, 15.0, >20, and 4.0 µg/ml. Hence, it is noted that among the cell lines examined, Meth-A cells exhibited high sensitivity for these lanostane-type compounds. The respective ED_{50} values for adriamycin, the reference standard, were evaluated to be 0.06, 0.02, 0.11, and 0.13 µg/ml [1].

Niedermeyer et al. [2] evaluated the antiviral activity of lucialdehyde B against influenza A virus (MDCK cells) and herpes simplex virus type I (HSV) (Vero cells) at noncytotoxic concentrations; the test compound was found to possess moderate activity against influenza A virus with an IC_{50} value of 3.0 µg/ml, but it exhibited strong inhibitory activities against HSV with an IC_{50} value of 0.075 µg/ml. The positive control, amantadine sulfate, showed the inhibitory activity against influenza A virus with an IC_{50} value of 15 µg/ml, while the another control, aciclovir, was found to be active against herpes simplex virus with an IC_{50} value of 0.1 µg/ml.

References

[1] Gao, J., Min, B., Ahn, E., Nakamura, N., Lee, H., and Hattori, M. (2002) *Chem. Pharm. Bull.*, **50**, 837.
[2] Niedermeyer, T.H.J., Lindequist, U., Mentel, R., Gordes, D., Schmidt, E., Thurow, K., and Lalk, M. (2005) *J. Nat. Prod.*, **68**, 1728.

Lucidenic acid N

Physical data: $C_{27}H_{40}O_6$, colorless powder (CHCl$_3$), mp 202–204 °C, $[\alpha]_D$ +119.5° (CHCl$_3$, c 0.23)

Lucidumoside C

Structure:

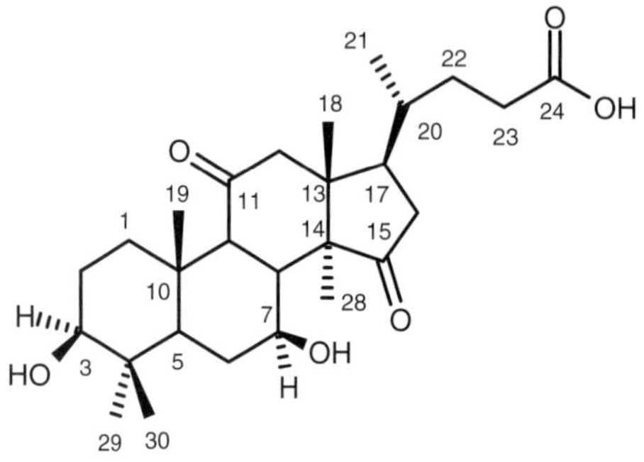

Lucidenic acid N

Compound class: Triterpene

Source: *Ganoderma lucidum* (Fr.) Krast (fruiting bodies; family: Polyporaceae)

Pharmaceutical potential: *Cytotoxic*
Lucidenic acid N was evaluated to display cytotoxicity against the growth of human Caucasian hepatocyte carcinoma HepG2, HepG2.2.15, lymphocytic leukemia P-388, nasopharyngeal carcinoma KB, and cerebral cavernous malformation CCM2 cells with IC_{50} values of 2.06×10^{-4}, 1.66×10^{-3}, 1.20×10^{-2}, 26.69, and 35.49 µM, respectively.

Reference

Wu, T.-S., Shi, L.-S., and Kuo, S.-C. (2001) *J. Nat. Prod.*, **64**, 1121.

Lucidumoside C

Physical data: $C_{27}H_{36}O_{14}$, amorphous powder, $[\alpha]_D^{15}$ −112° (MeOH, *c* 0.22)

Structure:

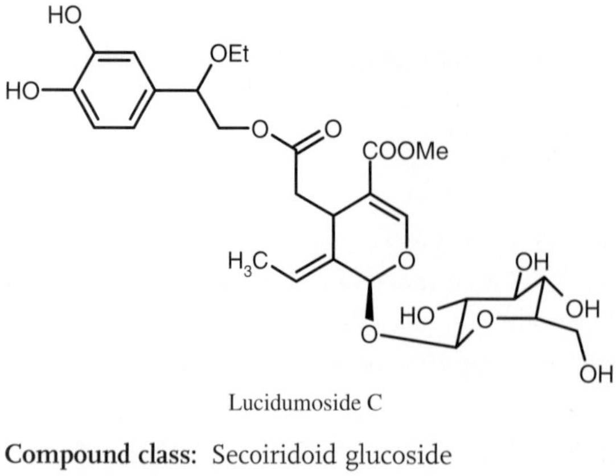

Lucidumoside C

Compound class: Secoiridoid glucoside

Source: *Ligustrum lucidum* Ait. (fruits; family: Oleaceae)

Pharmaceutical potential: *Antioxidant*
The isolate was evaluated to possess strong antioxidant potential as accessed from its efficacy to inhibit hemolysis of red blood cells (RBC) induced by 2,2'-azo-bis-(2-amidinopropane) dihydrochloride; the IC$_{50}$ value for the test compound was determined to be 9.3 µM. The antioxidant efficacy was noted to be almost four times greater than that of trolox used as a positive control. The investigators suggested that such hemolysis inhibitory effect of the compound might be associated with the free phenolic hydroxyl functions present within the molecule.

Reference

He, Z.-D., But, P.P.-H., Chan, T.-W.D., Dong, H., Xu, H.-X., Lau, C.-P., and Sun, H.-D. (2001) *Chem. Pharm. Bull.*, **49**, 780.

Lupulin A

Physical data: $C_{30}H_{46}O_{11}$, colorless needles (*n*-hexane–acetone), mp 172–174 °C

Structure:

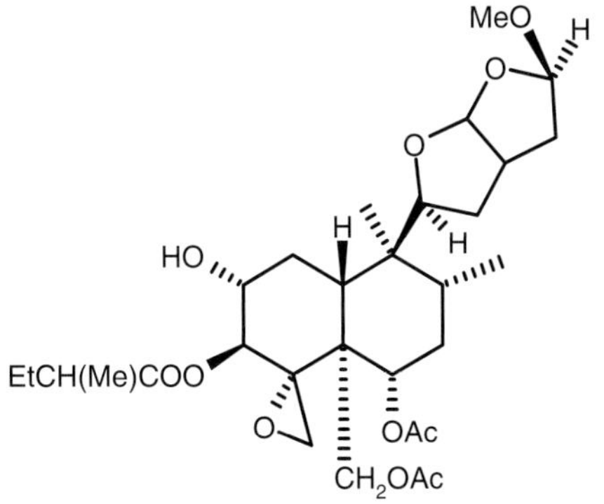

Lupulin A

Compound class: Diterpenoid

Source: *Ajuga lupulina* (Maxim.) (family: Lamiaceae)

Pharmaceutical potential: *Antibacterial*
Lupulin A was reported to display *in vitro* strong antibacterial activity against *Pseudomonas aeruginosa* and *Escherichia coli* (inhibitory zone 3–5 mm) and weak activity against *Staphylococcus aureus* (1.5 mm) as evaluated using the paper diffusion method.

Reference

Chen, H., Tan, R.X., Liu, Z.L., and Zhang, Y. (1996) *J. Nat. Prod.*, **59**, 668.

Luzonial A

Physical data: $C_{19}H_{20}O_7$, yellow oil, $[\alpha]_D^{21}$ $-7.1°$ (MeOH, c 1.04)

Structure:

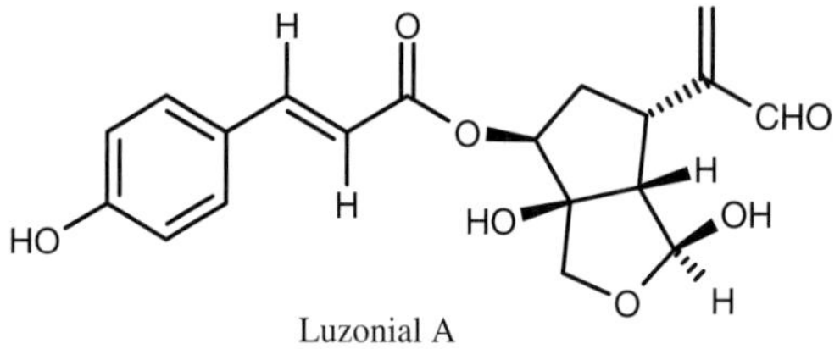

Luzonial A

Compound class: Iridoid (iridoid aldehyde)

Source: *Viburnum luzonicum* (leaves; family: Caprifoliaceae)

Pharmaceutical potential: *Cytotoxic*
In a cytotoxicity assay with the human epithelial cancer (HeLa S3) cell line, the test compound displayed moderate inhibitory activity with an IC_{50} value of 3.50 µM (positive controls used were cisplatin (IC_{50} = 2.46 µM) and fluorouracil (IC_{50} = 5.40 µM)).

Reference

Fukuyama, Y., Minoshima, Y., Kishimoto, Y., Chen, I.-S., Takahashi, H., and Esumi, T. (2005) *Chem. Pharm. Bull.*, **53**, 125.

Luzonial B and luzonidial B

Physical data:
- Luzonial B: $C_{19}H_{20}O_7$, yellow oil, $[\alpha]_D^{21}$ $-1.9°$ (MeOH, c 1.17)
- Luzonidial B: $C_{19}H_{18}O_6$, yellow oil, $[\alpha]_D^{21}$ $-183.4°$ (CHCl$_3$, c 0.94)

Structures:

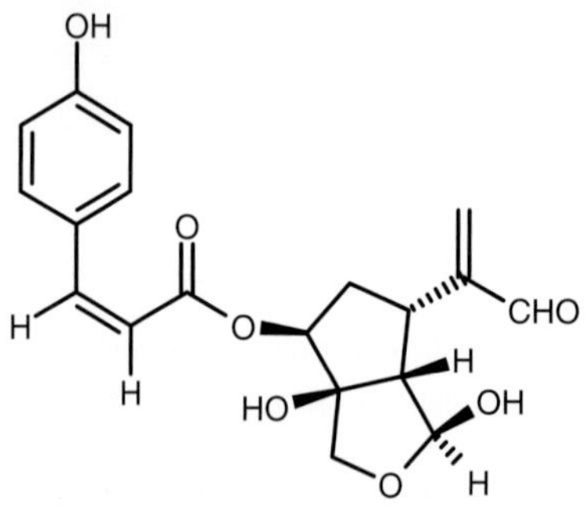

Luzonial B

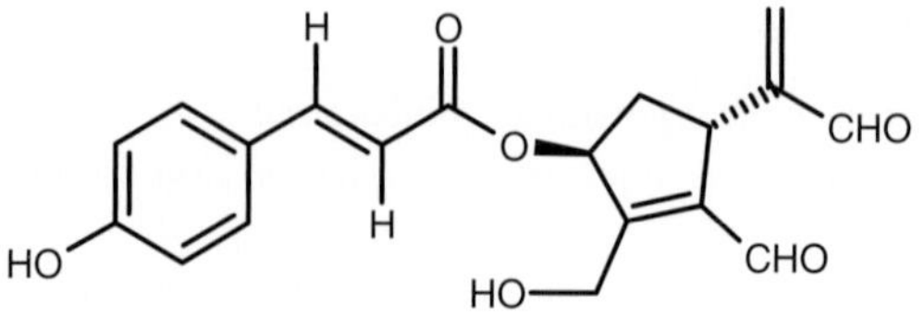

Luzonidial B

Compound class: Iridoids (iridoid aldehydes)

Source: *Viburnum luzonicum* (leaves; family: Caprifoliaceae)

Pharmaceutical potential: *Cytotoxic*

In a cytotoxicity assay with the human epithelial cancer (HeLa S3) cell line, luzonial B was found to have relatively stronger inhibitory activity with an IC_{50} value of 1.90 µM than those of luzonidial B (IC_{50} = 24.5 µM) and the reference standards, cisplatin (IC_{50} = 2.46 µM) and fluorouracil (IC_{50} = 5.40 µM).

Reference

Fukuyama, Y., Minoshima, Y., Kishimoto, Y., Chen, I.-S., Takahashi, H., and Esumi, T. (2005) *Chem. Pharm. Bull.*, **53**, 125.

Lychnostatins 1 and 2

Physical data:
- Lychnostatin 1: $C_{21}H_{28}O_8$, crystals (acetone–hexane), mp 228–230 °C, $[\alpha]_D^{24}$ +89° (CHCl$_3$, c 1.0)
- Lychnostatin 2: $C_{21}H_{28}O_7$, fine needles (acetone–hexane), mp 190–193 °C, $[\alpha]_D^{30}$ +20.9° (CHCl$_3$, c 0.67)

Structures:

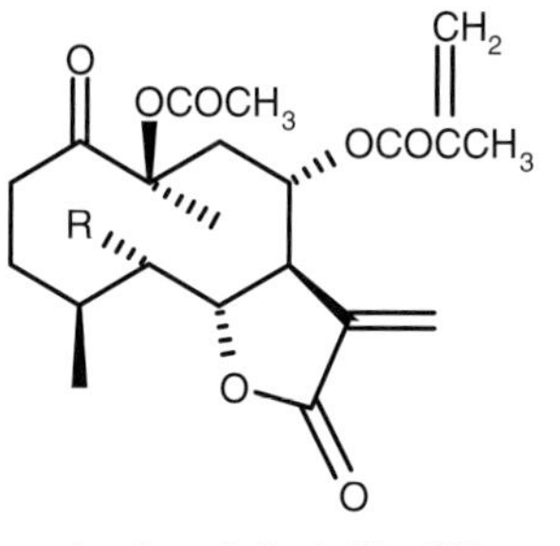

Lychnostatin 1: R = OH
Lychnostatin 2: R = H

Compound class: Germacranolides

Source: *Lychnophora antillana* (stems and leaves; family: Compositae)

Pharmaceutical potential: *Anticancerous*

Both lychnostatins 1 and 2 inhibited growth of P388 lymphocytic leukemia cells *in vitro* with ED_{50} values of 2.0 and 0.19 µg/ml, respectively.

Reference

Pettit, G.R., Herald, D.L., Cragg, G.M., Rideout, J.A., and Brown, P. (1990) *J. Nat. Prod.*, **53**, 382.

Lycojapodine A

Physical data: $C_{16}H_{22}NO_3$, colorless crystals (MeOH), mp 167–168 °C, $[\alpha]_D^{24.7}$ −140.98° (CHCl$_3$, c 0.2)

Structure:

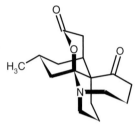

Lycojapodine A

Compound class: Alkaloid

Source: *Lycopodium japonicum* (club moss; family: Lycopodiaceae)

Pharmaceutical potentials: *Acetylcholinesterase inhibitor; anti-HIV-1*
Lycojapodine A was found to exhibit moderate inhibitory efficacy against acetylcholinesterase with an IC$_{50}$ value of 90.3 µM, which was comparable to that of (−)-huperzine A; the isolate also showed weak anti-HIV-1 activity (MTT method) with an EC$_{50}$ value of 85 µg/ml.

Reference

He, J., Chen, X.-Q., Li, M.-M., Zhao, Y., Xu, G., Cheng, X., Peng, L.-Y., Xie, M.-J., Zheng, Y.-T., Wang, Y.-P., and Zhao, Q.-S. (2009) *Org. Lett.*, **11**, 1397.

Lycoperine A

Physical data: $C_{31}H_{49}N_3O_2$, colorless solid, $[\alpha]_D^{24}$ −238° (MeOH, c 0.1)

Structure:

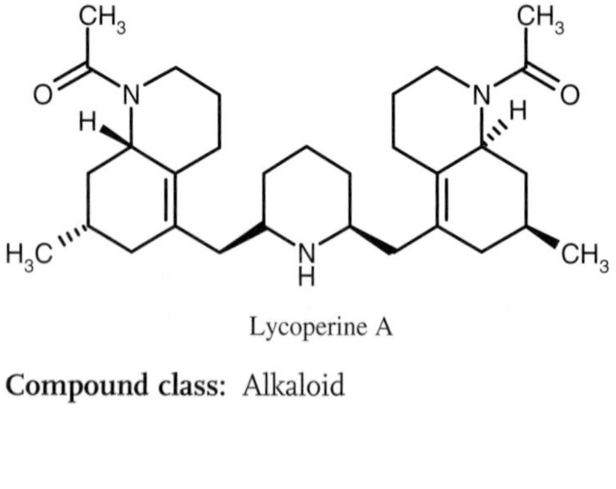

Lycoperine A

Compound class: Alkaloid

Source: *Lycopodium hamiltonii* (club moss; family: Lycopodiaceae)

Pharmaceutical potential: *Acetylcholinesterase inhibitor*
The alkaloid was found to inhibit acetylcholinesterase (from bovine erythrocyte) activity with an IC_{50} value of 60.9 µM.

Reference

Hirasawa, Y., Kobayashi, J., and Morita, H. (2006) *Org. Lett.*, 8, 123.

(+)-Lyoniresinol 4,4′-bis-*O*-β-D-glucopyranoside

Physical data: $C_{34}H_{48}O_{18}$, colorless gummy solid, $[\alpha]_D^{25}$ +18.2° (MeOH, *c* 0.21)

Structure:

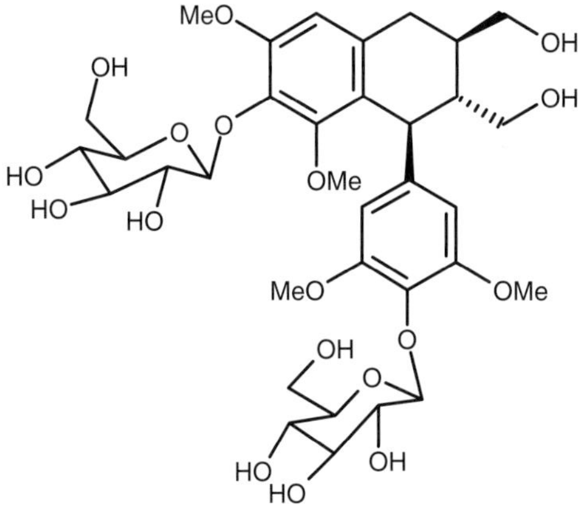

(+)-Lyoniresinol 4,4′-bis-*O*-β-D-glucopyranoside

Compound class: Lignan

Source: *Indigofera heterantha* (whole plants; family: Leguminosae) [1]

Pharmaceutical potential: *Lipoxygenase enzyme inhibitor*
The lignoid derivative exhibited significant inhibitory potential against lipoxygenase enzyme in a concentration-dependent manner; at 25, 50, and 100 µM concentrations, the inhibition percentages (%) were measured to be 44.5, 53.0, and 84.5, respectively [1]. The IC_{50} value for the compound was determined to be 41.5 ± 1.7 µM (baicalein was used as a positive control; IC_{50} = 22.6 ± 0.05 µM) [1].

Lipoxygenases are regarded as key enzymes for developing a variety of disorders such as bronchial asthma, inflammation [2], and several human cancers [3]. Hence, lipoxygenases are potential targets for the rational drug design and discovery of mechanism-based inhibitors for the treatment of such disease manifestations.

References

[1] Rehman, A.U., Malik, A., Riaz, N., Ahmad, H., Nawaz, S.A., and Choudhary, M.I. (2005) *Chem. Pharm. Bull.*, **53**, 263.
[2] Steinhilber, D.A. (1999) *Curr. Med. Chem.*, **6**, 71.
[3] Ding, X.Y., Tong, W.G., and Adrian, T.E. (2001) *Pancreatology*, **1**, 291.

(+)-Lyoniresinol-3-α-*O*-β-D-glucopyranoside

Physical data: $C_{28}H_{38}O_{13}$, amorphous powder, $[\alpha]_D^{24}$ +26° (MeOH, *c* 0.5)

Structure:

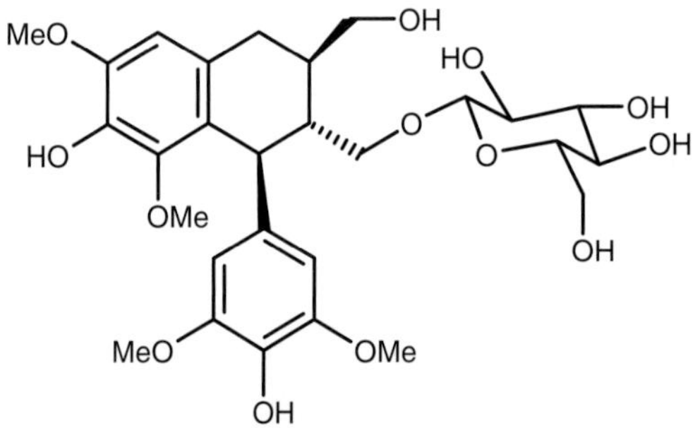

(+)-Lyoniresinol-3-α-*O*-β-D-glucopyranoside

Compound class: Lignan glycoside

Sources: *Stemmadenia minima* A. Gentry (stem barks; family: Apocynaceae) [1]; *Lycium chinense* (root barks; family: Solanaceae) [2]

Pharmaceutical potentials: *Antibacterial; antifungal*
The lignan glycoside exhibited potent antibacterial activity against the methicillin-resistant *Staphylococcus aureus* strains such as MRSA-1, MRSA-2, MRSA-3, and MRSA-4 with MIC values of 5, 2.5–5, 2.5, and 2.5 μg/ml, respectively, whereas the respective MIC values for the positive control, cefotaxime, were observed to be >40 μg/ml; in addition, the isolate was found to have potent and effective antifungal activity against *Candida albicans* [2]. The present investigators [2] evaluated that the test compound induced accumulation of intracellular trehalose on *C. albicans* as stress-response to the drug and thus disrupted the dimorphic transition that forms pseudohyphae caused by the pathogenesis. The compound was also found not to have any hemolytic effect on human erythrocytes; hence, this lignoid derivative might be a promising lead for developing antibiotic drugs.

References

[1] Achenbach, H., Lowel, M., Waibel, R., Gupta, M., and Solis, P. (1992) *Planta Med.*, **58**, 270.
[2] Lee, D.G., Jung, H.J., and Woo, E.-R. (2005) *Arch. Pharm. Res.*, **28**, 1031.

Lyratols C and D

Physical data:
- Lyratol C: $C_{15}H_{26}O_4$, colorless prismatic crystals, mp 207–208 °C, $[\alpha]_D^{25}$ +9.8° (CHCl$_3$, c 0.82)
- Lyratol D: $C_{15}H_{20}O_3$, pale yellow lamellar crystals, mp 164–165 °C, $[\alpha]_D^{25}$ −22.5° (CHCl$_3$, c 0.93)

Structures:

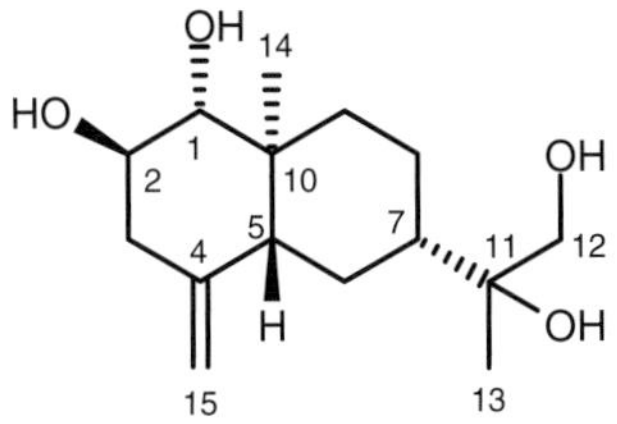

Lyratol C 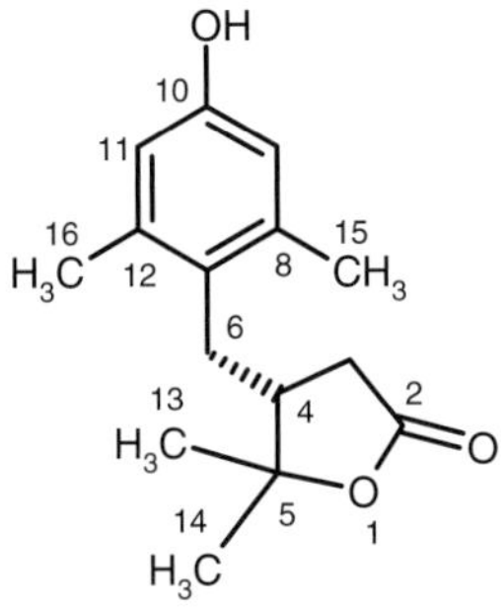   Lyratol D

Compound class: Sesquiterpenoids

Source: *Solanum lyratum* Thunb (whole plants; family: Solanaceae)

Pharmaceutical potential: *Cytotoxic*
Both the isolates exhibited significant *in vitro* cytotoxic activities against three human cancer cell lines, namely, HONE-1 nasopharyngeal, KB oral epidermoid carcinoma, and HT29 colorectal carcinoma cells, with respective IC$_{50}$ values of 3.7 ± 2.1, 6.1 ± 2.4, and 6.5 ± 1.6 for lyratol C; 5.6 ± 3.0, 8.1 ± 1.8, and 6.4 ± 2.3 for lyratol D; 3.8 ± 0.4, 4.5 ± 0.7, and 5.8 ± 1.2 for cisplatin (positive control); 0.6 ± 0.4, 0.9 ± 0.3, and 1.9 ± 0.6 for etoposide (positive control).

Reference

Ren, Y., Shen, L., Zhang, D.-W., and Dai, S.-J. (2009) *Chem. Pharm. Bull.*, **57**, 408.

M

Maackiaflavanones A and B

Systematic names:

- Maackiaflavanone A: (2*S*)-5,2′-dihydroxy-7-methoxy-8-(3-methylbut-2-enyl)-2‴,2‴-dimethylpyrano[5‴,6‴: 5′,4′] flavanone
- Maackiaflavanone B: (2*S*)-5,7-dihydroxy-8,3′-di(3-methylbut-2-enyl)-2‴,2‴-dimethylpyrano [5‴,6‴: 5′,4′] flavanone

Physical data:

- Maackiaflavanone A: $C_{26}H_{28}O_6$, white amorphous powder ($CHCl_3$), $[\alpha]_D^{20}$ −67.3° (MeOH, *c* 0.1)
- Maackiaflavanone B: $C_{30}H_{34}O5_6$, pale yellow gum ($CHCl_3$), $[\alpha]_D^{20}$ −34.7° (MeOH, *c* 0.2)

Structures:

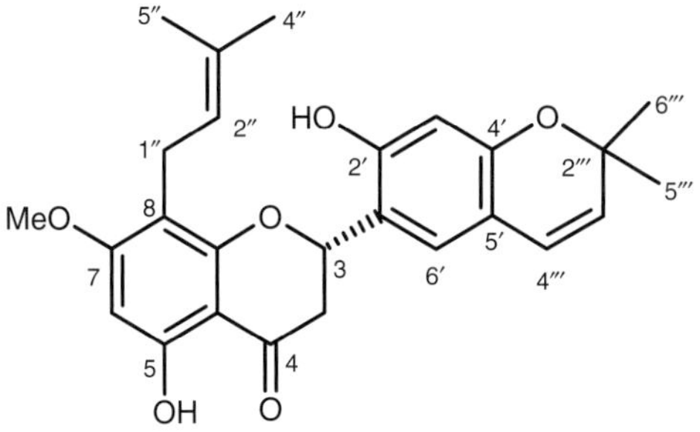

Maackiaflavanone A

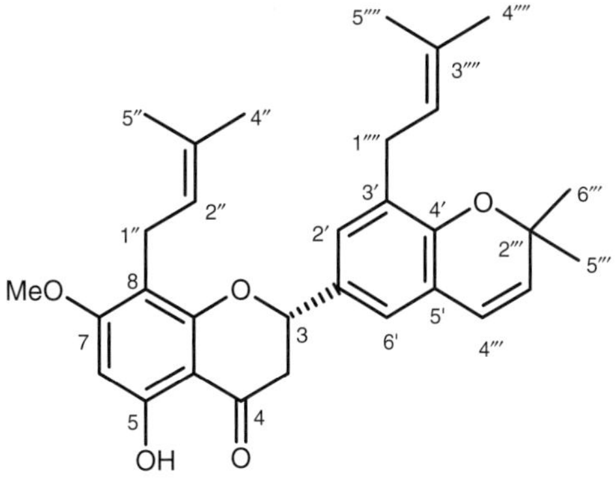

Maackiaflavanone B

Compound class: Flavonoids (prenylated flavanones)

Source: *Maackia amurensis* Rupr. et Maxim. (stem barks; family: Leguminosae)

Pharmaceutical potential: ***Cytotoxic***

Both the prenylated flavanones were found to have significant *in vitro* cytotoxicities against the four human cancer cell lines, namely, A375S2, HeLa, MCF-7, and HepG2, with respective IC_{50} values of 35.1 ± 0.8, 27.7 ± 1.1, 39.2 ± 0.9, and 40.3 ± 0.7 μM for maackiaflavanone A and 7.8 ± 1.4, 36.8 ± 1.0, 16.8 ± 1.2, and 37.4 ± 0.9 μM for maackiaflavanone B (5-FU was used as a positive control with respective IC_{50} values of 9.6 ± 0.2, 50.9 ± 0.4, 48.2 ± 0.2, and 33.6 ± 0.5 μM).

Handbook of Pharmaceutical Natural Products, Volume 2.
Goutam Brahmachari

ISBN: 978-3-527-32148-3

Maackiapterocarpan A

Reference

Li, X., Wang, D., Xia, M.-Y., Wang, Z.-H., Wang, W.-N., and Cui, Z. (2009) *Chem. Pharm. Bull.*, 57, 302.

Maackiapterocarpan A

Systematic name: 3-Hydroxy-8,9-methylenedioxy-[2′,2′-dimethyl-3′,4′-dihydropyrano-(5′,6′: 1,2)] [6aR,11aR]-pterocarpan

Physical data: $C_{21}H_{20}O_6$, white amorphous powder (MeOH), $[\alpha]_D^{20}$ −240° (MeOH, c 1.5)

Structure:

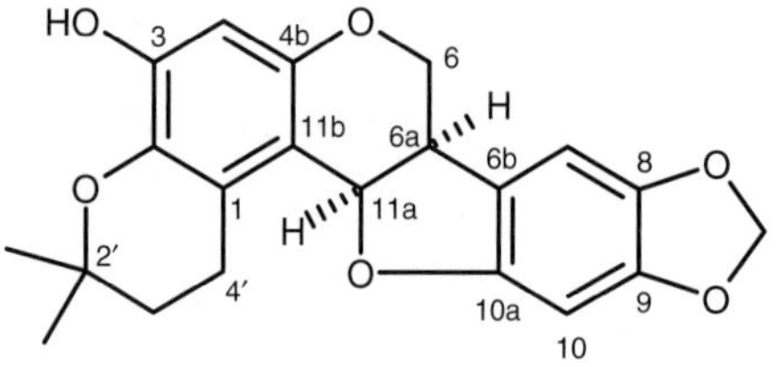

Maackiapterocarpan A

Compound class: Flavonoid (prenylated pterocarpan)

Source: *Maackia amurensis* Rupr. et Maxim. (stem barks; family: Leguminosae)

Pharmaceutical potential: *Cytotoxic*
The prenylated flavonoid was found to have significant *in vitro* cytotoxicity against four human cancer cell lines, namely, A375S2, HeLa, MCF-7, and HepG2, with respective IC$_{50}$ values of 12.2 ± 1.3, 7.6 ± 1.0, 66.2 ± 1.7, and 85.6 ± 1.4 μM (5-FU was used as a positive control with respective IC$_{50}$ values of 9.6 ± 0.2, 50.9 ± 0.4, 48.2 ± 0.2, and 33.6 ± 0.5 μM).

Reference

Li, X., Wang, D., Xia, M.-Y., Wang, Z.-H., Wang, W.-N., and Cui, Z. (2009) *Chem. Pharm. Bull.*, 57, 302.

Macabarterin

Physical data: $C_{42}H_{37}O_{32}$, brown amorphous powder, $[\alpha]_D^{30}$ +60.6° (MeOH, c 0.033)

Structure:

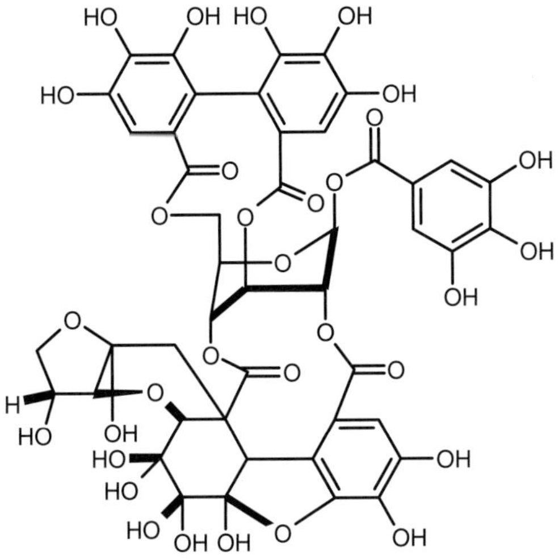

Macabarterin

Compound class: Ellagitannin

Source: *Macaranga barteri* Muell. Arg. (stem barks; family: Euphorbiaceae)

Pharmaceutical potential: *Inhibitor to human neutrophil respiratory burst activity (anti-inflammatory)*

The ellagitannin derivative was evaluated for its *in vitro* anti-inflammatory potential in a cell-based respiratory burst assay and was found to exert inhibition against superoxides (ROS) produced in the cellular system (activated by opsonized zymosan). The compound, thus by inhibiting human neutrophil respiratory burst activity, may have potential as a nonsteroidal anti-inflammatory agent. At a concentration of 1000 µg/ml, the test compound inhibited respiratory burst activity in human neutrophils by 73.23% (indomethacin and aspirin, used as positive controls, showed the inhibition, respectively, by 58.82 and 70.45% at the same concentration). The respective IC_{50} values in regard to suppression of superoxide production for the isolate, indomethacin, and aspirin were determined to be 821.1 ± 73.3, 757.99 ± 5.9, and 279.44 ± 4.42 µg/ml.

Reference

Ngoumfo, R.M., Ngounou, G.E., Tchamadeu, C.V., Qadir, M.I., Mbazoa, C.D., Begum, A., Ngninzeko, F.N., Lontsi, D., and Choudhary, M.I. (2008) *J. Nat. Prod.*, **71**, 1906.

Macaflavanone G

Physical data: $C_{30}H_{36}O_6$, off-white amorphous powder, $[\alpha]_D^{30}$ +75.9° (CHCl$_3$, *c* 1.40)

Structure:

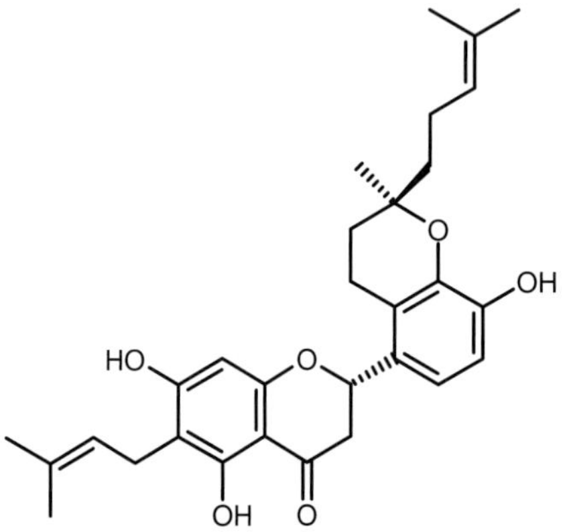

Macaflavanone G

Compound class: Flavonoid (prenylated flavanone)

Source: *Macaranga tanarius* (L.) Benth. Mull.-Arg. (leaves; family: Euphorbiaceae)

Pharmaceutical potential: *Cytotoxic*

Macaflavanone G was reported to exhibit cytotoxicity against KB and A549 cells with IC$_{50}$ values of 12.3 ± 3.0 and 13.4 ± 2.1 µM, respectively.

Reference

Kawakami, S., Harinantenaina, L., Matsunami, K., Otsuka, H., Shinzato, T., and Takeda, Y. (2008) *J. Nat. Prod.*, **71**, 1872.

Macaranone A

Physical data: $C_{30}H_{34}O_7$, yellow amorphous solid

Structure:

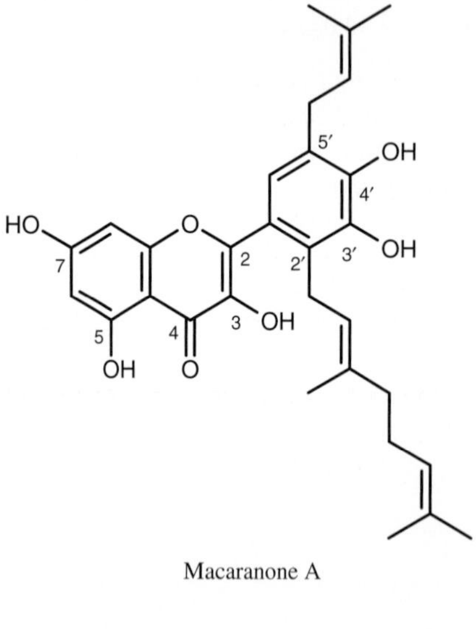

Macaranone A

Compound class: Flavonoid (prenylated flavonol)

Source: *Macaranga sampsonii* (leaves; family: Euphorbiaceae)

Pharmaceutical potential: *Cytotoxic*
The isolate showed *in vitro* cytotoxicity against the HepG2 cancer cell line with an IC_{50} value of 6.9 µg/ml.

Reference

Li, X., Xu, L., Wu, P., Xie, H., Huang, Z., Ye, W., and Wei, X. (2009) *Chem. Pharm. Bull.*, **57**, 495.

Macluraxanthones B and C

Systematic names:
- Macluraxanthone B: 1,3,6,7-tetrahydroxy-2-(1,1'-dimethylallyl)-4-(3-methyl-but-2-enyl)xanthone
- Macluraxanthone C: 1,3,5,7-tetrahydroxy-2-(1,1'-dimethylallyl)-4-(3-methyl-but-2-enyl)xanthone

Physical data: $C_{23}H_{24}O_6$ (both yellow amorphous solids) [1]; recrystallization of macluraxanthone B from a $CHCl_3$–MeOH solution as yellow prisms, mp 152 C [2]

Structures:

Macluraxanthone B   Macluraxanthone C

Compound class: Prenylated tetraoxygenated xanthones

Sources: Macluraxanthone B: *Maclura tinctoria* (family: Moraceae) [1]; *Cudrania tricuspidata* (root bark; family: Moraceae) [2]; macluraxanthone C: *Maclura tinctoria* (family: Moraceae) [1]

Pharmaceutical potentials: *Anti-HIV; cytotoxic*
Macluraxanthones B and C were reported to exhibit significant HIV-inhibitory activity with EC_{50} values of 1–2 and 1.3–2.2 µg/ml, respectively [1]; both the compounds showed very high toxicity toward the CEM-SS host cells with the respective IC_{50} values of 2.2–3.3 and 3.7µg/ml. It is assumed that the catechol unit present in both the structures proffers the enhanced HIV-inhibitory activity [1].

Macluraxanthone B exhibited *in vitro* cytotoxic activity [2] against human lung cancer (A549) and human ovarian cancer (SKOV3) cell lines with respective IC_{50} values of 2.88 ± 0.53 and 4.24 ± 0.40 µM; taxol was used as a positive control (IC_{50} values: $1.25 \pm 0.0.19$ µM (A549); 1.35 ± 0.29 µM (SK OV3)) [2].

References

[1] Groweiss, A., Cardellina, J.H., II, and Boyd, M.R. (2000) *J. Nat. Prod.*, **63**, 1537.
[2] Lee, B.W., Gal, S.W., Park, K.-M., and Park, K.H. (2005) *J. Nat. Prod.*, **68**, 456.

Madecassoside

Systematic name: O-6-Deoxy-α-L-mannopyranosyl-(1→4)-O-β-D-glucopyranosyl-(1→6)-β-D-glucopyranosyl (2α,3β,4α,6β)-2,3,6,23-tetrahydroxyurs-12-en-28-oate

Physical data: $C_{48}H_{78}O_{20}$

Structure:

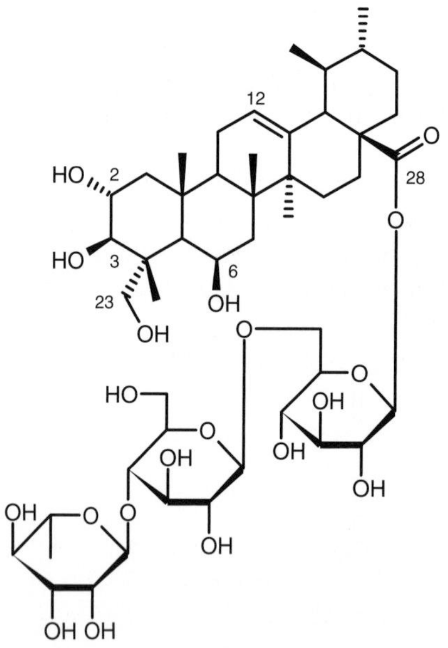

Madecassoside

Compound class: Triterpenoid saponin

Source: *Centella asiatica* L. Urban (syn. *Hydrocotyle asiatica* L. family: Umbelliferae) [1]

Pharmaceutical potentials: *Wound healing; protective against myocardial ischemia–reperfusion injury; anti-inflammatory; antirheumatoid arthritic; skin-care agent; antioxidant*

Medecassoside was reported to find useful applications in wound healing and preventing cicatrisation [2]. The saponin was found to promote *in vitro* synthesis of type I and type III collagens, the major components of skin dermis, and fibroblast proliferation in cultured human fibroblasts [3, 4]. Recently, Liu *et al.* [5] also showed that the test compound facilitates burn wound healing in mice through increased antioxidative activity and enhanced collagen synthesis and angiogenesis in a dose- and time-dependent manner – an oral administration of madecassoside (6, 12, and 24 mg/kg body weight) promoted wound closure, and almost completely wound closure was noted on day 20 in the group receiving the highest dose of 24 mg/kg b.w. of the saponin. Their experimental results revealed that the drug not only mitigated infiltration of inflammatory cells as well as enhanced epithelization resulting from dermal proliferation of fibroblasts, but also decreased nitric oxide (NO) levels and malondialdehyde (MDA) content in the burn skin tissue. A recent study demonstrated that it could accelerate healing of gastric ulcers also; antinociceptive and anti-inflammatory efficacies of medecassoside were evaluated by Somchit et al. [6].

Liu and his group [7] also evaluated the therapeutic potential and underlying mechanisms of madecassoside on collagen II (CII)-induced arthritis (CIA) in mice; the drug on oral administration (10, 20, and 40 mg/kg b.w.) from the day of the antigen challenge for twenty consecutive days, dose dependently prevented mouse CIA based on the reduced clinical scores, and elevated the body weights of mice. Madecassoside was noted to alleviate infiltration of inflammatory cells and synovial hyperplasia as well as to protect joint destruction. Moreover, it reduced the serum level of anti-CII IgG, suppressed the delayed type hypersensitivity against CII in ears as well as CII-stimulated proliferation of lymphocytes from popliteal lymph nodes in CIA mice [7].

Antilipid peroxidative, the enhancement of superoxide dismutase (SOD) activity, anti-inflammatory, and antidepressant effects of madecassoside were also reported [8, 9]. Li et al. [9] showed that it could exert protective effect against myocardial reperfusion injury in rabbit heart *in vivo*; treatment with the test compound (3.2, 1.6, and 0.8 mg/kg b.w. i.v.) during ischemia–reperfusion injury dose dependently attenuated myocardial damage [9]. Recently, the compound was also found to reduce significantly the myocardial infarction induced by ischemia–reperfusion injury in rat heart *in vivo* in a dose-dependent manner (at a dose of 2, 20, and 50 mg/kg b.w.); the protective effect on myocardial ischemia–reperfusion injury is supposed to be related to its roles as an antioxidative, anti-inflammatory, and antiapoptosis agents [10].

Madecassoside has been reported to have a variety of applications in skin-care including treatment of psoriasis [11].

References

[1] Pinhas, H. and Bondiou, J.-C. (1967) *Bull. Soc. Chim. Fr.*, 1888.
[2] Suguna, L., Sivakumar, P., and Chandrakasan, G. (1996) *Indian J. Exp. Biol.*, **34**, 1208.
[3] Bont,é, F., Dumas, M., Chaudagne, C., and Meybeck, A. (1995) *Ann. Pharm. Fr.*, **53**, 38.
[4] Lv, L., Wei, S.M., Lin, H.F., Ma, L.J., Ying, K., and Mao, Y.M. (2003) *China Surfact. Det. Cosmet.*, **33**, 9.
[5] Liu, M., Dai, Y., Li, Y., Luo, Y., Huang, F., Gong, Z., and Meng, Q. (2008) *Planta Med.*, **74**, 809.
[6] Somchit, M.N., Sulaiman, M.R., Zuraini, A., Samsuddin, L., Somchit, N., Israf, D.A., and Moin, S. (2004) *Indian J. Pharmacol.*, **36**, 377.
[7] Liu, M., Dai, Y., Yao, X., Li, Y., Luo, Y., Xia, Y., and Gong, Z. (2008) *Int. Immunopharmacol.*, **8**, 1561.
[8] Liu, M.R., Han, T., Chen, Y., Qin, L.P., Zheng, H.C., and Rui, Y.C. (2004) *Zhong Xi Yi Jie He Xue Bao*, **2**, 40.
[9] Li, G.-G., Bian, G.-X., Ren, J.-P., Wen, L.-Q., Zhang, M., and Li, Q.-J. (2007) *Yao Xue Xue Bao*, **42**, 475.
[10] Bian, G.-X., Li, G.G., Yang, Y., Liu, R.-T., Ren, J.-P., Wen, L.-Q., Guo, S.-M., and Lu, Q.-J. (2008) *Biol. Pharm. Bull.*, **31**, 458.
[11] Rougier, A. (2007) *J. Am. Acad. Dermatol.*, **56** (Suppl. 2), AB28 (Poster abstract, American Academy of Dermatology 65th Annual Meeting).

Madhucosides A and B

Systematic names:
- Madhucoside A: 3-O-β-D-apiofuranosyl-(1→2)-β-D-glucopyranosyl-28-O-{β-D-apiofuranosyl-(1→2)-[α-L-rhamnopyranosyl-(1→3)]-β-D-xylopyranosyl-(1→2)-α-L-rhamnopyranosyl-(1→2)-α-L-arabinopyranosyl}protobassic acid
- Madhucoside B: 3-O-β-D-apiofuranosyl-(1→2)-β-D-glucopyranosyl-28-O-{β-D-xylopyranosyl(1→2)-[α-L-rhamnopyranosyl(1→4)]-β-D-glucopyranosyl(1→3)-α-L-rhamnopyranosyl(1→2)-α-L-arabinopyranosyl}protobassic acid

Physical data:
- Madhucoside A: $C_{68}H_{110}O_{35}$, colorless amorphous yellow powder, $[\alpha]_D^{27}$ −38.8° (MeOH, c 1.0)
- Madhucoside B: $C_{69}H_{112}O_{36}$, colorless amorphous yellow powder, $[\alpha]_D^{27}$ −25.4° (MeOH, c 1.0)

Structures:

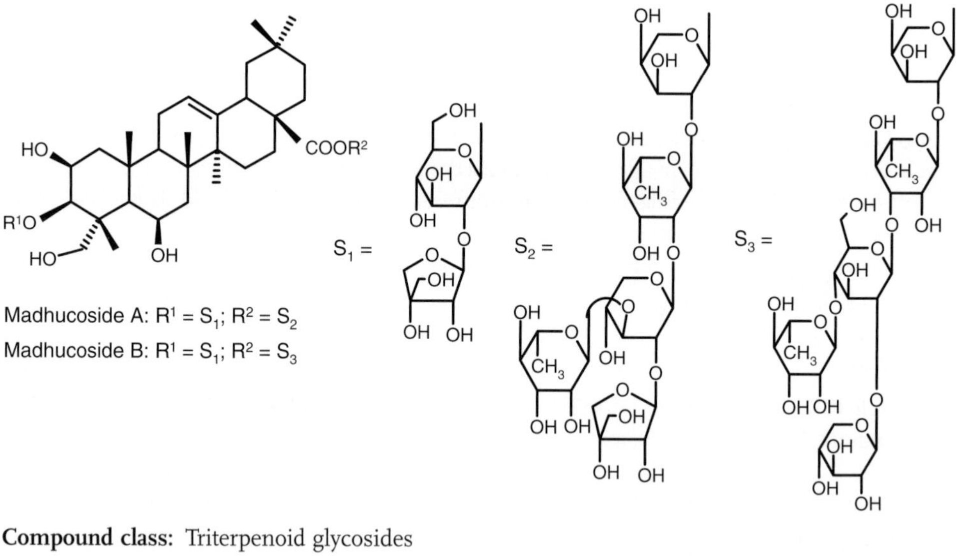

Madhucoside A: $R^1 = S_1$; $R^2 = S_2$
Madhucoside B: $R^1 = S_1$; $R^2 = S_3$

Compound class: Triterpenoid glycosides

Source: *Madhuca indica* J.F. Gmel (syn. *Madhuca latifolia, Bassia latifolia*) (barks; family: Sapotaceae)

Pharmaceutical potential: *Antioxidant*
The present investigators reported that both the compounds inhibit the production of hypochlorous acid by human neutrophils with IC$_{50}$ values of 21.9 and 144 µg/ml, respectively, studied in an *in vitro* luminol-enhanced chemiluminescence assay (using bacoside A$_3$ as a positive control, IC$_{50}$ = 19.84 µg/ml). Further, in the nitroblue tetrazolium (NBT) reduction assay used for the quantification of superoxide production from polymorphonuclear cells, madhucoside A was found to have an inhibitory effect with an IC$_{50}$ value of 138.3 µg/ml compared to the IC$_{50}$ values of 111.0 and 14.1 µg/ml for the two reference standards, quercetin and ascorbic acid; however, only 16% inhibition was achieved in the case of madhucoside B at the highest tested concentration of 200 µg/ml.

Reference

Pawar, R.S. and Bhutani, K.K. (2004) *J. Nat. Prod.*, **67**, 668.

(+)-Makassaric acid

Physical data: $C_{27}H_{38}O_3$, pale yellow oil, $[\alpha]_D^{25}$ +7.3° (MeOH, *c* 5.4)

Structure:

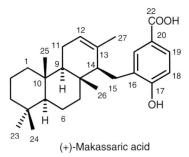

(+)-Makassaric acid

Compound class: Meroterpenoid

Source: *Acanthodendrilla* sp. (marine sponge; order: Dendroceratida, family: Dictyodendrillidae) [1]

Pharmaceutical potential: ***MAPKAP kinase 2 (MK2) inhibitor***
The isolate was found to possess moderate *in vitro* inhibitory activity against MAPKAP kinase 2 (MK2) enzyme with an IC_{50} value of 20 μM [1]. TNF-α plays a key role in developing various inflammatory diseases such as rheumatoid arthritis, and MK2 is reported to have a critical role in the regulation of TNF-α production; hence, MK2 kinase inhibitors represent potential therapeutic agents to treat inflammatory diseases [2].

References

[1] Williams, D.E., Telliez, J.-B., Liu, J., Tahir, A., van Soest, R., and Andersen, R.J. (2004) *J. Nat. Prod.*, **67**, 2127.
[2] Kotlyarov, A., Neininger, A., Schubert, C., Eckert, R., Birchmeier, C., Volk, H.-D., and Gaestel, M. (1999) *Nat. Cell Biol.*, **1**, 94.

Mallotophilippens A and B

Systematic names:

- Mallotophilippen A: 1-[5,7-dihydroxy-2,2-dimethyl-6-(2,4,6-trihydroxy-3-isobutyryl-5-methyl-benzyl)-2*H*-chromen-8-yl]-2-methyl-butan-1-one
- Mallotophilippen B: 1-[6-(3-acetyl-2,4,6-trihydroxy-5-methyl-benzyl)-5,7-dihydroxy-2,2-dimethyl-2*H*-chromen-8-yl]-2-methyl-butan-1-one

Mallotophilippens C–E

Physical data:
- Mallotophilippen A: $C_{28}H_{34}O_8$, yellow powder, $[\alpha]_D^{23}$ ±0° (MeOH, c 0.1)
- Mallotophilippen A: $C_{26}H_{30}O_8$, yellow powder, $[\alpha]_D^{23}$ ±0° (MeOH, c 0.1)

Structures:

Mallotophilippen A

Mallotophilippen B

Compound class: Phloroglucinol derivatives

Source: *Mallotus philippensis* (fruits; family: Euphorbiaceae)

Pharmaceutical potentials: *Inhibitors to NO production; anti-inflammatory; antiallergic*

Mallotophilippens A and B were found to exert potent inhibitory effect on nitric oxide (NO) production stimulated by lipopolysaccharide (LPS) in a murine macrophage-like cell line (RAW 264.7) with IC_{50} values of 4.2 and 3.2 µM, respectively; the efficacies were found to be more potent than that of quercetin used as a positive control (IC_{50} = 26.8 µM). The compounds also inhibited inducible nitric oxide synthase (iNOS) mRNA expression by RAW 264.7, which was activated by LPS and recombinant mouse interferon-γ (IFN-γ) in a dose-dependent fashion. Furthermore, mallotophilippens A and B were evaluated to have inhibitory activity against histamine release from rat peritoneal mast cells induced by compound 48/80, (histamine releasers; at a dose of 5 µg/ml) with respective IC_{50} values of 8.6 and 13.8 µM (indomethacin was used as a positive control; IC_{50} = 250 µM). Hence, the isolates may find useful application in the treatment of diseases related to inflammation and allergy.

Reference

Daikonya, A., Katsuki, S., Wu, J.-B., and Kitanaka, S. (2002) *Chem. Pharm. Bull.*, **50**, 1566.

Mallotophilippens C–E

Systematic names:
- Mallotophilippen C: 1-[6-(3,7-dimethyl-octa-2,6-dienyl)-5,7-dihydroxy-2,2-dimethyl-2H-chromen-8-yl]-3-(4-hydroxy-phenyl)-propenone
- Mallotophilippen D: 3-(3,4-dihydroxyphenyl)-1-[6-(3,7-dimethyl-octa-2,6-dienyl)-5,7-dihydroxy-2,2-dimethyl-2H-chromen-8-yl]-propenone
- Mallotophilippen E: 1-[5,7-dihydroxy-2-methyl-6-(3-methyl-but-2-enyl)-2-(4-methyl-pent-3-enyl)-2H-chromen-8-yl]-3-(3,4-dihydroxyphenyl)-propenone

Physical data:
- Mallotophilippen C: $C_{30}H_{34}O_5$, reddish yellow plate
- Mallotophilippen D: $C_{30}H_{34}O_6$, reddish yellow plate
- Mallotophilippen E: $C_{30}H_{34}O_6$, reddish yellow plate

Structures:

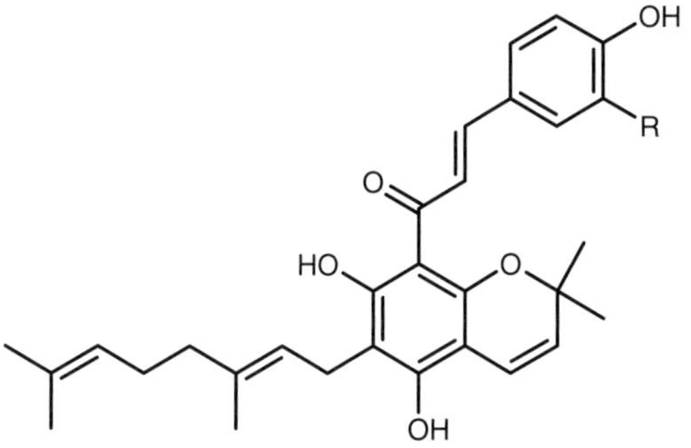

Mallotophilippen C: R = H
Mallotophilippen D: R = OH

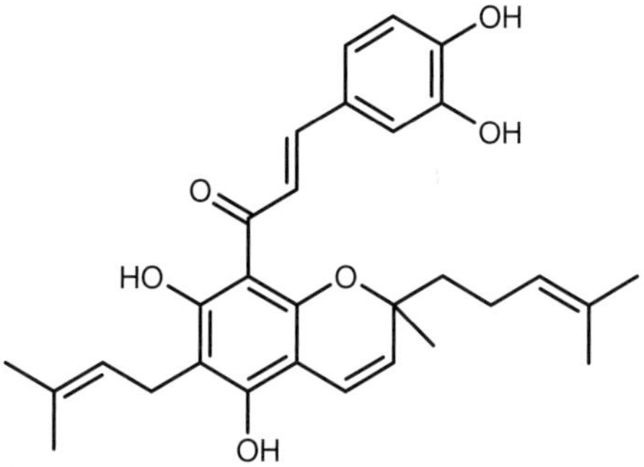

Mallotophilippen E

Compound class: Chalcones (flavonoids)

Source: *Mallotus philippinensis* (fruits; family: Euphorbiaceae)

Pharmaceutical potentials: *Inhibitors to NO production and iNOS gene expression*
The chalcone derivatives were found to possess potent inhibitory activities against lipopolysaccharide (LPS)- and recombinant mouse interferon-γ (IFN-γ)-induced nitric oxide (NO) production along with inducible NO synthase (iNOS) gene expression as evaluated in a murine macrophage-like cell line (RAW 264.7). The IC_{50} values for mallotophilippens C–E in inhibiting NO production by RAW 264.7 stimulated with LPS/IFN-γ were determined to be 7.6, 9.5, and 38.6 µM, respectively (quercetin was used as a positive control; IC_{50} = 26.8 µM). Furthermore, all of them downregulated cyclooxygenase-2 (COX-2) gene, interleukin-6 (IL-6) gene, and interleukin-1β (IL-1β) gene expression, thereby suggesting their possible anti-inflammatory and immunoregulatory effects.

Reference
Daikonya, A., Katsuki, S., and Kitanaka, S. (2004) *Chem. Pharm. Bull.*, **52**, 1326.

Mallotus benzopyrans 1 and 2

Systematic names:
- Benzopyran 1: 6-[1′-oxo-3′(R)-hydroxy-butyl]-5,7-dimethoxy-2,2-dimethyl-2H-1-benzopyran
- Benzopyran 2: 6-[1′-oxo-3′(R)-methoxy-butyl]-5,7-dimethoxy-2,2-dimethyl-2H-1-benzopyran

Physical data:
- 1: $C_{17}H_{22}O_5$, colorless oil, $[\alpha]_D^{25}$ −3.5° (CHCl$_3$, c 0.5)
- 2: $C_{18}H_{24}O_5$, colorless oil, $[\alpha]_D^{25}$ −3° (CHCl$_3$, c 0.5)

6α-Malonyloxymanoyl oxide

Structure:

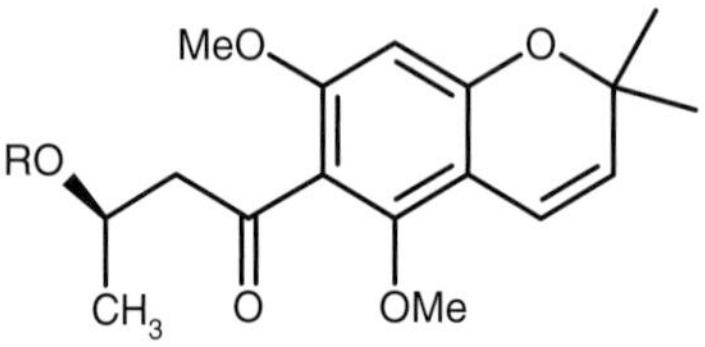

1: R = H
2: R = Me

Compound class: Benzopyrans

Source: *Mallotus apelta* Muell.-Arg. (leaves; family: Euphorbiaceae)

Pharmaceutical potential: *Cytotoxic*

The benzopyran **1** was reported to display strong cytotoxicity against two human cancer cell lines such as human hepatocellular carcinoma (Hep2) and rhabdosarcoma (RD) with respective IC_{50} values of 0.49 and 0.54 µg/ml, while compound **2** showed only moderate cytotoxic activity against the Hep2 cell line with an IC_{50} value of 4. 22 µg/ml.

Reference

Kiem, P.V., Dang, N.H., Bao, H.V., Huong, H.T., Minh, C.V., Huong, L.M., Lee, J.J., and Kim, Y.H. (2005) *Arch. Pharm. Res.*, **28**, 1131.

6α-Malonyloxymanoyl oxide

Physical data: $C_{23}H_{36}O_5$, white amorphous crystal, $[\alpha]_D^{30}$ +46.95° (CHCl$_3$, *c* 1.51)

Structure:

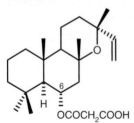

6α -Malonyloxymanoyl oxide

Compound class: Diterpenoid (labdane type)

Source: *Stemodia foliosa* Benth. (aerial parts; family: Scrophulariaceae)

Pharmaceutical potential: *Antibacterial*

The compound displayed moderate antibacterial activity against a bacteria panel consisting of *Staphylococcus aureus*, *Bacillus cereus*, *B. subtilis*, *B. anthracis*, *Micrococcus luteus*, *Mycobacterium smegmatis*, and *M. phlei* with respective MIC values of 15, 15, 15, 20, 17, 7, and 9 µg/ml (clarithromycin at a dose ranging from 0.5 to 2.0 µg/ml was used as a positive control).

Manassantins A and B

Reference

da Silva, L.L.D., Nascimento, M.S., Cavalheiro, A.J., Silva, D.H.S., Castro-Gamboa, I., Furlan, M., and da Silva Bolzani, V. (2008) *J. Nat. Prod.*, **71**, 1291.

Manassantins A and B

Physical data: Manassantin A: $C_{42}H_{52}O_{11}$, $[\alpha]_D - 100°$; Manassantin B: $C_{41}H_{48}O_{11}$, $[\alpha]_D - 99°$

Structures:

Manassantin A: $R^1 = R^2 = OCH_3$
Manassantin B: $R^1 = R^2 = -OCH_2O-$

Compound class: Dineolignoids

Sources: *Saururus cernuus* (family: Saururaceae) [1], *S. chinensis* Baill. (roots) [2]

Pharmaceutical potentials: *Inhibitor of transcription factor NF-κB and tumor necrosis factor-α (TNF-α); anti-inflammatory; anticancerous; neuroleptic; antiplasmodial*

Both the compounds were reported to possess anti-inflammatory potential; they were evaluated as potent inhibitors of transcription factor NF-κB [3] and also of tumor necrosis factor-α (TNF-α) that mediates systematic inflammation and immune responses by enhancing adhesion molecule expression and secretion of inflammatory mediators [4–6]. NF-κB is regarded as an important and attractive therapeutic target for drugs to treat many inflammatory diseases, including arthritis, asthma, and the autoimmune diseases – NF-κB has been established to have controlling effect on the expression of genes encoding the proinflammatory cytokines (e.g., IL-1, IL- 2, IL-6, TNF-α, etc.), chemokines (e.g., IL-8, MIP-1α, MCP1, RANTES, eotaxin, etc.), adhesion molecules (e.g., ICAM, VCAM, and E-selectin), inducible enzymes (COX-2 and iNOS), growth factors, some of the acute phase proteins, and immune receptors, all of which play critical roles in controlling most inflammatory processes [7]. Lee *et al*. [8] showed that manassantins A and B inhibit NF-κB activation by the suppression of transcriptional activity of RelA/p65 subunit of NF-κB; the compounds significantly inhibited the induced expression of NF-κB reporter gene by LPS or TNF-α in a dose-dependent manner – however, these compounds did not prevent the DNA-binding activity of NF-κB assessed by electrophoretic mobility shift assay as well as the induced-degradation of 1κB-α protein by LPS or TNF- α [8]. Further analysis revealed that manassantins A and B dose-dependently suppressed not only the induced NF-κB activation by overexpression of RelA/p65, but also transactivation activity of RelA/p65. Furthermore, treatment of cells with these compounds prevented the TNF-α-induced expression of antiapoptotic NF-κB target genes Bfl-1/A1, a prosurvival Bcl-2 homologue, and resulted in sensitizing HT-1080 cells to TNF-α-induced cell death; hence, manassantins A and B could be valuable candidates for the intervention of NF-κB-dependent pathological condition such as inflammation and cancer [8].

From their detailed study with manassantin B, Son et al. [9] showed that it induces attenuation of the expression of proinflammatory mediators probably results primarily from inhibition of NF-IL6, although inhibition of NF-κB may also play a part. The lignan inhibited phorbol 12-myristate 13-acetate (PMA)-induced expression of IL-1β, IL-1β mRNA, and IL-1β promoter activities in U937 cells with IC_{50} values of approximately 50 nM; it also inhibited NF-IL6- and NF-κB induced activation of IL-1β, with IC_{50} values of 78 nM and 1.6 µM, respectively, revealing a potent inhibitory effect on NF-IL6. Electrophoretic mobility shift assays showed that manassantin B had an inhibitory effect on DNA binding by NF-IL6, but not by NF-κB. Further analysis revealed that manassantin B suppresses expression of IL-1β in promonocytic cells by inhibiting not only NF-κB activity but also NF-IL6 activity. The present investigators [9] suggested that Manassantins suppress the transcriptional activity of NF-IL6 and NF-κB by modifying their phosphorylation status.

Rho et al. [2] reported that both the compounds inhibited PMA-induced intercellular adhesion molecule-1 (ICM-1) expression of HL-60 cells; hence, the same group of investigators [4] studied the effect of the compounds on the interaction of monocyte and human umbilical vein endothelial cells (HUVEC) and TNF-α-induced expression of ICAM-1, VCAM-1 (vascular cell adhesion molecule-1) and E-selection in HUVEC. It was found that when HUVEC were pretreated with manassantins A and B followed by stimulation with TNF-α, adhesion of THP-1 cells to HUVEC decreased with a dose-dependent manner with IC_{50} values of 5.0 and 7.0 ng/ml, respectively, without cytotoxicity. From their present study, the investigators suggested that both the compounds prevent monocyte adhesion to HUVEC through the inhibition of ICAM-1, VCAM-1, and E-selection expression stimulated by TNF-α, and might be useful for the prevention of atherosclerosis relevant to endothelial activation [4].

Hypoxia-inducible factor-1 (HIF-1) represents an important tumor-selective therapeutic target for solid tumors; Chowdhury et al. [10] evaluated both the compounds as potent inhibitors of hypoxia-induced HIF-1 activation with identical IC_{50} value of 3.0 nM, studied in a T47D cell-based reporter assay. It was also established that these compounds inhibited HIF-1 by blocking hypoxia-induced nuclear HIF-1α protein accumulation without affecting HIF-1α mRNA levels. Among other biological potentials, both the dineolignoids were also reported to have human ACAT inhibitory [11], murine neuroleptic [12], and in vitro antiplasmodial activities [13].

References

[1] Rao, K.V. and Alvarez, F.M. (1983) Tetrahedron Lett., **24**, 4947.
[2] Rho, M.C., Kwon, O.E., Kim, K., Lee, S.W., Chung, M.Y., Kim, Y.H., Hayashi, M., Lee, H.S., and Kim, Y.K. (2003) Planta Med., **69**, 1147.
[3] Hwang, B.Y., Lee, J.-H., Nam, J.B., Hong, Y.-S., and Lee, J.J. (2003) Phytochemistry, **64**, 765.
[4] Kwon, O.E., Lee, H.S., Lee, S.W., Chung, M.Y., Bae, K.H., Rho, M.-C., and Kim, Y.-K. (2005) Arch. Pharm. Res., **28**, 55.
[5] Pfeffer, K., Matsuyama, T., Kundig, T.M., Wakeham, A., Kishihara, K., Shahinian, A., Wiegmann, K., Ohashi, P.S., Kronke, M., and Mak, T.W. (1993) Cell, **73**, 457.
[6] Modur, V., Zimmerman, G.A., Prescott, S.M., and Mcintyre, T.M. (1996) J. Biol. Chem., **271**, 13094.
[7] Nam, N.-H. (2006) Mini Rev. Med. Chem., **6**, 945.

[8] Lee, J.-H., Hwang, B.Y., Kim, K.-S., Nam, J.B., Hong, Y.-S., and Lee, J.J. (2003) *Biochem. Pharmacol.*, **66**, 1925.
[9] Son, K.-N., Song, I.-S., Shin, Y.-H., Pai, T.-K., Chung, D.-K., Baek, N.-I., Lee, J.J., and Kim, J. (2005) *Mol. Cells*, **20**, 105.
[10] Chowdhury, F.H., Kim, Y.-P., Baerson, S.R., Zhang, L., Bruick, R.K., Mohammed, K.A., Agarwal, A.K., Nagle, D.G., and Zhou, Y.-D. (2005) *Biochem. Biophys. Res. Commun.*, **333**, 1026.
[11] Lee, W.S., Lee, D.-W., Baek, Y.-I., An, S.-J., Cho, K.-H., Choi, Y.-K., Kim, H.-C., Park, H.-Y., Bae, K.-H., and Jeong, T.-S. (2004) *Bioorg. Med. Chem. Lett.*, **14**, 3109.
[12] Rao, K.V., Puri, V.N., Diwan, P.K., and Alvarez, F.M. (1987) *Pharmacol. Res. Commun.*, **19**, 629.
[13] Kraft, C., Jenett-Siems, K., Kohler, I., Tofern-Reblin, B., Siems, K., Bienzle, U., and Eich, E. (2002) *Phytochemistry*, **60**, 167.

Mangiferin

Systematic name: 2-C-β-D-Glucopyranosyl-1,3,6,7-tetrahydroxyxanthone

Physical data: $C_{19}H_{18}O_{11}$, yellow crystalline powder, mp 270–274 °C

Structure:

Mangiferin

Compound class: *C*-Glucoxanthone

Sources: Mangiferin is by far the most widespread in nature; particularly, Leguminosae and Gentianaceae appeared to be the main centers of distribution for this type of compound [1]. Few are mentioned here. *Mangifera indica* (family: Anacardiaceae) [2–5]; *Canscora decussata* (family: Gentianaceae) [6]; *Cratoxylum pruniflorium* (family: Clusiaceae/Guttiferae) [7], *C. cochinchinense* [8]; *Cyclopia genistoides* R. Br. (family: Leguminosae) [9], *C. intermedia* [9], *C. maculate* [9], *C. sessiliflora* [9], *C. subternata* [10]; *Gentiana campestris* (family: Gentianaceae) [11], *G. cruciata* [12], *G. lutea* [13]; *Gentianopsis* (12 species) [14]; *Hypericum perforatum* (family: Clusiaceae/Guttiferae) [15], *H. sampsonii* [16]; *Salacia chinensis* (family: Hippocrateaceae) [17], *S. reticulate* [18]; *Swertia chirata* (family: Gentianaceae) [19], *S. cordata* [20], *S. corymbosa* [21], *S. elongate* [22], *S. franchetiana* [23], *S. punctata* [24]

Pharmaceutical potentials: *Antidiabetic; antihyperlipidemic; antiatherogenic; antitubercular; antioxidant; anti-inflammatory; immunomodulatory; antitumor, anticancerous, analgesic; hepatoprotective; cardioprotective; antiviral*

It is interesting to note that mangiferin was the first natural xanthone to be investigated for pharmacological purposes, and already established itself a highly promising natural polyphenolic compound possessing a variety of pharmacological properties [25]. When tested against *Mycobacterium tuberculosis* in mice, it displayed a very strong antituberculosis activity with no obvious toxicity to rats being detected on prolonged administration for four weeks [6].

Both mangiferin [26–31] and its glycoside (mangiferin-7-O-β-glucoside) showed antidiabetic property in mice [26, 29]; lowering in blood glucose level is responsible for increasing insulin sensitivity [26] and/or decreasing insulin resistance [29]. Mechanism studies revealed the inhibitory activity of mangiferin and its glycoside against several carbohydrate metabolizing enzymes [18, 32, 33]. In addition, mangiferin improved hyperinsulinemia and, on insulin tolerance test, reduced blood glucose levels of mice [28, 34, 35]. The compound at dose levels of 10–20 mg/kg exhibited antidiabetic, antihyperlipidemic, and antiatherogenic activities in streptozotocin (STZ)-induced diabetic rats and also showed the improvement in oral glucose tolerance in glucose-loaded normal rats without inducing hypoglycemic state [36]. Mangiferin and *Salacia reticulata* extract displayed antiobesity/lipolytic effects, thus, supporting their use as a supplementary food in Japan to prevent obesity and diabetes [30]. This natural chemical component having a number of advantageous properties, namely, antidiabetic, antihyperlipidenic, antiatherogenic, and antioxidant potentials without causing hypoglycemia would be of greater therapeutic benefit in the management of diabetes mellitus associated with abnormalities in lipid profiles; hence it demands merits for further detailed investigation to find out its exact mechanism of action and to establish its therapeutic potential in the treatment of diabetes and diabetic complications.

Pauletti et al. [37] investigated the antioxidant efficacies of mangiferin and mangiferin-derivatives [2-(2'-O-trans-cafeoyl)-C-β-D-glucopyranosyl-1,3,6,7-tetrahydroxyxanthone, 2-(2'-O-trans-cinnamoyl)-C-(-D-glucopyranosyl-1,3,6,7-tetrahydroxyxanthone, 2-(2'-O-trans-coumaroyl)-C-(-D-glucopyranosyl-1,3,6,7-tetrahydroxyxanthone, and 2-(2'-O-benzoyl)-C-(-D-glucopyranosyl-1,3,6,7-tetrahydroxyxanthone and muraxanthone], and found that all the compounds display free radical scavenging activity against DPPH, thereby establishing their antioxidant potential as evidenced by redox properties on EICD-HPLC. The antioxidant efficacy is, thus, due to their hydrogen-donating ability that enhances with the increase in number of hydroxyl or catechol units resulting in a more efficient radical scavenging potential [38]. Other workers [39–43] also proved the ability of mangiferin to scavenge free radicals involved in lipid peroxidation initiation [39, 40, 44]. It has also been found to protect the liver of rats from high altitude hypoxia [45]. On investigation in hepatic systems, it was found that extracts of *Mangifera indica* [46–48], *Cratoxylum cochinchinensis* [8], *Polygala elongate* [49], and *Salacia reticulate* [50] have the antioxidant profile similar to that of mangiferin, which is the principal polyphenolic component of those plant extracts. Besides, mangiferin was also found to exert its protective effect on liver [51], renal [52], cardiac [52], and brain tissues [51] from oxidative damage caused by reactive oxygen species (ROS), which are overproduced by peritoneal macrophages [50].

The anti-inflammatory and immunomodulatory properties of mangiferin were evidenced from its reductive effect on the expression of inflammation-related genes in macrophages such as cytokines, interleukins-1β [53], transforming growth factor-β (TGF-β) [42], TNF-α [42, 54, 55], colony-stimulating factor, NF-ϰB [53, 56] and secondary mediators such as NO synthase [42, 54, 55, 57], cyclooxygenase-2 (COX-2) [54, 56, 57], and intercellular adhesion molecule-1(ICAM-1) [56]. Furthermore, mangiferin is capable to inhibit TNF-induced p65 phosphorylation and translocation to nucleus, TNF-induced reactive oxygen intermediate generation, as well as NFϰB activation induced by other inflammatory agents [56]. Immunomodulatory activity of mangiferin was characterized from the observations of its effects on the expression, by activated mouse macrophages, of diverse genes related to the NFκB signaling pathway [58]. The inhibition of the Jun N-terminal kinase-1 (JNK1), together with stimulation of the Jun oncogene (c-JUN) [58] along with the previously reported superoxide-scavenging activity of mangiferin [41, 42, 53, 59, 60] clearly suggested that the test-compound might protect cells from oxidative damage and mutagenesis. Hence, mangiferin modulated the expression of a large number of genes critical to the regulation of apoptosis, inflammation, tumorigenesis, viral replication, and other various autoimmune diseases – the phenomena indicate the possibility of mangiferin to be used in the treatment of inflammatory diseases and/or cancer [58, 61]. The effect of mangiferin in rat colon carcinogenesis induced by the chemical carcinogen azoxymethane has been successfully examined [61]. It was also shown to have an antiproliferative effect on leukemia cells by induction of apoptosis, probably through downregulation of the bcr/abl gene expression [62]. Recently, Tang et al. [8] reported that a semipurified extract from *Cratoxylum cochinchinense*, which mostly contains mangiferin, is selectively toxic to certain tumor cell types, causing intense oxidative stress and ultimately cell death [63]. Besides, mangiferin showed *in vitro* and *in vivo*, growth inhibitory activity against ascitic fibrosarcoma [64], and enhanced tumor cell cytotoxicity of the splenic cells and peritoneal macrophages of normal and tumor-bearing mice [64].

Mangiferin was also found to antagonize the *in vitro* cytopathic effect of human immunodeficiency virus (HIV) [64]. It was found to be the active component in *Mangifera indica* extract for its activity against Herpes simplex virus [65–67], and it was suggested that mangiferin inhibited the late event in this virus replication [65, 68]. Bhattacharya et al. [69] found a remarkable CNS-stimulating effect of mangiferin in doses of 50–100 mg/kg, which could be blocked by pretreatment of chlorpromazine. This effect could be evidenced by hyperactivity, fine tremors, piloerection, increased spontaneous motility, sedation, potentiation of the analgesia, and also by potentiation of amphetamine toxicity in aggregated mice. Lin et al. [70] also reported the CNS stimulating effect of mangiferin. From the experimental observations of Ya et al. [71] mangiferin has been emerged as a potential hepatoprotective agent when tested against liver damage induced by acetaminophen, CCl_4, and D-GalN in rat.

References

[1] Richardson, P.M. (1983) *Biochem. Syst. Ecol.*, **11**, 371 (review); Jensen, S.R. and Schripsema, J. (2002) Chemotaxonomy and pharmacology of Gentianaceae, in Gentianaceae: Systematics and Natural History (eds L. Struwe and V. Albert), Cambridge University Press, pp. 573–631.
[2] Iseda, S. (1957) *Bull. Chem. Soc. Japan*, **30**, 625.
[3] Haynes, L.J. and Taylor, D.R. (1966) *J. Chem. Soc. (C)*, 1685.
[4] Bhatia, V.K. and Seshadri, T.R. (1968) *Tet. Lett.*, 1741.
[5] Nott, P.E. and Roberts, J.C. (1967) *Phytochemistry*, **7**, 741; Yoshimi, N., Matsunaga, K., Katayama, M., Yamada, Y., Kuno, T., Qiao, Z., Hara, A., Yamahara, J., and Mori, H. (2001) *Cancer Lett.*, **163**, 163.
[6] Ghosal, S. and Chaudhuri, R.K. (1975) *J. Pharm. Sci.*, **64**, 888.
[7] Kitanov, G.M., Assenov, I., and Van, A.T. (1988) *Pharmazie*, **43**, 879.
[8] Tang, S.Y., Whiteman, M., Peng, Z.F., Jenner, A., Yong, E.L., and Halliwell, B. (2004) *Free Radical Biol. Med.*, **36**, 1575.
[9] Joubert, E., Otto, F., Gruner, S., and Weinreich, B. (2003) *Eur. Food Res. Technol.*, **216**, 270.
[10] Kamara, B.I., Brand, D.J., Brandt, E.V., and Joubert, E. (2004) *J. Agric. Food Chem.*, **52**, 5391.
[11] Kaldas, M., Hostettmann, K., and Jacot-Guillarmod, A. (1975) *Helv. Chim. Acta*, **58**, 2188.

[12] Goetz, M., Hostettmann, K., and Jacot-Guillarmod, A. (1976) *Phytochemistry*, **15**, 2015.
[13] Menkovic, N., Savikin-Fudulovic, K., and Savin, K. (2000) *Planta Med.*, **66**, 178;Menkovic, N., Savikin-Fudulovic, K., Momcilovic, I., and Grubisic, D. (2000) *Planta Med.*, **66**, 96.
[14] Massias, M., Carbonnier, J., and Molho, D. (1982) *Biochem. Syst. Ecol.*, **10**, 319.
[15] Jurgenliemk, G. and Nahrstedt, A. (2002) *Planta Med.*, **68**, 88;Dias, A.C.P., Seabra, R.M., Andrade, P.B., Ferreres, F., and Ferreira, M.F. (2001) *J. Plant Physiol.*, **158**, 821.
[16] Don, M.-J., Huang, Y.-J., Huang, R.-L., and Lin, Y.-L. (2004) *Chem. Pharm. Bull.*, **52**, 866;Hong, D., Yin, F., Hu, L.-H., and Lu, P. (2004) *Phytochemistry*, **65**, 2595.
[17] Kishi, A., Morikkawa, T., Masuda, H., and Yoshikawa, M. (2003) *Chem. Pharm. Bull.*, **51**, 1051.
[18] Yoshikawa, M., Nishida, N., Shimoda, H., Takada, M., Kawahara, Y., and Matsuda, H. (2001) *Yakugaku Zasshi*, **121**, 371.
[19] Ghosal, S., Sharma, P.V., Chaudhuri, R.K., and Bhattacharya, S.K. (1973) *J. Pharm. Sci.*, **62**, 926;Mandal, S. and Chatterjee, A. (1987) *Tetrahedron Lett.*, **28**, 1309.
[20] Atta-ur-Rahman, Feroz, P.A.M., Choudhary, M.I., Qureshi, M.M., Perveen, S., Mir, I. Khan, M.I. (1994) *J. Nat. Prod*, **57**, 134.
[21] Ramesh, N., Viswanathan, M.B., Saraswathy, A., Balakrishna, K., Brindha, P., and Laksmanaperumalsamy, P. (2002) *Fitoterapia*, **73**, 160.
[22] Kong, D., Jiang, Y., Yao, Y., Luo, S., and and Li, H. (1995) *Zhongcaoyao*, **26**, 7–10.
[23] Yang, H., Duan, Y., Hu, F., and Liu, J. (2004) *Biochem. Syst. Ecol.*, **32**, 861.
[24] Menkovic, N., Savikin-Fodulovic, K., Bulatovic, V., Aljancic, I., Juranic, N., Mesquita, S., Vajs, V., and Milosavljevic, S. (2002) *Phytochemistry*, **61**, 415.
[25] Finnegan, R.A., Stephani, G.M., Ganguli, G., and Bhattacharya, S.K. (1968) *J. Pharm. Sci.*, **57**, 1039;Pinto, M.M.M., Sousa, M.E., and Nascimento, M.S.J. (2005) *Curr. Org. Chem.*, **12**, 2517; Fotie, J. and Bhole, D.S. (2006) *Anti-Infective Agents in Med. Chem.*, **5**, 15.
[26] Ichiki, H., Miura, T., Kubo, M., Ishihara, E., Komatsu, Y., Tanigawa, K., and Okada, M. (1998) *Biol. Pharm. Bull.*, **21**, 1389.
[27] Miura, T., Iwamoto, N., Kato, M., Ichiki, H., Kubo, M., Komatsu, Y., Ishida, T., Okada, M., and Tanigawa, K. (2001) *Biol. Pharm. Bull.*, **24**, 1091.
[28] Miura, T., Ichiki, H., Hashimoto, I., Iwamoto, N., Kato, M., Kubo, M., Ishihara, E., Komatsu, Y., Okada, M., Ishida, T., and Tanigawa, K. (2001) *Phytomedicine*, **8**, 85.
[29] Miura, T., Ichiki, H., Iwamoto, N., Kato, M., Kubo, M., Sasaki, H., Okada, M., Ishida, T., Seino, Y., and Tanigawa, K. (2001) *Biol. Pharm. Bull.*, **24**, 1009.
[30] Yoshikawa, M., Shimoda, H., Nishida, N., Takada, M., and Matsuda, H. (2002) *J. Nutr.*, **132**, 1819.
[31] Li, Y., Peng, G., Li, Q., Wen, S., Huang, T.H.-W., Roufogalis, B.D., and Yamahara, J. (2004) *Life Sci.*, **75**, 1735.
[32] Li, W.L., Zheng, H.C., Bukuru, J., and De Kimpe, N. (2004) *J. Ethnopharmacol.*, **92**, 1.
[33] Morikawa, T., Kishi, A., Pongpiriyadacha, Y., Matsuda, H., and Yoshikawa, M. (2003) *J. Nat. Prod.*, **66**, 1191.
[34] Matsuda, H., Tokunaga, M., Hirata, N., Iwahashi, H., Naruto, S., and Kubo, M. (2004) *Nat. Med.*, **58**, 278.
[35] Karunanayake, E.H. and Sirimanne, S.R. (1985) *J. Ethnopharm.*, **13**, 227.
[36] Muruganandan, S., Srinivasan, K., Gupta, S., Gupta, P.K., and Lal, J. (2005) *J. Ethnopharm.*, **97**, 497.
[37] Pauletti, P.M., Castro-Gamboa, I., Silva, D.H.S., Young, M.C.M., Tomazela, D.M., Eberlin, M.N., and Bolzani, V.S. (2003) *J. Nat. Prod.*, **66**, 1384.

[38] Son, S. and Lewis, B.A. (2002) *J. Agric. Food Chem.*, **50**, 468.
[39] Sato, T., Kawamoto, A., Tamura, A., Tatsumi, Y., and Fujii, T. (1992) *Chem. Pharm. Bull.*, **40**, 721.
[40] Ghosal, S., Rao, G., Saravana, V., Misra, N., and Rana, D. (1996) *Indian J. Chem. B*, **35**, 561.
[41] Calliste, C.A., Trouillas, P., Allais, D.P., Simon, A., and Duroux, J.L. (2001) *J. Agric. Food Chem.*, **49**, 3321.
[42] Leiro, J.M., Álvarez, E., Arranz, J.A., Siso, I.G., and Orallo, F. (2003) *Biochem. Pharmacol.*, **65**, 1361.
[43] Salvi, A., Brhlmann, C., Migliavacca, E., Carrupt, C.A., Hostettmann, K., and Testa, B. (2002) *Helv. Chim. Acta*, **85**, 867.
[44] Born, M., Carrupt, P.A., Zini, R., Brée, F., Tillement, J.P., Hostettmann, K., and Testa, B. (1996) *Helv. Chim. Acta*, **79**, 1147.
[45] Jizeng, D., Quingfen, L., and Chen, X. (1983) *Yaoxue Xuebao*, **18**, 174;*Chem. Abst.*, 1983, **99**, 16378.
[46] Sánchez, G.M., Rodríguez, M.A., Giuliani, A., Núñez Sellés, A.J., Rodríguez, N.P., Fernndez, H., Leon, O.S., and Re, L. (2003) *Phytother. Res.*, **17**, 197.
[47] Martinez, G., Delgado, R., Perez, G., Garrido, G., Nunez Selles, A.J., and Leon, O.S. (2000) *Phytother. Res.*, **14**, 424.
[48] Martinez, G., Giuliani, A., Leon, O.S., Perez, G., and Nunez Selles, A.J. (2001) *Phytother. Res.*, **15**, 581.
[49] Shirwaikar, A., Padma, R., Kumar, A.V., and Rao, L. (2002) *Indian J. Pharm. Sci.*, **64**, 345.
[50] Yoshikawa, M., Ninomiya, K., Shimoda, H., Nishida, N., and Matsuda, H. (2002) *Biol. Pharm. Bull.*, **25**, 72.
[51] Sanchez, G.M., Re, L., Giuliani, A., Nunez-Selles, A.J., Davison, G.P., and Leon-Fernandez, O.S. (2000) *Pharmacol. Res.*, **42**, 565.
[52] Muruganandan, S., Gupta, S., Kataria, M., Lal, J., and Gupta, P.K. (2002) *Toxicology*, **176**, 165.
[53] Garcia, D., Leiro, J., Delgado, R., Sanmartin, M.L., and Ubeira, F.M. (2003) *Phytother. Res.*, **17**, 1182.
[54] Leiro, J., Garcya, D., Arranz, J.A., Delgado, R., Sanmartyn, M.L., and Orallo, F. (2004) *Int. Immunopharmacol.*, **4**, 991.
[55] Garrido, G., Delgado, R., Lemus, Y., Rodriguez, J., Garcia, D., and Nunez-Selles, A. (2004) *J. Pharmacol. Res.*, **50**, 165.
[56] Sarkar, A., Sreenivasan, Y., Ramesh, G.T., and Manna, S.K. (2004) *J. Biol. Chem.*, **279**, 33768.
[57] Beltran, A.E., Alvarez, Y., Xavier, F.E., Hernanz, R., Rodriguez, J., Nunez, A.J., Alonso, M.J., and Salaices, M. (2004) *Eur. J. Pharmacol.*, **499**, 297.
[58] Leiro, J., Arranz, J.A., Yanez, M., Ubeira, F.M., Sanmartin, M.L., and Orallo, F. (2004) *Int. Immunopharmacol.*, **4**, 763.
[59] Garcia, D., Delgado, R., Ubeira, F.M., and Leiro, J. (2002) *Int. Immunopharmacol.*, **2**, 797.
[60] Makare, N., Bodhankar, S., and Rangari, V. (2001) *J. Ethnopharmacol.*, **78**, 133.
[61] Yoshimi, N., Matsunaga, K., Katayama, M., Yamada, Y., Kuno, T., Qiao, Z., Hara, A., Yamahara, J., and Mori, H. (2001) *Cancer Lett.*, **163**, 163.
[62] Peng, Z.G., Luo, J., Xia, L.H., Chen, Y., and Song, S.J. (2004) *Zhongguo Shi Yan Xue Ye Xue Za Zhi*, **12**, 590.
[63] Tang, S.Y., Whiteman, M., Jenner, A., Peng, Z.F., and Halliwell, B. (2004) *Free Radic. Biol. Med.*, **36**, 1588.

[64] Cao, Z., Baguley, B.C., and Ching, L.M. (2001) *Cancer Res.*, **61**, 1517.
[65] Zhu, X.M., Song, J.X., Huang, Z.Z., Wu, Y.M., and Yu, M.J. (1993) *Zhongguo Yao Li Xue Bao*, **14**, 452.
[66] Yoosook, C., Bunyapraphatsara, N., Boonyakiat, Y., and Kantasuk, C. (2000) *Phytomedicine*, **6**, 411.
[67] Vichkanova, S.A., Shipulina, L.O., Glyzin, V.I., Bankovskii, A.I., Pimanov, M.G., and Boryaev, K.I.Ger. Pat. DE 3 141 970/1983.
[68] Zheng, M.S. and Lu, Z.Y. (1990) *Chin. Med. J. Engl.*, **103**, 160.
[69] Bhattacharya, S.K., Ghosal, S., Chaudhuri, R.K., and Sanyal, A.K. (1972) *J. Pharm. Sci.*, **61**, 1838;Ghosal, S., Sharma, P.V., Chaudhuri, R.K., and Bhattachrya, S.K. (1975) *J. Pharm. Sci.*, **64**, 80.
[70] Lin, C., Chung, M.I., Ariswa, M., Shumizu, M., and Morita, N. (1984) *Shoyakugaki Yasshi*, **38**, 80;*Chem. Abstr.* 1984, **101**, 216263.
[71] Ya, B.Q., Nian, L.C., and Gen, X.P. (1998) *Pharm. Pharmacol. Commun.*, **4**, 597.

Mangiferin 7-O-β-glucoside

See Mangiferin

Mannioside A

Systematic name: 3β,17α-Dihydroxyspirost-5-ene-3-O-α-L-rhamnopyranosyl-(1→3)-β-D-glucopyranoside; alternatively, pennogenin-3-O-α-L-rhamnopyranosyl-(1→3)-β-D-glucopyranoside

Physical data: $C_{39}H_{62}O_{13}$, white amorphous powder

Structure:

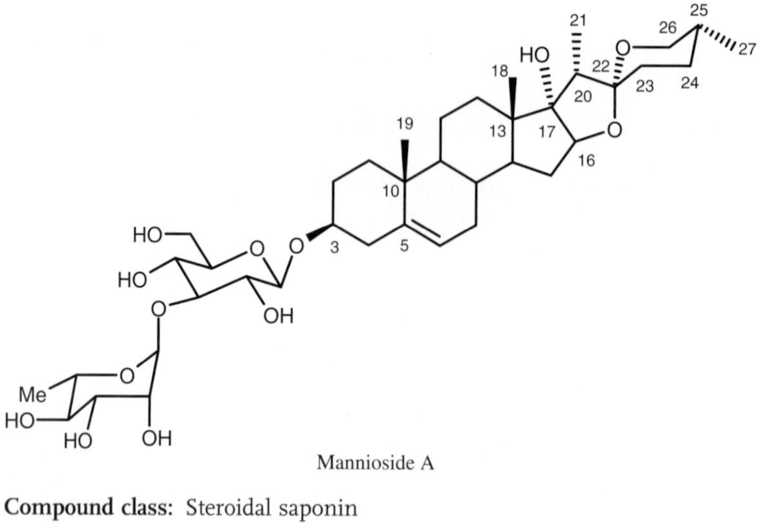

Mannioside A

Compound class: Steroidal saponin

Source: *Dracaena mannii* (stem barks; family: Dracaenaceae/Agavaceae)

Pharmaceutical potential: *Anti-inflammatory*

The steroidal saponin was evaluated to exert significant inhibition activity against carrageenan-induced paw edema in rats; the test compound, 1 h after carrageenan injection at a dose of 10 mg/kg body weight, showed a maximum inhibitory activity of 80.57%; indomethacin (used as a reference standard) produced an anti-inflammatory activity that increased gradually to reach a maximum inhibition of 62.36% at the fourth hour.

Reference

Tapondjou, L.A., Ponou, K.B., Teponno, R.B., Mbiantcha, M., Djoukeng, J.D., Nguelefack, T.B., Watcho, P., Cadenas, A.G., and Park, H.-J. (2008) *Arch. Pharm. Res.*, **31**, 653.

Manoalide

Physical data: $C_{25}H_{36}O_5$, colorless amorphous solid

Structure:

Manoalide

Compound class: Sesquiterpenoid antibiotic

Source: *Luffariella variabilis* (marine-derived sponge) [1]

Pharmaceutical potentials: *Anti-inflammatory; antibacterial; inhibitor to cobra venom and bee venom*

Manoalide, a sesterterpenoid with an α,β-unsaturation along with a δ-lactone and a hemiacetal ring, was first isolated as an antibacterial agent from the marine sponge *Luffariella variabilis* [1]. Manoalide is also well known for its potent anti-inflammatory activity as a selective inhibitor of phospholipase A2 (PLA2) [2]; from the experiments of Jacobs and his group [3, 4] it was revealed that manoalide antagonized phorbol ester induced local inflammation in murine epidermis but not that induced by arachidonic acid, thereby, suggesting that the test compound acts prior to the cyclooxygenase step in prostaglandin synthesis, possibly by inhibiting phospholipase A2 (PLA2). Manoalide has also been shown to be a potent inhibitor of both cobra ($IC_{50} = 2\,\mu M$) [2, 5, 6] and bee venom ($IC_{50} = 0.05\,\mu M$) [7–9]. It has been suggested that modification of phospholipase A2 (PLA2) involves a Michael addition of lysine residues to one or both of the α,β-unsaturated aldehydes (generated from opening of the δ-lactone and hemiacetal rings of manoalide molecule) [5, 6]; this phenomenon received from the fact that aplysinoplides A and B did not inhibit bovine pancreas PLA2 at a concentration of 100 μM, whereas manoalide exhibited a potent activity in a parallel experiment confirming the important role of the C-24 aldehyde for inhibition of PLA2 [10].

Manoalide has been developed as a novel lead for its anti-inflammatory properties [11–13]. However, it also possesses other activities, such as acting as calcium channel blocker in several cell types [13, 14]. Wheeler *et al.* [14] showed that in A431 cells the increase in epidermal growth factor receptor-mediated Ca^{2+} entry and release from intracellular Ca^{2+} stores were blocked by manoalide ($IC_{50} = 0.4$ µM) in a time-dependent manner, and in GH3 cells, the test compound blocked the thyrotropin-releasing hormone-dependent release of Ca^{2+} from intracellular stores without inhibition of the formation of inositol phosphates from phosphatidylinositol 4,5-bisphosphate. In addition, manoalide also inhibited the Ca^{2+} influx induced by concanavalin A in mouse spleen cells in a time- and temperature-sensitive manner with an IC_{50} value of 0.07 µM [14].

References

[1] de Silva, E.D. and Scheuer, P.J. (1980) *Tetrahedron Lett.*, **21**, 1611.
[2] Reynolds, L.J., Morgan, B.P., Hite, G.A., Mihelich, E.D., and Dennis, E.A. (1988) *J. Am. Chem. Soc.*, **110**, 5172.
[3] Jacobs, R.S., Culver, P., Langdon, R., O'Brien, T., and White, S. (1985) *Tetrahedron*, **41**, 981.
[4] Burley, E.S., Smith, B., Cutter, G., Ahlem, J.K., and Jacobs, R.S. (1982) *Pharmacologist*, **24**, 117.
[5] Lombardo, D. and Dennis, E.A. (1985) *J. Biol. Chem.*, **260**, 7234.
[6] Deems, R.A., Lombardo, D., Morgan, B.P., Mihelich, E.D., and Dennis, E.A. (1981) *Biochem. Biophys. Acta*, **917**, 258.
[7] Glaser, K.B. and Jacobs, R.S. (1986) *Biochem. Pharmacol.*, **35**, 449.
[8] Glaser, K.B. and Jacobs, R.S. (1987) *Biochem. Pharmacol.*, **36**, 2079.
[9] Potts, B.C.M., Faulkner, D.J., De Carvalho, M.S., and Jacobs, R.S. (1992) *J. Am. Chem. Soc.*, **114**, 5093.
[10] Glaser, K.B., De Carvalho, M.S., Jacobs, R.S., Kernan, M.R., and Faulkner, D.J. (1989) *Mol. Pharm.*, **36**, 782.
[11] Potts, B.C., Faulkner, D.J., and Jacobs, R.S. (1992) *J. Nat. Prod.*, **55**, 1701.
[12] Lombardo, D., Morgan, B.P., Mihelich, E.D., and Dennis, E.A. (1987) *Biochem. Biophys. Acta*, **917**, 258.
[13] De Vries, G.W., Mclaughlin, A., Wenzel, M.B., Perez, J., Harcourt, D., Lee, G., Garst, M., and Wheeler, L.A. (1995) *Inflammation*, **19**, 261.
[14] Wheeler, L.A., Sachs, G., De Vries, G., Goodrum, D., Woldemussie, E., and Muallem, S. (1987) *J. Bio. Chem.*, **262**, 6531.

Manzamine A and its hydroxy derivatives

Physical data:
- Manzamine A: isolated as hydrochloride salt (positive charge on pyrrolidinium nitrogen), $C_{36}H_{44}N_4O \cdot HCl$, colorless crystals (MeOH), mp >240 °C, $[\alpha]_D^{20}$ +50° ($CHCl_3$, *c* 0.28) [1]
- 6-Hydroxymanzamine A: $C_{36}H_{44}N_4O_2$ [2]
- 8-Hydroxymanzamine A: $C_{36}H_{44}N_4O_2 \cdot HCl$, pale yellow crystals, mp >230 °C, $[\alpha]_D$ +118.5° ($CHCl_3$, *c* 1.94) [3]

Structures:

Manzamine A: $R^1 = R^2 = H$
6-Hydroxymanzamine A: $R^1 = OH$, $R^2 = H$
8-Hydroxymanzamine A: $R^1 = H$, $R^2 = OH$

Compound class: Alkaloids (complex β-carboline alkaloids)

Sources: Manzamine A: *Haliclona* sp. (Okinawan sponge) [1]; *Pachypellina* sp. (sponge) [3], *Acanthostrongylophora* (common Indonesian sponge) [4]; 6-hydroxymanzamine A: *Haliclona* sp. [2], *Acanthostrongylophora* (common Indonesian sponge) [4]; 8-hydroxymanzamine A: *Pachypellina* sp. (sponge) [3], *Acanthostrongylophora* (common Indonesian sponge) [4]

Pharmaceutical potentials: *Antitumor; anti-HSV-II; GSK-3 and CDK-5inhibitor; antibacterial; antifungal; anti-HIV*

Manzamine A was reported to inhibit the growth of P388 mouse leukemia cells with an IC_{50} value of 0.07 µg/ml [1]. The compound and its 8-hydroxy isomer also showed antitumor activity against the KB and LoVo cell lines with respective IC_{50} values of 0.05 and 0.15 µg/ml for manzamine A; 0.30 and 0.26 µg/ml for 8-hydroxymanzamine A [3]; the same investigators also noted that both of them exhibited activity against the herpes simplex virus-II (HSV-II) with respective MIC values of 0.05 and 0.1 µg/ml [3].

Glycogen synthase kinase-3 (GSK-3), a serine–threonine kinase, has already emerged as one of the most attractive therapeutic targets for the development of selective inhibitors as new promising drugs for severe unmet pathologies, such as type 2 diabetes, bipolar disorders, stroke, and Alzheimer's disease, and different tau pathologies, such as Pick's disease, supranuclear palsy, and front-temporal dementia [5]. GSK-3 and also CDK-5 are known to phosphorylate the microtubule-associated protein tau in mammalian cells [6] – these two kinases are the key players in the hyperphosphorylation of tau protein in Alzheimer's disease [7]. Hamann *et al.* [4] evaluated that manzamine A, and its 6- and 8- hydroxy derivatives inhibit specifically human GSK-3β *in vitro*, respectively, by 73.2, 74.3, and 86.7% at a concentration of 25 µM; the respective IC_{50} values were determined as 10.2, 16.6, and 4.8 µM. From detailed experimental observations, the investigators suggested that manzamine A acts as a noncompetitive inhibitor of ATP binding, because the inhibition was not interfered on enhancing the ATP concentration [4]. Moreover, manzamine A also inhibited CDK-5 and proved to be effective in decreasing tau phosphorylation on human neuroblastoma cell lines, a demonstration of its ability to enter cells and interfere with tau pathology [4]. These results suggest that manzamine A constitutes a new scaffold from which more potent and selective GSK-3 inhibitors could be designed as potential therapeutic agents for the treatment of diseases mediated by GSK-3, such as the Alzheimer's disease.

Manzamine A and its 8-hydroxy derivative were also reported to display useful antibacterial and antifungal activities [8]; the respective IC_{50} values for antibacterial activity against *Staphylococcus aureus* and methicillin-resistant *Staph. aureus* were determined as: 0.5 and 0.7 μg/ml for manzamine A, and 0.9 and 4.0 μg/ml for 8-hydroxymanzamine A. The respective IC_{50} values of manzamine A and 8-hydroxymanzamine A against the targeted fungi were evaluated as: 3.0 μg/ml each against *Cryptococcus neoformans* and *Mycobacterium intracellulare*; 3.0 and 1.0 μg/ml, respectively, against *C. neoformans* and *M. intracellulare*. The same investigators [8] also reported anti-HIV potential of both the compounds having respective EC_{50} values of 4.2 and 0.59 mM.

References

[1] Sakai, R., Higa, T., Jefford, C.W., and Bernardinelli, G. (1986) *J. Am. Chem. Soc.*, **108**, 6404.
[2] Kobayashi, M., Chen, Y.J., Aoki, S., In, Y., Ishida, T., and Kitagawa, I. (1995) *Tetrahedron*, **51**, 3727.
[3] Ichiba, T., Corgiat, J.M., Scheuer, P.J., and Kelly-Borges, M. (1994) *J. Nat. Prod.*, **57**, 168.
[4] Hamann, M., Alonso, D., Martín-Aparicio, E., Fuertes, A., Pérez-Puerto, M.J., Castro, A., Morales, S., Navarro, M.L., del Monte-Millán, M., Medina, M., Pennaka, H., Balaiah, A., Peng, J., Cook, J., Wahyuono, S., and Martínez, A. (2007) *J. Nat. Prod.*, **70**, 1397.
[5] Martinez, A., Castro, A., Dorronsoro, I., and Alonso, M. (2002) *Med. Res. Rev.*, **22**, 373;Cohen, P. and Frame, S. (2001) *Nat. Rev. Mol. Cell Biol.*, **2**, 769.
[6] Lovestone, S., Reynolds, C.H., Latimer, D., Davis, D.R., Anderton, B.H., Gallo, J.M., Hanger, D., Mulot, S., and Marquardt, B. (1994) *Curr. Biol.*, **4**, 1077;Takashima, A., Murayama, M., Yasutake, K., Takahashi, H., Yokoyama, M., and Ishiguro, K. (2001) *Neurosci. Lett.*, **306**, 37.
[7] Imahori, K. and Uchida, T. (1997) *J. Biochem.*, **121**, 179;Flaherty, D.B., Soria, J.P., Tomasiewicz, H.G., and Wood, J.G. (2000) *J. Neurosci. Res.*, **62**, 463.
[8] Yousaf, M., Hammond, N.L., Peng, J., Wahyuono, S., McIntosh, K.A., Charman, W.N., Mayer, A.M.S., and Hamann, M.T. (2004) *J. Med. Chem.*, **47**, 3512.

Maoecrystal Z

Systematic name: 6β-Hydroxy-11α-acetoxy-6,7: 8,15-di-*seco*-7,20-olide-6,8-cyclo-*ent*-kaur-16-en-15-aldehyde

Physical data: $C_{22}H_{30}O_6$, colorless cubes (Me$_2$CO–MeOH–H$_2$O 90: 9: 1), $[\alpha]_D^{18.5}$ −119.05° (MeOH, *c* 0.98)

Structure:

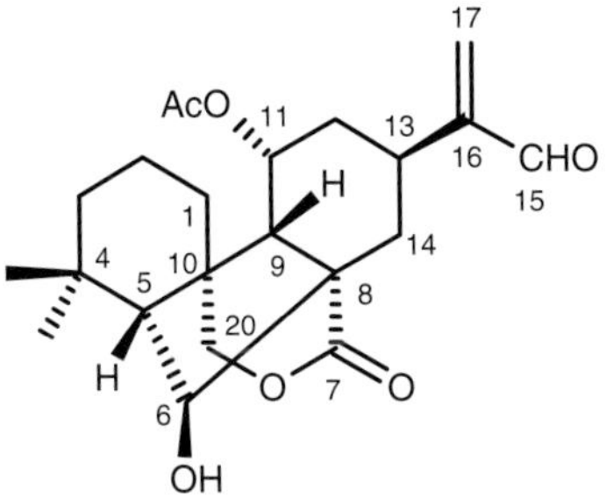

Maoecrystal Z

Compound class: Diterpenoid

Source: *Isodon eriocalyx* (Dunn.) Hara (leaves; family: Labiatae)

Pharmaceutical potential: *Cytotoxic (antitumor)*
The diterpenoid showed significant inhibitory efficacy against human K562 leukemia, MCF7 breast, and A2780 ovarian tumor cells with IC_{50} values of 2.90 ± 0.28, 1.63 ± 0.34, and 1.45 ± 0.03 μg/ml, respectively (camptothecin and paclitaxel were used as positive controls).

Reference

Han, Q.-B., Cheung, S., Tai, J., Qiao, C.-F., Song, J.-Z., Tso, T.-F., Sun, H.-D., and Xu, H.-X. (2006) *Org. Lett.*, **8**, 4727.

Maprounic acid derivatives

Systematic names:
- Compound 1: 1β-hydroxymaprounic acid 3-*p*-hydroxybenzoate
- Compound 2: 2α-hydroxymaprounic acid 2,3-bis-*p*-hydroxybenzoate

Physical data:
- Compound 1: $C_{37}H_{52}O_6$, mp 266–267 °C, $[\alpha]_D$ +7.8° (Py, *c* 0.103)
- Compound 2: $C_{44}H_{57}O_8$, mp 275–276 °C, $[\alpha]_D$ −38.6° (Py, *c* 0.13)

Structures:

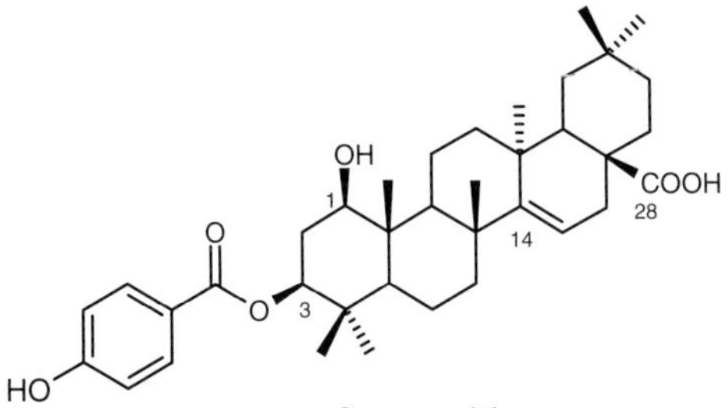

Compound **1**

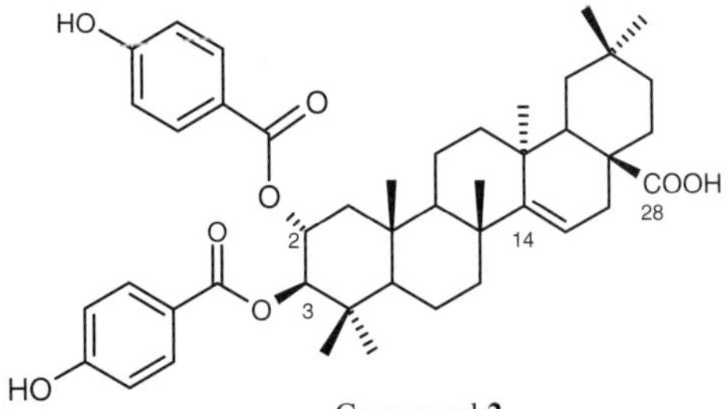

Compound **2**

Compound class: Pentacyclic triterpenoids

Source: *Maprounea africana* Muell.-Arg. (roots; family: Euphorbiaceae)

Pharmaceutical potential: *Anti-HIV*
The investigators tested both the compounds against HIV-1 reverse transcriptase (HIV-1 RT) and HIV-2 reverse transcriptase (HIV-2 RT); both of them were found to be potent inhibitors of HIV-1 RT with an equal IC_{50} value of 3.7 µM. However, both of them were less active toward HIV-2 RT compared to HIV-1 RT with respective IC_{50} values of 59.0 and 28.9 µM.

Reference

Pengsuparp, T., Cai, L., Fong, H.H.S., Kinghorn, A.D., Pezzuto, J.M., Wani, M.C., and Wall, M.E. (1994) *J. Nat. Prod.*, **57**, 415.

Marianine and marianosides A and B

Systematic names:
- Marianine: 24-methylene-7-oxo-lanosta-8(9)-ene-3β,25-diol
- Marianoside A: 24-methylene-lanoesta-8(9)-ene-25,28-diol 3-*O*-β-D-glucopyranoside
- Marianoside B: 24-methylenelanosta-8(9)-ene 3-*O*-β-D-glucopyranoside

Physical data:
- Marianine: $C_{31}H_{50}O_3$, colorless crystals, mp 160–161 °C, $[\alpha]_D^{20}$ +73° (CHCl$_3$, *c* 1.0)
- Marianoside A: $C_{37}H_{62}O_7$, amorphous solid, mp 271–273 °C, $[\alpha]_D^{25}$ −28° (CHCl$_3$, *c* 1.0)
- Marianoside B: $C_{37}H_{63}O_6$, amorphous solid, mp 240–242 °C, $[\alpha]_D^{25}$ −45.5° (CHCl$_3$, *c* 0.1)

Structures:

Marianine

Marianoside A: R = OH
Marianoside B: R = H

Compound class: Lanostane triterpenoid and its glycosides

Source: *Silybum marianum* Gaerth (whole plants; family: Compositae) [1]

Pharmaceutical potential: *Chymotrypsin inhibitory (anti-hepatitis C virus)*
Hepatitis C virus (HCV) is thought to replicate with the help of NS3 protease, a chymotrypsin-like serine protease; hence, inhibition to such protease enzymes, which are somewhat essential for the viral replication, has become a target for anti-HCV drugs [2–4]. Ahmed *et al.* [1] evaluated these lanostane terpenoids for their *in vitro* inhibitory potential against chymotrypsin. All the three

isolates were found to have significant inhibition against this enzyme – marianine and marianosides A and B showed the inhibitory activity in a dose-dependent manner with respective IC_{50} values of 9.4 ± 0.02, 22.6 ± 0.1, and $28.2 \pm 0.8\,\mu M$ (chymostatin was used as a positive control, $IC_{50} = 7.01 \pm 0.1\,\mu M$) [1].

References

[1] Ahmed, E., Malik, A., Ferheen, S., Afza, N., Azhar-Ul-Haq, Lodhi, M.A., and Choudhary, M.I. (2006) *Chem. Pharm. Bull.*, **54**, 103.
[2] Starkey, P.M. (1977) *Acta Biol. Med. Germ.*, **36**, 1549.
[3] Patrick, A.K. and Potts, K.E. (1998) *Clin. Microb. Rev.*, **11**, 614.
[4] Hussain, S., Ahmed, E., Malik, A., Jabbar, A., Ashraf, M., Lodhi, M.A., and Choudhary, M.I. (2006) *Chem. Pharm. Bull.*, **54**, 623.

Marineosins A and B

Physical data:
- Marineosin A: $C_{25}H_{35}N_3O_2$, colorless oil, $[\alpha]_D$ $-101.7°$ (MeOH, c 0.06)
- Marineosin B: $C_{25}H_{35}N_3O_2$, colorless oil, $[\alpha]_D$ $+143.5°$ (MeOH, c 0.09)

Structures:

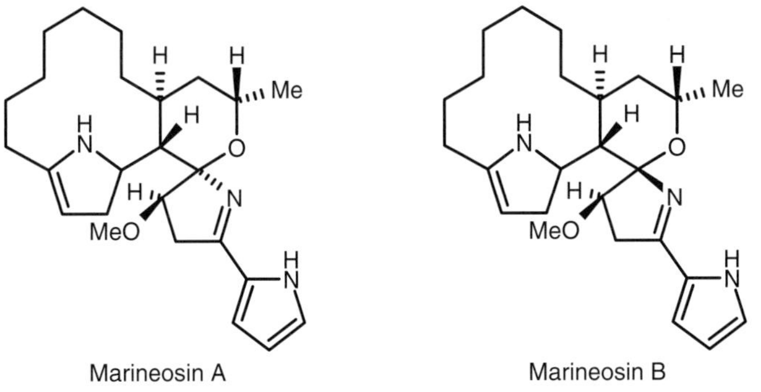

Marineosin A   Marineosin B

Compound class: Spiroaminals

Source: *Streptomyces* sp. (a marine sediment-derived actinomycete; culture broth)

Pharmaceutical potential: *Cytotoxic*
Marineosins A and B were evaluated for their ability to arrest cell division in the HCT-116 human colon tumor cell line; the former was found to be more active in the HCT-116 cytotoxicity assay ($IC_{50} = 0.5\,\mu M$) than the latter ($IC_{50} = 46\,\mu M$). The present investigators suggested that the difference in configuration at the spiroaminal center and in the tetrahydropyran conformation appears to significantly affect their bioactivities.

Reference

Boonlarppradab, C., Kauffman, C.A., Jensen, P.R., and Fenical, W. (2008) *Org. Lett.*, **10**, 5505.

Marinopyrroles A and B

Physical data:
- Marinopyrrole A: $C_{22}H_{12}Cl_4N_2O_4$
- Marinopyrrole B: $C_{22}H_{11}BrCl_4N_2O_4$

Structures:

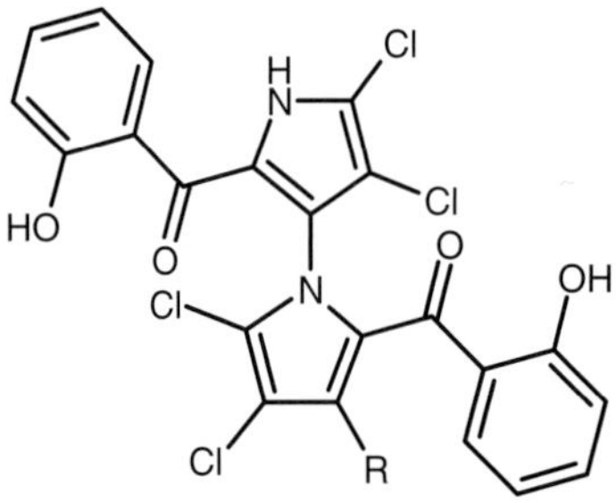

Marinopyrrole A: R = H
Marinopyrrole A: R = Br

Compound class: Antibiotics (halogenated bipyrrole derivatives)

Source: *Streptomyces* sp. strain CNQ-418 (culture broth)

Pharmaceutical potentials: *Antibacterial; cytotoxic*
Marinopyrroles A and B exhibited growth inhibitory activity against methicillin-resistant *Staphylococcus aureus* with MIC_{90} values of 0.61 and 1.1 µM, respectively (MIC_{90} value of vancomycin (positive control) = 0.14–0.27 µM). In addition, the compounds displayed moderate *in vitro* cytotoxicity against human colon cancer (HCT-116) cell lines with respective IC_{50} values of 8.8 and 9.0 µM (IC_{50} value of etoposide (positive control) = 0.49–4.9 µM).

Reference

Hughes, C.C., Prieto-Davo, A., Jensen, P.R., and Fenical, W. (2008) *Org. Lett.*, **10**, 629.

Marsupsin

Physical data: $C_{16}H_{14}O_6$, mp 193–195 °C, $[\alpha]_D^{26} -4°$ (MeOH, *c* 0.5)

Structure:

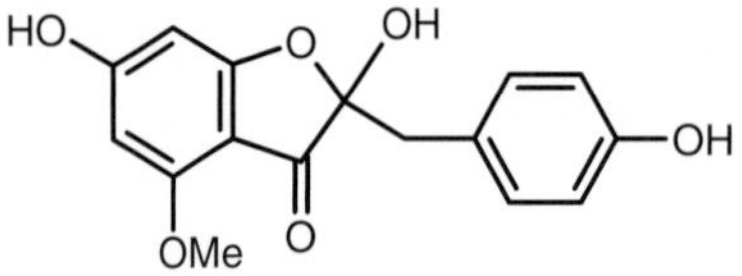

Marsupsin

Compound class: Phenolic

Source: *Pterocarpus marsupium* Roxb. (heartwoods; family: Leguminosae) [1, 2]

Pharmaceutical potential: *Antidiabetic*

Manickam et al. [3] evaluated antihyperglycemic activity of the phenolic compound against streptozotocin (STZ)-induced hyperglycemic rats; marsupsin was found to decrease significantly the plasma glucose level of STZ-induced diabetic rats, and the efficacy is almost comparable to that of the reference standard, metformin. The test compound at a dose of 20 mg/kg body weight (administered jp for 3 days) lowered 33% in plasma glucose level, while metformin lowered the plasma glucose level by 48% at a dose of 30 mg/kg body weight (administered jp for 3 days). Marsupsin did not alter the basal plasma glucose level in nondiabetic animals; furthermore, it significantly decreased the body weight in comparison to the vehicle-treated animals.

Marsupsin might be useful in noninsulin-dependent *diabetes mellitus*. The investigators [3] supposed that the test compound may have insulin-like effects on several tissues as in the case of the oral hypoglycemic agents such as metformin [4–7]; however, detailed investigations are needed to elucidate the exact mode of action.

References

[1] Maurya, R., Ray, A.B., Duah, F.K., Slatkin, D.J., and Schiff, P.L. Jr, (1984) *J. Nat. Prod.*, **47**, 179.
[2] Maurya, R., Ray, A.B., Duah, F.K., Slatkin, D.J., and Schiff, P.L. Jr, (1982) *Heterocycles*, **19**, 2103.
[3] Manickam, M., Ramanathan, M., Farboodniay Jahromi, M.A., Chansouria, J.P.N., and Ray, A.B. (1997) *J. Nat. Prod.*, **60**, 609.
[4] Jackson, R.A., Hawa, M.I., Japan, J.B., Sim, B.M., Silvio, D., Featherbe, L., and Kurtz, D. (1987) *Diabetes*, **36**, 632.
[5] Hermann, L.S. (1979) *Diabetic Metab.*, **5**, 233.
[6] Lord, J.M., Atkins, T.W., and Bailey, C.J. (1983) *Diabetologia*, **25**, 108.
[7] Puah, J.A. and Bailey, C.J. (1984) *Diabetologia*, **27**, 322.

Martefragin A

Physical data: $C_{20}H_{25}N_3O_3$, white powder, mp 147–148 °C

Structure:

Martefragin A

Compound class: Indole alkaloid

Source: *Martensia fragilis* (red alga)

Pharmaceutical potential: *Antioxidant*
Martefragin A displayed promising antioxidant efficacy by inhibiting NADPH-dependent lipid peroxidation in rat liver microsomes with an IC_{50} value of 2.8 µM; the activity was found to be more potent than that exhibited by the positive controls, α-tocopherol (IC_{50} = 87 µM) and ascorbic acid (IC_{50} = 200 µM). The alkaloid displayed no chelation with ferrous ion; hence, the mechanism of inhibition of NADPH-dependent lipid peroxidation is assumed to be due to radical stabilization through its extended conjugated system. However, detailed studies on the mode of action are suggested by the present investigators.

Reference

Takahashi, S., Matsunaga, T., Hasegawa, C., Saito, H., Fujita, D., Kiuchi, F., and Tsuda, Y. (1998) *Chem. Pharm. Bull.*, **46**, 1527.

Maslinic acid and *epi*-maslinic acid

Systematic name:
- 2α,3β-Dihydroxyolean-12-en-28-oic acid (Maslinic acid)
- 2α,3α-Dihydroxyolean-12-en-28-oic acid (*epi*-maslinic acid)

Physical data: $C_{30}H_{48}O_4$, white crystals

Structures:

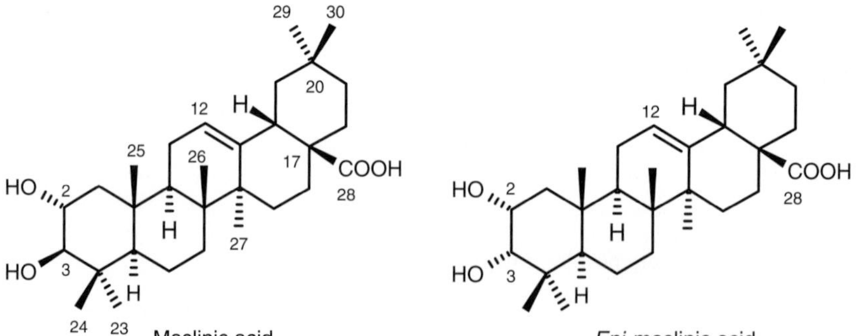

Maslinic acid

Epi-maslinic acid

Compound class: Triterpenoids (olean type)

Sources: Maslinic acid: *Eucalyptus viminalis* (family: Myrtaceae) [1]; *Euptelea polyandra* Sieb. et Zucc. (barks; family: Eupteleaceae) [2]; *Geum japonicum* Thunb. (whole plants; family: Rosaceae) [3]; *Luehea divaricata* (leaves; family: Tiliaceae) [4]; *Olea europaea* (pomaceous Olive; family: Oleaceae) [5, 6]; *Salvia canariensis* L. (aerial parts; family: Labiatae) [7]

Epi-maslinic acid: *Prunella vulgaris* (leaves and stems; family: Lamiaceae) [8]; tissue cultures derived from *Isodon japonicus* Hara (Family: Lamiaceae) [9]

Pharmaceutical potentials: *Maslinic acid: Antioxidant; anti-inflammatory; anticarcinogenic; hypolipidemic; growth promoting; anti-HIV*
Epi-maslinic acid: Antiproliferative
Montilla et al. [5] evaluated the antioxidative efficacy (in rat model) of maslinic acid obtained from pomaceous olive on the susceptibility of plasma (*in vivo* study) as well as hepatocyte membranes (*in vitro* study) to lipid peroxidation; endogenous plasma lipoperoxide levels and susceptibility to lipid peroxidation were found to decrease in the animals treated with maslinic acid, after exposure to hydroxyl radical (generated from Fe^{2+}/H_2O_2). The investigators observed that coincubation with maslinic acid prevented hepatocyte membrane lipid peroxidation as shown by the reduction of TBARS; hence, maslinic acid was considered to offer some advantages in the resistance of oxidative stress in the animals. Besides attenuating intracellular oxidative stress via inhibition of NO and H_2O_2 production, maslinic acid also reduced proinflammatory cytokine generation in murine macrophages [10]; the investigators tested the effect of this hydroxyl-pentacyclic triterpene derivative obtained from the non glyceride fraction of pomace olive oil upon oxidative stress and cytokine production using peritoneal murine macrophages. It significantly inhibited the enhanced production of nitric oxide (NO) induced by lypopolysaccharide (LPS) with an IC_{50} value of 25.4 µM. This inhibiting effect seems to be the consequence of an action at the level of the LPS-induction of the inducible nitric oxide synthethase (iNOS) gene enzyme expression rather than to a direct inhibitory action on enzyme activity. The investigators also observed that the secretion of the inflammatory cytokines interleukine-6 and TNF-α from LPS-stimulated murine macrophages was significantly reduced ($p < 0.05$ and 0.01) to 50 and 100 µM of masnilic acid, respectively. Besides, pretreatment with maslinic acid reduced the generation of hydrogen peroxide from stimulated macrophages in a dose-dependent manner ($IC_{50} = 43.6$ µM) as assayed from the oxidation of the peroxidase enzyme; however, no inhibitory effect on superoxide release, measured by the reduction of ferricytochrome c, was observed after the pretreatment with the test compound in the culture medium [10].

Liu et al. [11] demonstrated by means of their experiment in rat model (male Sprague-Dawley rat) that maslinic acid reduces diet-induced hyperlipidemia as well as modulates the abnormalities arisen out of such symptom; the natural triterpene exerts therapeutic effects on diet-induced hyperlipidemia by inhibiting the intestinal absorption and storage of cholesterol [11]. Fernández-Navarro et al. [12] also reported that maslinic acid can act as a growth factor when added to trout diet. Five groups of 120 fish of a mean body mass of 20 g were fed for 225 days with diets containing 0, 1, 5, 25, and 250 mg of the triterpene per kg diet. The highest weight increase was registered for the group fed 250 mg/kg, representing a 29% increase over controls. The total hepatic DNA or liver cell hyperplasia levels in trout fed with 25 and 250 mg of maslinic acid/kg were 37 and 68% higher than controls. Besides, in these same groups of trout fractional and absolute hepatic protein-synthesis rates were significantly higher than in control as well as significant increments in hepatic protein-synthesis efficiency and protein-synthesis capacity along with a larger rough-endoplasmic reticulum and larger glycogen stores than controls were also reported [12].

Maslinic acid showed antiproliferative activities against Caco-2 cancer cells ($EC_{50} = 15$ µM), HT-29 human colon cancer cells ($EC_{50} = 74$ µM), 1321N1 astrocytoma cells ($IC_{50} = 25$ µM), and human leukemia (CCRF-CEM and CEM/ADR5000) cells ($IC_{50} = 7$ and 9 µM, respectively) [13–16]; the antiproliferative activity likely develops from the induction of an oxidative apoptotic pathway by the drug, causing cell cycle and cytoskeleton alterations. From their detailed studies Reyes-Zurita et al. [17] demonstrated that treatment with maslinic acid results in a significant inhibition of cell proliferation in a dose-dependent manner and causes apoptotic death in colon-cancer cells; it inhibits considerably the expression of Bcl-2 whilst increasing that of Bax; it also stimulates the release of mitochondrial cytochrome-c and activates caspase-9 and caspase-3. Maslinic acid, thus, is able to induce caspase-dependent apoptosis in human colon-cancer cells via the intrinsic mitochondrial pathway [17]. Recently maslinic acid has been found to inhibit the spread of the HIV virus by inhibiting the replication of a primary HIV-1 isolate as well as decreased the cytopathic effect and p24 antigen levels in MT2 cells [18]. Xu et al. [7] reported that maslinic acid displayed potent inhibitory activity (100%) against HIV-1 protease at a concentration of 17.9 μg/ml.

The antiproliferative activity of *epi*-maslinic acid was determined against MK-1, HeLa, B16F10 cells; the respective GI_{50} (50% growth inhibition) values were determined as 55, 38, and 36 μM [19].

References

[1] Savina, A.A., Sokol'skaya, T.A., and Fesenko, D.A. (1983) *Chem. Nat. Comp.*, **19**, 114.
[2] Konoshima, T., Matsuda, T., Takasaki, M., Yamahara, J., Kozuka, M., Sawada, T., and Shingu, T. (1985) *J. Nat. Prod.*, **48**, 683.
[3] Savona, G. and Bruno, M. (1983) *J. Nat. Prod.*, **46**, 593.
[4] Tanaka, J.C.A., Vidotti, G.J., and da Silva, C.C. (2003) *J. Braz. Chem. Soc.*, **14**, 475.
[5] Montilla, M.P., Agil, A., Navarro, M.C., Jiménez, M.I., García-Granados, A., Parra, A., and Cabo, M.M. (2003) *Planta Med.*, **69**, 472.
[6] Garcia-Granados, A., Martinez, A., Moliz, J.N., Parra, A., and Rivas, F. (1998) *Molecules*, **3**, M88.
[7] Xu, H.-X., Zeng, F.-Q., Wan, M., and Sim, K.-Y. (1996) *J. Nat. Prod.*, **59**, 643.
[8] Kojima, H. and Ogura, H. (1989) *Phytochemistry*, **28**, 1703.
[9] Kojima, H. and Ogura, H. (1986) *Phytochemistry*, **25**, 729.
[10] Marquez, M.A., De La Puerta, V.R., Fernandez-Arche, A., and Ruiz-Gutierrez, V. (2006) *Free Rad. Res.*, **40**, 295.
[11] Liu, J., Sun, H., Wang, X., Mu, D., Liao, H., and Zhang, L. (2007) *Drug Dev. Res.*, **68**, 261.
[12] Fernández-Navarro, M., Peragón, J., Esteban, F.J., de la Higuera, M., and and Lupiáñez, J.A. (2006) *Comp. Biochem. Physiol. C Toxicol. Pharmacol.*, **144**, 130.
[13] He, X. and Liu, R.H. (2007) *J. Agric. Food Chem.*, **55**, 4366.
[14] Juan, M.E., Wenzel, U., Ruiz-Gutierrez, V., Daniel, H., and Planas, J.M. (2006) *J. Nutr.*, **136**, 2553.
[15] Martin, R., Carvalho, J., Ibeas, E., Hernández, M., Ruiz-Gutierrez, V., and Nieto, M.L. (2007) *Cancer Res.*, **67**, 3741.
[16] Wang, Y.-F., Lai, G.-F., Efferth, T., Cao, J.-X., and Luo, S.-D. (2006) *Chem. Biodivers.*, **3**, 1023.
[17] Reyes-Zurita, F.J., Rufino-Palomares, E.E., Lupiáñez, J.A., and Cascante, M. (2009) *Cancer Lett.*, **273**, 44.

[18] García, A."Compound from olive-pomace oil inhibits HIV spread", http://en.wikipedia.org/wiki/Maslinic_acid (accessed 17 July 2007).
[19] Yoshida, M., Fuchigama, M., Nagao, T., Okabe, H., Matsunaga, K., Takata, J., Karube, Y., Tsuchihashi, R., Kinjo, J., Mihashi, K., and Fujioka, T. (2005) *Biol. Pharm. Bull.*, **28**, 173.

Massadine

Physical data: $C_{22}H_{26}Br_4N_{10}O_5$, yellow powder, $[\alpha]_D^{17} -12°$ (MeOH, *c* 0.1)

Structure:

Massadine

Compound class: Alkaloid

Source: *Stylissa* aff. *massa* (marine sponge) [1]

Pharmaceutical potential: *Geranylgeranyltransferase type I (GGTase I) inhibitors*
It is regarded that GGTase I might be a potential antifungal drug target [2–7]; massadine was found to inhibit GGTase I from *Candida albicans* with an IC_{50} value of 3.9 µM [1]. Although the isolate inhibited the growth of *Cryptococcus neoformans* with an MIC value of 32 µM, it did not inhibit the growth of *C. albicans* up to a concentration of 64 µM [1].

References

[1] Nishimura, S., Matsunaga, S., Shibazaki, M., Suzuki, K., Furihata, K., van Soest, R.W.M., and Fusetani, N. (2003) *Org. Lett.*, **5**, 2255.
[2] Kelly, R., Card, D., Register, E., Mazur, P., Kelly, T., Tanaka, K.-I., Onishi, J., Williamson, J.M., Fan, H., Satoh, T., and and Kurtz, M. (2000) *J. Bacteriol.*, **182**, 704.
[3] Casey, P.J., Thissen, J.A., and Moormaw, J.F. (1991) *Proc. Natl. Acad. Sci. USA*, **88**, 8631.
[4] Shafer, W.R. and and Rine, J. (1992) *Annu. Rev. Genet.*, **30**, 209.
[5] Drgonova, J., Drgon, T., Tanaka, K., Kollar, R., Chen, G.-Ch., Ford, R.A., Chan, C.S.M., Takai, Y., and Cabib, E. (1996) *Science*, **272**, 277.
[6] Mazur, P., Register, E., Bonfiglio, C.A., Yuan, X., Kurtz, M.B., Williamson, J.M., and Kelly, R. (1999) *Microbiology*, **145**, 1123.
[7] Sunami, S., Ohkubo, M., Sagara, T., Ono, J., Asahi, S., Koito, S., and Morishima, H. (2002) *Bioorg. Med. Chem. Lett.*, **12**, 629.

Matsudone A

Systematic name: 7,8-(2″,2″-Dimethylpyrano)-5,3′,4′-trihydroxyflavone-3-O-β-D-glucopyranoside

Physical data: $C_{26}H_{26}O_{12}$, pale yellow amorphous powder

Structure:

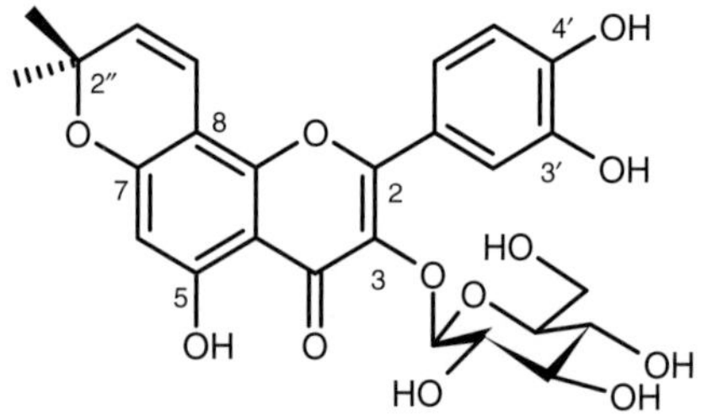

Matsudone A

Compound class: Flavonoid

Source: *Salix matsudana* Koidz (leaves; family: Salicaceae)

Pharmaceutical potential: *Cyclooxygenase inhibitory (anti-inflammatory)*
The flavonoid was tested for its inhibitory activities against cyclooxygenase (COX-1 and COX-2) enzymes; it was found to have potent inhibitory effect on COX-2 with an IC_{50} value of 27.3 μM but a weak effect on COX-1 (IC_{50} = 153.1 μM). Hence, considerable COX-2 inhibition activity of the isolated compound compared to the positive control aspirin (IC_{50} = 19.0 μM) prompts to screen it as a good candidate for further consideration as an anti-inflammatory prodrug.

Reference

Li, X., Liu, Z., Zhang, X.-F., Wang, L.-J., Zheng, Y.-N., Yuan, C.-C., and Sun, G.-Z. (2008) *Molecules*, 13, 1530.

Maytenfolone A

Physical data: $C_{30}H_{46}O_4$, cubic crystalline solid (EtOAc), mp 166–168 °C, $[\alpha]_D$ −85° (CHCl$_3$, c 0.1)

Structure:

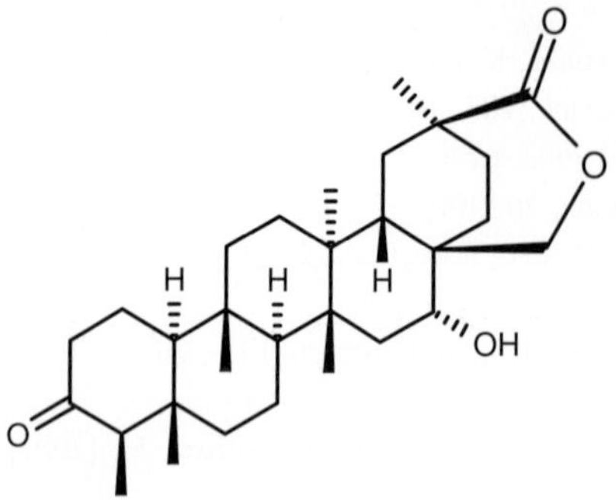

Maytenfolone A

Compound class: Triterpenoid

Source: *Celastrus hindsii* (stems; family: Celastraceae)

Pharmaceutical potentials: *Anti-HIV; cytotoxic*
Maytenfolone A was reported to possess moderate anti-HIV potential; the friedelin triterpene inhibited HIV replication in H9 lymphocyte cells with an EC_{50} value of 1.8 μg/ml and a lower toxicity at 7.0 μg/ml, whereas the standard agent, AZT, showed an EC_{50} value of 0.01 μg/ml and toxicity at 500 μg/ml.

Maytenfolone A also demonstrated cytotoxicity against hepatoma (HEPA-2B) cells with an ED_{50} value of 2.3 μg/ml and against nasopharynx carcinoma cells with an ED_{50} value of 3.8 μg/ml; the ED_{50} value of the reference compound, mitomycin C, was determined to be <0.04 μg/ml.

Reference

Kuo, Y.H. and Yang-Kuo, L.M. (1997) *Phytochemistry*, **44**, 1275.

Maytenonic acid

See Polpunonic acid

Mediomycins A and B

Physical data:
- Mediomycin A: $C_{62}H_{98}NO_{18}S$, amorphous yellow powder, $[\alpha]_D^{25}$ −10.4° (MeOH, *c* 0.28)
- Mediomycin B: $C_{62}H_{98}NO_{15}$, amorphous yellow powder, $[\alpha]_D^{25}$ −15.4° (MeOH, *c* 0.21)

Structures:

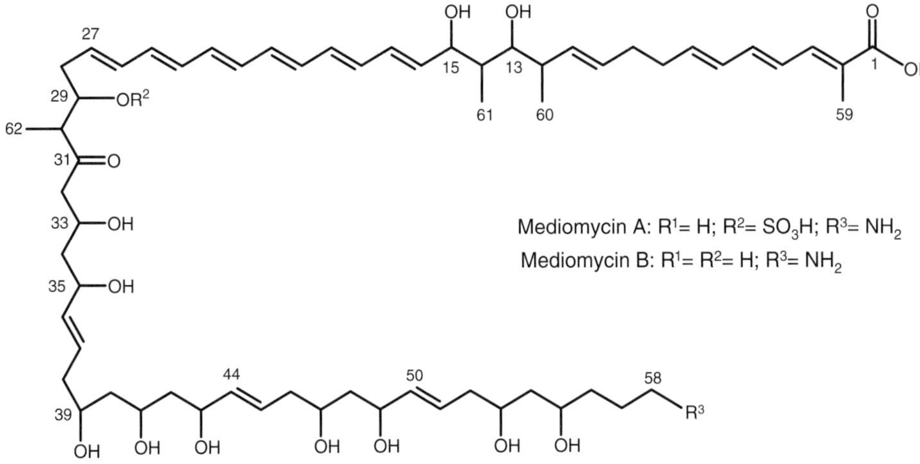

Mediomycin A: R^1= H; R^2= SO_3H; R^3= NH_2
Mediomycin B: R^1= R^2= H; R^3= NH_2

Compound class: Polyene antibiotics

Source: *Streptomyces mediocidicus* ATCC23936

Pharmaceutical potential: *Broad-spectrum antifungal*

The present investigators demonstrated that both the isolates could act as broad-spectrum antifungal agents *in vitro* against a variety of yeasts and fungi including pathogenic yeasts and filamentous fungi; mediomycin B was found to be more potent than mediomycin A. Amphotericin B, a representative polyene antibiotic, was used as a reference standard. The respective sets of MIC values (in μg/ml) of mediomycins A and B and amphotericin B (reference) against *Candida albicans* (GC 3064), *C. albicans* (GC 3065), *C. albicans* (GC 3066), *C. parapsilosis* (GC 3074), *C. parapsilosis* (GC 3075), *C. parapsilosis* (GC 3076), *C. pseudotropicalis* (GC 3070), *C. tropicalis* (GC 3080), *C. tropicalis* (GC 3081), *C. krussii* (GC 3067), *C. lusitaniae* (GC 3068), *C. rugosa* (GC 3077), *Aspergillus fumigatus* (GC 3092), and *A. fumigatus* (GC 3091) were determined to be 2, 1, 0.50; 2, 1, 0.25; 4, 1, 0.25; 16, 2, 0.25; 16, 4, 0.50; 16, 2, 0.50; 1, 1, 0.25; 16, 2, 0.50; 8, 2, 0.50; 2, 1, 0.50; 2, 2, 0.25; 2, 1, 0.50; 128, 16, 0.25; and 1, 1, 0.25 μg/ml.

Reference

Cai, P., Kong, F., Fink, P., Ruppen, M.E., Williamson, R.T., and Keiko, T. (2007) *J. Nat. Prod.*, **70**, 215.

Megathyrin A

Systematic name: 1α,7β,14β-Trihydroxy-*ent*-7α,20-epoxy-kaur-16-en-15-one

Physical data: $C_{20}H_{28}O_5$, colorless crystals, mp 180–182 °C, $[\alpha]_D$ −22.5° (MeOH, *c* 0.082)

Structure:

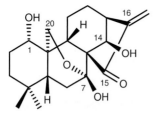

Megathyrin A

Compound class: Diterpenoid

Source: *Isodon megathyrsus* (leaves; family: Labiatae)

Pharmaceutical potential: *Cytotoxic*

The isolate displayed significant cytotoxic activity *in vitro* against a panel of cancer cell lines such as Lu-1 (human lung carcinoma), KB (human oral epidermoid carcinoma), KB-V (vinblastine-resistant KB), LNCaP (hormone-dependent human prostate carcinoma), and ZR-75-1 (hormone-dependent human breast carcinoma) with ED_{50} values of 3.4, 1.1, 19.1, 1.8, and 1.8 μg/ml, respectively.

Reference

Sun, H.-D., Lin, Z.-W., Niu, F.-D., Lin, L.-Z., Chai, H., Pezzuto, J.M., and Cordell, G.A. (1994) *J. Nat. Prod.*, **57**, 1424.

Melemeleone B

Physical data: $C_{23}H_{33}NSO_5$, red amorphous solid, mp 190–200 °C, $[\alpha]_D^{20}$ −22° (CH_2Cl_2, c 0.01)

Structure:

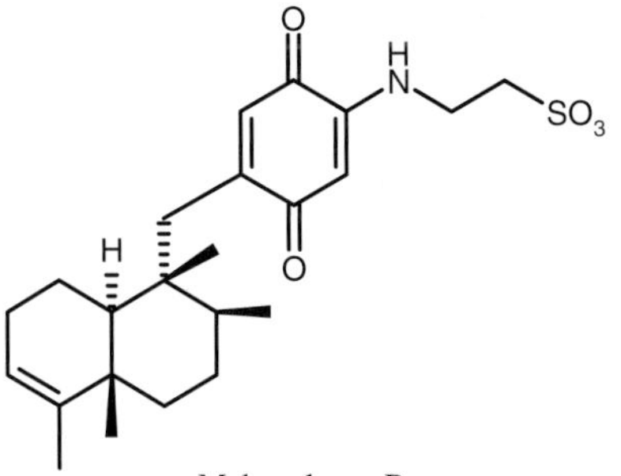

Melemeleone B

Compound class: Sesquiterpene quinone

Source: *Dysidea avara* (marine sponge)

Pharmaceutical potential: *Protein tyrosine kinase (PTK) inhibitor*
The metabolite showed moderate inhibitory activity against protein tyrosine kinase enzyme with an IC_{50} value of approximately 28 µM.

Reference

Alvi, K.A., Diaz, M.C., and Crews, P. (1992) *J. Org. Chem.*, **57**, 6604.

Melophlins P–S

Physical data:
- Melophlin P: $C_{22}H_{39}NO_3$, yellowish oil
- Melophlin Q: $C_{21}H_{37}NO_3$, yellowish oil
- Melophlin R: $C_{21}H_{37}NO_3$, yellowish oil
- Melophlin S: $C_{21}H_{37}NO_3$, yellowish oil

Structures:

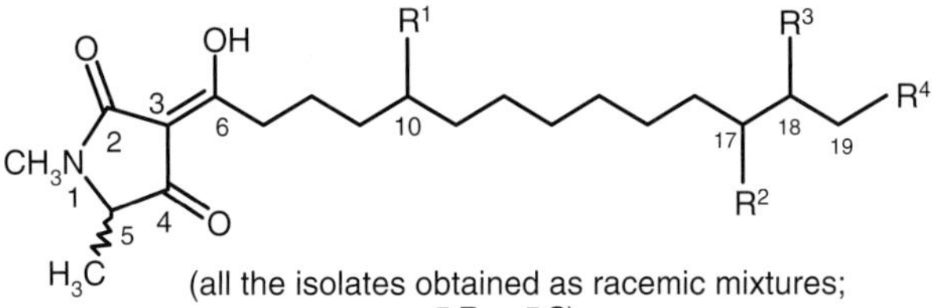

(all the isolates obtained as racemic mixtures; 5R or 5S)

Melophlin P: $R^1 = R^2 = R^3 = H$; $R^4 = CH_3CH_2$
Melophlin Q: $R^1 = R^2 = R^4 = H$; $R^3 = CH_3$
Melophlin R: $R^1 = R^3 = R^4 = H$; $R^2 = CH_3$
Melophlin S: $R^1 = CH_3$; $R^2 = R^3 = R^4 = H$

Compound class: Tetramic acid derivatives

Sources: *Melophlus* cf. *sarasinorum* and *Melophlus* sp. (marine sponges collected, respectively, in Palau and Palauan)

Pharmaceutical potential: *Cytotoxic*
Melophlins P, Q, R, and S were found to show growth inhibitory activity against L1210 cells with IC_{50} values of 5.13, 0.85, 10.5, and 20.0 µM, respectively.

Reference

Xu, J., Hasegawa, M., Harada, K., Kobayashi, H., Nagai, H., and Namikoshi, M. (2006) *Chem. Pharm. Bull.*, **54**, 852.

Metachromins S and T

Physical data:
- Metachromin S: $C_{26}H_{39}NO_3$, purple oil, $[\alpha]_D^{22}$ +45.1° ($CHCl_3$, *c* 0.2)
- Metachromin T: $C_{23}H_{32}O_4$, colorless oil, $[\alpha]_D^{22}$ −41.8° ($CHCl_3$, *c* 0.5)

Structures:

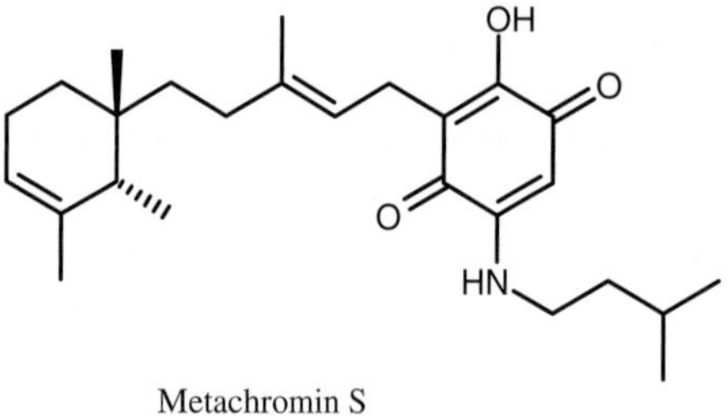

Metachromin S

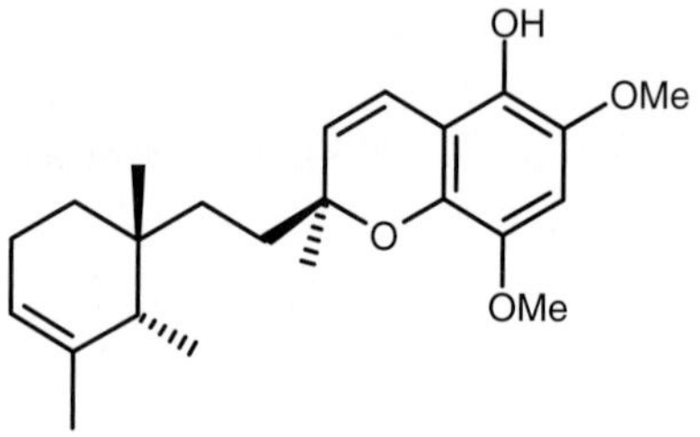

Metachromin T

Compound class: Sesquiterpenoids

Source: *Spongia* sp. (SS-1037; an Okinawan sponge)

Pharmaceutical potential: *Cytotoxic*
The sesquiterpenoids, metachromins S and T, were reported to show moderate *in vitro* cytotoxicity against L1210 murine leukemia and KB human epidermoid carcinoma cells with respective IC_{50} values of 5.2 and >10 µg/ml for metachromin S and 3.0 and 5.6 µg/ml for metachromin T.

Reference

Takahashi, Y., Yamada, M., Kubota, T., Fromont, J., and Kobayashi, J. (2007) *Chem. Pharm. Bull.*, **55**, 1731.

5-Methoxy-8-(2'-hydroxy-3'-buthoxy-3'-methylbutyloxy)-psoralen

Physical data: $C_{21}H_{26}O_7$, yellowish amorphous powder, $[\alpha]_D^{20}$ +11.1° (MeOH, c 0.27)

Structure:

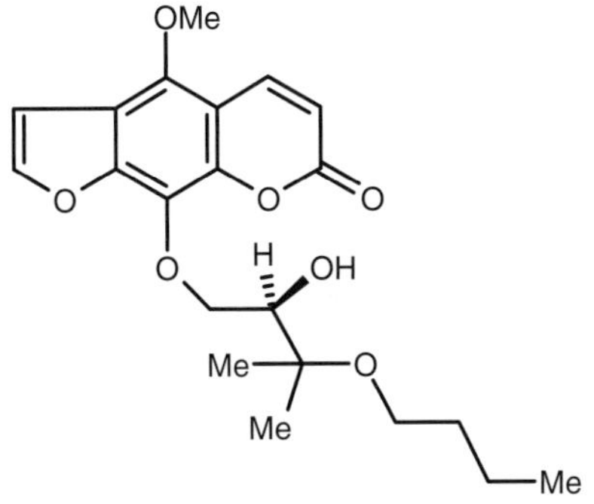

5-Methoxy-8-(2'-hydroxy-3'-buthoxy-3'-methylbutyloxy)-psoralen

Compound class: Furanocoumarin

Source: *Angelica dahurica* Benth. et Hook. (roots; family: Umbelliferae) [1, 2]

Pharmaceutical potential: *Inhibitor of COX-2 and 5-LOX (anti-inflammatory)*
Hua et al. [2] evaluated the furanocoumarin as a potent inhibitor of cyclooxygenase-2 (COX-2)-dependent phase of prostaglandin D_2 (PGD_2) generation in bone marrow-derived mast cells with an IC_{50} value of 23.5 µM; in addition, this compound consistently inhibited the production of leukotriene C_4 in a dose-dependent manner (IC_{50} = 2.5 µM). From their detailed studies, the present investigators [2] demonstrated that the test compound not only inhibited both cyclooxygenase-2 and 5-lipoxygenase activities but also inhibited the degranulation reaction (IC_{50} = 4.1 µM). Hence, this compound might provide a basis for novel anti-inflammatory drug development [2].

References

[1] Lee, S.-H., Li, G., Kim, H.-J., Kim, J.-Y., Chang, H.-W., Jhang, Y., Woo, M.-H., Song, D.-K., and Son, J.-K. (2003) *Bull. Korean Chem. Soc.*, **24**, 1699.
[2] Hua, J.M., Moon, T.C., Hong, T.G., Park, K.M., Son, J.K., and Chang, H.W. (2008) *Arch. Pharm. Res.*, **31**, 617.

(3R,4R)-(−)-6-Methoxy-1-oxo-3-*n*-pentyl-3,4-dihydro-1*H*-isochromen-4-yl acetate

Alternative name: (3R,4R)-(−)-6-Methoxy-3,4-dihydro-3-*n*-pentyl-4-acethoxy-1*H*-2-benzopyran-1-one

Physical data: $C_{15}H_{22}O_5$

(3R,4R)-(−)-6-Methoxy-1-oxo-3-n-pentyl-3,4-dihydro-1H-isochromen-4-yl acetate

Structure:

(3R,4R)-(−)-6-Methoxy-1-oxo-3-n-pentyl-3,4-dihydro-1H-isochromen-4-yl acetate

Compound class: Dihydroisocoumarin

Source: *Xyris pertygoblephara* (family: Xyridaceae) [1, 2]

Pharmaceutical potentials: *Antifungal; aromatase inhibitor; antiproliferative*
The dihydroisocoumarin displayed potent *in vitro* antifungal activity (formed significant inhibition zones at 100 μg/disc in agar diffusion method) against clinical isolates of the dermatophyte-fungi *Epidermophyton floccosum*, *Trichophyton mentagrophytes*, and *T. rubrum* [1].
The test compound exhibited moderate aromatase inhibitory activity [2]. Aromatase (CYP19), a key cytochrome P450 (CYP) enzyme, plays a crucial role in regulating estrogen synthesis in postmenopausal women and that's why the regulation and inhibition of aromatase activity are regarded as a well-established target for the chemoprevention of breast cancer [3–6]. Aromatase inhibitors (e.g., anastrozole and letrozole) have shown improved efficacy and reduced side effects against advanced as well as early stage breast cancer in comparison with the estrogen antagonist tamoxifen [3, 4, 7–10]. The specificity of aromatase inhibition by the compound was evaluated through the assays carried out with xenobiotic-metabolizing cytochrome P450 enzymes CYP1A1, CYP2C8, and CYP3A4—the test compound inhibited CYP1A1 with an IC_{50} value of 38.0 ± 2.0 μM, but showed no significant inhibition to CYP2C8 and CYP3A4 enzymes [2]. It also exhibited weak antiproliferative activity against MCF-7 ($IC_{50} = 66.9 \pm 2.3$ μM) and LNCaP ($IC_{50} = 57.5 \pm 2.0$ μM) cells and remained inactive against LU-1 and HepG2 cells in culture; hence this compound can hopefully be regarded as a prototype for future development, which may result in a new chemopreventive agent for the prevention or treatment of human breast cancer [2].

References

[1] Guimaraes, K.G., Souza Filho, J.D., Mares-Guia, T.R., and Braga, F.G. (2008) *Phytochemistry*, **69**, 439.
[2] Endringer, D.C., Guimaraes, K.G., Kondratyuk, T.P., Pezzuto, J.M., and Braga, F.G. (2008) *J. Nat. Prod.*, **71**, 1082.
[3] Mastro, L.D., Clavarezza, M., and Venturini, M. (2007) *Cancer Treat. Rev.*, **33**, 681.
[4] Briest, S. and Wolff, A.C. (2007) *Expert Rev. Anticancer Ther.*, **7**, 1243.
[5] Maiti, A., Cuendet, M., Croy, V.L., Endringer, D.C., Pezzuto, J.M., and Cushman, M. (2007) *J. Med. Chem.*, **50**, 2799.
[6] Smith, I.E. and Dowsett, M. (2003) *N. Engl. J. Med.*, **348**, 2431.
[7] Wong, Z.W. and Ellis, M.J. (2004) *Br. J. Cancer*, **90**, 20.

[8] Milla-Santos, A., Milla, L., Portella, J., Rallo, L., Pons, M., Rodes, E., Casanovas, J., and Puig-Gali, M. (2003) *Am. J. Clin. Oncol.*, **26**, 317.
[9] Arora, A. and Potter, J.F. (2004) *J. Am. Geriatr. Soc.*, **52**, 611.
[10] Goss, P.E., (1999) *Endocr. Relat. Cancer*, **6**, 325.

9-Methoxycanthin-6-one

Physical data: $C_{15}H_{10}N_2O$, yellow crystals (acetone), mp 281–283 °C

Structure:

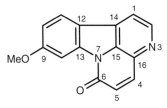

9-Methoxycanthin-6-one

Compound class: Alkaloid

Sources: *Simaba multiflora* A. Juss. (woods; family: Simaroubaceae) [1, 2], *S. cuspidata* Spruce ex Engl. [3]; *Eurycoma longifolia* Jack (roots; family: Simaroubaceae) [4]

Pharmaceutical potential: *Cytotoxic (antitumor)*
The alkaloid was evaluated against a panel of cell lines comprising a number of human cancer cell types such as breast, colon, fibrosarcoma, lung, melanoma, KB, KB-V1 (a multidrug-resistant (vincristine-resistant) cell line derived from KB), and murine lymphocytic leukemia (P-388) and found to be cytotoxic to all tumor cell lines tested, with the exception of the KB-V1 cell line – the respective ED_{50} values were determined to be 2.5, 4.2, 3.5, 4.0, 4.5, 2.1, >20, and 1.4 μg/ml [4].

References

[1] Polonsky, J., Gallas, J., Varenne, J., Prange, T., Pascard, C., Jacquernin, H., and Moretti, C. (1982) *Tetrahedron Lett.*, **23**, 869.
[2] Arisawa, M., Kinghorn, A.D., Cordell, G.A., and Farnsworth, N.R. (1983) *J. Nat. Prod.*, **46**, 222.
[3] Giesbrecht, A.M., Gottlieb, H.E., Gottlieb, O.R., Goulart, M.O.F., De Lama, R.A., and Santana, A.E.G. (1980) *Phytochemistry*, **19**, 313.
[4] Kardono, L.B.S., Angerhofer, C.K., Tsauri, S., Padmawinata, K., Pezzuto, J.M., and Kinghorn, A.D. (1991) *J. Nat. Prod.*, **54**, 1360.

1-Methoxycanthinone

Physical data: $C_{15}H_{10}N_2O_2$, pale yellow needles, mp 250–250.5 °C

Structure:

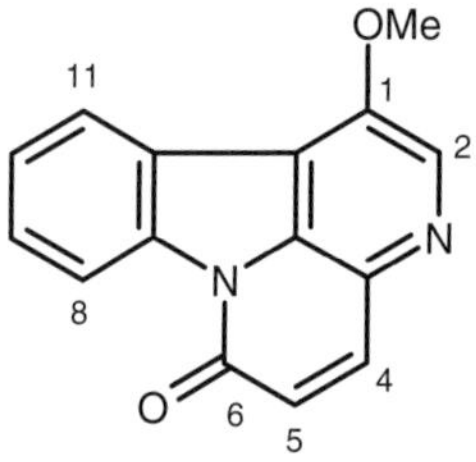

1-Methoxycanthinone

Compound class: Alkaloid (β-carboline)

Sources: *Ailanthus altissima* Swingle (dried woods; family: Simaroubaceae) [1]; *Leitneria floridana* Chapman (aerial parts; family: Leitneriaceae) [2]

Pharmaceutical potentials: *Anti-HIV; antitumor*
Xu et al. [2] evaluated that the alkaloid possesses potent *in vitro* anti-HIV efficacy by significantly suppressing HIV-infected H9 cell growth with an impressive therapeutic index (TI: IC_{50}/EC_{50}) of >391. The corresponding IC_{50} (concentration that inhibited uninfected H9 cell growth by 50%) and EC_{50} (concentration that inhibited viral replication by 50%) values were determined to be >100 and 0.256 μg/ml, respectively. Furthermore, the test compound inhibited the growth of a panel of human tumor cell lines such as KB, A-549, HCT-8, CAKI-1, and MCF-7 with respective ED_{50} values of 3.6, 4.3, 3.6, 3.6, and 13.5 μg/ml [2].

References

[1] Ohmoto, T., Tankaka, R., and Nikado, T. (1976) *Chem. Pharm. Bull.*, **24**, 1532.
[2] Xu, Z., Chang, F.-R., Wang, H.-K., Kashiwada, Y., McPhail, A.T., Bastow, K.F., Tachibana, Y., Cosentino, M., and Lee, K.-H. (2000) *J. Nat. Prod.*, **63**, 1712.

(+)-*N*-(Methoxycarbonyl)-*N*-norboldine

Physical data: $C_{20}H_{21}NO_6$, colorless crystals (MeOH), mp 140–142 °C, $[\alpha]_D^{20}$ + 269.2° (MeOH, *c* 0.19)

Structure:

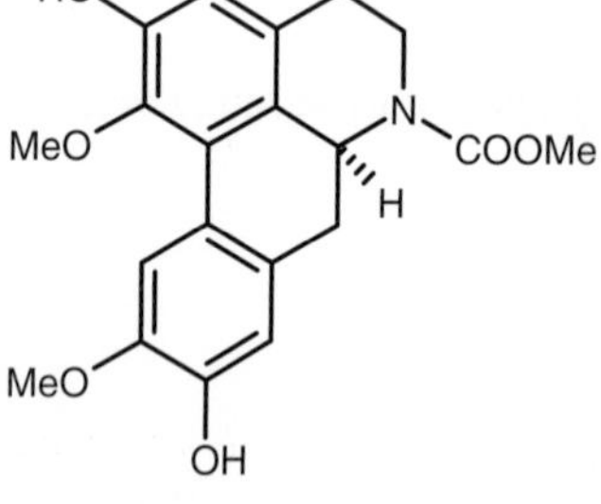

(+)-*N*-(Methoxycarbonyl)-*N*-norboldine

Compound class: Isoquinoline alkaloid

Source: *Litsea cubeba* (Lour.) Pers. (aerial parts; family: Lauraceae)

Pharmaceutical potential: *Antibacterial*
The alkaloid compound displayed significant *in vitro* antibacterial activity against *Staphylococcus aureus*; it was found to create a zone of inhibition of 9 mm diameter against the bacterial growth at a concentration of 50 μg/disk (rifampicin, the standard reference, showed the zone of inhibition of 40 mm diameter at the same concentration).

Reference

Feng, T., Xu, Y., Cai, X.-H., Du, Z.-Z., and Luo, X.-D. (2009) *Planta Med.*, **75**, 76.

25-*O*-Methoxycimigenol 3-*O*-α-L-arabinopyranoside

Physical data: $C_{36}H_{58}O_9$, solid

Structure:

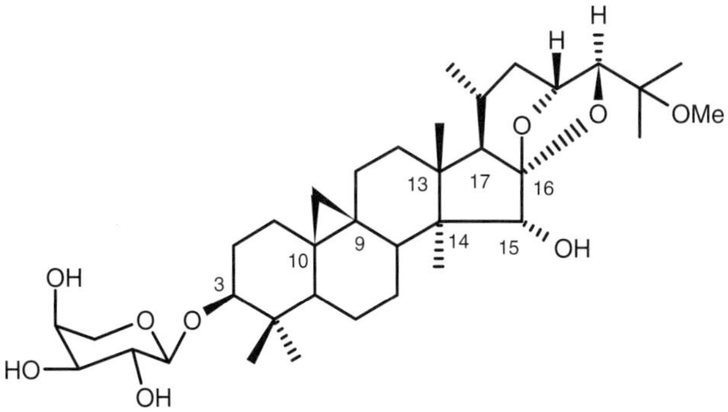

25-*O*-Methoxycimigenol 3-*O*-α-L-arabinopyranoside

Compound class: Triterpenoid (cycloartane glycoside)

Source: *Cimicifuga racemosa* Nutt. (rhizomes; family: Ranunculaceae) [1, 2]

Pharmaceutical potential: *Cytotoxic*
Watanabe *et al.* [2] evaluated that the isolate exhibits *in vitro* cytotoxic activity against human oral squamous cell carcinoma (HSC-2) cells in a dose-dependent manner with an IC_{50} value of 30 μM.

References

[1] Bedir, E. and Khan, I.A. (2001) *Pharmazie*, **56**, 268.
[2] Watanabe, K., Mimaki, Y., Sakagami, H., and Sashida, Y. (2002) *Chem. Pharm. Bull.*, **50**, 121.

(R)-4''-Methoxydalbergione

Physical data: $C_{16}H_{14}O_3$

Structure:

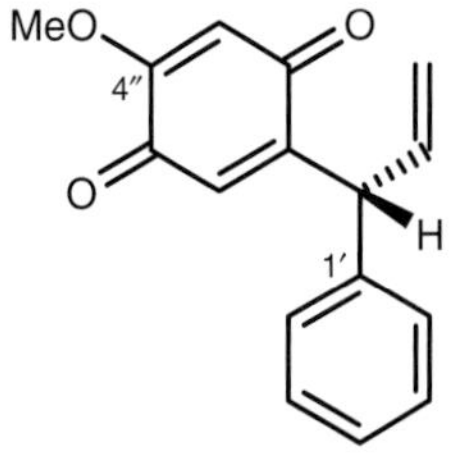

(R)-4''-Methoxydalbergione

Compound class: 3,3-Diarylpropene

Source: *Dalbergia louvelii* R. Viguier (heartwoods; family: Fabaceae) [1–3]

Pharmaceutical potential: *Antimalarial*
The isolate was reported to possess potent *in vitro* antimalarial activity; it inhibited the growth of the chloroquine-resistant strain FcB1 of *Plasmodium falciparum* with an IC_{50} value of 5.8 ± 0.15 µM (chloroquine used as a positive control; IC_{50} = 0.13 ± 0.03 µM) [1]. The *para*-quinone moiety present in the molecule has been reported to play a vital role in exhibiting antiplasmodial property as established from the fact that the reduced derivative, obtusaquinol, and the *ortho*-methylated derivative, 9-hydroxy-6,7-dimethoxydalbergiquinol, showed a two- and eightfold decrease, respectively, in antimalarial potency compared to the parent compound (R-4''-methoxydalbergione) [1].

References

[1] Beldjoudi, N., Mambu, L., Labaied, M., Grellier, P., Ramanitrahasimbola, D., Rasoanaivo, P., Martin, M.T., and Frappier, F. (2003) *J. Nat. Prod.*, **66**, 1447.
[2] Eyton, W.B., Ollis, W.D., Sutherland, I.O., Jackman, L.M., Gottlieb, O.R., and Magalhâes, M.T. (1962) *Proc. Chem. Soc.*, 301.
[3] Eyton, W.B., Ollis, W.D., Sutherland, I.O., Gottlieb, O.R., Magalhâes, M.T., and Jackman, L.M. (1965) *Tetrahedron*, **21**, 2683.

7-Methoxydeoxymorrellin

Physical data: $C_{34}H_{40}O_7$, orange crystals (CH_2Cl_2–MeOH), mp 129–130 °C, $[\alpha]_D^{30}$ −442.0° ($CHCl_3$, *c* 0.1)

Structure:

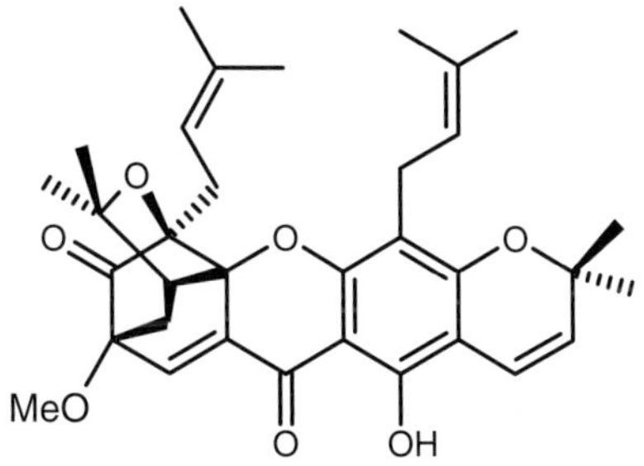

7-Methoxydeoxymorrellin

Compound class: Xanthone (caged)

Source: *Garcinia hanburyi* Hook. F. (resin and fruits; family: Guttiferae)

Pharmaceutical potential: *Cytotoxic*
7-Methoxydeoxymorrellin was reported to have cytotoxic property as evaluated against a panel of cell lines – the ED_{50} (µg/ml) values were determined to be 2.13 (P-388, murine lymphocytic leukemia), 7.71 (KB, human oral nasopharyngeal carcinoma), 1.44 (Col-2, human colon cancer), 2.59 (BCA-1, human breast cancer), 2.61 (Lu-1, human lung cancer), and 0.78 (ASK, rat glioma).

Reference

Reutrakul, V., Anantachoke, N., Pohmakotr, M., Jaipetch, T., Sophasan, S., Yoosook, C., Kasisit, J., Napaswat, C., Santisuk, T., and Tuchinda, P. (2007) *Planta Med.*, **73**, 33.

4-Methoxymagnaldehyde B

Systematic name: 5′-Allyl-2′-hydroxyphenyl-4-methoxy-3-cinnamic aldehyde

Physical data: $C_{19}H_{18}O_3$, yellowish oil

Structure:

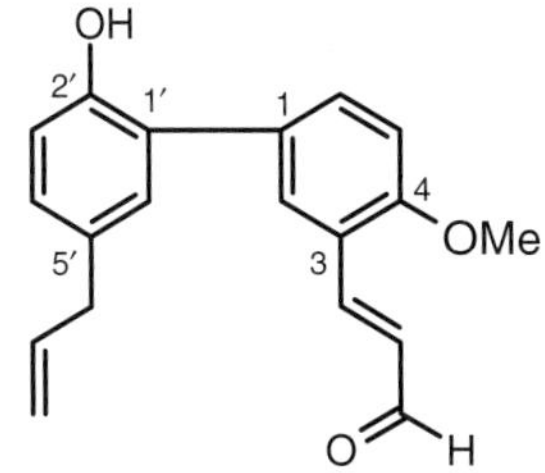

4-Methoxymagnaldehyde B

Compound class: Lignan

Source: *Magnolia obovata* Thunb. (stem barks; family: Magnoliaceae)

Pharmaceutical potential: *Cytotoxic*

Youn et al. [1] evaluated the compound for its cytotoxic potential against HeLa, A549, and HCT116 cancer cell lines; the compound showed the strongest cytotoxic activity against the HCT116 cancer cells. The IC_{50} values against HeLa, A549, and HCT116 cancer cell lines were observed to be 8.02 ± 0.7, 7.3 ± 1.1, and 1.3 ± 0.3 µg/ml, respectively [1]. The compound was also found to be more efficacious than its parent compound, magnaldehyde B [1, 2].

References

[1] Youn, U., Chen, Q.C., Lee, I.S., Kim, H., Yoo, J.-K., Lee, J., Na, M., Min, B.-S., and Bae, K. (2008) *Chem. Pharm. Bull.*, **56**, 115.
[2] Yahara, S., Nishiyori, T., Kohda, A., Nohara, T., and Nishioka I. (1991) *Chem. Pharm. Bull.*, **39**, 2024.

7-Methoxypraecansone B

Systematic names: 7,2′,6′-Trimethoxy-6″,6″-dimethylpyrano-(3′,4′:2″,3″)-chalcone

Physical data: $C_{23}H_{24}O_5$, yellow oil

Structure:

7-Methoxypraecansone B

Compound class: Flavonoid (coumarin)

Source: *Pongamia pinnata* (L.) Pierre (stem barks; family: Leguminosae)

Pharmaceutical potential: *Quinone reductase inducer*

The coumarin was found to induce quinone reductase (QR) activity; it was evaluated for its ability to act as phase II enzyme inducer using cultured Hepa 1clc7 mouse hepatoma cells by the present investigators – the experimental results for the compound were obtained as follows: CD (concentration required to double QR activity) = 1.2 µg/ml; IC_{50} (concentration inhibiting cell growth by 50%) = 9.6 µg/ml; CI (chemoprevention index, IC_{50}/CD) = 8.0 (sulforaphane used as a positive control: CD = 0.087 µg/ml; IC_{50} = 2.1 µg/ml; CI = 24.1).

Reference

Carcache-Blanco, E.J., Kang, Y.H., Park, E.J., Su, B.N., Kardono, L.B.S., Riswan, S., Fong, H.H.S., Pezzuto, J.M., and Kinghorn, A.D. (2003) *J. Nat. Prod.*, **66**, 1197.

6-Methoxyspirotryprostatin B

Physical data: $C_{22}H_{23}N_3O_4$, pale yellow amorphous powder, $[\alpha]_D^{21}$ −47.7° (CHCl$_3$, c 0.26)

Structure:

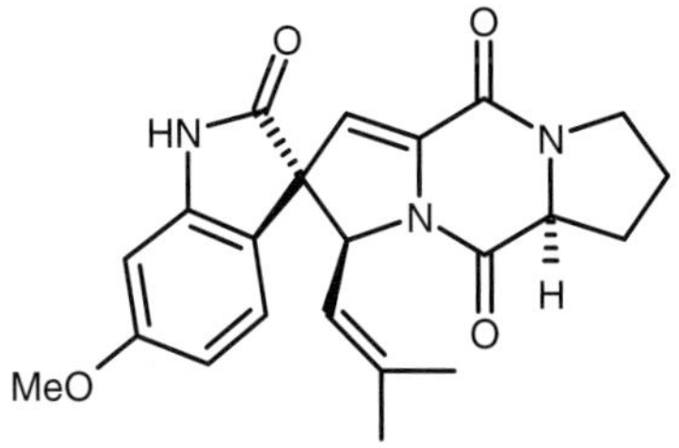

6-Methoxyspirotryprostatin B

Compound class: Alkaloid (diketopiperazine alkaloid)

Source: *Aspergillus sydowii* PFW1-13 strain (marine-derived fungus)

Pharmaceutical potential: *Cytotoxic*
The alkaloid exhibited moderate cytotoxicity against A-549 (lung cancer) and HL-60 (human leukemia) cells with IC$_{50}$ values of 8.29 and 9.71 μM, respectively.

Reference

Zhang, M., Wang, W.-L., Fang, Y.-C., Zhu, T.-J., Gu, Q.-Q., and Zhu, W.-M. (2008) *J. Nat. Prod.*, **71**, 985.

Methyl 3,5-di-*O*-caffeoyl quinate

Physical data: $C_{26}H_{26}O_{12}$, yellowish gum, $[\alpha]_D^{25}$ −187.3° (MeOH, c 0.02), −194.3° (MeOH, c 0.20), −199.5° (MeOH, c 0.17)

Structure:

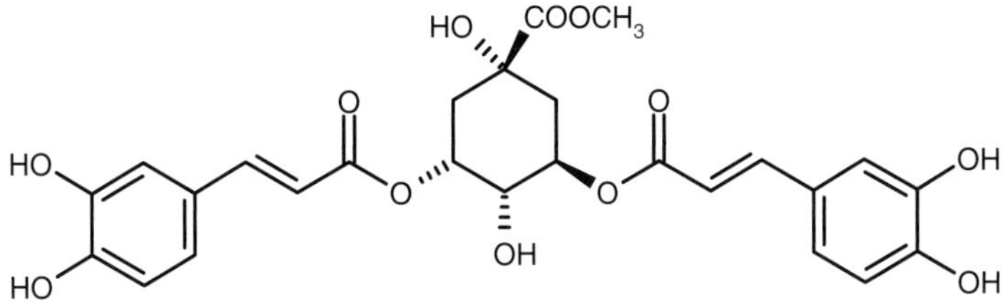

Methyl 3,5-di-*O*-caffeoyl quinate

Compound class: Phenolic compound

Sources: *Dichrocephala bicolor* (Roth) Schltdl. (whole plants; family: Asteraceae) [1]; *Solidago virga-aurea* var. *gigantea* (aerial parts; family: Compositae) [2]; *Suaeda glauca* (whole plants; family: Chenopodiaceae) [3]

Pharmaceutical potential: *Hepatoprotective*
The isolate was reported to exert hepatoprotective activity against tacrine-induced cytotoxicity in human liver-derived HepG2 cells with an EC_{50} value of 72.7 ± 6.2 µM (silybin was used as a positive control; EC_{50} = 82.4 ± 4.1 µM) [3].

References

[1] Lin, L.-C., Kuo, Y.-C., and Chou, C.-J. (1996) *J. Nat. Prod.*, **62**, 405.
[2] Choi, S.Z., Choi, S.U., and Lee, K.R. (2004) *Arch. Pharm. Res.*, **27**, 164.
[3] An, R.-B., Sohn, D.-H., Jeong, G.-S., and Kim, Y.-C. (2008) *Arch. Pharm. Res.*, **31**, 594.

Methyl 3,4-dihydroxy-5-(3'-methyl-2'-butenyl)benzoate

Physical data: $C_{13}H_{16}O_4$, amorphous solid

Structure:

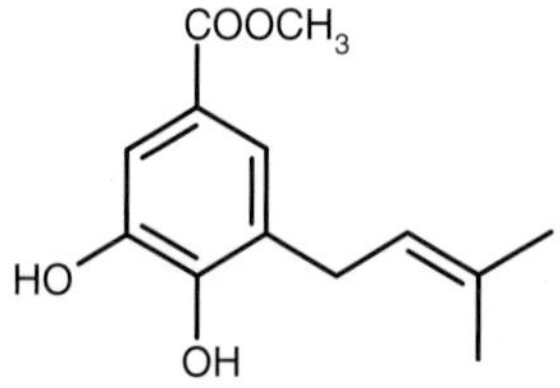

Methyl 3,4-dihydroxy-5-(3'-methyl-2'-butenyl)benzoate

Compound class: Benzoic acid derivative

Sources: *Piper glabratum* and *P. acutifolium* (leaves; family: Piperaceae)

Pharmaceutical potential: *Antiparasitic (leishmanicidal; trypanocidal; antiplasmodial)*
The benzoic acid derivative exhibited significant *in vitro* antiparasitic activities against the promastigote forms of *Leishmania amazonensis*, *L. braziliensis*, *L. donovani*, *Trypanosoma cruzi*, and *Plasmodium falciparum* with IC_{50} values of 18.2 ± 0.7, 13.8 ± 0.5, 18.5 ± 0.2, 18.5 ± 0.3, and 4.1 ± 0.5 µg/ml, respectively.

Reference

Flores, N., Jimenez, I.A., Gimenez, A., Ruiz, G., Gutierrez, D., Bourdy, G., and Bazzocchi, I.L. (2008) *J. Nat. Prod.*, **71**, 1538.

Methyl [2-(2,3-dihydroxy-3-methylbutyl),4-amino] benzoate

Physical data: $C_{13}H_{19}NO_4$, colorless gum, $[\alpha]_D^{27}$ −19.17° ($CHCl_3$, *c* 0.90)

Structure:

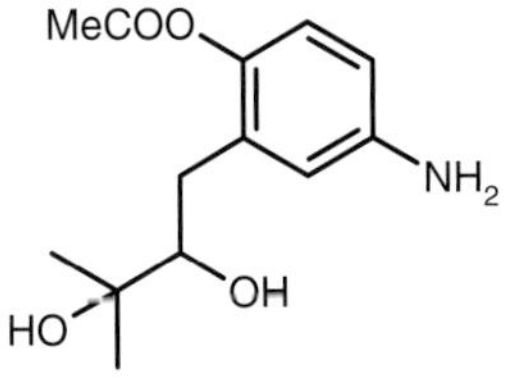

Methyl [2-(2,3-dihydroxy-3-methylbutyl),4-amino] benzoate

Compound class: Methylaminobenzoate

Source: *Xylaria* sp. BCC 9653 (wood-decayed fungus)

Pharmaceutical potential: *Cytotoxic*
The aminoester exhibited cytotoxic activity against standard Vero cells with an IC_{50} value of 26.13 µM.

Reference

Pongcharoen, W., Rukachaisirikul, V., Isaka, M., and Sriklung, K. (2007) *Chem. Bull. Pharm.*, **55**, 1647.

Methyl brevifolincarboxylate

Physical data: $C_{14}H_{10}O_8$, yellow needles, 198–200 °C

Structures:

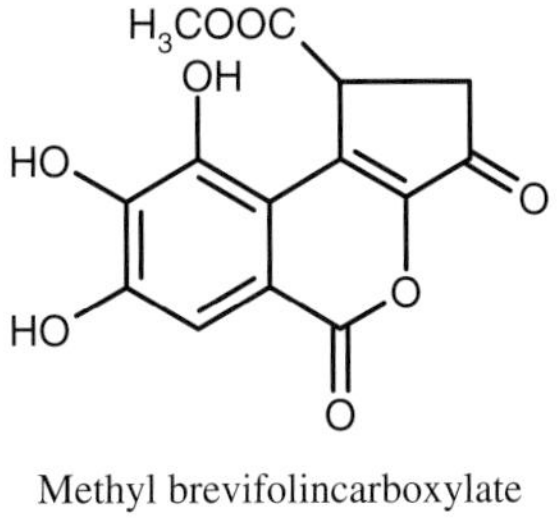

Methyl brevifolincarboxylate

Compound class: Benzopyran derivative

Sources: *Phyllanthus urinaria* L. (family: Euphorbiaceae) [1]; *Phyllanthus niruri* L. (leaves; family: Euphorbiaceae) [2]

Pharmaceutical potentials: *Inhibitor of platelet aggregation; vasorelaxant*
The investigators studied the inhibitory effects of the isolate on platelet aggregation induced by collagen or ADP at different concentrations using adenosine as the positive control; the test compound was found to be a potent inhibitor of platelet aggregation comparable to adenosine in spite of differences in the inhibitory mechanisms. At a concentration of 5.0 µM, it markedly inhibited platelet aggregation induced by collagen or ADP compared to the positive control.

The compound was also found to exhibit vasorelaxant effect on rat aortic rings [3]; it exhibited slow relaxation activity against norepinephrine (NE)-induced contractions of rat aorta with or without endothelium. The compound was found not to affect contractions induced by a high concentration (60 mM) of K^+, whereas it inhibited NE-induced vasocontraction in the presence of nicardipine. From their detailed studies, Iizuka et al. [3] established that such vasodilating property of the test compound is attributable to its inhibitory effects on receptor-operated Ca^{2+} channels (ROCs); hence, this compound might be a new type of vasodilator that does not belong to the inhibitors of voltage-dependent Ca^{2+} channels (VDCs).

References

[1] Yao, Q.-Q. and Zuo C.-X. (1993) *Acta Pharm. Sin.*, **28**, 829.
[2] Iizuka, T., Nagai, M., Taniguchi, A., Moriyama, H., and Hoshi, K. (2007) *Biol. Pharm. Bull.*, **30**, 382.
[3] Iizuka, T., Moriyama, H., and Nagai, M. (2006) *Biol. Pharm. Bull.*, **29**, 177.

3′-(3-Methyl-2-butenyl)-4′-O-β-D-glucopyranosyl-4,2′-dihydroxychalcone

Physical data: $C_{26}H_{30}O_9$, yellow amorphous powder, [α] +80.2° (MeOH, *c* 0.1)

Structure:

3′-(3-Methyl-2-butenyl)-4′-*O*-β-D-glucopyranosyl-4,2′-dihydroxychalcone

Compound class: Flavonoid (chalcone glycoside)

Source: *Maclura* (*Chlorophora*) *tinctoria* Linn. (Gaud.) Linn. (stem barks; family: Moraceae)

Pharmaceutical potential: *Antioxidant*
The chalcone glycoside was found to have significant antioxidant potential as determined by measuring free radical scavenging effects when studied in two different assays, such as the TEAC (i.e., Trolox equivalent antioxidant capacity) and the autooxidation assay (i.e., coupled oxidation of β-carotene and linoleic acid). In the TEAC assay, its efficacy was noted to be almost comparable to that of quercetin (used as a positive control), while in the autooxidation assay it exhibited antioxidant activity weaker than that of BHT (2,6-di-*tert*-butyl-4-methoxyphenol used as a positive control).

Reference

Cioffi, G., Escobar, L.M., Braca, A., and De Tommasi, N. (2003) *J. Nat. Prod.*, **66**, 1061.

12-Methyl-5-dehydrohorminone and 12-methyl-5-dehydroacetylhorminone

Systematic names:
- 12-Methyl-5-dehydrohorminone: 7-hydroxy-12-methoxy-11,14-dioxoabieta-5,8,12-triene
- 12-Methyl-5-dehydroacetylhorminone: 7-acetyl-12-methoxy-11,14-dioxoabieta-5,8,12-triene

Physical data:
- 12-Methyl-5-dehydrohorminone: $C_{21}H_{28}O_4$, amorphous powder
- 12-Methyl-5-dehydroacetylhorminone: $C_{23}H_{30}O_5$, amorphous powder

Structures:

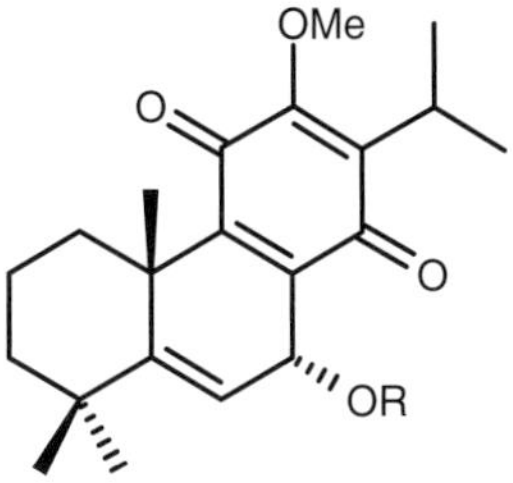

12-Methyl-5-denydrohorminone: R = H
12-Methyl-5-dehydroacetylhorminone: R = Ac

Compound class: Abietane diterpenoids

Source: *Salvia multicaulis* (roots; family: Labiatae)

Pharmaceutical potential: *Antitubercular*

12-Methyl-5-dehydrohorminone and its acetyl derivative showed strong antitubercular activity against *Mycobacterium tuberculosis* strain H37Rv with MIC values of 1.2 and 0.89 μg/ml, respectively.

Reference

Ulubelen, A., Topcu, G., and Johansson, C.B. (1997) *J. Nat. Prod.*, **60**, 1275.

6-*O*-Methyl-2-deprenylrheediaxanthone B and vieillardixanthone

Physical data:
- 6-*O*-Methyl-2-deprenylrheediaxanthone B: $C_{19}H_{18}O_6$, white amorphous solid, $[\alpha]_D^{25}$ 0° (CHCl$_3$, *c* 0.04)
- Vieillardixanthone: $C_{19}H_{18}O_6$, yellow oil

Structures:

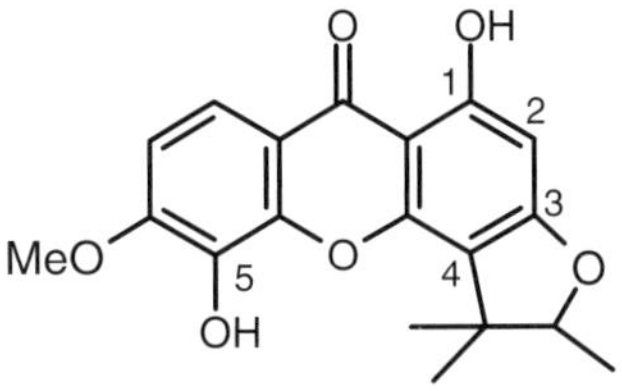

6-*O*-Methyl-2-deprenylrheediaxanthone

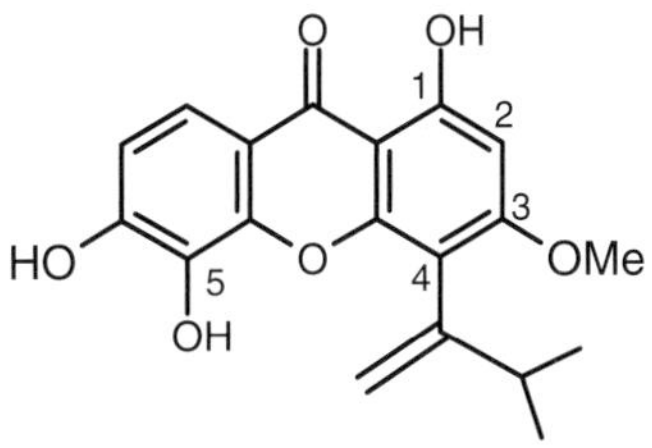

Vieillardixanthone

Compound class: Xanthones

Source: *Garciana vieillardii* (stem barks; family: Clusiaceae)

Pharmaceutical potential: *Antioxidant*
Both the isolates exhibited radical scavenging property (concentration-dependent) as studied in the DPPH assay using BHA (2,6-di-*tert*-butyl-4-hydroxyanisol) and α-tocopherol as reference standards. Vieillardixanthone was found to be more potent than the other, and its activity was noted to be almost similar/comparable to those of BHA and α-tocopherol. The free radical scavenging activity of such aromatic polyphenolics is believed to increase in the presence of hydroxyl groups or catechol or *para*-di-OH, and hence, the relative antioxidant activity of these two isolates is in accordance to this fact.

Reference

Hay, A.-E., Aumond, M.-C., Mallet, S., Dumontet, V., Litaudon, M., Rondeau, D., and Richomme, P. (2004) *J. Nat. Prod.*, **67**, 707.

Methyl rocaglate

See Aglafolin

4-Methylaeruginoic acid

Systematic name: 2-(*o*-Hydroxyphenyl)-4-methyl-2-thiazoline-4-carboxylic acid

Physical data: $C_{11}H_{11}NO_3S$, yellow powder

Structure:

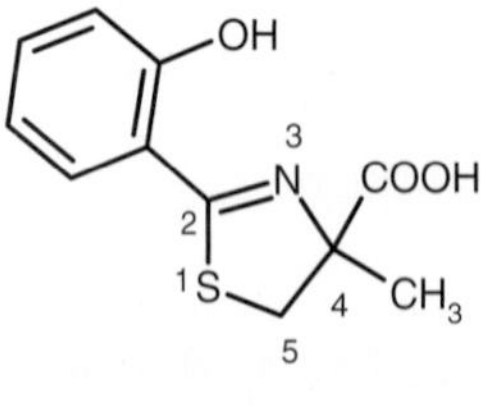

4-Methylaeruginoic acid

Compound class: Phenylthiazoline derivative

Source: *Streptomyces* sp. KCTC 9303 (culture broth)

Pharmaceutical potential: *Cytotoxic (antitumor)*
The isolate exhibited potent cytotoxic activities against a number of tumor cell lines such as CRL 1579 (skin), SNB-75 (CNS), M0LT-4F (leukemia), NCI-H522 (lung), and PC-3 (prostate) using adriamycin as a positive control. The respective ED_{50} values were determined to be 0.89, 0.02, 0.10, 8.68, and 0.04 for the test compound and 0.16, 0.62, 0.02, 0.14, and 1.09 for adriamycin. Hence, in comparison to adriamycin, 4-methylaeruginoic acid showed higher cytotoxicity against SNB-75 and PC-3.

Reference

Ryoo, I.-J., Song, K.-S., Kim, J.-P., Kim, W.-G., Koshino, H., and Yoo, I.-D. (1997) *J. Antibiot.*, **50**, 256.

Methylbellidifolin

See Swerchirin

3-*O*-Methylcalopocarpin

Systematic name: 9-Hydroxy-3-methoxy-2-prenylpterocarpan

Physical data: $C_{21}H_{22}O_4$, white amorphous solid, $[\alpha]_D$ −235° (CHCl$_3$, *c* 0.28)

Structure:

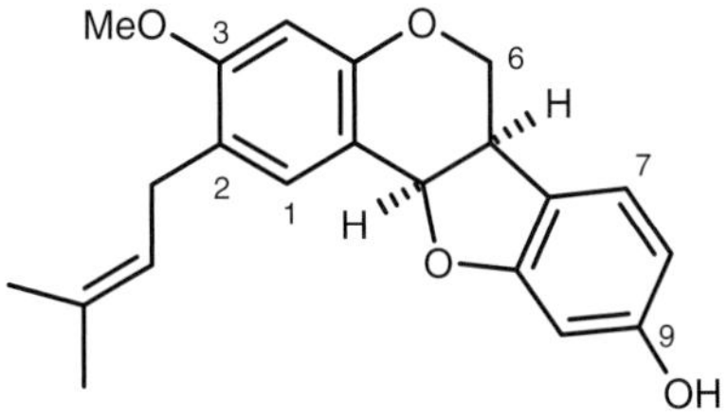

3-*O*-Methylcalopocarpin

Compound class: Pterocarpan

Sources: *Erythrina glauca* Willd. (syn. *Erythrina fusca* Lour.) (barks; family: Leguminosae) [1]; *E. burttii* (stem barks) [2]

Pharmaceutical potential: *Anti-HIV*
The pterocarpan was reported to significantly inhibit the cytopathic effects of *in vitro* HIV-1 infection in a human T-lymphoblastoid cell line (CEM-SS) with EC$_{50}$ and IC$_{50}$ values of 0.2 and 3.0 µg/ml, respectively, showing a maximum of 80–95% protection [2].

References

[1] McKee, T.C., Bokesch, H.R., McCormick, J.L., Rashid, M.A., Spielvogel, D., Kirk, R., Gustafson, K.R., Alavanja, M.M., Cardellina, J.H., II, and Boyd, M.R. (1997) *J. Nat. Prod.*, **60**, 431.
[2] Yenesew, A., Irungu, B., Derese, S., Midiwo, J.O., Heydenreich, M., and Peter, M.G. (2003) *Phytochemistry*, **63**, 445.

2-Methylchromone glycosides

Systematic names:
- **1**: 5-Hydroxy-6-methoxy-2-methylchromone-7-*O*-rutinoside
- **2**: 5-Hydroxy-2-methylchromone-7-*O*-rutinoside

Physical data:
- **1**: $C_{23}H_{30}O_{14}$, amorphous light yellow solid, $[\alpha]_D^{20}$ −67.1° (MeOH, *c* 0.136)
- **2**: $C_{22}H_{28}O_{13}$, amorphous light yellow solid, $[\alpha]_D^{20}$ −54.5° (MeOH, *c* 0.149)

Structures:

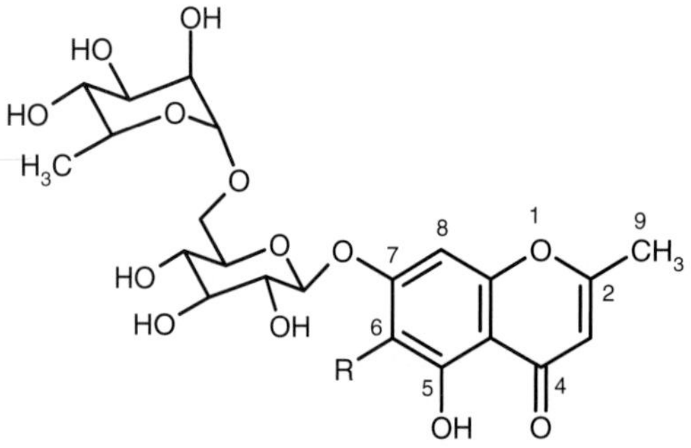

1: R = OCH$_3$
2: R = H

Compound class: Flavonoids (2-methylchromone glycosides)

Source: *Crossosoma bigelovii* S. Watt. (leaves, twigs, flowers and fruits; family: Crossosomataceae)

Pharmaceutical potential: *Cytotoxic*
The 2-methylchromone glycosides **1** and **2** showed significant cytotoxic activities against the three human cancer cell lines such as A-549 (human lung carcinoma), MCF-7 (human breast adenocarcinoma), and HT-29 (human colon adenocarcinoma), both having identical GI$_{50}$ (50% growth inhibitory) values of 0.4, 0.8, and 6.0 µM, respectively (adriamycin was used as a positive control; GI$_{50}$ values: 0.007, 0.1, and 0.06 µM, respectively).

Reference

Klausmeyer, P., Zhou, Q., Scudiero, D.A., Uranchimeg, B., Melillo, G., Cardellina, J.H., II, Shoemaker, R.H., Chang, C.-J., and McCloud, T.G. (2009) *J. Nat. Prod.*, **72**, 805.

N-Methylnarceimicine

Physical data: $C_{22}H_{22}NO_8$, yellow crystalline solid (MeOH), mp 194–196 °C

Structure:

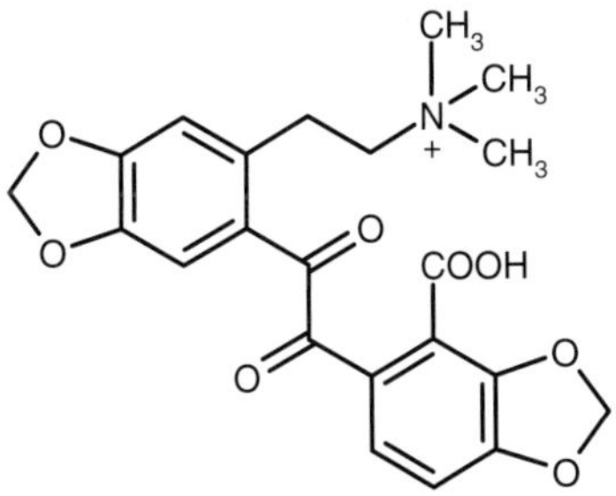

N-Methylnarceimicine

Compound class: Quaternary alkaloid

Source: *Corydalis saxicola* Bunting (roots; family: Fumariodeae)

Pharmaceutical potential: *Anti-hepatitis B virus*
The alkaloid showed inhibitory effect against secretions of HBsAg (hepatitis B virus surface antigen) and HBeAg (hepatitis B virus e antigen) in the HepG2.2.15 cell line with IC_{50} values of 1.24 µM (SI = 1.6) and 1.86 µM (SI = 1.0), respectively; lamivudine was used as a positive control (IC_{50} value against HBsAg: 15.37 µM (SI = 2.9); IC_{50} value against HBeAg: 44.85 µM (SI = 1.0)).

Reference

Wu, Y.-R., Ma, Y.-B., Zhao, Y.-X., Yao, S.-Y., Zhou, J., Zhou, Y., and Chen, J.J. (2007) *Planta Med.*, **73**, 787.

2-(Methyldithio)pyridine-*N*-oxide

Physical data: $C_6H_7NOS_2$, colorless waxy solid

Structure:

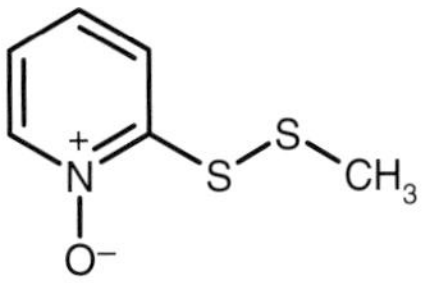

2-(Methyldithio)pyridine-*N*-oxide

Compound class: Pyridine-*N*-oxide alkaloid

Source: *Allium sativum* L. (bulbs; family: Liliaceae/Alliaceae)

Pharmaceutical potentials: *Antimicrobial; antiproliferative*

The isolate exhibited potent activity against *Mycobacterium bovis* BCG (with an MIC value of 0.1 μg/ml) and *Mycobacterium tuberculosis* H37Rv; it was found to be bactericidal against nonreplicating cells of *M. tuberculosis* with a concentration of 1.25 μg/ml, resulting in a 99% decrease in viable counts of anaerobically adapted cells after 1 week of exposure. The test compound also showed potent antibacterial activity against methicillin-resistant *Staphylococcus aureus* and multidrug-resistant (MDR) variants of *S. aureus*.

In a human cancer cell line antiproliferative assay, the dithioalkaloid displayed cytotoxic activity against human breast cancer (MCF7), human lung cancer (A549), human colorectal cancer (HT29), and lung fibroblast (WI38) cancer cells with IC_{50} values of 0.35, 0.22, 1.84, and 0.8 μM, respectively (cisplatin was used as a positive control; respective IC_{50} values were recorded as 0.76, 2.47, 3.27, and 1.0 μM).

Reference

O'Donnell, G., Poeschl, R., Zimhony, O., Gunaratnam, M., Moreira, J.B.C., Neidle, S., Evangelopoulos, D., Bhakta, S., Malkinson, J.P., Boshoff, H.I., Lenaerts, A., and Gibbons, S. (2009) *J. Nat. Prod.*, **72**, 360.

Michaolides A–K

Physical data:
- Michaolide A (1): $C_{24}H_{32}O_8$, colorless oil, $[\alpha]_D^{25}$ −29° (CHCl$_3$, *c* 0.1)
- Michaolide B (2): $C_{24}H_{32}O_8$, $[\alpha]_D^{25}$ −37° (CHCl$_3$, *c* 0.1)
- Michaolide C (3): $C_{24}H_{32}O_8$, colorless oil, $[\alpha]_D^{25}$ −33° (CHCl$_3$, *c* 0.1)
- Michaolide D (4): $C_{24}H_{30}O_8$, $[\alpha]_D^{25}$ −7° (CHCl$_3$, *c* 0.1)
- Michaolide E (5): $C_{24}H_{32}O_7$, $[\alpha]_D^{25}$ −7° (CHCl$_3$, *c* 0.1)
- Michaolide F (6): $C_{22}H_{30}O_6$, $[\alpha]_D^{25}$ −88° (CHCl$_3$, *c* 0.1)
- Michaolide G (7): $C_{26}H_{34}O_{10}$, $[\alpha]_D^{25}$ −10° (CHCl$_3$, *c* 0.1)
- Michaolide H (8): $C_{27}H_{39}O_{10}$, $[\alpha]_D^{25}$ −13° (CHCl$_3$, *c* 0.1)
- Michaolide I (9): $C_{26}H_{36}O_{10}$, $[\alpha]_D^{25}$ +10° (CHCl$_3$, *c* 0.1)
- Michaolide J (10): $C_{28}H_{38}O_{11}$, $[\alpha]_D^{25}$ +19° (CHCl$_3$, *c* 0.1)
- Michaolide K (11): $C_{24}H_{34}O_8$, $[\alpha]_D^{25}$ +61° (CHCl$_3$, *c* 0.1)

Michaolides A–K

Structures:

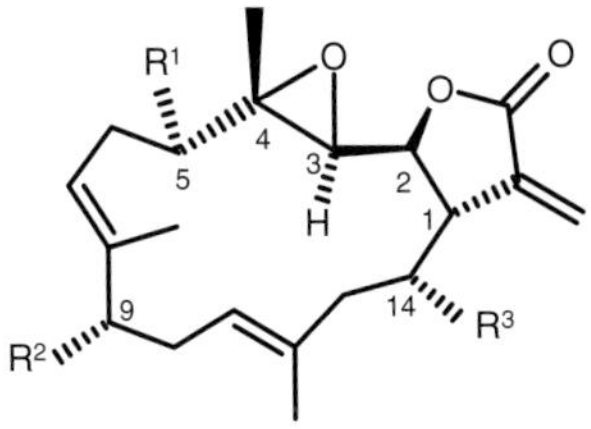

1: $R^1 = R^3 = OAc; R^2 = OH$
2: $R^1 = R^2 = OAc; R^3 = OH$
3: $R^1 = OH; R^2 = R^3 = OAc$
4: $R^1 = R^3 = OAc; R^2 = O$
5: $R^1 = R^3 = OAc; R^2 = H$
6: $R^1 = OAc; R^2 = H; R^3 = OH$

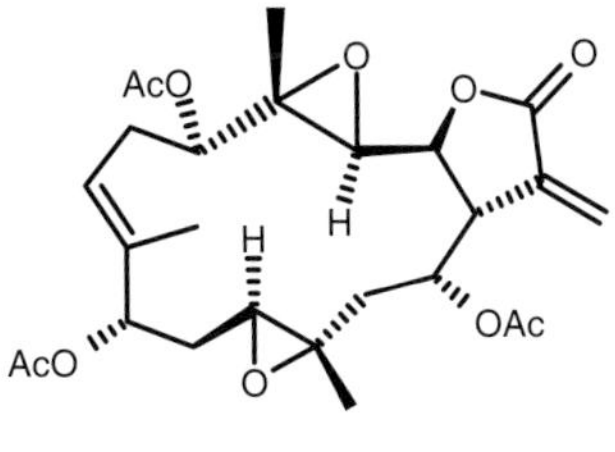

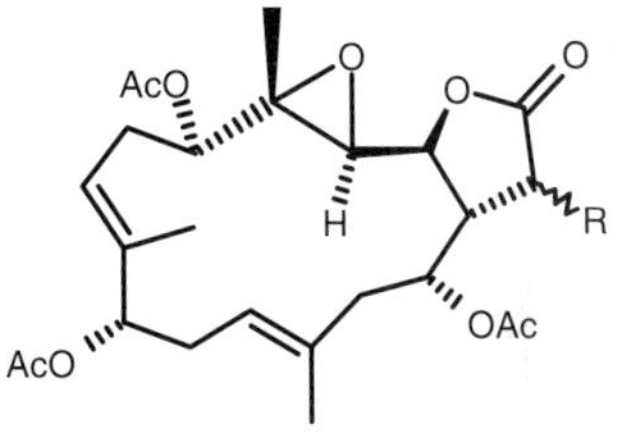

8: $R = \beta\text{-}CH_2OMe$

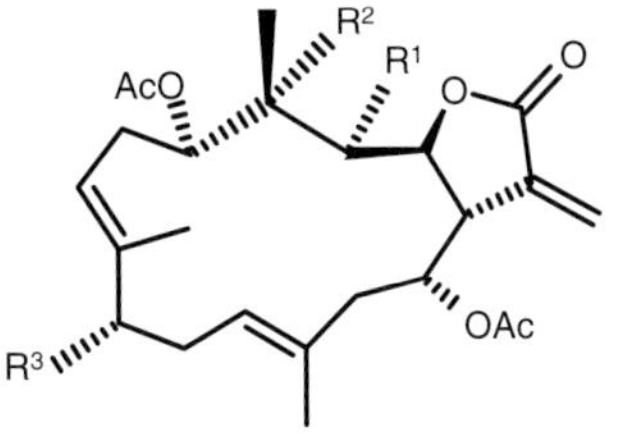

9: $R^1 = R^2 = OH; R^3 = OAc$
10: $R^1 = R^3 = OAc; R^2 = OH$
11: $R^1 = R^2 = OH; R^3 = H$

Compound class: Terpenoids (cembranolides)

Source: *Lobophytum michaelae* Tixier-Durivault (Formosan soft coral; family: Alcyoniidae)

Pharmaceutical potential: *Cytotoxic*

The present investigators studied *in vitro* cytotoxic potentials of all the isolates against HT-29 and P-388 cell lines. Michaolides B and F were found to have potent cytotoxicity against both HT-29 and P-388 cells; the others also showed moderate cytotoxicity against the target cell lines. The respective ED_{50} values (μg/ml) of the cembranolides **1–11** against HT-29 cells were determined to be 0.7, 0.03, 0.8, 1.5, 0.05, 0.001, 0.8, 3.1, 1.4, 0.8, and 9.9 μg/ml, whereas their ED_{50} values (μg/ml) against P-388 cells were evaluated as 0.1, 0.001, 0.4, 0.2, 0.007, 0.004, 0.3, 0.4, 0.8, 0.6, and 0.6 μg/ml, respectively. From structure–activity relationships, the investigators assumed that hydroxylation at C-14 together with an α-exo-methylene-γ-lactone and a 3,4-trisubstituted epoxy might be important for potent cytotoxicity.

Reference

Wang, L.-T., Wang, S.-K., Soong, K., and Duh, C.-Y. (2007) *Chem. Pharm. Bull.*, **55**, 766.

Michellamines A and B

Physical data:
- Michellamine A: $C_{46}H_{48}N_2O_8$, $[\alpha]_D$ −10.5°, $[\alpha]_{365}$ +65.7° (MeOH, c 0.38)
- Michellamine B: $C_{46}H_{48}N_2O_8$, $[\alpha]_D$ −14.8°, $[\alpha]_{365}$ −23.4° (MeOH, c 0.74)

Structures:

Michellamine A   (atropisomeric pair)   Michellamine B

Compound class: Naphthalene tetrahydroisoquinoline alkaloids

Source: *Ancistrocladus abreviatus* (aerial parts; family: Ancistrocladaceae)

Pharmaceutical potential: *Anti-HIV*

Both the alkaloids were evaluated as anti-HIV agents – the present investigators found that michellamine B completely inhibits the cytopathic effects of HIV-1 upon CEM-SS human lymphoblastoid target cells *in vitro* with an EC_{50} value of ∼20 µM; at higher concentrations, it exerted only a weak cytotoxic effect on uninfected target cells (IC_{50} value of ∼200 µM). Similar results were also obtained with michellamine A. Moreover, the water-soluble HBr and HOAc salts of both compounds were observed to be equally effective as the free bases.

The alkaloids inhibit HIV-1 during the early phase of viral infection of T lymphocytes by inhibiting reverse transcriptase as well as at later stages by inhibiting cellular fusion and syncytium formation. The investigators also noted that michellamine B, either as the free base or as the HBr salt, exhibited the same potency against the cytopathic effects of HIV-2 upon MT-2 target cells *in vitro* as it did against HIV-1 upon CEM-SS cells; michellamine A was somewhat less effective than the other against HIV-2, affording only partial protection at comparable concentrations. The inhibitory characteristics of the compounds, their range of activity, and the difficulty in cell membrane penetration as a whole are indicative of the supposition that the drugs might act at the cell surface.

Reference

Manfredi, K.P., Blunt, J.W., Cardellina, J.H.I., McMahon, J.B., Pannell, L.K., Cragg, G.M., and Boyd, M.R. (1991) *J. Med. Chem.*, **34**, 3402.

Microclavatin

Physical data: $C_{20}H_{28}O_3$, colorless prisms (EtOAc–petroleum ether 1:4), mp 145–146 °C, $[\alpha]_D^{20}$ −15.7° (CHCl$_3$, c 0.07)

Structure:

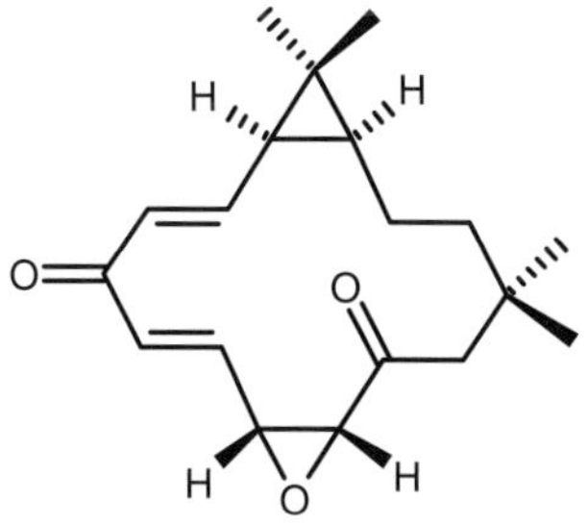

Microclavatin

Compound class: Diterpenoid

Source: Soft coral *Sinularia microclavata*

Pharmaceutical potential: Cytotoxic
Microclavatin was reported to display moderate cytotoxic activity against the KB and MCF tumor cell lines with IC$_{50}$ values of 5.0 and 20.0 µg/ml, respectively (MTT assay was used).

Reference

Zhang, C.-X., Yan, S.-J., Zhang, G.-W., Lu, W.-G., Su, J.-U., Zeng, L.-M., Gu, L.-Q., Yang, X.-P., and Lian, Y.-J. (2005) *J. Nat. Prod.*, **68**, 1087–1089.

Microspinosamide

Physical data: $C_{75}H_{109}BrN_{18}O_{22}S$, white amorphous powder, $[\alpha]_D$ +2.4° (MeOH, c 0.5)

Structure:

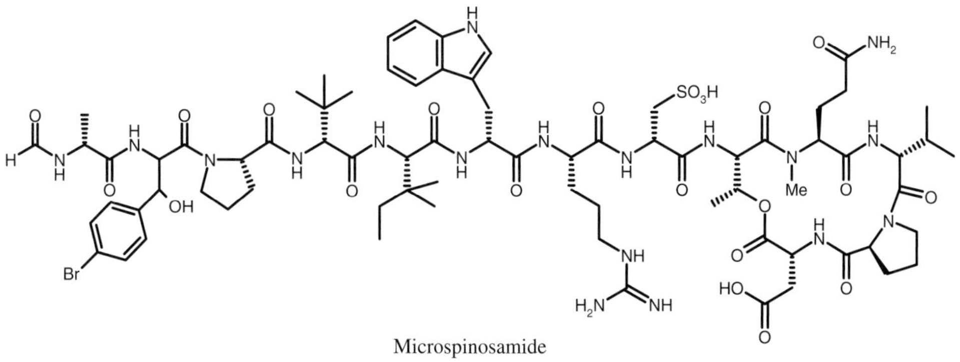

Microspinosamide

Compound class: Peptide (a cyclic depsidecapeptide)

Source: *Sidonops microspinosa* Wilson (Geodiidae; Indonesian marine sponge)

Pharmaceutical potential: *Anti-HIV*

Microspinosamide, the first naturally occurring peptide to contain a β-hydroxy-*p*-bromophenylalanine residue, was reported to display anti-HIV activity by inhibiting cytopathic effect of HIV-1 infection in CEM-SS target cells (XTT-based *in vitro* assay) with an EC_{50} value of ~0.2 µg/ml. However, the investigators showed that the peptide was only cytoprotective over a modest concentration range because of cytotoxic effects toward the target cells (IC_{50} value of ~3.0 µg/ml).

Reference

Rashid, M.A., Gustafson, K.R., Cartner, L.K., Shigematsu, N., Pannell, L.K., and Boyd, M.R. (2001) *J. Nat. Prod.*, **64**, 117.

Migrastatin

Physical data: $C_{27}H_{39}NO_7$, white powder, mp 54–55 °C, $[\alpha]_D^{27}$ +17.9° (MeOH, *c* 3.19)

Structure:

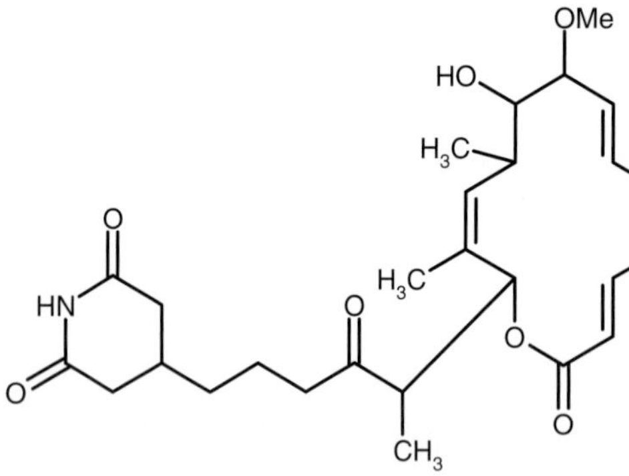

Migrastatin

Compound class: Antibiotic (a glutarimide antibiotic having a unique 14-membered lactone ring)

Source: *Streptomyces* sp. MK929-43F1 (culture broth) [1, 2]

Pharmaceutical potential: *Antitumor*

Nakae *et al.* [1] demonstrated that the glutarimide antibiotic acts as an inhibitor of cell migration of tumor cells; it was found to inhibit cell migration of human esophageal cancer EC17 cells (IC_{50} = 1.0 µg/ml) and mouse melanoma B16 cells (IC_{50} = 3.0 µg/ml) as estimated by both wound healing assay and chemotaxicell chamber assay. It was evaluated that migration inhibitory activity of migrastatin was not dependent on cytotoxicity or inhibition of protein synthesis. In contrast, other 14-membered ring macrolides such as clarithromycin and erythromycin did not inhibit cell migration of both EC17 and B16 cells up to 100 µg/ml [1].

References

[1] Nakae, N., Yoshimoto, Y., Ueda, M., Sawa, T., Takahashi, Y., Naganawa, H., Takeuhi, T., and Imoto, M. (2000) *J. Antibiot.*, **53**, 1130.
[2] Nakae, N., Yoshimoto, Y., Ueda, M., Sawa, T., Takahashi, Y., Naganawa, H., Takeuhi, T., and Imoto, M. (2000) *J. Antibiot.*, **53**, 1228.

Militarinone A

Physical data: $C_{26}H_{37}NO_6$, yellow solid, $[\alpha]_D^{25}$ −14.6° (MeOH, *c* 1.3)

Structure:

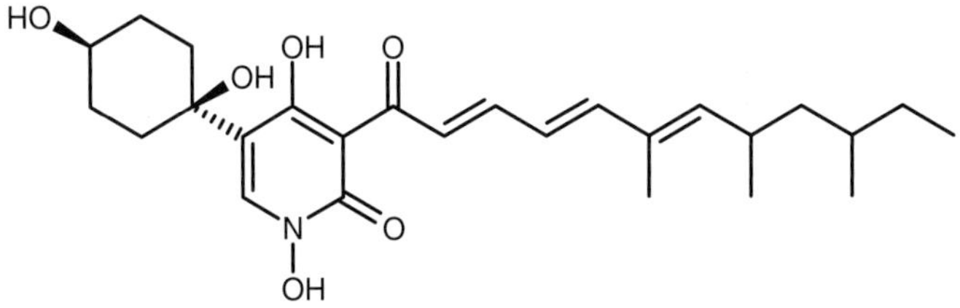

Militarinone A

Compound class: Alkaloid (pyridone alkaloid)

Source: *Paecilomyces militaris* (L.:Fr.) Link (mycelium of the entomogenous fungus) [1]

Pharmaceutical potential: *Neurotrophic*

Stimulation of neuronal differentiation of PC-12 cells is regarded as a well-established model for the study of neurotrophins [2]. The present investigators [1] evaluated such potential of the alkaloid studying at various concentrations – the test compound induced pronounced neurite sprouting by about 80 and 70% after 24 h at the concentrations of 33 and 10 µM, respectively. However, militarinone A displayed no significant cytotoxicity in PC-12 cells at concentrations up to 100 µM [1].

References

[1] Schmidt, K., Gnther, W., Stoyanova, S., Schubert, B., Li, Z., and Hamburger, M. (2002) *Org. Lett.*, **4**, 197.
[2] Greene, L.A. and Rukenstein, A. (1989) The quantitative bioassay of NGF with PC12 cells, in Nerve Growth Factors (ed. R.A. Rush), John Wiley & Sons, Ltd, Chichester, Chapter 7, pp. 139–147.

Millewanins G and H

Physical data:
- Millewanin G: $C_{25}H_{26}O_7$, colorless oil, $[\alpha]_D^{24}$ +7.0° (MeOH, *c* 0.108)
- Millewanin H: $C_{25}H_{26}O_7$, colorless oil, $[\alpha]_D^{24}$ +8.8° (MeOH, *c* 0.0057)

Structures:

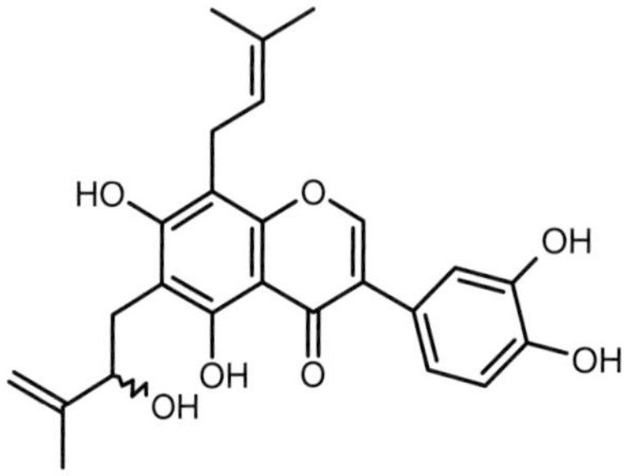

Millewanin G

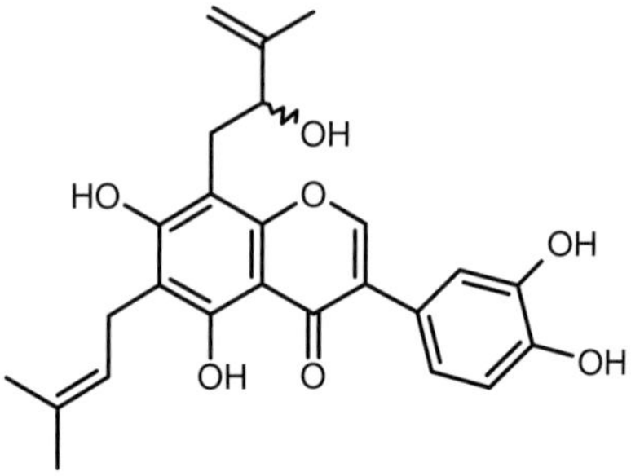

Millewanin H

Compound class: Flavonoids (isoflavonoids)

Source: *Millettia pachycarpa* Benth. (leaves; family: Leguminosae)

Pharmaceutical potential: *Antiestrogenic*
The prenylated isoflavonoids, millewanins G and H, showed significant antiestrogenic potential in a dose-dependant manner by inhibiting β-galactosidase activity induced by 17β-estradiol (studied in the yeast two-hybrid assay); the IC_{50} (a 50% inhibitory concentration of 1 nM 17β-estradiol) values were determined to be 29 and 18 μM, respectively. The values are slightly greater than that of 4-hydroxytamoxifen (OHT; IC_{50} = 4.4 μM), a typical estrogen antagonist. Antiestrogenic compounds inhibit estrogen-dependent breast cancer proliferation.

Reference

Ito, C., Itoigawa, M., Kumagaya, M., Okamoto, Y., Ueda, K., Nishihara, T., Kijima, N., and Furukawa, H. (2006) *J. Nat. Prod.*, **69**, 138.

Mimosifolenone

Physical data: $C_{16}H_{18}O_3$, colorless gum, $[\alpha]_D$ +81° (CHCl$_3$, *c* 0.25)

Structure:

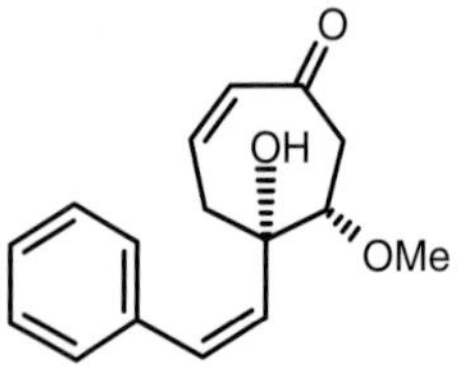

Mimosifolenone

Compound class: Aromatic compound (C_{16}-styrylcycloheptenone derivative)

Source: *Aeschynomene mimosifolia* Vatke (root woods; family: Leguminosae)

Pharmaceutical potential: *Cytotoxic*
The isolate showed moderate cytotoxicity against human oral epidermoid carcinoma (KB) cells with an ED_{50} value of 3.3 µg/ml.

Reference

Fullas, F., Kornberg, L.J., Wani, M.C., and Wall, M.E. (1996) *J. Nat. Prod.*, **59**, 150.

Minheryin G

Systematic name: 1α,3β,7α,14β-Tetrahydroxy-*ent*-kaur-16-en-15-one

Physical data: $C_{20}H_{30}O_5$, white powder, $[\alpha]_D^{25}$ −30.52° (pyridine, *c* 0.20)

Structure:

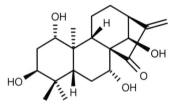

Minheryin G

Compound class: *ent*-Kaurane diterpenoid

Source: *Isodon henryi* (leaves; family: Labiatae)

Pharmaceutical potential: *Cytotoxic*
The isolate was found to possess potent cytotoxicity against K562 and HepG2 cell lines with IC_{50} values of 0.43 and 0.32 µg/ml, respectively (cisplatin was used as a positive control with respective IC_{50} values of 1.33 and 0.32 µg/ml).

Reference

Zhao, Y., Huang, S.-X., Yang, L.-B., Pu, J.-X., Xiao, W.-L., Li, L.-M., Lei, C., Weng, Z.-Y., Han, Q.-B., and Sun, H.-D. (2009) *Planta Med.*, **75**, 65.

Miquelianin

Systematic name: Quercetin 3-*O*-β-D-glucuronopyranoside

Physical data: $C_{21}H_{18}O_{13}$, yellow powder, mp 219–221 °C

Miquelianin

Structure:

Miquelianin

Compound class: Flavonoid glycoside (flavonol glycoside)

Sources: It occurs in various plants, such as *Salvia* sp. (family: Labiatae) [1]; *Foeniculum vulgare* (flowers; family: Apiaceae) Mill. [2, 3], *F. dulce* DC. (flowers) [3]; *Hypericum hirsutum* L. (family: Hypericaceae) [4], *H. perforatum* L. [5]; *Callistemon lanceolatus* DC. (flowers and leaves; family: Myrtaceae) [6], *Polygonum salicifolium* (family: Polygonaceae) [7], *Phaseolus vulgaris* L. (family: Fabaceae) [8], *Reynoutria sachalinensis* (Fr. Schm.) Nakai (flower; family: Polygonaceae) [9], *Schinus molle* L. (leaves; family: Anacardiaceae) [10]

Pharmaceutical potentials: *Antioxidant; antidepressant*
The flavonoid glycoside showed potent antioxidant activity as evaluated in DPPH, superoxide radical scavenging assays as well as in Cu^{2+}-mediated human plasma LDL oxidation assay; the compound displayed significant antioxidative effect in all these assays with IC_{50} values of 46.2 µM (DPPH radical scavenging), 4.4 µM (superoxide radical scavenging) and 5.4 µM (against LDL oxidation) [9]. Its strong radical scavenging properties on lipid peroxidation, hydroxyl radical, and superoxide anion generations in comparison with that of quercetin as a positive control *in vitro* were also reported by Marzouk *et al.* [10]. Its antidepressant activity was evidenced by the forced swimming test, an *in vivo* pharmacological model with rats [5].

References

[1] Lu, Y.R. and Foo, L.Y. (2002) *Phytochemistry*, **59**, 117.
[2] Krizman, M., Baricevic, D., and Prosek, M. (2006) *J. Pharm. Biomed. Anal.*, **43**, 481.
[3] Soliman, F.M., Shehata, A.H., Khaleel, A.E., and Ezzat, S.M. (2002) *Molecules*, **7**, 245.
[4] Kitanov, G.M. (1988) *Khim. Prirr. Soedinm*, **1**, 132; *Chem. Nat. Comp.*, (1988,) **24**, 119.
[5] Juergenliemk, G., Boje, K., Huewel, S., Lohmann, C., Galla, H.-J., and Nahrstedt, A. (2003) *Planta Med.*, **69**, 1013.
[6] Marzouk, M.S., (2008) *Phytochem. Anal.*, **19**, 541.
[7] Calis, I., Kuruüzüm, A., Demirezer, L.O., Sticher, O., Ganci, W., and Rüedi, P. (1999) *J. Nat. Prod.*, **62**, 1101.
[8] Price, K.R., Colquhoun, I.J., Barnes, K.A., and Rhodes, M.J. (1998) *J. Agric. Food Chem.*, **46**, 4898.

[9] Zhang, X., Thuong, P.T., Jin, W.Y., Su, N.D., Sok, D.E., Bae, K., and Kang, S.S. (2005) *Arch. Pharm. Res.*, **28**, 22.
[10] Marzouk, M.S., Moharram, F.A., Haggag, E.G., Ibrahim, M.T., and Badary, O.A. (2006) *Phytother. Res.*, **20**, 200.

Monodictyochromones A and B

Physical data:
- Monodictyochromone A: $C_{30}H_{30}O_{11}$, yellow solid, $[\alpha]_D^{24}$ $-67°$ (CHCl$_3$, c 0.42)
- Monodictyochromone B: $C_{30}H_{30}O_{11}$, yellow solid, $[\alpha]_D^{24}$ $+74°$ (CHCl$_3$, c 0.60)

Structures:

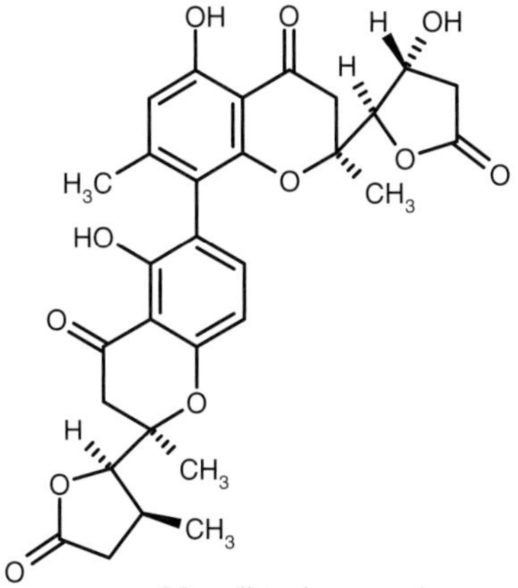

Monodictychromone A 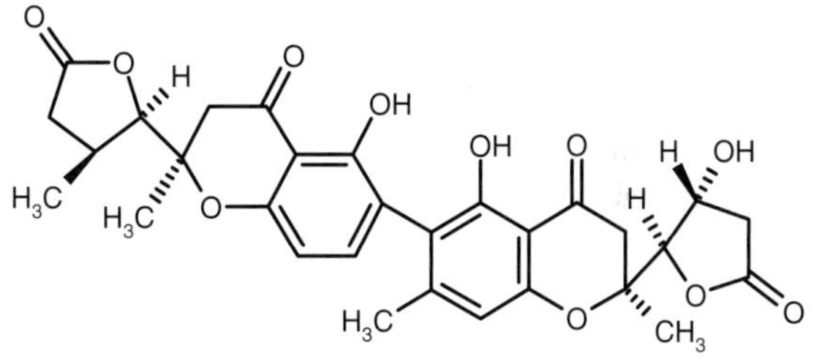   Monodictychromone B

Compound class: Chromanone derivatives (dimeric xanthonoids)

Source: *Monodictys putredinis* (marine-derived fungus)

Pharmaceutical potential: *Anticancerous*
Both the compounds were evaluated for their cancer chemopreventive potentials and were found to inhibit cytochrome P450 1A activity with IC$_{50}$ values of 5.3 ± 1.1 and 7.5 ± 0.9 µM, respectively. In addition, they also showed moderate activity as inducers of NAD(P)H:quinone reductase (QR) in cultured mouse Hepa 1c1c7 cells with respective CD values (concentration required to double the specific activity of QR) of 22.1 ± 1.0 and 24.8 ± 2.6 µM. In addition, monodictyochromones A and B showed inhibitory activity against aromatase enzyme with IC$_{50}$ values of 24.4 ± 4.8 and 16.5 ± 1.8 µM, respectively.

Reference

Pontius, A., Krick, A., Mesry, R., Kehraus, S., Foegen, S.E., Muller, M., Klimo, K., Gerhauser, C., and Konig, G.M. (2008) *J. Nat. Prod.*, **71**, 1793.

Monodictysins B and C

Systematic names:
- Monodictysin B: (1S,3S,4S,4aS,9aS)-1,4,8-trihydroxy-3,4a-dimethyl-1,2,3,4,4a,9a-hexahydro-9H-xanthen-9-one
- Monodictysin C: (1S,3S,4S,4aS,9aS)-1,4,8-trihydroxy-6-methoxy-3,4a-dimethyl-1,2,3,4,4a,9a-hexahydro-9H-xanthen-9-one

Physical data:
- Monodictysin B: $C_{15}H_{18}O_5$, white solid, $[\alpha]_D^{24}$ +80.5° (CHCl$_3$, c 0.2)
- Monodictysin C: $C_{16}H_{20}O_6$, white solid, $[\alpha]_D^{24}$ +102° (CHCl$_3$, c 0.5)

Structures:

Monodictysin B: R = H
Monodictysin C = R = OCH$_3$

Compound class: Xanthone derivatives

Source: *Monodictys putredinis* (fungus) [1]

Pharmaceutical potential: *Anticancerous*

Monodictysins B and C were evaluated for their cancer chemopreventive activity by analyzing their potential to modulate drug metabolizing enzymes, that is, inhibition of the phase I enzyme Cyp1A (cytochrome P450 1A) and induction of the phase II enzyme NAD(P)H:quinone reductase (QR). These mechanisms are most relevant during initiation phase of carcinogenesis [2] – Cyp1A isoenzyme is involved in the metabolic conversion of procarcinogens into carcinogens, while QR is carcinogen-detoxifying enzyme. Monodictysins B and C inhibited Cyp1A enzyme activity with IC$_{50}$ values of 23.3 ± 3.9 and 3.0 ± 0.7 µM, respectively [1]. Both the compounds also displayed moderate activity as inducers of NAD(P)H:quinone reductase in cultured mouse Hepa 1c1c7 cells with CD values (concentration required to double the specific activity of QR) of 12.0 ± 4.8 and 12.8 ± 2.6 µM, respectively, without exhibiting cytotoxic effects at concentrations up to 50 µM; in addition, monodictysin C was further identified as a weak inhibitor of aromatase activity essential for biosynthesis of estrogens [1].

References

[1] Krick, A., Kehraus, S., Gerhauser, C., Klimo, K., Nieger, M., Maier, A., Fiebig, H.-H., Atodiresei, I., Raabe, G., Fleischhauer, J., and Konig, G.M. (2007) *J. Nat. Prod.*, **70**, 353.
[2] Lewis, D.F. and Lake, B.G. (1996) *Xenobiotica*, **26**, 723.

Monotesone A

Systematic name: (2S)-2,3-Dihydro-5,7-dihydroxy-2-{3-hydroxy-4-[3-methylbut-2-enyloxy]phenyl}-4H-1-benzopyran-4-one

Physical data: $C_{20}H_{20}O_6$, mp 81–83 °C, $[\alpha]_D^{23}$ −24.7° (MeOH, c 0.32)

Structure:

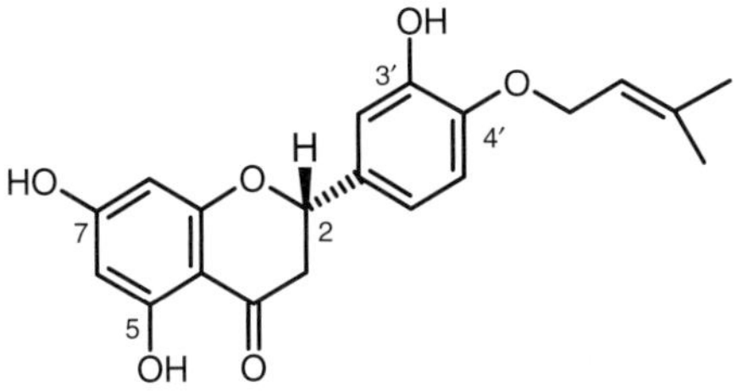

Monotesone A

Compound class: Flavonoid (flavanone)

Source: *Monotes engleri* (leaves; family: Dipterocarpaceae)

Pharmaceutical potential: *Antifungal*
The O-prenylated flavanone was found to possess antifungal activity against *Candida albicans* with an MIC value of 20 µg/ml.

Reference

Garo, E., Wolfender, J.-L., Hostettmann, K., Hiller, W., Antus, S., and Mavi, S. (1998) *Helv. Chim. Acta*, **81**, 754.

Morellic acid and 8,8a-epoxymorellic acid

Physical data:
- Morellic acid: $C_{33}H_{36}O_8$, amorphous powder
- 8,8a-Epoxymorellic acid: $C_{33}H_{36}O_9$, orange semisolid, $[\alpha]_D^{29}$ −138.0° (CHCl$_3$, c 0.1)

Structures:

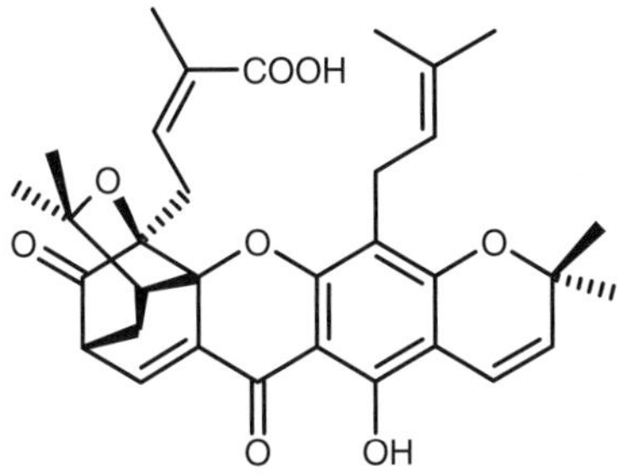

Morellic acid

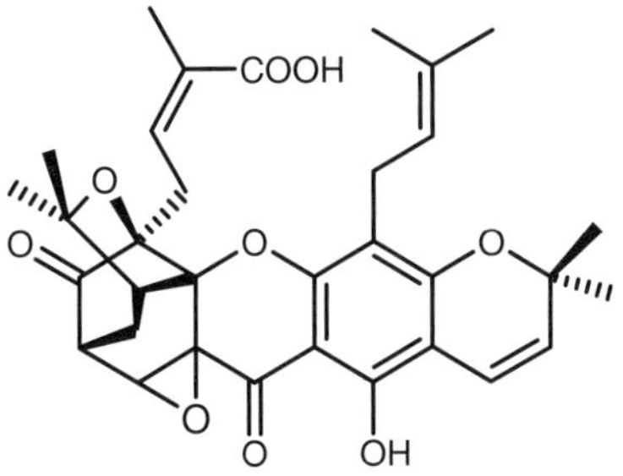

8,8a-Epoxymorellic acid

Compound class: Xanthones (caged)

Sources: Morellic acid: *Garcinia morella* (family: Guttiferae) [1], *G. hanburyi* Hook. F. (resin and fruits) [2–4]; 8,8a-epoxymorellic acid: *G. hanburyi* Hook. F. (resin and fruits) [2]

Pharmaceutical potentials: *Cytotoxic; antibacterial; anti-HIV*
Morellic acid was reported to have cytotoxic property [2–4]; it also showed moderate antibacterial activity against methicillin-resistant *Staphylococcus aureus* with an MIC value of 25 µg/ml [4]. Reutrakul *et al.* [2] studied the cytotoxicity of the compound against a panel of cell lines – the ED_{50} (µg/ml) values were determined to be 0.17 (P-388, murine lymphocytic leukemia), 10.29 (KB, human oral nasopharyngeal carcinoma), 2.52 (Col-2, human colon cancer), 9.40 (BCA-1, human breast cancer), 8.28 (Lu-1, human lung cancer), and 2.37 (ASK, rat glioma). The same investigators [2] also evaluated 8,8a-epoxymorellic acid as a potent cytotoxic agent – the ED_{50} (µg/ml) values were determined to be 0.38 (P-388), 2.80 (KB), 2.31 (Col-2), 2.10 (BCA-1), 2.58 (Lu-1), and 0.70 (ASK).

Besides the cytotoxicity of the compounds, Reutrakul *et al.* also reported that both of them possess anti-HIV activity [2]. Both the xanthones showed potent anti-HIV activity studied in the HIV RT assay; at a concentration of 200 µg/ml, morellic acid and its epoxy derivative inhibited the HIV-RT activity by 86.7 and 94.1%, respectively, with respective IC_{50} values of 101.8 and 186.2 µg/ml.

References

[1] Karanjgaonkar, C.G., Nair, P.M., and Venkataraman, K. (1966) *Tetrahedron Lett.*, **7**, 687.
[2] Reutrakul, V., Anantachoke, N., Pohmakotr, M., Jaipetch, T., Sophasan, S., Yoosook, C., Kasisit, J., Napaswat, C., Santisuk, T., and Tuchinda, P. (2007) *Planta Med.*, **73**, 33.
[3] Asano, J., Chiba, K., Tada, M., and Yoshii, T. (1996) *Phytochemistry*, **41**, 815.
[4] Sukpondma, Y., Rukachaisirkul, V., and Phongpaichit, S. (2005) *Chem. Pharm. Bull.*, **53**, 850.

Moromycins A and B

Physical data:
- Moromycin A: $C_{43}H_{46}O_{14}$, yellow solid, $[\alpha]_D^{25}$ +11.1° (MeOH, *c* 0.27)
- Moromycin B: $C_{31}H_{30}O_{10}$, yellow solid, $[\alpha]_D^{25}$ +44° (MeOH, *c* 0.25)

Structures:

Moromycin A: R¹ = X, R² = Y
Moromycin B: R¹ = X, R² = H

Compound class: Antibiotics (*C*-glycosylangucycline-type antibiotics)

Source: *Streptomyces sp.* KY002

Pharmaceutical potential: *Cytotoxic (anticancer)*
Both the antibiotics were found to exhibit cytotoxicity against H-460 human lung cancer and MCF-7 human breast cancer cells; however, moromycin B showed higher efficacy. Respective GI_{50} (50% inhibition of cell growth), TGI (total growth inhibition), and LC_{50} (50% lethal concentration) values for moromycin B were determined to be 4.1, 14.1, and 52.3 μM against H-460 human lung cancer cells, while those against MCF-7 human breast cancer cells were measured to be 5.6, 9.2, and 16.3 μM, respectively. Moromycin A showed relatively less potency; respective GI_{50}, TGI, and LC_{50} values for it were evaluated as 18.7, 51.2, and 69.1 μM (against H-460 human lung cancer cells); 25.1, 38.9, and 55.0 μM (against MCF-7 human breast cancer cells).

Reference

Abdelfattah, M.S., Kharel, M.K., Hitron, J.A., Baig, I., and Rohr, J. (2008) *J. Nat. Prod.*, **71**, 1569.

Moronic acid

Systematic name: 3-oxo-Olean-18-en-28-oic acid

Physical data: $C_{30}H_{46}O_3$, colorless powder, $[\alpha]_D$ +59.3° ($CHCl_3$, *c* 1.01)

Structure:

Moronic acid

Compound class: Triterpenoid

Sources: *Boronia inornata* (aerial parts and roots; family: Rutaceae) [1]; *Evodia meliafolia* (stem barks; family: Rutaceae) [2]; *Myrceugenia euosma* (O. Berg) Legrand (Brazilian propolis; family: Myrtaceae) [3]; *Rhus javanica* (family: Anacardiaceae) [4]; *Phoradendron reichenbachianum* (aerial parts; mistletoe; family: Loranthaceae) [5]

Pharmaceutical potentials: *Anti-HIV; anti-HSV; cytotoxic*
Ito and coworkers [3] established moronic acid as a potent HIV inhibitory agent; they determined that at a concentration of 18.6 μg/ml (IC_{50}) it became cytotoxic to 50% of the H9 cells and at a concentration of <0.1 μg/ml (EC_{50}) it inhibited viral replication in the H9 cells by 50% (therapeutic index >186).

Moronic acid was also found to exhibit potent *in vitro* and *in vivo* (in mice) anti-herpes simplex virus (anti-HSV) activity [4]; the EC_{50} value of plaque reduction of moronic acid for wild-type HSV type 1 (HSV-1) was determined to be 3.9 µg/ml with a therapeutic index (TI) of 10.3–16.3. The investigators [4] reported that the susceptibility of acyclovir-phosphonoacetic acid-resistant HSV-1, thymidine kinase-deficient HSV-1, and wild-type HSV-2 to moronic acid was similar to that of the wild-type HSV-1; on oral administration (three times daily) to mice infected cutaneously with HSV-1, it significantly retarded the development of skin lesions and/or prolonged the mean survival times of infected mice without toxicity compared to the control. The terpenoid drug molecule suppressed the virus yields in the brain more efficiently than those in the skin – the phenomenon was found to be consistent with the prolongation of mean survival times [4]. Moronic acid also has cytotoxic property [5].

References

[1] Ahsan, M., Armstrong, J.A., Gray, A.I., and Waterman, P.G. (1995) *Phytochemistry*, **38**, 1275.
[2] Rao, E.V., Rao, Vadlamudi Rao, V.S.V., Sridhar, P., Venkata Rao, E., and Rao, V.S. (1997) *Indian J. Pharm. Sci.*, **59**, 251.
[3] Ito, J., Chang, F.-R., Wang, H.-K., Park, Y.K., Ikegaki, M., Kilgore, N., and Lee, K.-H. (2001) *J. Nat. Prod.*, **64**, 1278.
[4] Kurokawa, M., Basnet, P., Ohsugi, M., Hozumi, T., Kadota, S., Namba, T., Kawana, T., and Shiraki, K. (1999) *J. Pharmacol. Exp. Ther.*, **289**, 72.
[5] Rios, M.Y., Salina, D., and Villarreal, M.L. (2001) *Planta Med.*, **67**, 443.

Motualevic acids A and F

Systematic names:
- Motualevic acid A: glycyl conjugate of ω-brominated lipid (*E*)-14,14-dibromotetradeca-2,13-dienoic acid
- Motualevic acid F: (*E*)-3-(13,13-dibromotrideca-1,12-dienyl)-2*H*-azirine-2(*R*)-carboxylic acid

Physical data:
- Motualevic acid A: $C_{16}H_{25}Br_2NO_3$, colorless solid
- Motualevic acid F: $C_{16}H_{23}Br_2NO_2$, $[\alpha]_D$ −75°

Structures:

Motualevic acid A

Motualevic acid F

Compound class: Brominated long-chain acids

Source: *Siliquariaspongia* sp. (marine sponge)

Pharmaceutical potential: *Antibacterial*
Motualevic acids A and F exhibited antibacterial activities against *Staphylococcus aureus* with respective MIC values of 10.9 and 1.2 µg/ml and against methicillin-resistant *S. aureus* (MRSA) with MIC values of 9.3 and 3.9 µg/ml, respectively.

Reference

Keffer, J.L., Plaza, A., and Bewley, C.A. (2009) *Org. Lett.*, **11**, 1087.

Multicaulin and 12-demethylmulticaulin

Systematic names:
- Multicaulin: 12-methoxyabieta-1,3,5(10),6,8,11,13-heptaene
- 12-Demethylmulticaulin: 12-hydroxyabieta-1,3,5(10),6,8,11,13-heptaene

Physical data:
- Multicaulin: $C_{20}H_{22}O$, dark orange amorphous powder
- 12-Demethylulticaulin: $C_{19}H_{20}O$, dark orange amorphous powder

Structures:

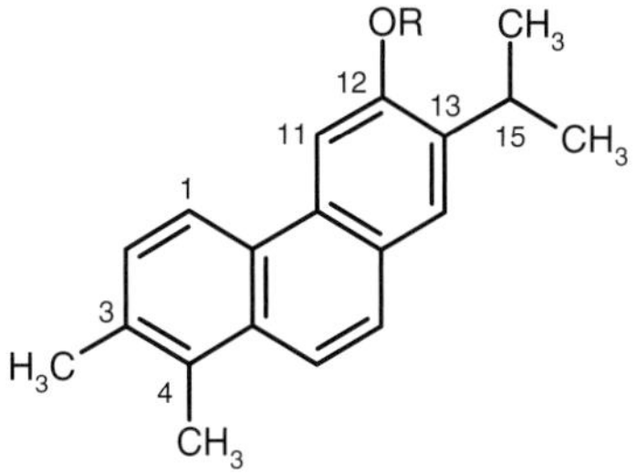

Multicaulin: R = CH_3
12-Demethylmulticaulin: R = H

Compound class: Norditerpenoids

Source: *Salvia multicaulis* (roots; family: Labiatae)

Pharmaceutical potential: *Antitubercular*
Multicaulin and 12-demethylmulticaulin showed strong antitubercular activity against *Mycobacterium tuberculosis* strain H37Rv with MIC values of 5.6 and 0.46 µg/ml, respectively.

Reference

Ulubelen, A., Topcu, G., and Johansson, C.B. (1997) *J. Nat. Prod.*, **60**, 1275.

Multiorthoquinone and 2-demethylmultiorthoquinone

Systematic names:
- Multiorthoquinone: 2-methoxy-11,12-dioxoabieta-1,3,5(10),6,8,13-hexaene
- 2-Demethylmultiorthoquinone: 2-hydroxy-11,12-dioxoabieta-1,3,5(10),6,8,13-hexaene

Physical data:
- Multiorthoquinone: $C_{20}H_{20}O_3$, amorphous powder
- 2-Demethylmultiorthoquinone: $C_{19}H_{18}O_3$, amorphous powder

Structure:

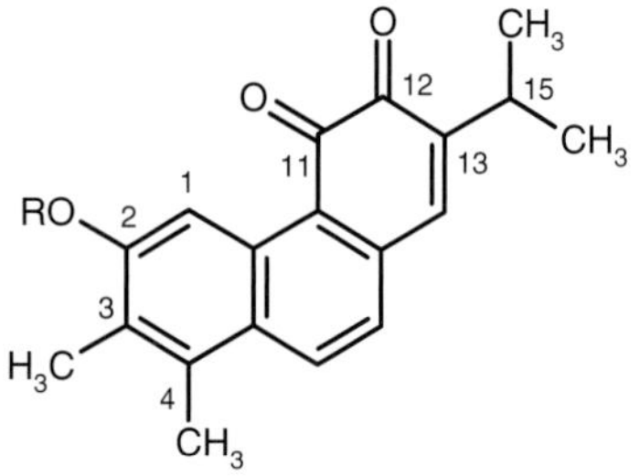

Multiorthoquinone: R = CH_3
2-Demethylmultiorthoquinone: R = H

Compound class: Norditerpenoids

Source: *Salvia multicaulis* (roots; family: Labiatae)

Pharmaceutical potential: *Antitubercular*
Multiorthoquinone and 2-demethylmultiorthoquinone showed strong antitubercular activity against *Mycobacterium tuberculosis* strain H37Rv with MIC values of 2.0 and 1.2 µg/ml, respectively.

Reference

Ulubelen, A., Topcu, G., and Johansson, C.B. (1997) *J. Nat. Prod.*, **60**, 1275.

Multipolides A and B

Physical data:
- Multipolide A: $C_{10}H_{14}O_5$, colorless oil, $[\alpha]_D^{29}$ +6.7° ($CHCl_3$, *c* 0.18)
- Multipolide B: $C_{14}H_{18}O_6$, colorless oil, $[\alpha]_D^{29}$ +24.5° ($CHCl_3$, *c* 0.40)

Structures:

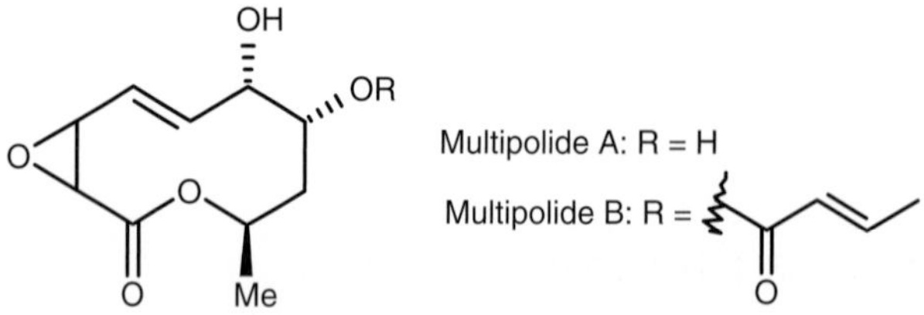

Multipolide A: R = H
Multipolide B: R =

Compound class: 10-Membered lactones

Source: *Xylaria multiplex* BCC 1111 (fungus; culture broth)

Pharmaceutical potential: *Antifungal*
Multipolides A and B displayed antifungal activity against *Candida albicans* with IC_{50} values of 7.0 and 2.0 µg/ml, respectively; both of them were also evaluated as noncytotoxic to BC-1 and KB cell lines (at 20 µg/ml).

Reference

Boonphong, S., Kittakoop, P., Isaka, M., Pittayakhajonwut, D., Tanticharoen, M., and Thebtaranonth, Y. (2001) *J. Nat. Prod.*, **64**, 965.

Muntingia biflavans 1 and 2

Systematic names:
- Both of them were isolated as a mixture of two diastereoisomers
- *Muntingia* biflavan 1: [(M)(2S),(2″S)- and (P),(2S),(2″S)- of 8,8″-5′-trihydroxy-7,7″-3′,3″-4′,4″-5″-heptamethoxy-5,5″-biflavan
- *Muntingia* biflavan 2: [(M)(2S),(2″S)- and (P),(2S),(2″S)- of 8,8″-5′,5″-tetrahydroxy-7,7″-3′,3″-4′,4″-hexamethoxy-5,5″-biflavan

Physical data:
- **1:** $C_{37}H_{40}O_{12}$, colorless prism (MeOH), mp >228 °C, $[\alpha]_D^{22}$ +11.3° (CHCl$_3$, c 0.31)
- **2:** $C_{36}H_{38}O_{12}$, colorless prism (MeOH), mp >211 °C, $[\alpha]_D^{22}$ +18.9° (CHCl$_3$, c 0.22)

Structures:

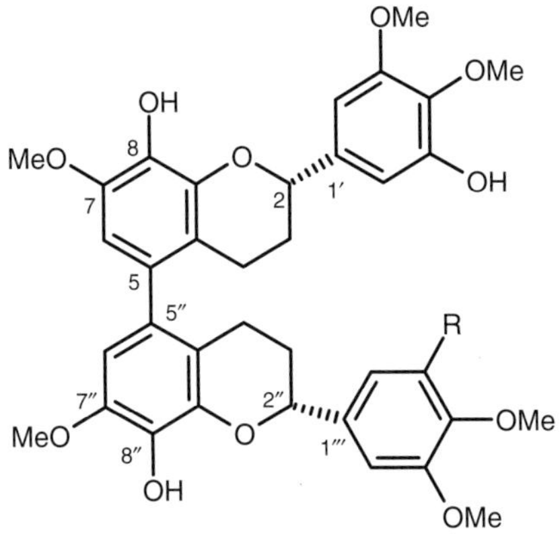

Compound class: Flavonoids (biflavans)

Source: *Muntingia calabura* L. (roots; family: Elaeocarpaceae)

Pharmaceutical potential: *Cytotoxic*
The *Muntingia* biflavans **1** and **2** displayed significant cytotoxicities against a series of cancer cell lines such as BC1 (human breast cancer), HT-1080 (human fibrosarcoma), Lu1 (human lung cancer), Me-12 (human melanoma), Co-12 (human colon cancer), KB (human nasopharyngeal carcinoma), KB-V (vincristine resistant KB), and P-388 (murine lymphocytic leukemia); the present investigators determined their respective ED_{50} values to be 12.0, 5.5, 12.4, 10.2, 6.2, 2.2, 8.3, and 3.7 µg/ml (**1**), 16.0, 5.0, 15.6, 8.7, 9.0, 5.2, 12.6, and 4.8 µg/ml (**2**).

Reference
Kaneda, N., Pezzuto, J.M., Soejarto, D.D., Kinghorn, A.D., and Farnsworth, N.R. (1991) *J. Nat. Prod.*, **54**, 196.

Muqubilone

Physical data: $C_{24}H_{40}O_6$, colorless oil, $[\alpha]_D^{25}$ +48° (CHCl$_3$, *c* 0.1)

Structure:

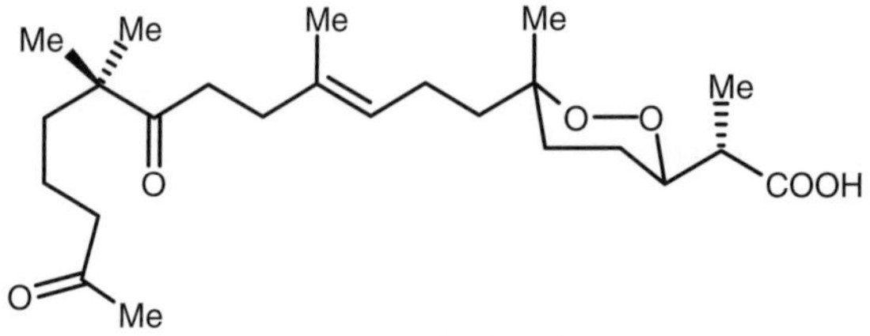

Muqubilone

Compound class: Norsesterterpene acid

Source: *Diacarnus erythraeanus* (Red Sea sponge)

Pharmaceutical potential: *Antiviral*
The isolate displayed *in vitro* antiviral activity against herpes simplex type 1 (HSV-1) by protecting a confluent nonproliferating monolayer of Vero African green monkey kidney cells from the cytopathic effect of the virus with an ED_{50} value of 30 µg/ml.

Reference
El Sayed, K.A., Hamann, M.T., Hashish, N.E., Shier, W.T., Kelly, M., and Khan, A.A. (2001) *J. Nat. Prod.*, **64**, 522.

Mustakone

Physical data: $C_{15}H_{22}O$, oil, bp 128–129 °C/1 mm, $[\alpha]_D$ +0.34° (CHCl$_3$, *c* 0.026)

Structure:

Mustakone

Compound class: Sesquiterpenoid

Sources: *Cyperus rotundus* Linn. (essential oil; family: Cyperaceae) [1]; *C. articulatus* Linn. (rhizomes; family: Cyperaceae) [2]

Pharmaceutical potential: *Antimalarial*
Mustakone was reported to exhibit strong antiplasmodial activity against the *Plasmodium falciparum* strains NF54 and ENT30 with respective IC_{50} values of 0.14 ± 0.03 and 0.25 ± 0.01 µg/ml (IC_{50} value of chloroquine phosphate (positive control) = 0.03 ± 0.005 µg/ml) [2].

References

[1] Kapadia, V.H., Nagasampangi, B.A., Naik, V.G., and Sukh Dev (1963) *Tetrahedron Lett.*, **4**, 1933.
[2] Rukunga, G.M., Muregi, F.W., Omar, S.A., Gathirwa, J.W., Muthaura, C.N., Peter, M.G., Heydenreich, M., and Mungai, G.M. (2008) *Fitoterapia*, **79**, 188.

Myrciacitrins I–V

Systematic names:
- Myrciacitrin I: (2S)-6,8-dimethyl-5,7,2′,5′-tetrahydroxyflavanone 7-O-β-D-glucopyranoside
- Myrciacitrin II: (2S)-6,8-dimethyl-5,7,2′- trihydroxy-5′-methoxyflavanone 7-O-β-D-glucopyranoside
- Myrciacitrin III: (2S)-6,8-dimethyl-5,7,2′,5′-tetrahydroxyflavanone 2′-O-β-D-glucopyranoside
- Myrciacitrin IV: (2S)-6,8-dimethyl-5,7,2′,5′-tetrahydroxyflavanone 7-O-(6″-O-p-coumaroyl)-β-D-glucopyranoside
- Myrciacitrin V: (2S)-6,8-dimethyl-5,7,2′,5′-tetrahydroxyflavanone 7-O-(6″-O-p-hydroxybenzoyl)-β-D-glucopyranoside

Physical data:
- Myrciacitrin I: $C_{23}H_{26}O_{11}$, yellow powder, $[\alpha]_D^{27}$ −51° (MeOH, *c* 0.17)
- Myrciacitrin II: $C_{24}H_{28}O_{11}$, yellow powder, $[\alpha]_D^{24}$ −20.7° (pyridine, *c* 0.1)
- Myrciacitrin III: $C_{23}H_{26}O_{11}$, yellow powder, $[\alpha]_D^{25}$ −104.2° (EtOH, *c* 0.16)
- Myrciacitrin IV: $C_{32}H_{32}O_{13}$, yellow powder, $[\alpha]_D^{26}$ −99.2° (MeOH, *c* 0.68)
- Myrciacitrin V: $C_{30}H_{30}O_{13}$, yellow powder, $[\alpha]_D^{26}$ +108.2° (MeOH, *c* 0.028)

Structures:

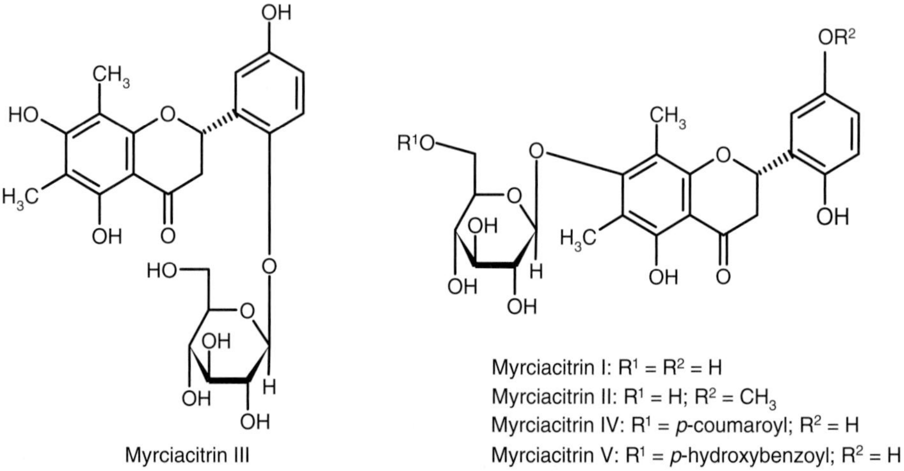

Myrciacitrin I: R¹ = R² = H
Myrciacitrin II: R¹ = H; R² = CH₃
Myrciacitrin IV: R¹ = *p*-coumaroyl; R² = H
Myrciacitrin V: R¹ = *p*-hydroxybenzoyl; R² = H

Compound class: Flavonoids (flavanone glycosides)

Source: *Myrcia multiflora* DC (leaves; family: Myrtaceae) [1, 2]

Pharmaceutical potential: *Aldose reductase inhibitors (antidiabetics)*
All the isolates were evaluated to possess significant aldose reductase inhibition activity; aldose reductase is reported to catalyze reduction of glucose to sorbitol in the polyol pathway, which is eventually associated with several chronic complications such as peripheral neuropathy, retinopathy, and cataracts. Myrciacitrins I–V displayed significant inhibitory activity against rat lens aldose reductase with IC_{50} values of 3.2, 15, 46, 0.79, and 16 μM, respectively [1, 2]. Myrciacitrin IV appeared to be the most potent in this series, but was less active than epalrestat (positive control), a commercially available synthetic aldose reductase inhibitor (IC_{50} = 0.072 μM) [2, 3].

References

[1] Yoshikawa, M., Shimada, H., Nishida, N., Li, Y., Toguchida, I., Yamahara, J., and Matsuda, H. (1998) *Chem. Pharm. Bull.*, **46**, 113.
[2] Matsuda, H., Nishida, N., and Yoshikawa, M. (2002) *Chem. Pharm. Bull.*, **50**, 429.
[3] Terashima, H., Hama, K., Yamamoto, R., Tsuboshima, M., Kikkawa, R., Hatanaka, I., and Shigeta, Y. (1984) *J. Pharmacol. Exp. Ther.*, **229**, 226.

Myriaporones 3 and 4

Physical data: $C_{19}H_{32}O_7$, $[\alpha]_D^{26}$ +44.6° (MeOH, *c* 0.06)

Structures: Myriaporones 3 and 4 were isolated as a mixture of two isomers in equilibrium; the major isomer was the hemiketal (myriaporone 3) and the minor isomer was the hydroxy ketone (myriaporone 4) with the relative ratio 3: 1 (determined by ¹H NMR). These two compounds were inseparable under normal conditions.

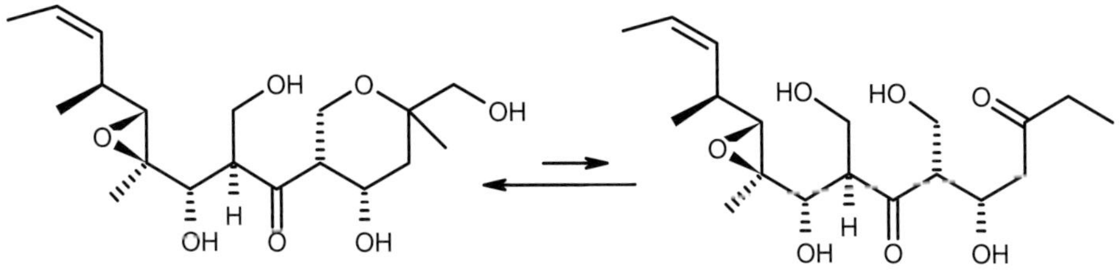

Myriaporone 3 ⇌ Myriaporone 4

Compound class: Polyketide-derived metabolites

Source: *Myriapora truncata* (Mediterranean bryozoan)

Pharmaceutical potential: *Cytotoxic*

The investigators reported that the inseparable equilibrium mixture of myriaporones 3 and 4 displayed potent inhibition (88%) against the growth of L1210 murine leukemia cells at 0.2 μg/ml.

Reference

Cheng, J.-F., Lee, J.-S., Sakai, R., Jares-Erijman, E.A., Silva, M.V., and Rinehart, K.L. (2007) *J. Nat. Prod.*, **70**, 332.

Myricitrin-5-methyl ether

Systematic name: 3,7,3′,4′,5′-Pentahydroxy-5-methoxyflavone-3-*O*-α-L-rhamnopyranoside

Physical data: $C_{22}H_{22}O_{12}$, yellow powder, mp 170–172 °C, $[\alpha]_D$ −139° (MeOH, *c* 0.2)

Structure:

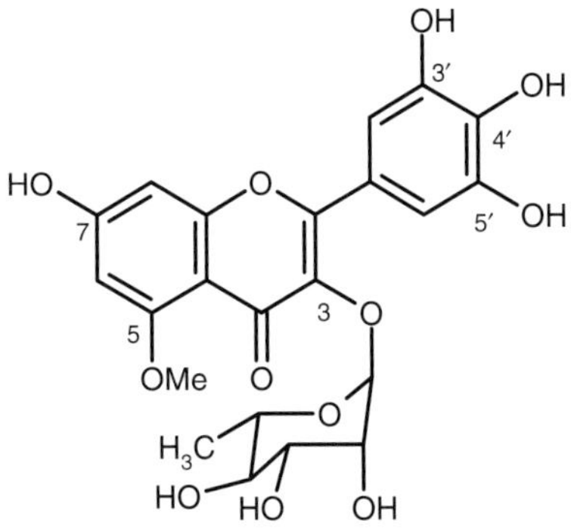

Myricitrin-5-methyl ether

Compound class: Flavonoid glycoside

Source: *Rhododendron yedoense* var. *poukhanense* (flowers; family: Ericaceae)

Pharmaceutical potentials: *Antioxidant*
The flavonoid glycoside was found to possess significant antioxidant potential as studied using DPPH, TBARS (thiobarbituric acid reactive substance), and superoxide anion radical (O_2^-) in the xanthine/xanthine oxidase assay systems. In the DPPH scavenging assay, the compound showed about twofold greater antioxidant activity (with an IC_{50} value of 4.5 ± 0.48 μM) than α-tocopherol (reference standard; $IC_{50} = 9.8 \pm 0.94$ μM). The flavonoid also exhibited greater antioxidant activity than L-ascorbic acid in the TBRAS assay (IC_{50} (flavonoid glycoside) $= 1.7 \pm 0.22$ μM; IC_{50} (L-ascorbic acid) $= 7.4 \pm 0.63$ μM). In addition, the compound displayed high efficacy in inhibiting the activity of xanthine oxidase ($IC_{50} = 1.1 \pm 0.21$ mM) and also in the activation of superoxide scavenging.

Reference

Jung, S.J., Kim, D.-H., Hong, Y.-H., Lee, J.-H., Song, H.-N., Rho, Y.-D., and Baek, N.-I. (2007) *Arch. Pharm. Res.*, **30**, 146.

Myrothenone A

Systematic name: 5(*R*)-5-Ethenyl-3-formamido-5-hydroxy-2-cyclopenten-1-one

Physical data: $C_8H_9NO_3$, colorless oil, $[\alpha]_D^{20}$ +61° (MeOH, *c* 0.6)

Structure:

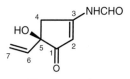

Myrothenone A

Compound class: Cyclopentenone derivative

Source: *Myrothecium* sp. (marine-derived fungus) [1]

Pharmaceutical potential: *Tyrosinase inhibitory activity*
Li *et al.* [1] evaluated myrothenone A as a promising tyrosinase enzyme inhibitory agent; this cyclopentenone derivative was found to be more active ($ED_{50} = 6.6$ μM) than kojic acid ($ED_{50} = 7.7$ μM) currently being used as a functional personal care compound.

Tyrosinase, a key enzyme involved in the metabolism of melanin in melanocytes, is associated with skin coloring and local hyperpigmentation such as melasma, ephelis, and lentigo [2, 3]. Hence, tyrosinase inhibitory compounds find immense applications in formulating functional personal care products for skin whitening effects and for preventive and therapeutic effects on the local hyperpigmentation diseases.

References

[1] Li, X., Kim, M.K., Lee, U., Kim, S.-K., Kang, J.S., Choi, H.D., and Son, B.W. (2005) *Chem. Pharm. Bull.*, **53**, 453.

[2] Shin, N.H., Lee, K.S., Kang, S.H., Min, K.R., Lee, S.H., and Kim, Y. (1997) *Nat. Prod. Sci.*, **3**, 111.
[3] Iwata, M., Corn, T., Iwata, S., Everett, M.A., and Fuller, B.B. (1990) *J. Invest. Dermatol.*, **95**, 9.

Myrsine saponin

Systematic name: 3-β-*O*-β-D-Glucopyranosyl-(1→2)-[α-L-rhamnopyranosyl-(1→2)-β-D-glucopyranosyl-(1→4)]-α-L-arabinopyranosyl-13β,28-epoxy-16α-hydroxyoleanane

Physical data: $C_{53}H_{88}O_{21}$, colorless powder, mp 268–275 °C, $[\alpha]_D^{20}$ −10° (MeOH, *c* 0.2)

Structure:

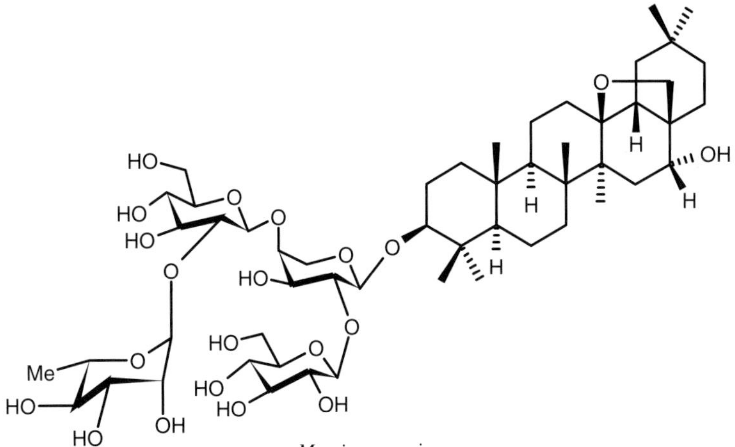

Myrsine saponin

Compound class: Triterpenoid saponin

Sources: *Myrsine australis* (leaves and twigs; family: Myrsinaceae) [1]; *Ardisia japonica* (Thunb.) Bl. (whole plants; family: Myrsinaceae) [2]

Pharmaceutical potential: *Cytotoxic*

The isolate was evaluated to possess significant cytotoxic activity against a panel of human cancer cell lines such as HL-60 myeloid leukemia, KATO-III stomach adenocarcinoma, and A549 lung adenocarcinoma cells with IC_{50} values of 1.9 ± 0.5, 0.4 ± 0.3, and 3.7 ± 0.3 μM, respectively [2].

References

[1] Bloor, S. and Qi, L. (1994) *J. Nat. Prod.*, **57**, 1354.
[2] Chang, X., Li, W., Jia, Z., Satou, T., Fushiya, S., and Koike, K. (2007) *J. Nat. Prod.*, **70**, 179.

Myrsinoic acids A, B, C, and F

Systematic names:
- Myrsinoic acid A: 3-geranyl-4-hydroxy-5-(3'-methyl-2'-butenyl)benzoic acid
- Myrsinoic acid B: 5-carboxy-2,3-dihydro-2-(1',5'-dimethyl-1'-hydroxy-4'-hexenyl)-7-(3''-methyl-2''-butenyl)benzofuran
- Myrsinoic acid C: (2S), (3S)-6-carboxy-2,3-dihydro-3-hydroxy-2-methyl-2(4'-methylpenta-3'-enyl)-8-(3''-methyl-2''-butenyl)chroman
- Myrsinoic acid F: 5-carboxy-2,3-dihydro-2-(1',5'-dimethyl-1'E,4'-hexadienyl)-7-(3''-methyl-2''-butenyl)benzofuran

Physical data:
- Myrsinoic acid A: $C_{22}H_{30}O_3$
- Myrsinoic acid B: $C_{22}H_{30}O_4$, oil, $[\alpha]_D^{17}$ $-41.5°$ (MeOH, c 0.524)
- Myrsinoic acid C: $C_{22}H_{30}O_4$, oil, $[\alpha]_D^{17}$ $+19.7°$ (MeOH, c 0.311)
- Myrsinoic acid F: $C_{22}H_{28}O_3$, oil

Structures:

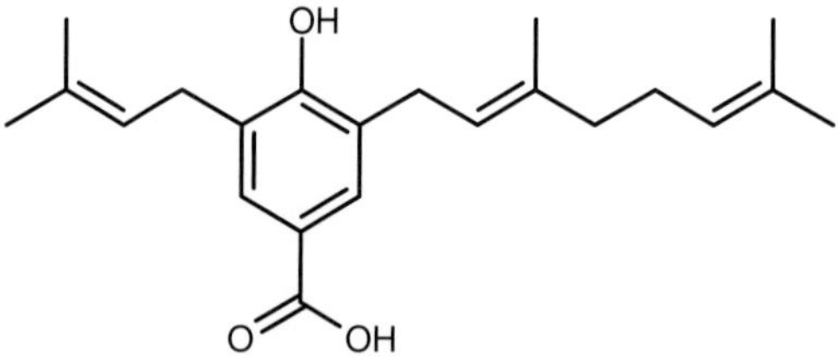

Myrsinoic acid A

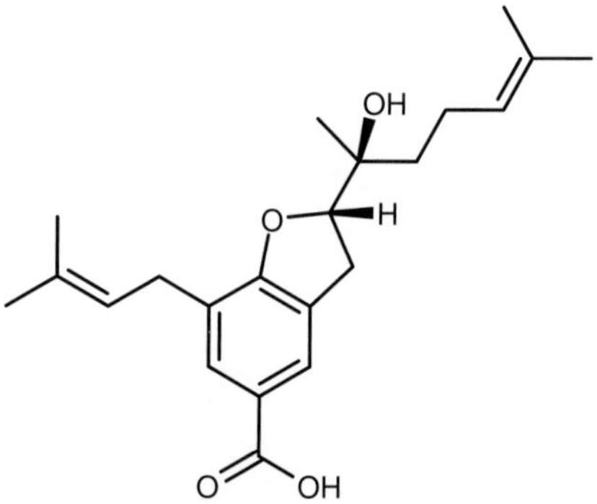

Myrsinoic acid B

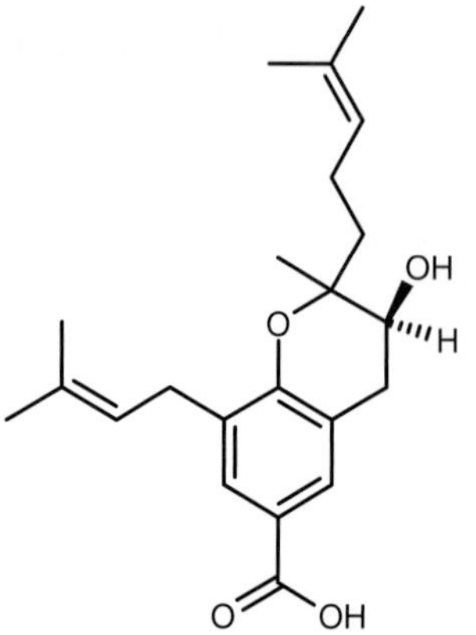

Myrsinoic acid C

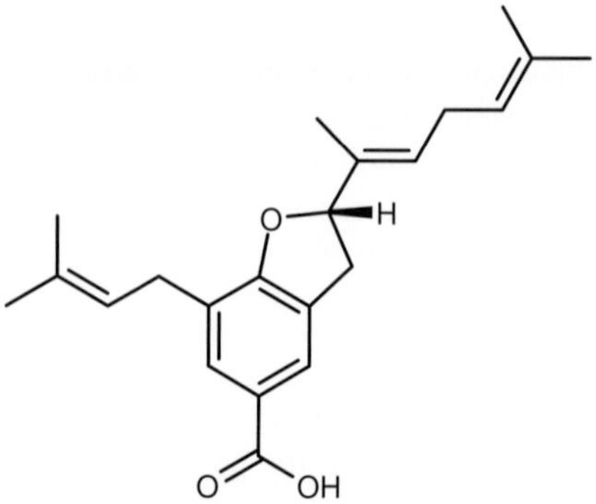

Myrsinoic acid F

Compound class: Hydroxybenzoic acid derivatives

Source: *Myrsine seguinii* (fresh leaves and twigs; family: Myrsinaceae) [1, 2]

Pharmaceutical potential: *Anti-inflammatory*

Hirota et al. [2] evaluated anti-inflammatory activity of all these isolates using 12-O-tetradecanoylphorbol-13-acetate (TPA)-induced edema of mouse ear assay – at a dose of 1.4 µM/ear, myrsinoic acids A, B, and C suppressed TPA-induced inflammation by up to 65, 83, and 68%, respectively; these activities were found to be almost comparable to that of indomethacin (78% suppression at a dose of 1.4 µM/ear). Dong et al. [1] also reported anti-inflammatory efficacy of myrsinoic acid A, which suppressed TPA-induced inflammation on mouse ears by 65% at a dose of 500 µg. The investigators [2] established that among these isolates myrsinoic acid F showed the highest efficacy; it inhibited 77% inflammation at a dose of 0.56 µM/ear; hence, this compound appeared as a stronger anti-inflammatory chemotype than indomethacin, a widely used anti-inflammatory agent.

References

[1] Dong, M., Nagaoka, M., Miyazaki, S., Iriye, R., and Hirota, M. (1999) *Biosci. Biotechnol. Biochem.*, **63**, 1650.

[2] Hirota, M., Miyazaki, S., Minakuchi, T., Takagi, T., and Shibata, H. (2002) *Biosci. Biotechnol. Biochem.*, **66**, 655.

N

Nalanthalide

Physical data: $C_{30}H_{44}O_5$, amorphous powder, mp 96.5–98 °C, $[\alpha]_D^{25}$ −58.2° ($CHCl_3$, *c* 0.275)

Structure:

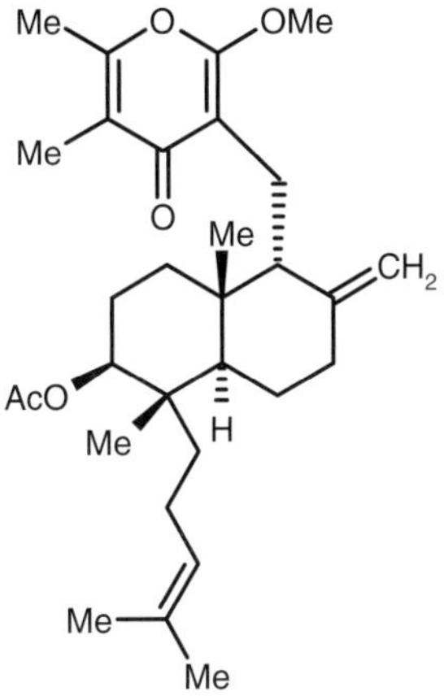

Nalanthalide

Compound class: Pyrone diterpenoid

Sources: *Nalanthamala* sp. MF 5638 (fungal culture) [1]; *Chaunopycnis alba* MF 6799 (culture broth) [1]

Pharmaceutical potential: ***Voltage-gated potassium channel Kv1.3 blocker (immunosuppressive)***
Kv1.3 has appeared as a potential novel target for immunosuppression [2]; blockers of Kv1.3 channel were found to suppress lymphokine production and to cause membrane depolarization in T cells that limits the Ca^{2+} entry that normally occurs upon activation of the T-cell receptor [3, 4]; hence, such compounds suppress activation and proliferation of human T cells.

Goetz *et al.* [1] evaluated that nalanthalide inhibits the binding of charybdotoxin (ChTX) to Jurkat membranes with an IC_{50} value of 3 μM; the pyrone diterpenoid also blocked the $^{86}Rb^+$ efflux in CHO-Kv1.3 cells with an IC_{50} value of 3.9 μM. In electrophysiological measurements, the investigators demonstrated that it depolarizes human T cells to the same extent as margatoxin (MgTX) with an EC_{50} value of 500 nM. It was also suggested that the acetate group at C-3 plays a role in the ChTX binding activity and has less effect on Rb^+ efflux activity.

References

[1] Goetz, M.A., Zink, D.L., Dezeny, G., Dombrowski, A., Polishook, J.D., Felix, J.P., Slaughter, R. S., and Singh, S.B. (2001) *Tetrahedron Lett.*, **42**, 1255.

Handbook of Pharmaceutical Natural Products, Volume 2.
Goutam Brahmachari

ISBN: 978-3-527-32148-3

[2] Lin, C.S., Boltz, R.C., Blake, J.T., Nguyen, M., Talento, A., Fischer, P.A., Springer, M.S., Sigal, N.H., Slaughter, R.S., Garcia, M.L., Kaczorowski, G.J., and Koo, G.C. (1993) *J. Exp. Med.*, **177**, 637.
[3] Felix, J.P., Bugianesi, R.M., Schmalhofer, W.A., Borris, R.P., Goetz, M.A., Hensens, O.D., Bao, J.-M., Kayser, F., Parsons, W.H., Rupprecht, K., Garcia, M.L., Kaczorowski, G.J., and Slaughter, R.S. (1999) *Biochemistry*, **38**, 4922.
[4] Koo, G.C., Blake, J.T., Shah, K., Staruch, M.J., Dumont, F., Wunderler, D., Sanchez, M., McManus, O.B., Sirotina-Meisher, A., Fischer, P.A., Boltz, R.C., Goetz, M.A., Baker, R., Bao, J.-M., Kayser, F., Rupprecht, K.M., Parsons, W.H., Tong, X.-C., Ita, I.E., Pivnichny, J., Vincent, S., Cunningham, P., Hora, D., Feeney, W., Kaczorowski, G.J., and Springer, M.S. (1999) *Cell Immunol.*, **197**, 99.

Nardoperoxide and isonardoperoxide

Physical data:
- Nardoperoxide: $C_{15}H_{22}O_4$, solid, mp 129–130 °C, $[\alpha]_D$ +31° (MeOH, c 0.33)
- Isonardoperoxide: $C_{15}H_{22}O_4$, yellowish oil, $[\alpha]_D$ +6.8° (MeOH, c 0.38)

Structures:

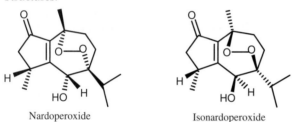

Nardoperoxide   Isonardoperoxide

Compound class: Sesquiterpenoids (guaiane-type endoperoxides)

Source: *Nardostachys chinensis* Batalin (roots; family: Valerianaceae)

Pharmaceutical potential: *Antimalarial*
The sesquiterpenoids, nardoperoxide and isonardoperoxide, were found to be promising antimalarial agents; both of them showed the activity against *Plasmodium falciparum* with EC_{50} values of 1.5 and 0.6 µM, respectively. The compounds nardoperoxide and isonardoperoxide exhibited cytotoxicities against FM3A and KB cells with respective EC_{50} values of 3.4×10^{-5} and $>6.4 \times 10^{-5}$ for nardoperoxide and 4.5×10^{-5} and 1.0×10^{-4} for isonardoperoxide; it was found that the activities as well as the selectivities (cytotoxicity/antimalarial activity) of isonardoperoxide were comparable to those of quinine, a clinically used drug. Nardoperoxide and isonardoperoxide seem to be promising lead compounds for antimalarial drugs.

Reference

Takaya, Y., Kurumada, K., Takeuji, Y., Kim, H.S., Shibata, Y., Ikemoto, N., Wataya, Y., and Oshima, Y. (1998) *Tetrahedron Lett.*, **39**, 1361.

Negundin B

Physical data: $C_{20}H_{22}O_6$, white amorphous solid, $[\alpha]_D^{26}$ $-56°$ (MeOH, c 0.11)

Structure:

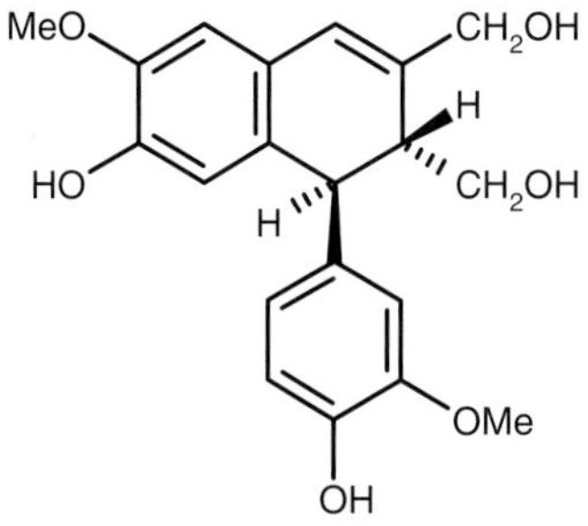

Negundin B

Compound class: Lignoid

Source: *Vitex negundo* Linn. (syn. *V. inesia* Lam.) (roots; family: Verbenaceae)

Pharmaceutical potential: *Lipoxygenase (LOX) enzyme inhibitor*
The lignoid showed promising *in vitro* inhibitory activity against lipoxygenase (LOX) enzyme with an IC_{50} value of $6.25 \pm 0.5\,\mu M$, which was found to be more potent than that of baicalein ($IC_{50} = 22.5 \pm 0.25\,\mu M$) used as a positive control in the experiment.

Reference

Azhar-Ul-Haq, Malik, A., Anis, I., Khan, S.B., Ahmed, E., Ahmed, Z., Nawaz, S.A., and Choudhary, M.I. (2004) *Chem. Pharm. Bull.*, **52**, 1269.

Neobavaisoflavone

Systematic name: 3-[4-Hydroxy-3-(3-methylbut-2-enyl)phenyl]-7-hydroxychroman-4-one

Physical data: $C_{20}H_{18}O_4$, yellow solid

Structure:

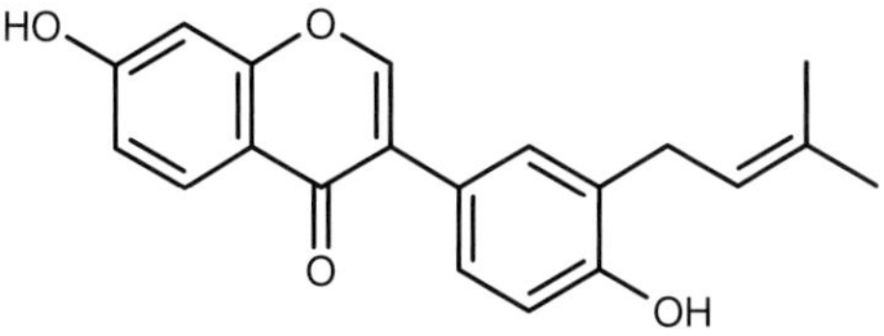

Neobavaisoflavone

Compound class: Flavonoid (isoflavone)

Source: *Psoralea corylifolia* L. (seeds; family: Leguminosae) [1–4]

Pharmaceutical potential: *Platelet aggression inhibitor*

Tsai et al. [4] evaluated antiplatelet activity of the isoflavone, and it was found to inhibit platelet aggregation (rabbit platelets) induced by collagen, arachidonic acid, and platelet activating factor (PAF) with the respective IC$_{50}$ values of 62.4 ± 10.4, 7.8 ± 2.5, and $2.5 \pm 0.3\,\mu$M. The IC$_{50}$ values for aspirin (positive control) were measured to be 30.5 ± 5.7 and $32.7 \pm 6.4\,\mu$M, respectively, against the collagen- and arachidonic acid-induced platelet aggregation, while the IC$_{50}$ value for CV-3988 (positive control) was measured to be $1.1 \pm 0.3\,\mu$M against PAF-induced platelet aggregation [4].

References

[1] Bhalla, V.K., Kayak, U.R., and Dev, S. (1968) *Tetrahedron Lett.*, **9**, 2401.
[2] Bajwa, B.S., Khanna, P.L., and Seshadri, T.R. (1974) *Indian J. Chem.*, **12**, 15.
[3] Nakayama, M., Eguchi, S., Hayashi, S., Tsukayama, M., Horie, T., Yamada, T., and Masmura, M. (1978) *Bull. Chem. Soc. Jpn*, **51**, 2398.
[4] Tsai, W.-J., Hsin, W.-C., and Chen, C.-C. (1996) *J. Nat. Prod.*, **59**, 671.

Neopyrrolomycin B

Physical data: $C_{10}H_6Cl_6NO$, light yellow solid, mp 117–120 °C, $[\alpha]_D$ +4.5° (CHCl$_3$, c 0.001)

Structure:

Neopyrrolomycin B

Compound class: Antibiotic

Source: *Streptomyces* sp. AMRI-33844 (fermentation broth)

Pharmaceutical potential: *Antibacterial*

Neopyrrolomycin B displayed broad-spectrum antibacterial activities against a panel of pathogens (including a variety of drug-susceptible and -resistant phenotypes) as listed below, with corresponding MIC values (μg/ml) in parentheses: *Staphylococcus aureus* 100 (MSSA; 0.25), *S. aureus* 1137 (MRSA; 0.12), *S. aureus* 2170 (MRSA; ≤0.06), *S. aureus* 2012 (VISA; ≤0.06), *S. aureus* 2018 (VISA; ≤0.06), *S. aureus* 1725 (LRSA; ≤0.06), *S. aureus* 1651 (LRSA; ≤0.06), *S. aureus* 2144 (CA; 0.12), *S. epidermidis* 1597 (MSSE; ≤0.06), *S. epidermidis* 1452 (MRSE; ≤0.06), *S. epidermidis* 495 (≤0.06), *Enterococcus faecalis* 846 (VRE; ≤0.06), *E. faecium* 843 (VSE; ≤0.06), *E. faecium* 1254 (VRE; ≤0.06), *Streptococcus pneumoniae* 866 (PSSP; 16.0), *S. pneumoniae* 940 (PRSP; 1.0), *S. pneumoniae* 748 (PRSP; 1.0), *S. pneumoniae* 376 (Quin-R; 2.0), *S. pneumoniae* 379 (Quin-R; 16.0), *S. pneumoniae* 933 (MDR; 1.0), *S. pneumoniae* 1195 (ATCC 49619; 2.0), *S. pyogenes* 723 (>16.0), *S. agalactiae* 2033 (16.0), *Moraxella catarrhalis* 557 (≤0.06), *Haemophilus influenzae* 1742 (ampR; 0.5), *H. influenzae* 1224 (ATCC 49247; 0.25), *H. parainfluenzae* 2319 (ATCC 7901; 1.0), *Escherichia coli* 102 (ATCC 25922, QC strain; 8.0), *E. coli* 2269 (ESBL-prod; 8.0), *Klebsiella pneumoniae* 2239 (16.0), *K. pneumoniae* 2262 (ampC, MDR; 8.0), *Serratia marcescens* 1635 (16.0), and *Pseudomonas aeruginosa* 1473 (16.0). (MRSA: methicillin-resistant *Staph. aureus*; MSSA: methicillin-susceptible *Staph. aureus*; VISA: vancomycin-intermediate *Staph. aureus*; LRSA: linezolid-resistant methicillin-resistant *Staph. aureus*; MSSE: methicillin-susceptible *Staph. epidermidis*; MRSE: methicillin-resistant *Staph. epidermidis*; VRE: vancomycin-resistant *Enterococcus*; PSSP: penicillin-susceptible *S. pneumoniae*; PRSP: penicillin-resistant *S. pneumoniae*; Quin-R: quinolone-resistant *S. pneumoniae*; MDR: multiple-drug resistant.)

Reference

Hopp, D.C., Rhea, J., Jacobsen, D., Romari, K., Smith, C., Rabenstein, J., Irigoyen, M., Clarke, M., Francis, L., Luche, M., Carr, G.J., and Mocek, U. (2009) *J. Nat. Prod.*, 72, 276.

Nepalensinols D–F

Physical data:
- Nepalensinol D: $C_{42}H_{34}O_{10}$, reddish brown powder, mp 230 °C, $[\alpha]_D$ −82° (MeOH, *c* 0.3)
- Nepalensinol E: $C_{56}H_{45}O_{13}$, brown powder, mp 250 °C, $[\alpha]_D$ −307.8° (MeOH, *c* 0.5)
- Nepalensinol F: $C_{56}H_{42}O_{12}$, brown powder, mp >300 °C, $[\alpha]_D$ +26.3° (MeOH, *c* 0.4)

Structures:

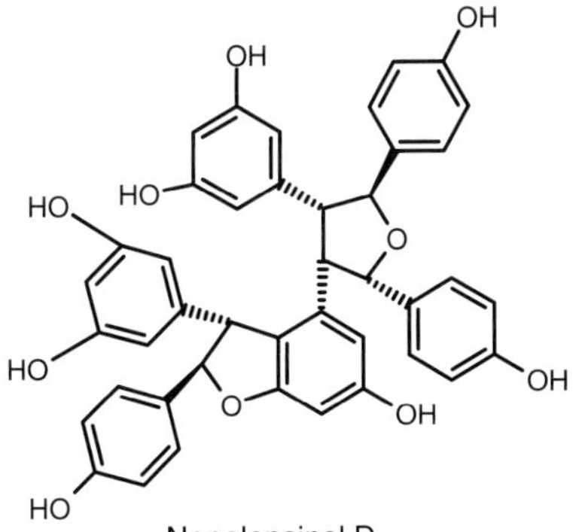

Nepalensinol D

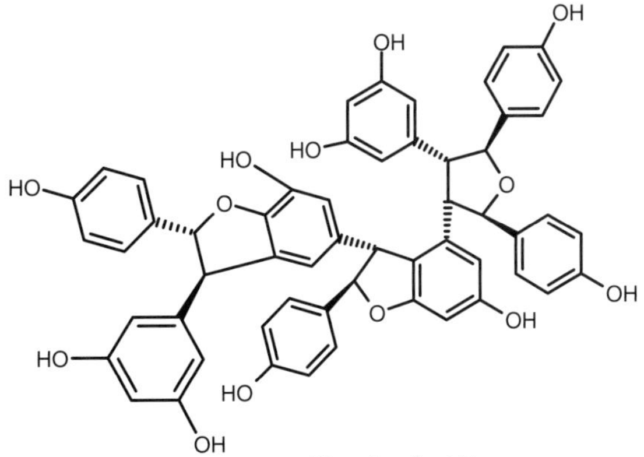

Nepalensinol E

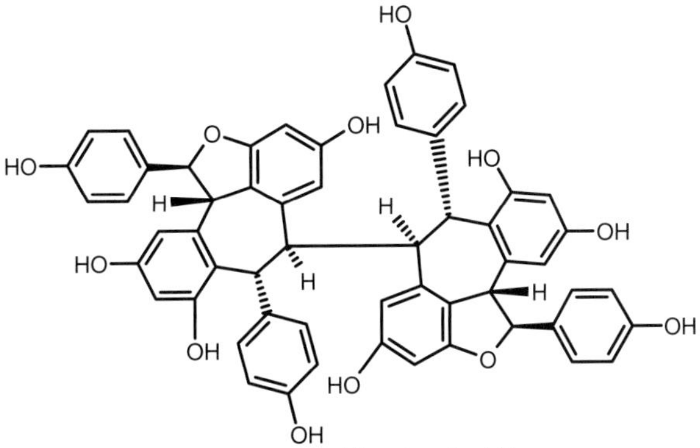

Nepalensinol F

Compound class: Stilbenoids (resveratrol oligomers)

Source: *Kobresia nepalensis* (stems; family: Cyperaceae) [1]

Pharmaceutical potential: *Human topoisomerase II inhibitor (antitumor)*
Nepalensinols A, E, and F were evaluated for their inhibitory activity against human topoisomerase II, a potential target of antitumor agents, in terms of their activity against the decatenation activity of topoisomerase II on kinetoplast DNA [1]; daunorubicin (a potent inhibitor of topoisomerase II and a clinically valuable anticancer drug [2]) was used as a positive control. Nepalensinols D and E exhibited almost the same activity (IC_{50} = 14.8 and 11.7 µM, respectively) as daunorubicin (IC_{50} = 9.1 µM), while nepalensinol F exhibited more potent inhibitory activity (IC_{50} = 5.5 µM) than the reference standard.

References

[1] Yamada, M., Hayashi, K., Hayashi, H., Tsuji, R., Kakumoto, K., Ikeda, S., Hoshino, T., Tsutsui, K., Tsutsui, K., Ito, T., Iinuma, M., and Nozaki, H. (2006) *Chem. Pharm. Bull.*, **54**, 354.
[2] Zunino, F. and Capranico, G. (1990) *Anticancer Drug Des.*, **5**, 307.

Nerolidol glycoside

Systematic name: Nerolidol-3-*O*-{α-L-rhamnopyranosyl-(1 → 4)-α-L-rhamnopyranosyl-(1 → 2)-[α-L-rhamnopyranosyl-(1 → 6)]-β-D-glucopyranoside}

Physical data: $C_{39}H_{66}O_{18}$, solid, $[\alpha]_D^{25}$ −50° (MeOH, *c* 1)

Structure:

R = Rha (1→4) Rha
R' = Rha

Nerolidol glycoside

Compound class: Sesquiterpene glycoside

Source: *Eriobotrya japonica* (Thunb.) Lindl. (leaves; family: Rosaceae) [1, 2]

Pharmaceutical potential: *Antihyperglycemic*
Chen et al. [2] evaluated the sesquiterpene glycoside as an antihyperglycemic agent; it produced a significant antihypoglycemic effect when administered orally to alloxan-diabetic mice. At doses of 25 and 75 mg/kg, the test compound exerted significant hypoglycemic ($p < 0.05$) effect; however, the dose of 75 mg/kg showed more efficacy than that of 25 mg/kg. In addition, the dose

of 75 mg/kg was found to be more effective than that of 50 mg/kg at 2 and 4 h after the oral administration of gliclazide.

References

[1] De Tommasi, N., De Simone, F., and Aquino, R. (1990) *J. Nat. Prod.*, **53**, 810.
[2] Chen, J., Li, W.L., Wu, J.L., Ren, B.R., and Zhang, H.Q. (2008) *Phytomedicine*, **15**, 98.

Neurolenins A–D

Physical data:
- Neurolenin A: $C_{20}H_{28}O_6$, colorless crystals, mp 127–128 °C, $[\alpha]_D^{25}$ −257.70 (CHCl$_3$, *c* 1.0)
- Neurolenin B: $C_{22}H_{30}O_8$, colorless crystals, mp 165–166 °C, $[\alpha]_D^{25}$ −350° (CHCl$_3$, *c* 0.76)
- Neurolenins C and D: $C_{20}H_{28}O_7$, crystalline mixture, mp 78 °C

Structures:

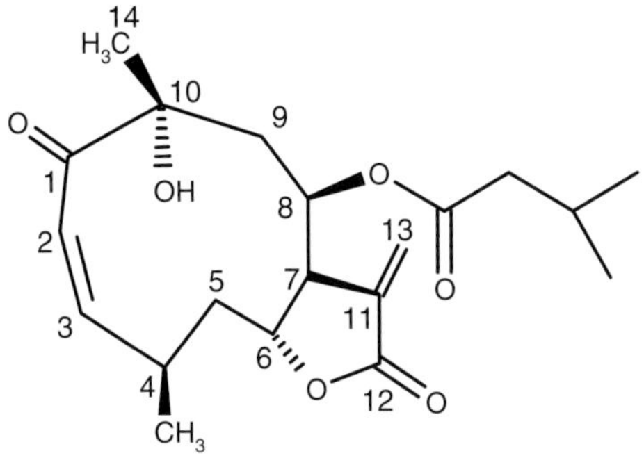

Neurolenin A

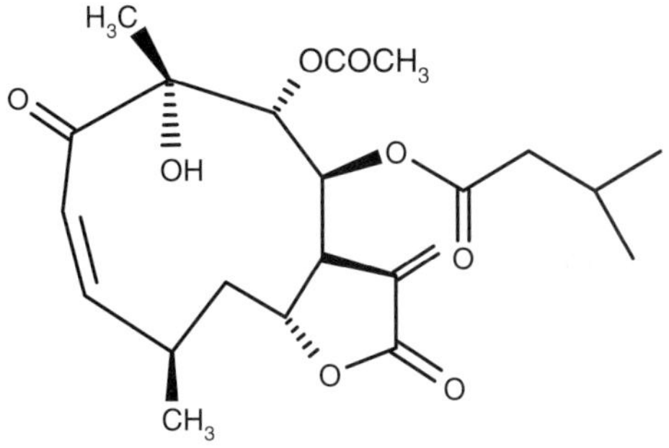

Neurolenin B

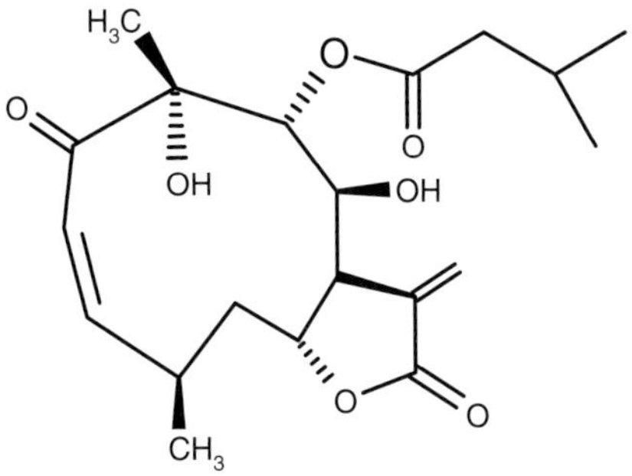

Neurolenin C

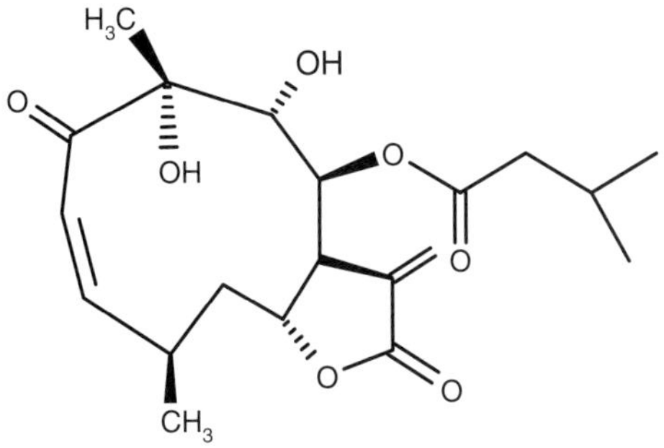

Neurolenin D

Compound class: Sesquiterpenoids (germacranolide type)

Sources: *Eupatorium inulaefolium* (*Austroeupatorium inulaefolium*) (stems and leaves; family: Asteraceae) [1]; *Neurolaena lobata* (L) R. Br. (aerial parts; family: Asteraceae) [2–7], *N. cobanensis* [8], *N. oaxacana* B.L. Turner (leaves) [9]

Pharmaceutical potentials: *Antimalarial; cytotoxic; antidysentric; antifeedant*

François et al. [3] tested antiplasmodial activity of the isolated sesquiterpene lactones; neurolenins A and B, were found to possess significant *in vitro* activity against *Plasmodium falciparum* with IC_{50} values of 0.336 µg/ml (0.92 µM) and 0.0.263 µg/mL (0.62 µM), respectively, comparable with those of the reference antimalarial agents artemisinin [IC_{50} 0.039 µg/ml (0.14 µM)] and quinine [IC_{50} 0.063 µg/ml (0.19 µM)]. In case of neurolenins C and D (available only as mixture), mixtures in two different ratios were evaluated – neurolenins C and D (1 : 3 mixture): IC_{50} 2.794 µg/ml (7.35 µM); neurolenins C and D (3 : 2 mixture): IC_{50} 0.236 µg/ml (0.62 µM). Among the germacranolides tested, neurolenin C appeared to possess the highest activity as derived from the comparison of IC_{50} values for the mixtures – increase in the proportion of neurolenin C in this mixture leads to a pronounced increase of activity, although this activity was less than that of the reference antimalarial agents, artemisinin and quinine. From structure–activity relationships, it may be suggested that one of the structural requirement for high antiplasmodial activity *in vitro* is an α,β-unsaturated carbonyl function; further, a free hydroxy group at C-8 increased the antiplasmodial activity, while a free hydroxyl at C-9 decreased the activity. It is noteworthy that the biological activity of α,β-unsaturated carbonyls, which is usually deduced from their conjugation, seems to be different in the case of neurolenins, because the keto function and the 2,3-double bond are to be out of the plane as evidenced from spectroscopic studies [10].

The same group of investigators [3] also tested cytotoxic activity of the neurolenins against two tumor cell lines, GLC_4 and COLO 320; all the four compounds showed some activity against the tested cell lines – the highest activity was displayed by neurolenin B with respective IC_{50} values (measured at 2 h incubation) of 4.4 ± 0.3 and 5.4 ± 0.2 µM (the respective values for the reference cisplatin were determined as 4.4 ± 0.3 and 8.5 ± 0.4 µM). From comparison of both the antiplasmodial and cytotoxic efficacies of neurolenins A–D, the investigators hypothesized that the antiplasmodial effects of the neurolenins are not due to their general cytotoxicity, caused by the alkylating properties of exocyclic methylene group fused to the lactone ring, but are rather dependent on a more specific mechanism of action [3].

Neurolenins B–D, isolated from *Neurolaena oaxacana*, were found to be active against *Entamoeba histolytica* and *Giardia intestinalis*; the investigators [9] recorded the MIC values for neurolenin B against *E. histolytica* as 7.6 µM, and for 1 : 1 mixture of neurolenins C and D as 8.40 µM while the value for the standard compound emetine was determined as 2.90 µM. Neurolenin B showed a slightly higher activity against *G. intestinalis* with MIC value of 3.8 µM, while the mixture of neurolenins C and D was again twofold less active; MIC value for the standard compound metronidazole was found at 1.17 µM. Although the neurolenins are less active than the standard compounds, emetine and metronidazole, they are probably involved in the antidysenteric activity of *N. lobata* [11, 12], as they are found in the leaves in high concentrations [13].

Neurolenin B, isolated from aerial parts of *Neurolaena lobata*, was found to be antifeedant against fifth instar larvae of *Spodoptera litura* (Noctuidae) with the efficacy less than parthenolide, but more than picrotoxinin, and comparable to strychnine [6]; the antifeedant effect of neurolenin B appears to result from interaction with $GABA_A$ receptors in the chemosensilla of *S. litura* as suggested for the known GABA antagonists strychnine and picrotoxinin.

References

[1] Blair, S., Mesa, J., Correa, A., Carmona-Fonseca, J., Granados, H., and Sáez, J. (2002) *Pharmazie*, **57**, 413.
[2] Passreiter, C.M., Wendisch, D., and Gondol, D. (1995) *Phytochemistry*, **39**, 133;Manchand, P.S. and Blount, J.F. (1978) *J. Org. Chem.*, **43**, 4352.
[3] François, G., Passreiter, C.M., Woerdenbag, H.J., and Van Looveren, M. (1996) *Planta Med.*, **62**, 126–129.
[4] Passreiter, C.M. (1998) *Phytochem. Anal.*, **9**, 67.
[5] Todorova, M.N., Ognyanov, I.V., and Navas, H. (1997) *Biochem. Syst. Ecol.*, **25**, 267.
[6] Passreiter, C.M. and Isman, M.B. (1997) *Biochem. Syst. Ecol.*, **25**, 371.
[7] Manchand, P.S. and Blount, J.F. (1978) *J. Org. Chem.*, **43**, 4352.
[8] Passreiter, C.M., Medinilla, B., Velasquez, R., and Moreno, P. (1998) *Pharm. Pharmacol. Lett.*, **8**, 119.
[9] Passreiter, C.M., Sandoval-Ramirez, J., and Wright, C.W. (1999) *J. Nat. Prod.*, **62**, 1093.
[10] Herz, W. and Kumar, N. (1980) *Phytochemistry*, **19**, 593.
[11] Giro'n, L.M., Freire, V., Alonzo, A., and Ca'ceres, A. (1991) *J. Ethnopharmacol.*, **34**, 173.
[12] Morton, J.F. (1981) Atlas of Medicinal Plants of Middle America: Bahamas to Yucatan, Charles C. Thomas, Springfield, IL, 949.
[13] Passreiter, C.M. and Medinilla Aldana, B. (1988) *Planta Med.*, **64**, 427.

NG-061

Systematic name: Phenylacetic acid 2-(2-methoxy-4-oxocyclohexa-2,4-dienylidene)hydrazide

Physical data: $C_{15}H_{14}N_2O_3$, pale yellow crystal, mp 189–190 °C

Structure:

NG-061

Compound class: Antibiotic (phenyl acetic acid hydrazide derivative)

Source: *Penicillium minioluteum* F-4627 (fermentation broth) [1, 2]

Pharmaceutical potential: *Potentiator of nerve growth factor (NGF)*
NG-061 was evaluated to show neurotrophic effect and to induce neurite outgrowth in PC12 cells at doses of 1–10 µg/ml; the maximal effect was similar in magnitude to that of NGF at 10 ng/ml [1]. Neurite outgrowth induced by the NG-061 appeared to be more effective in the presence of low dose of NGF; however, the fungal metabolite showed no protective and survival effects on hypoxic stress in the primary culture of mouse cerebral cortical neuron.

References

[1] Ito, M., Sakai, N., Ito, K., Mizobe, F., Hanada, K., Mizoue, K., Bhandari, R., Eguchi, T., and Kakinuma, K. (1999) *J. Antibiot.*, **52**, 224.
[2] Bhandari, R., Eguchi, T., Sekine, A., Ohashi, Y., Kakinuma, K., Ito, M., and Mizobe, F. (1999) *J. Antibiot.*, **52**, 231.

6,7-Di-*O*-nicotinoylscutebarbatine G

See Scutebarbatines G and H

7-*O*-Nicotinoylscutebarbatine H

See Scutebarbatines G and H

Nidulalin A (F390)

Physical data: $C_{16}H_{14}O_6$, orange prisms (*n*-hexane); mp 119–121 °C, $[\alpha]_D^{25}$ −426° (CHCl$_3$, *c* 0.47)

Structure:

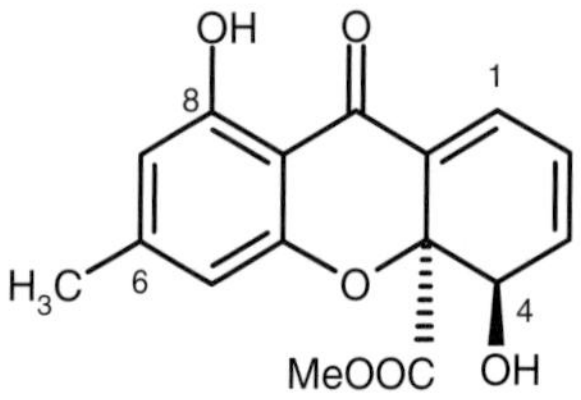

Nidulalin A (F390)

Compound class: Antibiotic (dihydroxanthone derivative)

Sources: *Emericella nidulans* (Eidam) vull. var. *lata* (Thom et Raper) Subram. (anamorph. *Aspergillus nidulans* Samson et W. Gams) [1]; *Penicillium* sp. AJ117291 (culture broth) [2]

Pharmaceutical potential: *Antitumor*
The isolate was evaluated to possess potent *in vitro* antitumor potential against both human and murine tumor cell lines HCT-116, K562, and P388 with IC$_{50}$ values of 0.042, 0.096, and 0.0072 µg/ml, respectively [2]. In addition, the fungal metabolite showed potent cytotoxicity against adriamycin-resistant FM3A mouse mammary carcinoma cells (FM3A/ADR) and its parent cells (FM3A/S) with respective IC$_{50}$ values of 0.13 and 0.09 µg/ml [2].

References

[1] Kawahara, N., Sekita, S., Satake, M., Udagawa, S.-I., and Kawai, K.-I. (1994) *Chem. Pharm. Bull.*, **42**, 1720.

[2] Sato, S., Nakagawa, R., Fudo, R., Fukuda, Y., Yoshimura, T., Kaida, K.-I., Ando, T., Kameyama, T., and Tsui, T. (1997) *J. Antibiot.*, **50**, 614.

Nigranoic acid

Systematic name: 3,4-Secocycloarta-4(28),24-(Z)-diene-3,26-dioic acid

Physical data: $C_{30}H_{46}O_4$, white needles, mp 128–130 °C, $[\alpha]_D$ +61.5° (MeOH, *c* 1.15) [1], $[\alpha]_D$ +55.8° (MeOH, *c* 2.4) [2]; colorless crystals, mp 123–124 °C [3]

Structure:

Nigranoic acid

Compound class: A ring-secocycloartene triterpenoid

Sources: *Schisandra sphaerandra* Stapf. (stems; family: Schisandraceae) [1], *S. henryi* (stems) [3], *S. propinqua* (stems) [4]; *Kadsura heteroclite* (Roxb) Craib (stems; family: Schisandraceae) [5, 6]

Pharmaceutical potentials: *Anti-HIV; cytotoxic*
Sun et al. [1] evaluated nigranoic acid as a significant anti-HIV agent; it was found to be active in several reverse transcriptase and polymerase assays, using fagaronine chloride as a positive control. Its inhibition percentage at 200 µM against HIV-1 RT and HIV-2 RT is a 99.4 and 76.7%, respectively.

Nigranoic acid showed weak cytotoxicity against leukemia and HeLa cells *in vitro* with equal IC_{50} value of 45.6 µg/ml (or 0.097 µmol/ml) [3]. It also showed significant cytotoxic effect against human decidual cells and rat luteal cells *in vitro* [4] as well as moderate efficacy against the human tumor cell lines Bel-7402, BGC-823, MCF-7, and HL-60 [6].

References

[1] Sun, H.-D., Qiu, S.-X., Lin, L.-Z., Wang, Z.-Y., Lin, Z.-W., Pengsuparp, T., Pezzuto, J.M., Fong, H.H.S., Cordell, G.A., and Farnsworth, N.R. (1996) *J. Nat. Prod.*, **59**, 525–527.
[2] Kikuchi, M. and Yoshikoshi, A. (1972) *Chem. Lett.*, 725.
[3] Chen, Y.-G., Wu, Z.-C., Lv, Y.-P., Gui, S.-H., Wen, J., Liao, X.-R., Yuan, L.-M., and Halaweish, F. (2003) *Arch Pharm. Res.*, **26**, 912.
[4] Chen, Y.-G., Qin, G.-W., Cao, L., Leng, Y., and Xie, Y.-Y. (2001) *Fitoterapia*, **72**, 435.

[5] Wang, W., Liu, J., Yang, M., Sun, J., Wang, X., Liu, R., and Guo, D. (2006) *Chromatographia*, **64**, 297.
[6] Wang, W., Liu, J., Han, J., Xu, Z., Liu, R., Liu, P., Wang, W., Ma, X., Guan, S., and Guo, D. (2006) *Planta Med.*, **72**, 450.

Niruriside

Physical data: $C_{38}H_{42}O_{17}$, white amorphous powder

Structure:

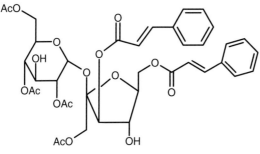

Niruriside

Compound class: Carbohydrate

Source: *Phyllanthus niruri* L. (dried leaves; family: Euphorbiaceae) [1]

Pharmaceutical potentials: *Anti-HIV*
For replication of human immunodeficiency virus (HIV), it is essential to regulate the transport of viral RNA to the cytoplasm by REV (regulation of virion expression), an HIV protein – such an interaction of the basic domain of REV with the REV-responsive element (RRE), a stem-loop RNA structure in the envelope region, is required for REV function [1–3]; hence, specific inhibitors of this interaction could be of use as antiviral agents. Niruriside was found to a specific inhibitor to the binding of REV protein to [^{33}P]-labeled RRE RNA with an IC_{50} value of 3.3 µM; however, the compound did not protect CEM-SS cells from acute HIV infection at concentrations up to 260 µM using an XTT dye reduction assay [1].

References

[1] Qian-Cutrone, J., Huang, S., Trimble, J., Li, H., Lin, P.-F., Alam, M., Klohr, S.E., and Kadow, K.F. (1996) *J. Nat. Prod.*, **59**, 196.
[2] Daly, T.J., Cook, K.S., Gary, G.S., Malone, T.E., and Rusche, J.R. (1989) *Nature*, **342**, 816.
[3] Bartel, D.P., Zapp, M.L., Green, M.R., and Szostak, J.W. (1991) *Cell*, **67**, 529.

Nitensidine E

Physical data: $C_{11}H_{19}N_3$, colorless oil, $[\alpha]_D^{25}$ +140° (MeOH, *c* 0.01)

Structure:

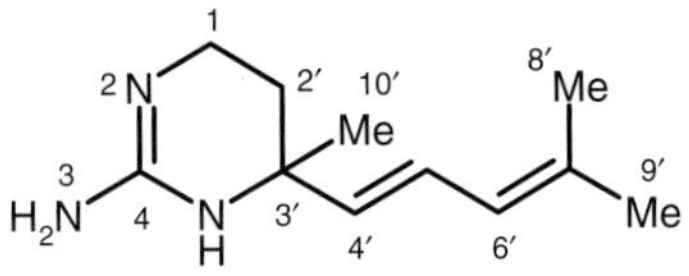

Nitensidine E

Compound class: Alkaloid (guanidine alkaloid)

Source: *Pterogyne nitens* Tul. (leaves; family: Fabaceae)

Pharmaceutical potential: *Cytotoxic (anticancerous)*
Nitensidine E displayed cytotoxicity against HL-60 (human myeloblastic leukemia) and SF-245 (human glioblastoma) cells with IC_{50} values of 3.6 and 4.9 μg/ml, respectively.

Reference

Regasini, L.O., Castro-Gamboa, I., Silva, D.H.S., Furlan, M., Barreiro, E.J., Ferreira, P.M.P., Pessoa, C., Lotufo, L.V.C., de Moraes, M.O., Young, M.C.M., and da Silva Bolzani, V. (2009) *J. Nat. Prod.*, **72**, 473.

NK154183A and B

Physical data:
- NK154183A: $C_{41}H_{70}O_{13}$, colorless prism
- NK154183B: $C_{49}H_{85}NO_{14}$, colorless prism

Structures:

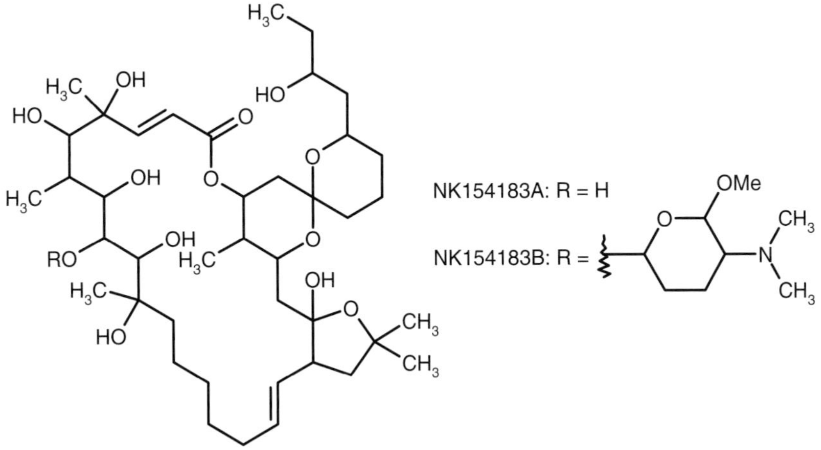

NK154183A: R = H

NK154183B: R =

Compound class: Antibiotics

Source: *Streptomyces* sp. NK154183 (fermentation broth)

Pharmaceutical potentials: *Antitumor; antifungal*

NK154183A and B were evaluated to possess *in vitro* antitumor activity against the human colon adenocarcinoma (SW1116) and NIH Swiss mouse embryo cells (NIH 3T3); the respective IC_{50} values were determined to be 0.89 and 1.21 µg/ml for NK154183A and 5.22 and 8.19 µg/ml for NK154183B. The acute toxicities (when administered intravenously) for mice of NK154183A and B were found to be 1.5 mg/kg (LD_{50}) and 1.25 mg/kg (LD_{50}), respectively.

In addition, both the isolates exhibited antifungal activities against some species of phytopathogens such as *Pyricularia oryzae* (MIC values: NK154183A, 0.78 µg/ml; NK154183B, 0.39 µg/ml) and *Collerichum frgariae* (MIC values: NK154183A, 0.78 µg/ml; NK154183B, 1.56 µg/ml), but they were inactive against Gram-positive and Gram-negative bacteria.

Reference

Tsuchiya, K., Kimura, C., Nishikawa, K., Harada, T., Nishikiori, T., and Tatsuta, K. (1996) *J. Antibiot.*, **49**, 1281.

NK372135A, B, and C

Physical data:
- NK372135A: $C_{20}H_{18}N_2O_2$, yellowish powder
- NK372135B: $C_{21}H_{20}N_2O_3$, yellowish powder, $[\alpha]_D^{20}$ +188.7° (MeOH, *c* 1.0)
- NK372135C: $C_{20}H_{18}N_2O_3$, yellowish powder

Structures:

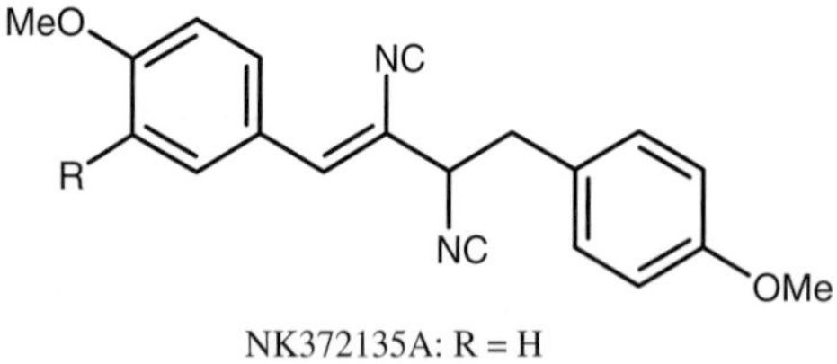

NK372135A: R = H
NK372135B: R = OMe
NK372135C: R = OH

Compound class: Antibiotics

Source: *Neosartorya fischeri* var. *glabra* IFO9857 (fungal strain; culture broth)

Pharmaceutical potential: *Antifungal*

All the microbial metabolites, NK372135A, B, and C, showed strong *in vitro* growth inhibition activity against *Candida albicans* with IC_{50} values of 2.12, 0.53, and 0.27 µg/ml, respectively (nystatin was used as a positive control; $IC_{50} = 0.8$ µg/ml).

Reference

Morino, T., Nishimoto, M., Itou, N., and Nishikiori, T. (1994) *J. Antibiot.*, **47**, 1546.

Nocardione A

Physical data: $C_{13}H_{10}O_4$, reddish brown powder, mp 115–120 °C, $[\alpha]_D^{26}$ −85.4° (CHCl$_3$, c 1.0)

Structure:

Nocardione A

Compound class: Tricyclic polyketide *ortho*-quinone antibiotic (naphtha[1,2-*b*]furan-4,5-dione derivative)

Source: *Nocardia* sp. TC-A0248 (culture broth) [1]

Pharmaceutical potentials: *Cdc25B tyrosine phosphatase inhibitor; antifungal; cytotoxic*

It has been demonstrated that protein tyrosine phosphatases (PTPase) and dual specific phosphatases (DSPase) are key enzymes in signal transduction pathway for a wide range of cellular processes [2]; Cdc25 is known to be DSPase involved in both cell cycle regulation and response to the stimulation of growth factors [3], usually responsible for various cancerous evolutions [4–7]. The inhibitory activity (IC$_{50}$) of the compound against Cdc25B, PTP1B, and FAP-1 protein tyrosine phosphatases was found to be 17, 14, and 89 µM, respectively [1].

The isolate showed cytotoxic efficacy against HeLa and SBC-5 cells with IC$_{50}$ values of 0.38 and 0.54 µM, respectively; in addition, it induced apoptosis in U937 human myeloid leukemia cell lines in RPMI 1640 medium at concentrations of less than 6.25 µg/ml [1]. Besides, the antibiotic compound was found to display moderate antifungal activity against a number of yeasts and fungi with MIC values at a range of 6.25–12.5 µg/ml [1].

References

[1] Otani, T., Sugimoto, Y., Aoyagi, Y., Igarashi, Y., Furumai, T., Satto, N., Yamada, Y., Asao, T., and And Oki, T. (2000) *J. Antibiot.*, **53**, 337.
[2] Walton, K.M. and Dixson, J.E. (1993) *Ann. Rev. Biochem.*, **62**, 101.
[3] Dunphy, W.D. and Kumagai, A. (1991) *Cell*, **67**, 189.
[4] Honda, R., Ohba, Y., Nagata, A., Okayama, H., and Yasuda, H. (1993) *FEBS Lett.*, **318**, 331.
[5] Nagata, A., Igarashi, M., Junno, S., Sato, K., and Okayama, H. (1991) *New Biol.*, **3**, 959.
[6] Galaktionov, K., Lee, A.K., Eckstein, J., Draetta, G., Meckler, J., Lods, M., and Beach, D. (1995) *Science*, **29**, 1575.
[7] Gasparotto, D., Maestro, R., Piccinin, S., Vukosavljevic, T., Barzan, L., and Sulfaro, S. (1997) *Cancer Res.*, **57**, 2366.

Norathyriol

Systematic name: 1,3,6,7-Tetrahydroxyxanthone

Physical data: $C_{13}H_8O_6$

Structure:

Norathyriol

Compound class: Xanthone (simple tetraoxygenated; aglycon of magiferin)

Sources: *Allanblackia floribunda* (family: Clusiaceae/Guttiferae) [1]; *Anaxagorea luzonensis* (family: Annonaceae) [2]; *Anthyrium mesosorum* (family: Polypodiaceae) [3]; *Canscora decussata* (family: Gentianaceae) [4]; *Chlorophora tinctoria* (family: Moraceae) [5]; *Cratoxylum pruniflorum* (family: Clusiaceae/Guttiferae) [6]; *Garcinia echinocarpa* (family: Clusiaceae/Guttiferae) [7], *G. mangostana* [8], *G. multiflora* [9], *G. pedunculata* [10], *G. terpnophylla* [7]; *Gymnocarpium robertianum* (family: Dryopteridaceae) [11]; *Hypericum androsaemum* (family: Clusiaceae/Guttiferae) [12], *H. aucheri* [13], *H. japonicum* [14], *H. sampsonii* [15]; *Iris nigricans* (family: Iridaceae) [16]; *Maclura pomifera* (family: Moraceae) [17]; *Mammea Africana* (family: Clusiaceae/Guttiferae) [18]; *Morus tinctoria* (family: Moraceae) [19]; *Ochrocarpus odoratus* (family: Clusiaceae/Guttiferae) [20]; *Pentaphalangium solomonse* (family: Clusiaceae/Guttiferae) [21]; *Symphonia globulifera* (family: Clusiaceae/Guttiferae) [22]; *Tripterospermum lanceolatum* (family: Gentianaceae) [31, 32].

Pharmaceutical potentials: *Enzyme inhibitor; anti-inflammatory; antioxidant; analgesic*
Norathyriol alone has been found to be a noncompetitive inhibitor of xanthine oxidase [23]; it was also found to inhibit angiotensin-I-converting-enzyme activity [24], and platelet aggregation [25]. The terahydroxyxanthone was reported to induce Ca^{2+} release in a dose-dependent manner from the actively loaded sarcoplasmic reticulum vesicles that was blocked by ruthenium red, a specific Ca^{2+} release inhibitor [26]. The same compound also increased, dose dependently, apparent [3H]-ryanodine binding; it produced a synergistic effect on binding activation when added concurrently with caffeine [26]. In the presence of Mg^{2+}, which inhibits ryanodine binding, both caffeine and norathyriol could restore the binding to the level observed in the absence of Mg^{2+}. In addition, norathyriol was reported to relax the rat thoracic aorta [27], and also to suppress pleurisy and cutaneous plasma extravasation caused by inflammatory mediators in mice [28].

Norathyriol suppressed thromboxane B_2 (TXB_2) and leukotriene B_4 (LTB_4) formation in calcium ionophore (A23187)- and formyl-methionyl-leucyl-phenylalanine (fMLP)-stimulated rat neutrophils. Norathyriol was three to four times more active against LTB_4 formation ($IC_{50} = 2.8\,\mu M$) than against TXB_2 formation ($IC_{50} = 10.0\,\mu M$). Norathyriol also inhibited prostaglandin D_2 (PGD_2) formation in A23187-stimulated rat mast cells ($IC_{50} = 3.0 \pm 1.2\,\mu M$) and in arachidonic acid (AA)-activated mast cell lysate. Moreover, norathyriol inhibited COX-2 and 12-lipoxygenase (12-LOX) with IC_{50} values of 19.6 ± 1.5 and $1.2 \pm 0.1\,\mu M$, respectively, similar to those required for the inhibition of COX-1 and 5-LO IC_{50} values of 16.2 ± 1.5 and $1.8 \pm 0.4\,\mu M$, respectively. Inhibition of 5-lipoxygenase (5-LOX) by norathyriol was slightly less active. Norathyriol had no effect on A23187-induced arachidonic acid (AA) release from neutrophils and did not affect phospholipase-2 (PLA_2) activity in a cell-free system. From their detailed experimental results, the investigators [29] concluded that norathyriol inhibits the formation of PGs and LTs in neutrophils probably through direct blockade of COX and 5-LOX activities; hence the xanthone, a single molecule with multiple targets, might provide a potential therapeutic benefit in the treatment of inflammatory diseases.

Norathyriol also suppressed inflammation related edema, probably due partly to suppression of mast cell degranulation and hence reduction in the release of chemical mediators that increase vascular permeability, and partly, at least in higher doses, to offer protection of the vasculature from challenge by various mediators [30]. The xanthone was reported to have *in vivo* anti-inflammatory and analgesic activities on A23187-induced pleurisy and analgesia in mice [31]. Norathyriol reduced the A23187-induced protein leakage ($ID_{50} \sim 30.6\,mg/kg$ i.p.) as well as A23187-induced PMN leukocytes accumulation ($ID_{50} \sim 16.8\,mg/kg$, i.p.). It also reduced both LTB_4 and PGE_2 production (ID_{50} determined as 18.6 and $29.1\,mg/kg$ i.p., respectively). The investigators [31] demonstrated the analgesic effect of norathyriol on the acetic acid-induced writhing response; acetic acid-induced writhing response was depressed by norathyriol ($ID_{50} \sim 27.9\,mg/kg$ i.p.) as also suppressed by the standard drugs, indomethacin and ibuprofen. On the basis of their detailed study, Wang *et al.* [31] suggested that norathyriol might be a dual, yet weak, cyclooxygenase and lipoxygenase pathway blocker; the inhibitory effect of norathyriol on the A 23 187-induced pleurisy and acetic acid-induced writhing response in mice is proposed to be dependent on the reduction of eicosanoids mediators formation in the inflammatory site.

Norathyriol exhibited potent inhibitory effects on superoxide formation by rat neutrophils [32]; it inhibited the formylmethionyl-leucyl-phenylalanine (fMLP)-induced superoxide anion (O_2^-) generation and O_2 consumption in rat neutrophils in a concentration dependent manner. In cell-free oxygen radical generating system, norathyriol inhibited the O_2^- generation during dihydroxyfumaric acid (DHF) autoxidation and in hypoxanthine-xanthine oxidase system. fMLP-induced transient elevation of Ca^{2+}, and the formation of inositol trisphosphate (IP3) were significantly inhibited by norathyriol ($30\,\mu M$) (approximately 30 and 46% inhibition, respectively). The test compound dose dependently suppressed the neutrophil cytosolic phospholipase-C (PLC). In contrast with the marked attenuation of fMLP-induced protein tyrosine phosphorylation (approximately 70% inhibition at $10\,\mu M$ norathyriol), it only slightly modulated the phospholipase D (PLD) activity as determined by the formation of phosphatidic acid (PA) and, in the presence of ethanol, phosphatidylethanol (PEt). Norathyriol did not modulate the intracellular cyclic AMP level. The investigators [32] also noted that in the presence of NADPH, the phorbol 12-myristate 13-acetate (PMA)-activated particulate NADPH oxidase activity was suppressed by norathyriol in a concentration-dependent manner and the inhibition was noncompetitive with respect to NADPH. Hence, the reduction in the formation of superoxide anion by the xathone was attributed to the scavenging of the generated superoxide anion, which is responsible for the blockade of the phospholipase C pathway, the reduction of protein tyrosine phosphorylation, and the suppression of NADPH oxidase activity through the interruption of electrons transport at the FAD redox center [32].

References

[1] Locksley, H.D. and Murray, I.G. (1971) *J. Chem. Soc. C*, 1332.
[2] Gonda, R., Takeda, T., and Akiyama, T. (2001) *Natural Medicines*, **55**, 316.
[3] Ueno, A. (1962a) *Yakugeku Zasshi*, **82**, 1482.
[4] Ghosal, S. and Chaudhuri, R.K. (1973) *Phytochemistry*, **12**, 2035.
[5] Gottlieb, O.R., Lima, R.A., Mendes, P.H., and Magalhaes, M.T. (1975) *Phytochemistry*, **14**, 1674.
[6] Kitanov, G.M., Assenov, I., and Van, A.T. (1988) *Pharmazie*, **43**, 879.
[7] Bandarnayake, W.M., Selliah, S.S., Sultanbawa, M.U.S., and Ollis, W.D. (1975) *Phytochemistry*, **14**, 1878.
[8] Holloway, D.M. and Scheinmann, F. (1975) *Phytochemistry*, **14**, 2517.
[9] Chen, F.-C., Lin, Y.-M., and Hung, J.-C. (1975) *Phytochemistry*, **14**, 300.
[10] Rama Rao, A.V., Sharma, M.R., Venkataraman, K., and Yemul, S.S. (1974) *Phytochemistry*, **13**, 1241.
[11] Takao, M., Nobutashi, T., Hiroaki, W., Yasushisa, S., and Chiu Ming, C. (1986) *Yakugaku Zasshi*, **106**, 378.
[12] Nielsen, H. and Arends, P. (1979) *J. Nat. Prod.*, **42**, 301; Dias, A.C.P., Seabra, R.M., Andrade, P.B., Ferreres, F., and Ferreira, M.F. (2000) *Plant Sci.*, **150**, 93.
[13] Kitanov, G.M. and Blinova, K.F. (1980) *Khimiya Prirodnykh Soedinenii*, **2**, 256.
[14] Fu, P., Li, T.-Z., Liu, R.-H., Zhang, W., Zhang, W.-D., and Chen, H.-S. (2004) *Nat. Prod. Res. Dev.*, **16**, 511.
[15] Don, M.-J., Huang, Y.-J., Huang, R.-L., and Lin, Y.-L. (2004) *Chem. Pharm. Bull.*, **52**, 866.
[16] Al-Khalil, S., Tosa, H., and Iinuma, M. (1995) *Phytochemistry*, **38**, 729.
[17] Wolfrom, M.L. and Bhat, H.B. (1965) *Phytochemistry*, **4**, 765.
[18] Carpenter, I., Locksley, H.D., and Scheinmann, F. (1969) *Phytochemistry*, **8**, 2013; Carpenter, I., Loeksly, H.D., and Scheinmann, F. (1969) *J. Chem. Soc. (C)*, 486.
[19] Jefferson, A. and Scheinmann, F. (1965) *Nature*, **207**, 1193.
[20] Locksley, H.D and Murray, I.G. (1971b) *Phytochemistry*, **10**, 3179.
[21] Owen, P.J. and Schienmann, F. (1974) *J. Chem. Soc. C*, 1018.
[22] Locksley, H.D., Moore, I., and Scheinmann, F. *J. Chem. Soc. C*, (1966) 430; Locksley, H.D. Moore, I., and Scheinmann, F. (1967) *Tetrahedron*, **23**, 2229.
[23] Noro, T., Ueno, A., Mizutani, M., Hashimoto, T., Miyase, T., Kuroyanage, M., and Fukushima, S. (1984) *Chem. Pharm. Bull.*, **32**, 4455.
[24] Chen, C.H., Lin, J.Y., Lin, C.N., and Hsu, S.Y. (1992) *J. Nat. Prod.*, **55**, 691.
[25] Teng, C.M., Lin, C.N., Ko, F.N., Cheng, K.L., and Huang, T.F. (1989) *Biochem. Pharmacol.*, **38**, 3791.
[26] Kang, J.J., Cheng, Y.W., Ko, F.N., Kuo, M.L., Lin, C.N., and Teng, C.M. (1996) *Braz. J. Pharmacol.*, **118**, 1736.
[27] Ko, F.N., Lin, C.N., Liou, S.S., Huang, T.F., and Teng, C.M. (1991) *Eur. J. Pharmacol.*, **192**, 133.
[28] Wang, J.P., Raung, S.L., Lin, C.N., and Teng, C.M. (1994) *Eur. J. Pharmacol.*, **251**, 35.
[29] Hsu, M.-F., Lin, C.-N., Lu, M.-C., and Wang, J.-P. *Naunyn-Schmiedeberg's Arch. Pharmacol.*, **369**, 507.
[30] Wang, J.P., Raung, S.L., Lin, C.N., and Teng, C.M. (1994) *Eur. J. Pharmacol.*, **251**, 35; Lin, C.N., Chung, M., Liou, S.J., Lee, T.H., and Wang, J.P. (1996) *J. Pharmaceu. Pharmacol.*, **48**, 532.
[31] Wang, J.-P., Ho, T.-F., Lin, C.N., and Teng, C.M. (1994) *Naunyn-Schmiedeberg's Arch. Pharmacol.*, **350**, 90.

[32] Hsu, M.F., Raung, S.L., Tsao, L.T., Lin, C.N., and Wang, J.P. (1997) *Free Radic. Biol. Med.*, **23**, 1035;Wanh, J.P., Raung, S.L., Tsao, L.T., and Lin, C.N. (1997) *Eur. J. Pharmacol.*, **336**, 81.

14-Norpseurotin A

Physical data: $C_{21}H_{23}NO_8$, pale yellow powder, $[\alpha]_D^{25}$ −6.55° (MeOH, *c* 0.12)

Structure:

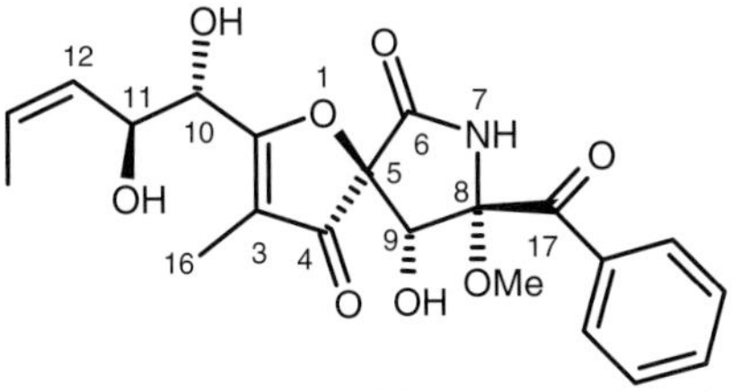

14-Norpseurotin A

Compound class: Alkaloid (diketopiperazine alkaloid)

Source: *Aspergillus sydowii* PFW1-13 strain (marine-derived fungus)

Pharmaceutical potential: ***Antimicrobial***
The test compound exhibited significant antimicrobial activities against *Escherichia coli, Bacillus subtilis,* and *Micrococcus lysodeikticus* with respective MIC values of 3.74, 14.97, and 7.49 μM.

Reference

Zhang, M., Wang, W.-L., Fang, Y.-C., Zhu, T.-J., Gu, Q.-Q., and Zhu, W.-M. (2008) *J. Nat. Prod.*, **71**, 985.

Norswertianolin

See Bellidin

Nostocarboline

Physical data: $C_{12}H_{10}N_2Cl$ (counterion was not determined)

Structure:

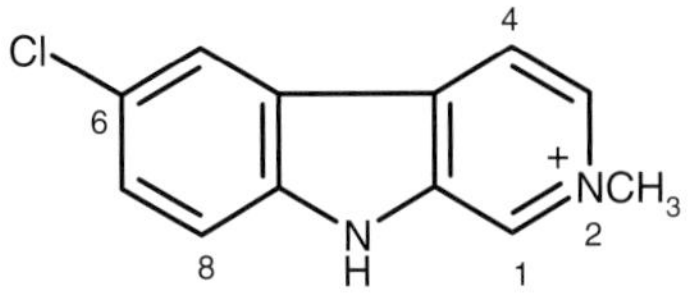

Nostocarboline

Compound class: Alkaloid (quaternary β-carboline alkaloid)

Source: Fresh-water cyanobacterium *Nostoc* 78-12A

Pharmaceutical potential: *Butyrylcholinesterase (BChE) inhibitor*
The investigators synthesized nostocarboline iodide and then evaluated both the natural (with unknown counterion) and the synthetic compound as inhibitors of butyrylcholinesterase (BChE); both the natural and synthetic alkaloids were found to exhibit potent inhibitory activity against butyrylcholinesterase with IC_{50} values of 19.4 ± 2.8 and $13.2 \pm 2.2\,\mu M$, respectively. The small difference in IC_{50} values between the synthetic and natural samples can be attributed to the different counterion that, however, remained unknown for the natural isolate. Such efficacy is comparable to that of galantamine hydrobromide ($IC_{50} = 16.9 \pm 0.9\,\mu M$), an approved drug against Alzheimer's disease. Nostocarboline can thus be regarded as a promising lead for the development of novel neuroprotectives.

Reference

Becher, P.G., Beuchat, J., Gademann, K., and Juttner, F. (2005) *J. Nat. Prod.*, **68**, 1793.

Nothramicin

Physical data: $C_{30}H_{37}NO_{11}$, red powder, mp 140–145 °C

Structure:

Nothramicin

Compound class: Antibiotic (anthraquinone derivative)

Source: *Nocardia* sp. MJ896-43F17 (culture broth)

Pharmaceutical potential: *Antimycobacterial*
Nothramicin exhibited significant antimycobacterial activity against several drug-resistant (paramomicin-resistant, capreomycin-resistant, viomycin-resistant, streptothricin-resistant, kanamycin-resistant, streptomycin-resistant, and rifampicin-resistant) *Mycobacterium smegmatis* strains with an equal MIC value of 6.25 μg/ml; the respective MIC values for the test compound against other *Mycobacterium* species such as *M. smegmatis* ATCC 607, *M. smegmatis phlei*, *M. vaccae* ATCC 15483, and *M. fortuitum* were determined to be 1.56, 6.25, 12.5, and 25.0 μg/ml. In addition,

actuate toxicity of nothramicin was found to be less as evaluated in mice ($LD_{50} \geq 100$ mg/kg, i.v.). However, it did not inhibit the growth of Gram-positive and Gram-negative bacteria and yeast.

Reference

Momosae, I., Kinoshita, N., Sawa, R., Naganawa, H., Iinuma, H., Hamada, M., and Takeuchi, T. (1998) *J. Antibiot.*, **51**, 130.

NP-101A

Systematic name: 2-Acetamidobenzamide

Physical data: $C_9H_{10}N_2O_2$, white amorphous powder, mp 160–163 °C

Structure:

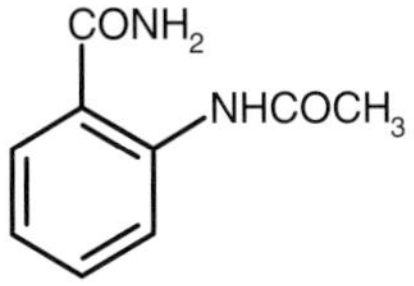

NP-101A

Compound class: Antibiotic (acetamidobenzamide)

Source: *Streptomyces aurantiogriseus* NPO-101 (fermentation broth)

Pharmaceutical potential: *Antifungal*
NP-101A exhibited antifungal activity against some microbial strains, namely, *Alternaria* sp. S-1, *Penicillium roqueforti* AHU8057, *Pyricularia oryzae* Ina 168, *Phytophthora infestans*, *Cladosporium herbarum* AHU9032, and *Aspergillus oryzae* AHU7134, with MIC values of 7.5, 3.75, 3.75, 7.5, 7.5, and 7.5 µg/ml, respectively; it showed weaker activity against *Rhizopus oryzae* AHU6536 with an MIC value of 15.0 µg/ml and against *Mucor javanicus* AHU6052, *Fusarium roseum* AHU9056, and *Rhizoctonia zeae* with an equal MIC value of 30.0 µg/ml.

Reference

Phay, N., Yada, H., Higashiyama, T., Yokota, A., Ichihara, A., and Tomita, F. (1996) *J. Antibiot.*, **49**, 703.

Nymphaeols A–C and isonymphaeol B

Physical data:
- Nymphaeol A: $C_{25}H_{28}O_6$
- Nymphaeol B: $C_{25}H_{28}O_6$
- Nymphaeol C: $C_{30}H_{36}O_6$
- Isonymphaeol B: $C_{25}H_{28}O_6$, yellow powder, mp 123–126 °C, $[\alpha]_D^{25}$ −17.8° (MeOH, *c* 0.2)

Structures:

Nymphaeol A: R^1 = R; R^2 = R^3 = H
Nymphaeol B: R^1 = R^3 = H; R^2 = R
Nymphaeol C: R^1 = R'; R^2 = R; R^3 = H
Isonymphaeol B: R^1 = R^2 = H; R^3 = R

Compound class: Flavonoids (prenylated flavanones)

Sources: Nymphaeols A–C: *Hernandia nymphaefolia* (presl) Kubitzki (family: Hernandiaceae) [1]; *Macaranga tanarius* (Linn.) Muell. Arg. (syn. *M. tomentosa* Bl.; leaves; family: Euphorbiaceae) [2, 3]; Nymphaeols A–C and isonymphaeol B: Propolis collected in Okinawa, Japan [4]; Nymphaeol A: *Schizolaena hystrix* (fruits; family: Sarcolaenaceae) [5]; Nymphaeol B: Taiwanese propolis [6].

Pharmaceutical potentials: *Cytotoxic; antioxidant; anti-inflammatory*
Phommart et al. [2] evaluated nymphaeols A–C for their cytotoxic, anti-inflammatory and antioxidant activities; only nymphaeol A showed cytotoxicity against human oral carcinoma (KB), human breast cancer (BC), human small cell lung cancer (NCI-H187), and Vero cell lines, with IC_{50} values of 4.6, 2.7, 1.5, and 3.8 μg/mL, respectively. On the other hand, nymphaeol B exhibited anti-inflammatory activity in a cyclooxygenase-2 (COX-2) inhibition assay, with an IC_{50} value of 3.7 μg/mL. The investigators [2] studied their antioxidant potential in DPPH radical-scavenging assay; nymphaeols A–C were found to show comparable radical-scavenging properties with IC_{50} values of 14 ± 1, 13 ± 2, and 15 ± 2 μM, respectively, and the activity was found stronger than 2,6-di(*tert*-butyl)-4-methylphenol (BHT; IC_{50} = 30 ± 1 μM). Strong DPPH radical-scavenging ability of nymphaeols A–C as well as isonymphaeol B was also reported by Kumazawa et al. [4]; the respective IC_{50} values were recorded as 6.5, 7.1, 9.8, and 8.5 μg/mL. Strong cytotoxic activity of nymphaeol A against A2780 human ovarian cancer cell line, with an IC_{50} value of 5.5 μg/mL, was further assessed by Murphy et al. [5].

References

[1] Yakushijin, K., Shibayama, K., Murata, H., and Furukawa, H. (1980) *Heterocycles*, **14**, 397.
[2] Phommart, S., Sutthivaiyakit, P., Chimnoi, N., Ruchirawat, S., and Sutthivaiyakit, S. (2005) *J. Nat. Prod.*, **68**, 927.
[3] Tseng, M.-H., Chou, C.-H., Chen, Y.-M., and Kuo, Y.-H. (2001) *J. Nat. Prod.*, **64**, 827.
[4] Kumazawa, S., Goto, H., Hamasaka, T., Fukumoto, S., Fujimoto, T., and Nakayama, T. (2004) *Biosci. Biotechnol. Biochem.*, **68**, 260.
[5] Murphy, B.T., Cao, S., Norris, A., Miller, J.S., Ratovoson, F., Andriantsiferana, R., Rasamison, V.E., and Kingston, D.G.I. (2005) *J. Nat. Prod.*, **68**, 417.
[6] Chen, C.-N., Wu, C.-L., Shy, H.-S., and Lin, J.-K. (2003) *J. Nat. Prod.*, **66**, 503.

O

Oblonganoside A

Systemic name: 3,23-*O*-Hydroxyethylidene-3β,23-dihydroxyurs-12,19(20)-dien-28-oic acid 28-β-D-glucopyranosyl ester

Physical data: $C_{38}H_{58}O_{10}$, colorless amorphous powder, $[\alpha]_D^{21} - 29°$ (MeOH, *c* 0.24)

Structure:

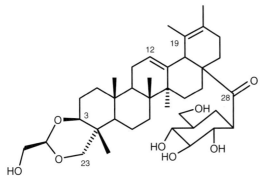

Oblonganoside A

Compound class: Triterpenoid saponin

Source: *Ilex oblonga* (leaves; family: Aquifoliaceae)

Pharmaceutical potential: ***Antiviral***
Oblonganoside A showed appreciable inhibitory activity against tobacco mosaic virus (TMV) replication with an EC_{50} value of 0.074 mg/ml. The investigators found that the activity is dose dependent; the compound was found to inhibit the TMV replication by 87.3, 80.7, 61.9, 30.1, and 10.1% at 0.4, 0.2, 0.1, 0.05, and 0.025 mg/ml concentrations, respectively.

Reference

Wu, Z.-J., Ouyang, M.-A., Wang, C.-Z., and Zhang, Z.-K. (2007) *Chem. Pharm. Bull.*, **55**, 422.

Obochalcolactone

Physical data: $C_{34}H_{30}O_7$, yellow powder, $[\alpha]_D^{25}$ −52.7° ($CHCl_3$, *c* 1.21)

Handbook of Pharmaceutical Natural Products, Volume 2.
Goutam Brahmachari

ISBN: 978-3-527-32148-3

Obolactone

Structure:

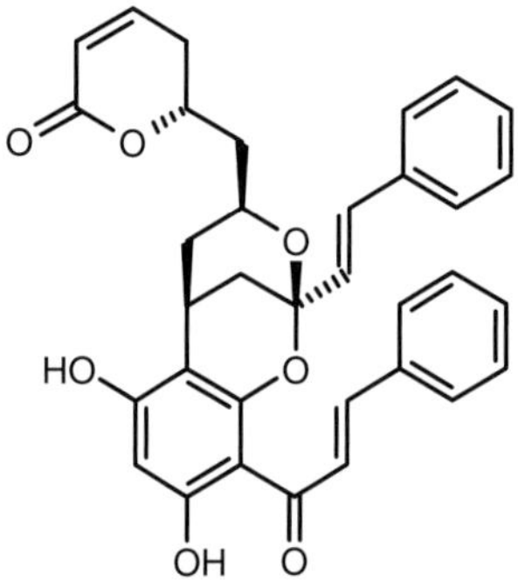

Obochalcolactone

Compound class: Flavonoid (chalcone)

Source: *Cryptocarya obovata* R. Br. Benth. (trunk barks; family: Lauraceae)

Pharmaceutical potential: *Cytotoxic*
The isolate showed moderate cytotoxicity against human oral epidermoid carcinoma (KB) cells with an IC_{50} value of 5 μM.

Reference

Dumontet, V., Van Hung, N., Adeline, M.-T., Riche, C., Chiaroni, A., Sevenet, T., and Gueritte, F. (2004) *J. Nat. Prod.*, **67**, 858.

Obolactone

Physical data: $C_{19}H_{18}O_4$, pale yellow needles (AcOEt), mp 116 °C, $[\alpha]_D^{25}$ +286° (CHCl$_3$, *c* 1.12)

Structure:

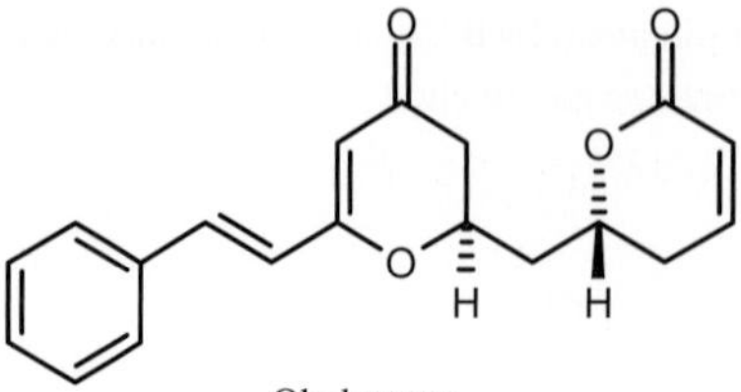

Obolactone

Compound class: α-Pyrone

Source: *Cryptocarya obovata* R. Br. Benth. (trunk barks; family: Lauraceae)

Pharmaceutical potential: *Cytotoxic*
The isolate showed moderate cytotoxicity against human oral epidermoid carcinoma (KB) cells with an IC_{50} value of 3 μM.

Obtusafuran

Systematic name: 5-Hydroxy-6-methoxy-3-methyl-2-phenyl-2,3-dihydrobenzofuran

Physical data: $C_{16}H_{16}O_3$

Structure:

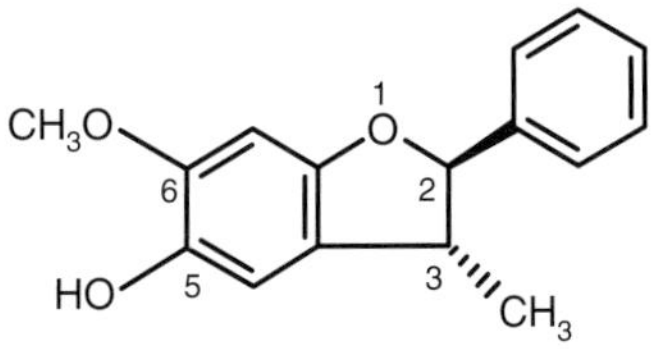

Obtusafuran

Compound class: Arylbenzofuran

Source: *Dalbergia louvelii* R. Viguier (heartwoods; family: Fabaceae) [1]

Pharmaceutical potential: *Antimalarial*
The compound was reported to possess potent *in vitro* antimalarial activity; it inhibited the growth of the chloroquine-resistant strain FcB1 of *Plasmodium falciparum* with an IC_{50} value of $8.7 \pm 0.6\ \mu M$ (chloroquine used as a positive control; $IC_{50} = 0.13 \pm 0.03\ \mu M$) [1]. The absence of C_2–C_3 double bond in the molecule has been found to enhance the antiplasmodial activity to a significant extent compared to its dehydro derivative, parvifuran (5-hydroxy-6-methoxy-3-methyl-2-phenyl-2,3-dihydrobenzofuran) [2], having a double bond at C-2 and C-3 positions.

References

[1] Beldjoudi, N., Mambu, L., Labaied, M., Grellier, P., Ramanitrahasimbola, D., Rasoanaivo, P., Martin, M.T., and Frappier, F. (2003) *J. Nat. Prod.*, **66**, 1447.
[2] Muangnoicharoen, N. and Frahm, A.N. (1981) *Phytochemistry*, **20**, 291.

Obyanamide

Physical data: $C_{30}H_{41}N_5O_6S$, white powder, $[\alpha]_D^{27}$ +20° (MeOH, *c* 0.04)

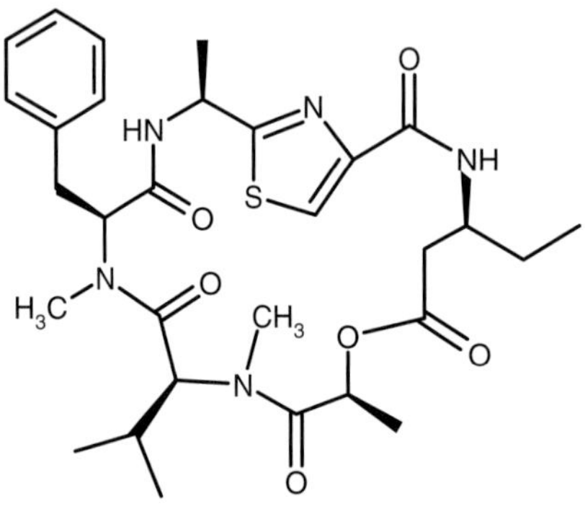

Obyanamide

Compound class: Depsipeptide

Source: *Lyngbya confervoides* (marine cyanobacterium)

Pharmaceutical potential: *Cytotoxic*
Obyanamide displayed potent *in vitro* cytotoxic activity against KB (human nasopharyngeal carcinoma) cells with an IC$_{50}$ value of 0.58 µg/ml.

Reference

Williams, P.G., Yoshida, W.Y., Moore, R.E., and Paul, V.J. (2002) *J. Nat. Prod.*, **65**, 29.

Oceanalin A

Physical data: $C_{41}H_{72}N_2O_9$, colorless amorphous solid, $[\alpha]_D$ −5.7° (EtOH, *c* 0.14)

Structure:

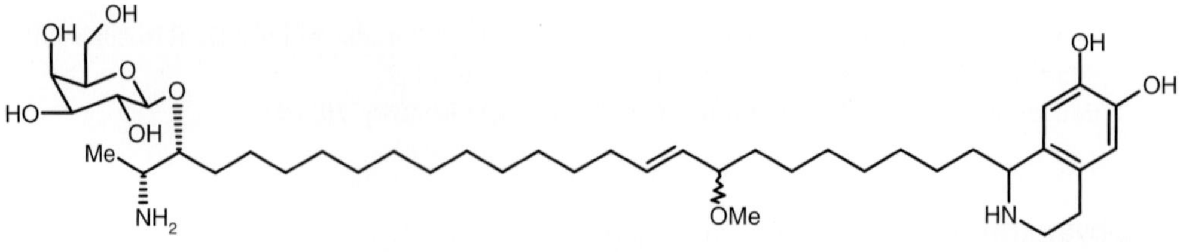

Oceanalin A

Compound class: Sphingolipid (tetrahydroquinoline-containing dimeric sphingolipid)

Source: *Oceanapia* sp. (marine sponge)

Pharmaceutical potential: *Antifungal*
The isolate was found to display *in vitro* antifungal activity against fluconazole-resistant yeast *Candida glabrata* with an MIC value of 30 µg/ml; the present investigators assumed that this

activity of the compound is due the blockade of phytosphingolipid biosynthesis in *C. glabrata* by inhibiting ceramide synthase.

Reference

Makarieva, T.N., Denisenko, V.A., Dmitrenok, P.S., Guzii, A.G., Santalova, E.A., Stonik, V.A., MacMillan, J.B., and Molinski, T.F. (2005) *Org. Lett.*, **7**, 2897.

Ochracenomicin A

Physical data: $C_{19}H_{16}O_6$, orange prism, mp 195–203 °C, $[\alpha]_D^{23}$ +14° (CHCl$_3$, *c* 0.2)

Structure:

Ochracenomicin A

Compound class: Antibiotic (benz[*a*]anthraquinone)

Source: *Amycolatopsis* sp. MJ950-89F4 (culture broth)

Pharmaceutical potential: *Antimicrobial*

The antibiotic displayed a broad spectrum of antimicrobial potential against a number of microbial strains such as *Staphylococcus aureus* FDA209P, *S. aureus* Smith, *S. aureus* MS9610, *S. aureus* MS 16526 (MRSA), *S. aureus* TY-04282 (MRSA), *Micrococcus luteus* IFO 3333, *M. luteus* PCI 1001, *Bacillus subtilis* NRRLB-558, *B. cereus* ATCC 10702, *Corynebacterium bovis* 1810, *Escherichia coli* NIHJ, *Shigella dysenteriae* JS 11910, *Salmonella enteritidis*, *Proteus mirabilis* IFM OM-9, *Providencia rettgeri* GN 466, *Klebsiella pneumoniae* PCI 602, *Mycobacterium smegmatis* ATCC 607, and *Candida albicans* 3147 with MIC values of 1.56, 0.78, 1.56, 1.56, 1.56, 0.39, 0.39, 1.56, 3.12, 6.25, 6.25, 3.12, 25, 6.25, 25, 25, 50, and 25 µg/ml, respectively.

Reference

Igarashi, M., Sasao, C., Yoshida, A., Naganawa, H., Hamada, M., and Takeuchi, T. (1995) *J. Antibiot.*, **48**, 335.

Ochrocarpinones A–C

Physical data:
- Ochrocarpinone A: $C_{33}H_{42}O_6$, viscous liquid, $[\alpha]_D$ +8.7° (CHCl$_3$, *c* 0.15)
- Ochrocarpinone B: $C_{33}H_{42}O_5$, viscous liquid, $[\alpha]_D$ −3.5° (CHCl$_3$, *c* 0.22)
- Ochrocarpinone C: $C_{33}H_{42}O_5$, viscous liquid, $[\alpha]_D$ +10.2° (CHCl$_3$, *c* 0.18)

Ochrocarpinones A–C

Structures:

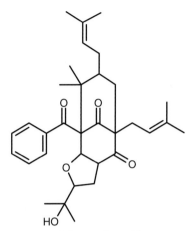

Ochrocarpinone A

Ochrocarpinone B

Ochrocarpinone C

Compound class: Benzophenones

Source: *Ochrocarpus punctatus* H. Perrier (barks; family: Clusiaceae)

Pharmaceutical potential: *Cytotoxic*

Ochrocarpinones A–C exhibited *in vitro* cytotoxicities against the A-2780 ovarian cancer cell line with IC_{50} values of 6.9 ± 0.3, 7.4 ± 0.2, and 5.2 ± 0.3 µg/ml, respectively.

Reference

Chaturvedula, V.S.P., Schilling, J.K., and Kingston, D.G.I. (2002) *J. Nat. Prod.*, **65**, 965.

Ochrocarpins A–G

Systematic names:
- Ochrocarpin A: 5-hydroxy-8-(1-hydroxy-1-methylethyl)-6-(2-methyl-1-oxobutyl)-4-phenyl-2H-furo[2′,3′:5,6]benzo[1,2-b]pyran-2-one
- Ochrocarpin B: 5-hydroxy-8-(1-hydroxy-1-methylethyl)-6-(3-methyl-1-oxobutyl)-4-phenyl-2H-furo[2′,3′:5,6]benzo[1,2-b]pyran-2-one
- Ochrocarpin C: 5-hydroxy-8-(1-hydroxy-1-methylethyl)-6-(2-methyl-1-oxopropyl)-4-phenyl-2H-furo[2′,3′:5,6]benzo[1,2-b]pyran-2-one
- Ochrocarpin D: 5-hydroxy-8-(1-methoxy-1-methylethyl)-6-(2-methyl-1-oxobutyl)-4-phenyl-2H-furo[2′,3′:5,6]benzo[1,2-b]pyran-2-one
- Ochrocarpin E: 2,3-dihydro-4-hydroxy-2-(1-hydroxy-1-methylethyl)-5-(2-methyl-1-oxopropyl)-9-phenyl-7H-furo[2′,3′:3,4]benzo[1,2-b]pyran-2-one
- Ochrocarpin F: 8,9-dihydro-5-hydroxy-8-(1-hydroxy-1-methylethyl)-6-(2-methyl-1-oxobutyl)-4-(1S-acetoxypropyl)-2H-furo[2′,3′:5,6]benzo[1,2-b]pyran-2-one
- Ochrocarpin G: 8,9-dihydro-5-hydroxy-8-(1-hydroxy-1-methylethyl)-6-(2-methyl-1-oxobutyl)-4-(1R-acetyloxypropyl)-2H-furo[2′,3′:5,6]benzo[1,2-b]pyran-2-one

Physical data:
- Ochrocarpin A: $C_{25}H_{24}O_6$, viscous liquid, $[\alpha]_D - 0.28°$ (CHCl$_3$, c 0.32)
- Ochrocarpin B: $C_{25}H_{24}O_6$, viscous liquid, $[\alpha]_D + 0.12°$ (CHCl$_3$, c 0.45)
- Ochrocarpin C: $C_{24}H_{22}O_6$, viscous liquid, $[\alpha]_D + 0.21°$ (CHCl$_3$, c 0.38)
- Ochrocarpin D: $C_{26}H_{26}O_6$, viscous liquid, $[\alpha]_D - 0.48°$ (CHCl$_3$, c 0.52)
- Ochrocarpin E: $C_{24}H_{24}O_6$, viscous liquid, $[\alpha]_D - 0.06°$ (CHCl$_3$, c 0.46)
- Ochrocarpin F: $C_{24}H_{30}O_8$, viscous liquid, $[\alpha]_D - 0.16°$ (CHCl$_3$, c 0.64)
- Ochrocarpin G: $C_{24}H_{30}O_8$, viscous liquid, $[\alpha]_D - 0.12°$ (CHCl$_3$, c 0.42)

Structures:

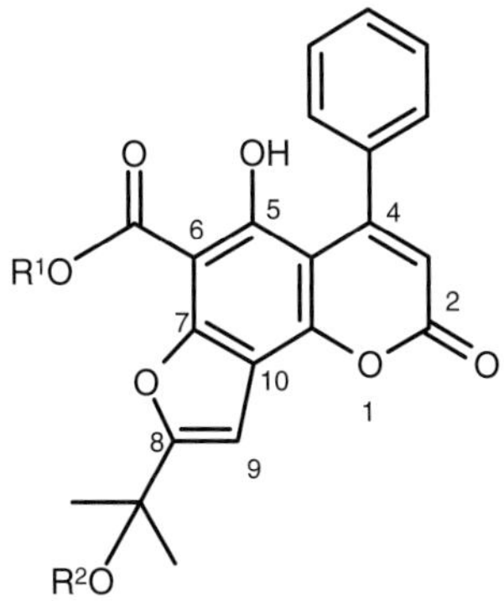

Ochrocarpin A: R^1 = CH(CH$_3$)CH$_2$CH$_3$; R^2 = H
Ochrocarpin B: R^1 = CH$_2$CH(CH$_3$)$_2$; R^2 = H
Ochrocarpin C: R^1 = CH(CH$_3$)$_2$; R^2 = H
Ochrocarpin D: R^1 = CH(CH$_3$)CH$_2$CH$_3$; R^2 = CH$_3$

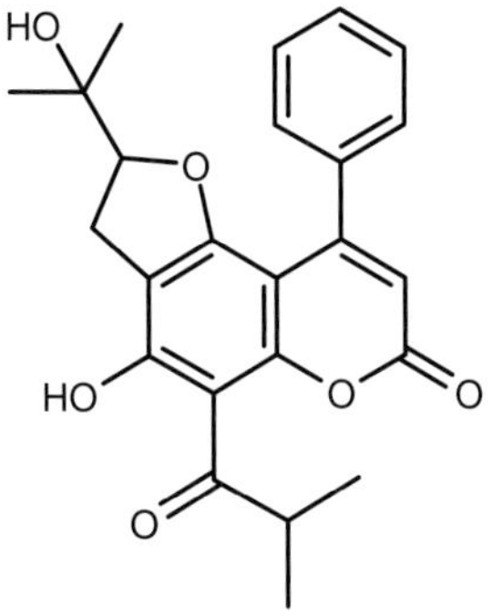

Ochrocarpin E

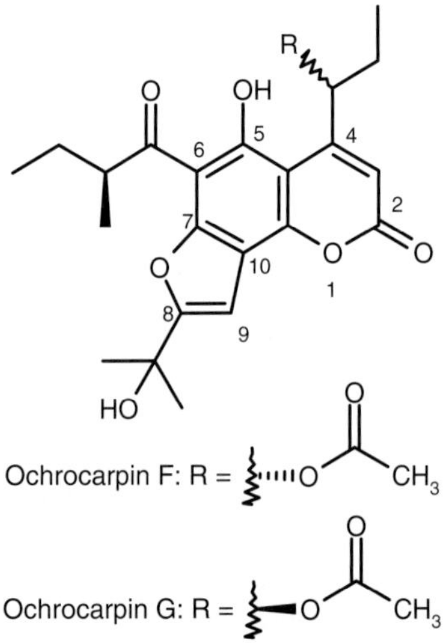

Ochrocarpin F: R = (stereobond)O-C(=O)-CH₃ → Ochrocarpin F: R = $\overset{\text{\tiny{\textbackslash\textbackslash\textbackslash}}}{}$O-CO-CH₃

Compound class: Flavonoids (coumarins)

Source: *Ochrocarpus punctatus* H. Perrier (barks; family: Clusiaceae)

Pharmaceutical potential: *Cytotoxic*
Ochrocarpins A–G were evaluated to possess *in vitro* cytotoxicities against the A-2780 ovarian cancer cell line with IC_{50} values of 5.2 ± 0.4, 3.8 ± 0.3, 3.7 ± 0.5, 6.3 ± 0.3, 4.9 ± 0.1, 8.9 ± 0.6, and 8.6 ± 0.3 µg/ml, respectively.

Reference

Chaturvedula, V.S.P., Schilling, J.K., and Kingston, D.G.I. (2002) *J. Nat. Prod.*, **65**, 965.

Oleanolic acid glycoside

Systematic name: Oleanolic acid 3-O-{O-β-D-glucopyranosyl-(1 → 4)-O-β-D-glucopyranosyl-(1 → 3)-O-α-L-rhamnopyranosyl-(1 → 2)-α-L-arabinopyranoside}

Physical data: $C_{53}H_{86}O_{21}$, amorphous solid, $[\alpha]_D^{27} - 6.0°$ (MeOH–CHCl₃ 1:1, *c* 0.1)

Structure:

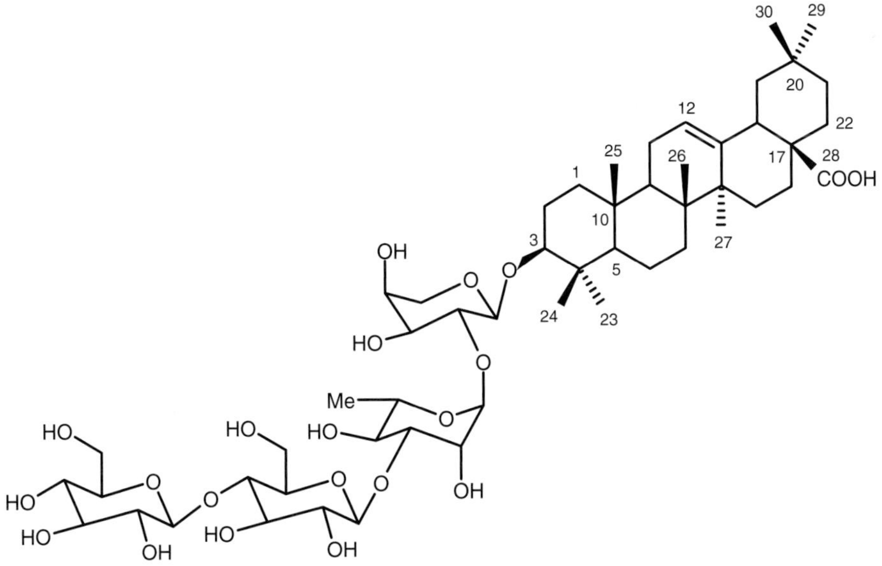

Oleanolic acid glycoside

Compound class: Triterpenoid saponin

Source: *Pulsatilla chinensis* (Burge) Regel (roots; family: Ranunculaceae)

Pharmaceutical potential: *Cytotoxic (antitumor)*
The triterpene saponin was evaluated for its *in vitro* cytotoxicity against human leukemia cells (HL-60) with an IC_{50} value of 2.6 µg/ml.

Reference

Yoshihiro, M., Akihito, Y., Minpei, K., Tomoki, A., and Yutaka, S. (1999) *J. Nat. Prod.*, **62**, 1279.

Oleoyl danshenxinkun A

Physical data: $C_{36}H_{48}O_5$, reddish oil, $[\alpha]_D^{25}$ − 78.8° (CHCl$_3$, *c* 0.25)

Structure:

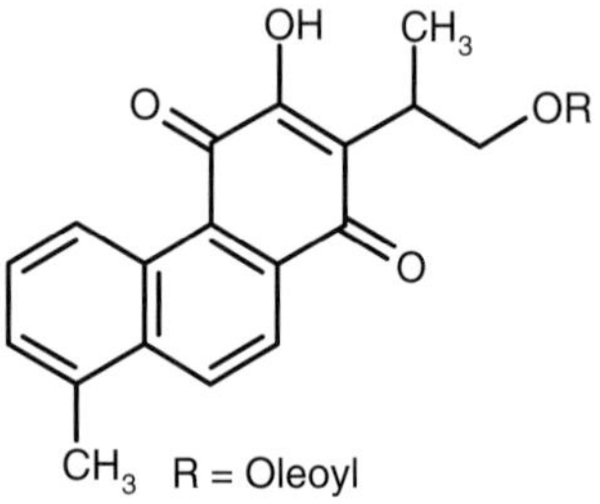

CH₃ R = Oleoyl

Oleoyl danshenxinkun A

Compound class: Abietane diterpenoid

Source: *Salvia miltiorrhiza* Bunge (roots; family: Labiatae)

Pharmaceutical potential: *Platelet aggregation inhibitor*
The test compound was found to inhibit platelet aggregation (rabbit platelets) induced by arachidonic acid and collagen with IC$_{50}$ values of 25.5 ± 1.9 and 60.5 ± 2.6 µM, respectively; however, it remained inactive against thrombin-induced platelet aggregation. Aspirin was used as a positive control for which the IC$_{50}$ value was measured to be 27.0 ± 1.1 µM against arachidonic acid-induced platelet aggregation.

Reference

Lin, H.-C., Ding, H.-Y., and Chang, W.-L. (2001) *J. Nat. Prod.*, **64**, 648.

Oleoyl neocryptotanshinone

Physical data: $C_{37}H_{54}O_5$, yellow oil, $[\alpha]_D^{25}$ + 14.3° (CHCl$_3$, *c* 0.35)

Structure:

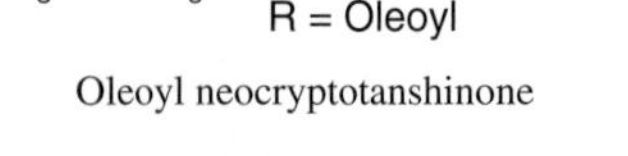

R = Oleoyl

Oleoyl neocryptotanshinone

Compound class: Abietane diterpenoid

Source: *Salvia miltiorrhiza* Bunge (roots; family: Labiatae)

Pharmaceutical potential: *Platelet aggregation inhibitor*
The test compound was found to inhibit platelet aggregation (rabbit platelets) induced by arachidonic acid and collagen with IC$_{50}$ values of 5.1 ± 0.8 and 50.4 ± 1.4 µM, respectively;

however, it remained inactive against thrombin-induced platelet aggregation. Aspirin was used as a positive control for which the IC_{50} value was measured to be $27.0 \pm 1.1\,\mu M$ against arachidonic acid-induced platelet aggregation.

Reference

Lin, F H.-C., Ding, F H.-Y., and Chang, F W.-L. (2001) *J. Nat. Prod.*, **64**, 648.

Onosmins A and B

Systematic names:
- Onosmin A: 2-[(4-methylbenzyl)amino]benzoic acid
- Onosmin B: methyl 2-[(4-methylbenzyl)amino]benzoate

Physical data:
- Onosmin A: $C_{15}H_{15}NO_2$, white amorphous solid, mp 185–187 °C
- Onosmin B: $C_{16}H_{17}NO_2$, white amorphous solid, mp 137–140 °C

Structures:

Onosmin A: R = H
Onosmin B: R = CH$_3$

Compound class: Aromatic compounds (*N*-aryl aminobenzoic acid derivatives)

Source: *Onosma hispida* (whole plants; family: Boraginaceae) [1]

Pharmaceutical potential: *Lipoxygenase (LOX) enzyme inhibitor*
Onosmins A and B were found to inhibit lipoxygenase (LOX) enzyme activity in a concentration-dependent manner with respective IC_{50} values of 24.2 and 36.0 µM (baicalein was used as a positive control with an IC_{50} value of 22 µM); from their detailed experimental studies, it was evident that the nature of inhibition is of noncompetitive type with dissociation constant (K_i) values of 22.0 and 31.1 µM, respectively [1]. The relatively greater inhibitory efficacy of onosmin A may be attributed to its stronger binding activity against LOX using free carboxylic acid group in the molecule; in addition, the carboxylic acid functionality being more electron withdrawing than the methyl benzoate group in onosmin B can convert the active state Fe^{2+} of LOX into inactive Fe^{3+} more efficiently [1].

LOX are key enzymes in the biosynthesis of several bioregulatory compounds; it has been shown that many LOX products play important role in developing a variety of disorders such as bronchial asthma and inflammation [2], including several human cancers [3]. Hence, LOX is a potential target for rational drug design and discovery of mechanism-based inhibitors for treatment of bronchial asthma, inflammation, cancer, and autoimmune diseases [1].

References

[1] Ahmad, I., Nawaz, S.A., Afza, N., Malik, A., Fatima, I., Khan, S.B., Ahmad, N. and Choudhary, M.I. (2005) *Chem. Pharm. Bull.*, **53**, 907.
[2] Steinhilber, D. (1999) *Curr. Med. Chem.*, **6**, 71.
[3] Ding, X.Y., Tong, W.G., and Adrian, T.E. (2001) *Pancreatology*, **4**, 291.

Ophioglonin

Physical data: $C_{16}H_{10}O_7$, greenish yellow needles (EtOH), mp 275–277 °C

Structure:

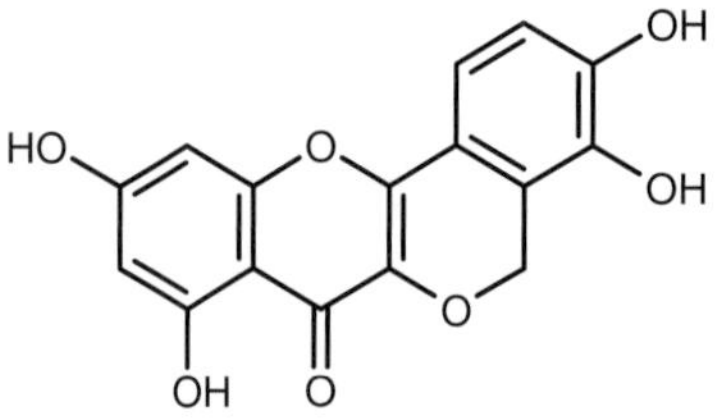

Ophioglonin

Compound class: Flavonoid (homoflavonoid)

Source: *Ophioglossum petiolatum* Hook (whole plants; family: Ophioglossaceae)

Pharmaceutical potential: *Anti-hepatitis B virus (anti-HBV)*
The homoflavonoid compound was found to possess weak antihepatitis B virus activity *in vitro* as studied using the MS-G2 hepatoma cell line; the test compound showed slight anti-HBV surface antigen secretion at 25 µM.

Reference

Lin, Y.-L., Shen, C.-C., Huang, Y.-J., and Chang, Y.-Y. (2005) *J. Nat. Prod.*, **68**, 381.

Orbiculins (A, D–I), celafolin A-1, ejap-2, 2β-benzoyloxyejap-2, and triptogelin C-1

Systematic names:
- Orbiculin A: 1β,2β-diacetoxy-6α,9α-dibenzoyloxy-β-dihydroagarofuran
- Orbiculin D: 1β-acetoxy-6α,9α-di(3-furoyloxy)-dihydro-β-agarofuran
- Orbiculin E: 1β,2β-diacetoxy-9α-benzoyloxy-6α-(3-furoyloxy)-dihydro-β-agarofuran
- Orbiculin F: 1β-acetoxy-9α-benzoyloxy-2β,6α-di(3-furoyloxy)-dihydro-β-agarofuran
- Orbiculin G: 1β-acetoxy-2β,6α,9α-tribenzoyloxy-dihydro-β-agarofuran
- Orbiculin H: 1β,8β-diacetoxy-6α,9α-di(3-furoyloxy)-dihydro-β-agarofuran
- Orbiculin I: 1β-acetoxy-2β,6α,9α-tri(3-furoyloxy)-dihydro-β-agarofuran
- Celafolin A-1: 1β-acetoxy-9α-benzoyloxy-6α-cinnamoyloxy-dihydro-β-agarofuran
- Ejap-2: 1β,6α,15-triacetoxy-9α-benzoyloxy-dihydro-β-agarofuran

- 2β-Benzoyloxyejap-2: 1β,6α,15-triacetoxy-2β,9α-dibenzoyloxy-dihydro-β-agarofuran
- Triptogelin C-1: 1β,2β,6α-triacetoxy-9α-benzoyloxy-dihydro-β-agarofuran

Physical data:
- Orbiculin A: $C_{33}H_{38}O_9$, white amorphous powder, mp 122–125 °C, $[\alpha]_D^{25} +9.52°$ (MeOH, c 0.21)
- Orbiculin D: $C_{27}H_{32}O_9$, white amorphous powder, mp 76–78 °C, $[\alpha]_D^{25} +16.87°$ (MeOH, c 2.49)
- Orbiculin E: $C_{31}H_{36}O_{10}$, white amorphous powder, mp 134–136 °C, $[\alpha]_D^{25} +34.67°$ (MeOH, c 1.24)
- Orbiculin F: $C_{34}H_{36}O_{11}$, white amorphous powder, mp 98–100 °C, $[\alpha]_D^{25} +45°$ (MeOH, c 0.80)
- Orbiculin G: $C_{38}H_{40}O_9$, white amorphous powder, mp 110–112 °C, $[\alpha]_D^{25} +47.06°$ (MeOH, c 0.85)
- Orbiculin H: $C_{29}H_{34}O_{11}$, white amorphous powder, mp 111–113 °C, $[\alpha]_D^{25} -19.5°$ (MeOH, c 1.0)
- Orbiculin I: $C_{32}H_{34}O_{12}$, white needles, mp 253–255 °C, $[\alpha]_D^{25} +39.2°$ (MeOH, c 0.63)
- Celafolin A-1: $C_{33}H_{38}O_7$, amorphous powder, $[\alpha]_D^{25} +23.2°$ (MeOH, c 0.43)
- Ejap-2: $C_{28}H_{34}O_9$, solid, mp 149–150 °C
- 2β-Benzoyloxyejap-2: $C_{35}H_{38}O_{11}$
- Triptogelin C-1: $C_{28}H_{36}O_9$, needles, mp 182–184 °C, $[\alpha]_D^{25} +41.7°$ (MeOH, c 1.0)

Structures:

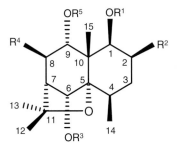

Orbiculin A: R¹ = Ac; R² = OAc; R³ = Bz; R⁴ = H; R⁵ = Bz
Orbiculin D: R¹ = Ac; R² = H; R³ = Fu; R⁴ = H; R⁵ = Fu
Orbiculin E: R¹ = Ac; R² = OAc; R³ = Fu; R⁴ = H; R⁵ = Bz
Orbiculin F: R¹ = Ac; R² = OFu; R³ = Fu; R⁴ = H; R⁵ = Bz
Orbiculin G: R¹ = Ac; R² = OBz; R³ = Bz; R⁴ = H; R⁵ = Bz
Orbiculin H: R¹ = Ac; R² = H; R³ = Fu; R⁴ = OAc; R⁵ = Fu
Orbiculin I: R¹ = Ac; R² = OFu; R³ = Fu; R⁴ = H; R⁵ = Fu

Celafolin A-1: R¹ = Ac; R² = H; R³ = Cinn; R⁴ = H; R⁵ = Bz
Triptogelin C-1: R¹ = Ac; R² = OAc; R³ = Ac; R⁴ = H; R⁵ = Bz

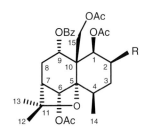

Ejap-2: R = H
2β-Benzoyloxyejap-2: R = OBz

Fu = 3-furoyl group   Cinn = cinnamoyl group

Compound class: Sesquiterpenes (β-agarofurans)

Sources: Orbiculin A: *Celastrus orbiculatus* (roots; family: Celastraceae) [1]; orbiculins D–F: *C. orbiculatus* (roots) [2]; orbiculin G: *C. orbiculatus* (roots) [2] and *Microtropis fokienensis* Dunn (roots; family: Celastraceae) [3]; orbiculins H and I: *C. orbiculatus* (roots) [4]; celafolin A-1: *C. stephanotiifolius* (seeds) [5], *C. orbiculatus* (roots) [1]; ejap-2: *Euonymus japonicus* (family: Celastraceae) [6], *C. orbiculatus* (roots) [2]; 2α-benzoyloxyejap-2: *C. orbiculatus* (roots) [2]; triptogelin C-1: *Tripterygium wilfordii* Hook fii var. *regelii* Makino (family: Celastraceae) [7]; *C. orbiculatus* (roots) [2]; *C. stephanotiifolius* (seeds) [5]

Pharmaceutical potentials: *Cytotoxic; antitubercular*

Jin et al. [4] evaluated the effects of the sesquiterpenes, orbiculins A, D–F, H, and I, on LPS-induced NF-χB activation in murine macrophage RAW 264.7 cells transfected with NF-χB-mediated reporter gene construct and on nitric oxide production in LPS-stimulated RAW 264.7 cells – orbiculins D, H, and I showed moderate inhibition in both NF-χB activation and nitric oxide production, thereby suggesting that the furoyloxy groups at C-6 and C-9 (as in orbiculins H, I, and D) are noticeable structural factors of dihydro-β-agarofuran sesquiterpenes in the modulation of NF-χB activity.

From the investigation on MDR reversing activity studied by Kim et al. [2], it was found that orbiculin A and celafolin A-1 partially or completely reversed resistance to adriamycin, vinblastine, and paclitaxel in multidrug-resistant KB-V1 and MCF7/ADR cells; the thorough research showed that ejap-2, orbiculin E and F, and triptogelin C-1 were more active than verapamil in reversing vinblastine resistance in multidrug-resistant KB-V1 cells [8]. Orbiculin A, orbiculin E, and triptogelin C-1, which had an acetoxy group at C-2, showed strong reversal activity, while orbiculin F, with a furoyloxy at C-2, showed half activity to that of orbiculin E, and orbiculin G, with a benzoyloxy at C-2, was the weakest; the compounds ejap-2 and 2β-benzoyloxyejap-2, which had two acetoxy groups at C-1 and C-15, exhibited strong activity irrespective of the presence of ester group at C-2. Hence, the polarity of C-1/C-2 or C-1/C-15 might be an important factor in MDR reversal activity [8].

Orbiculin G was also found to display potent *in vitro* antitubercular activity [9] against *Mycobacterium tuberculosis* 90-221387 with an MIC value of 14.6 µM; the investigators [9] used the clinical antitubercular agent ethambutol as the positive control, which showed an MIC value of 30.6 µM.

References

[1] Kim, S.E., Kim, Y.H., and Lee, J.J. (1998) *J. Nat. Prod.*, **61**, 108.
[2] Kim, S.E., Kim, H.S., Hong, Y.S., Kim, Y.C., and Lee, J.J. (1999) *J. Nat. Prod.*, **62**, 697.
[3] Chen, J.-J., Chou, T.-H., Peng, C.-F., Chen, I.-S., and Yang, S.-Z. (2007) *J. Nat. Prod.*, **70**, 202.
[4] Jin, H.Z., Hwang, B.Y., Kim, H.S., Lee, J.H., Kim, Y.H., and Lee, J.J. (2002) *J. Nat. Prod.*, **65**, 89.
[5] Takaishi, Y., Ohshima, S., Nakano, K., Tomimatsu, T., Tokuda, H., Nishino, H., and Iwashima, A. (1993) *J. Nat. Prod.*, **56**, 815.
[6] Rozsa, Z. and Perjesi, A. (1989) *J. Chem. Soc., Perkin Trans. 1*, 1079.
[7] Takaishi, Y., Tokura, K., Tamai, S., Ujita, K., Nakano, K., and Tornirnatsu, T. (1991) *Phytochemistry*, **30**, 1567.
[8] Kim, S.E., Kim, Y.H., and Lee, J.J. (1998) *J. Nat. Prod.*, **61**, 108.
[9] Chen, J.-J., Chou, T.-H., Peng, C.-F., Chen, I.-S., and Yang, S.-Z. (2007) *J. Nat. Prod.*, **70**, 202.

Oriciacridones C and F

Systematic names:
- Oriciacridone C: (+)-1,5-dihydroxy-2-isopropenyldihydrofuran[3,4-c]acridone
- Oriciacridone F: (+)-bis-5-hydroxy-(10H)-hydronoracromycine

Physical data:
- Oriciacridone C: $C_{18}H_{15}NO_4$, yellow needles (MeOH), mp 253–254 °C, $[\alpha]_D^{25} + 21.8°$ (MeOH, c 0.39)
- Oriciacridone F: $C_{36}H_{32}N_2O_8$, yellow crystals (MeOH), mp 187–189 °C, $[\alpha]_D^{25} + 35.6°$ (MeOH, c 0.62)

Structures:

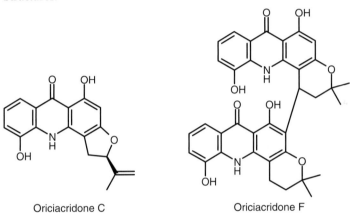

Oriciacridone C   Oriciacridone F

Compound class: Acridone alkaloids

Source: *Oriciopsis glaberrima* Engl. (stem barks; family: Rutaceae)

Pharmaceutical potentials: *α-Glucosidase inhibitory;antioxidant (radical scavenging)*
Both the alkaloids showed significant α-glucosidase inhibitory potential. Oriciacridone C showed 81.8 and 44.4% inhibition against the enzyme, respectively, at the concentrations of 100 and 50 μM; oriciacridone F exhibited slightly greater such activity by exhibiting 77.4, 54.2, and 25.5% inhibition, respectively, at 100, 50, and 25 μM concentrations. The respective IC_{50} values for oriciacridones C and F were calculated as 56 ± 5.4 and 34.05 ± 17 μM (IC_{50} value of deoxynojirimycin (positive control) $= 330 \pm 8.14$ μM).

Oriciacridones C and F also showed a moderate antioxidant activity (free radical scavengers; IC_{50} values were determined, respectively, to be 60.79 ± 1.23 and 482.0 ± 1.80 mM) in the DPPH assay with respect to the reference 3-*tert*-butyl-4-hydroxyanisole (BHA; $IC_{50} = 44.2 \pm 0.02$ mM).

Reference

Duplex Wansi, J., Wandji, J., Mbaze Meva'a, L., Kamdem Waffo, A.F., Ranjit, R., Khan, S.N., Asma, A., Choudhary, M.I., Lallemand, M.-C., Tillequin, F., and Tanee, Z.F. (2006) *Chem. Pharm. Bull.*, **54**, 292.

Oryzafuran

Systematic name: 2-(3,4-Dihydroxyphenyl)-4,6-dihydroxybenzofuran-3-carboxylic acid methyl ester

Physical data: $C_{16}H_{12}O_7$, pale brown needles (MeOH), mp 251–252 °C

Structure:

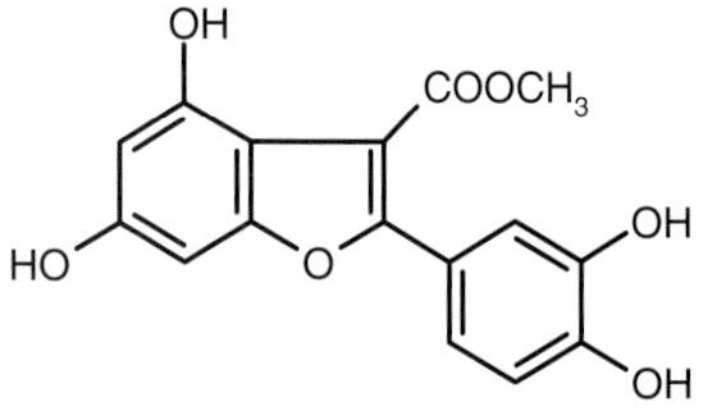

Oryzafuran

Compound class: 2-Arylbenzofuran

Source: *Oryza sativa* L. (rice bran; family: Poaceae)

Pharmaceutical potential: *Antioxidant*
The benzofuran derivative showed antioxidant efficacy when evaluated in the 1,1-diphenyl-2-picrylhydrazyl (DPPH) assay – it displayed free radical scavenging activity with an EC_{50} value of 1.58 ± 0.001 µg/ml; the activity was found to be more potent than that of ascorbic acid used as a positive control (EC_{50} = 3.35 ± 0.006 µg/ml).

Reference

Han, S.J., Ryu, S.N., and Kang, S.S. (2004) *Chem. Pharm. Bull.*, **52**, 1365.

7-Oxo-10α-cucurbitadienol

Systematic name: 7-Oxo-10α-cucurbita-5,24-dien-3β-ol

Physical data: $C_{30}H_{48}O_2$, solid, mp 158–160 °C

Structure:

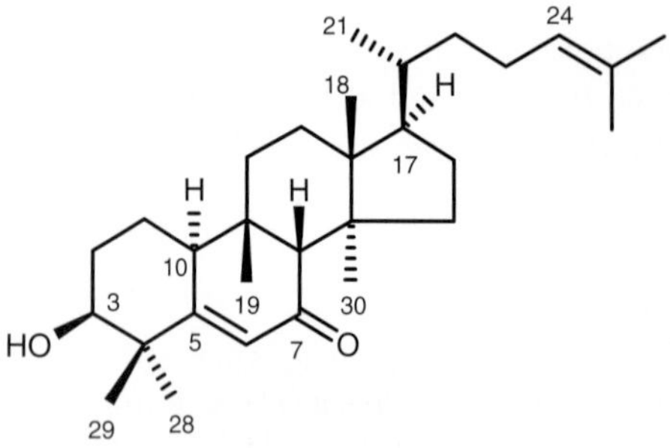

7-Oxo-10α-cucurbitadienol

Compound class: Triterpenoid (cucurbitane-type triterpene)

Source: *Trichosanthes kirilowii* Maxim. (seeds; family: Cucurbitaceae)

Pharmaceutical potential: *Anti-inflammatory*
7-Oxo-10α-cucurbitadienol and its acetyl and 24-dihydro derivatives were found to display marked inhibitory activity against 12-O-tetradecanoylphorbol-13-acetate (TPA)-induced ear inflammation in mice; the 50% inhibitory doses (ID$_{50}$) were determined, respectively, to be 0.7, 0.7, and 0.4 mg/ear (indomethacin was used as a positive control; ID$_{50}$ = 0.3 mg/ear).

Reference

Akihisa, T., Yasukawa, K., Kimura, Y., Takido, M., Kokke, W.C.M.C., and Tamura, T. (1994) *Phytochemistry*, **36**, 153.

7-Oxo-3,8,9-trihydroxystaurosporine and 7-oxo-8,9-dihydroxy-4′-N-demethylstaurosporine

Physical data:
- 7-Oxo-3,8,9-trihydroxystaurosporine (**1**): $C_{28}H_{24}N_4O_7$, red amorphous solid, $[\alpha]_D^{25}$ +37.1° (MeOH, *c* 0.040)
- 7-Oxo-8,9-dihydroxy-4′-*N*-demethylstaurosporine (**2**): $C_{27}H_{22}N_4O_6$, orange amorphous solid, $[\alpha]_D^{25}$ +47.7° (MeOH, *c* 0.068)

Structures:

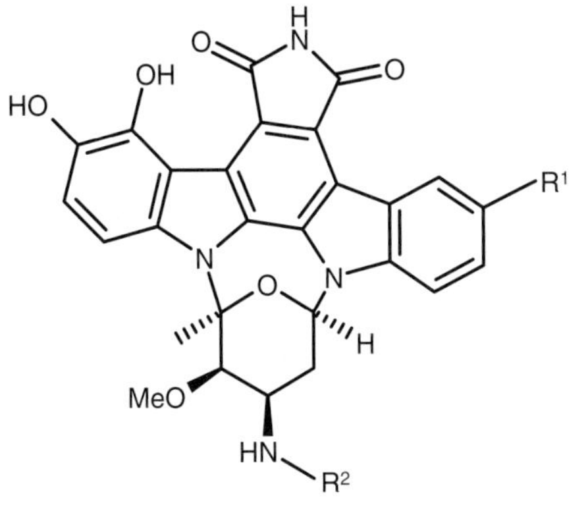

1: R^1 = OH, R^2 = CH$_3$
2: R^1 = R^2 = H

Compound class: Alkaloids (indolo[2,3-*a*]carbazole alkaloids)

Source: *Cystodytes solitus* Monniot (marine ascidian)

Pharmaceutical potential: *Cytotoxic*
Both the test compounds were found to exhibit strong *in vitro* growth inhibitory activity against three human tumor cell lines including lung (A549), colon (HT29), and breast

(MDA-MB-231); the respective GI_{50} values were determined to be 26.6 nM (A549), 68.1 nM (HT29), and 28.4 nM (MDA-MB-231) for compound **1** and 17.5 nM (A549), 90.3 nM (HT29), and 32.1 nM (MDA-MB-231) for compound **2**. The parent compound (staurosporine) was used as a reference standard for which GI_{50} values were determined to be 2.4 nM (A549), 10.9 nM (HT29), and 7.1 nM (MDA-MB-231).

Reference

Reyes, F., Fernández, R., Rodríguez, A., Bueno, S., de Eguilior, C., Francesch, A., and Cuevas, C. (2008) *J. Nat. Prod.*, **71**, 1046.

7-Oxohernangerine

Physical data: $C_{18}H_{11}NO_5$, orange prisms (MeOH), mp 256–258 °C, $[\alpha]_D^{24} \pm 0°$ (CHCl$_3$, *c* 0.12)

Structure:

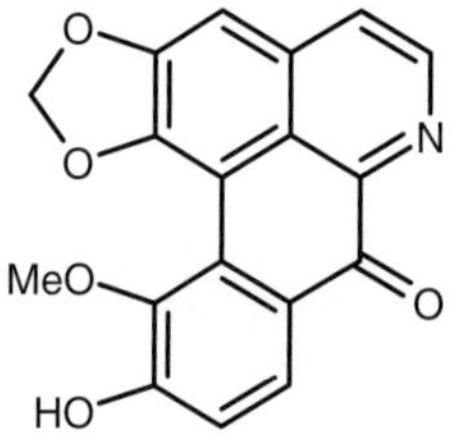

7-Oxohernangerine

Compound class: Alkaloid

Sources: *Hernandia nymphaeifolia* (Presl) Kubitzki (*Hernandia peltata* Meissn. (stem barks; family: Hernandiaceae) [1]; *Lindera chunii* Merr. (roots; family: Lauraceae) [2].

Pharmaceutical potential: *Anti-HIV*
The alkaloid was evaluated to possess significant antihuman immunodeficiency virus type-1 (HIV-1) integrase activity with an IC_{50} value of 18.2 µM (suramin was used as a positive control; IC_{50} = 2.4 µM) [2].

References

[1] Cheng, J.J., Tsai, L.L., and Chen, I.S. (1996) *J. Nat. Prod.*, **59**, 156.
[2] Zhang, C., Nakamura, N., Tewtrakul, S., Hattori, M., Sun, Q., Wang, Z., and Fujiwara, T. (2002) *Chem. Pharm. Bull.*, **50**, 1195.

18-Oxotryprostatin A

Physical data: $C_{22}H_{25}N_3O_4$, yellow amorphous powder, $[\alpha]_D^{21} - 31.5°$ (CHCl$_3$, *c* 0.14)

Structure:

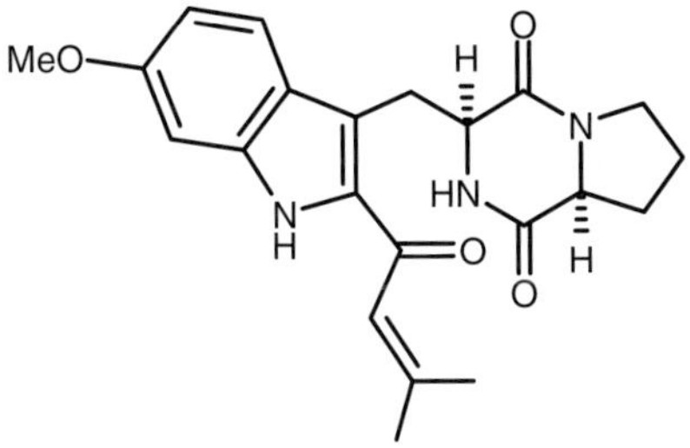

18-Oxotryprostatin A

Compound class: Alkaloid (diketopiperazine alkaloid)

Source: *Aspergillus sydowii* PFW1-13 strain (marine-derived fungus)

Pharmaceutical potential: *Cytotoxic*
The alkaloid exhibited moderate cytotoxicity against A-549 (lung cancer) cells with an IC_{50} value of 1.28 μM.

Reference

Zhang, M., Wang, W.-L., Fang, Y.-C., Zhu, T.-J., Gu, Q.-Q., and Zhu, W.-M. (2008) *J. Nat. Prod.*, **71**, 985.

Oxypeucedanin hydrate acetonide

Physical data: $C_{19}H_{20}O_6$, white crystals, mp 157–158 °C

Structure:

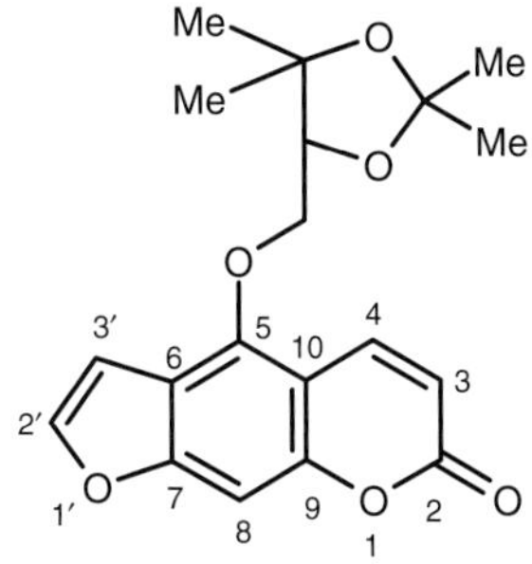

Oxypeucedanin hydrate acetonide

Compound class: Coumarin derivative

Sources: *Peucedanum turcomanicum* (family: Apiaceae) [1]; *Angelica dahurica* Benth et Hook (roots; family: Umbelliferae) [2]

Pharmaceutical potential: *Cytotoxic (antitumor)*
The coumarin isolate displayed significant *in vitro* cytotoxicity against the four tumor cell lines such as L1210, HL-60, K562, and B16F10 with IC_{50} values of 9.4, 9.5, 8.6, and 9.8 µg/ml, respectively [2].

References

[1] Abyshev, A.Z., Azhdarov, B., and Gashimov, N.F. (1979) *Khim. Prir. Soedin.*, **6**, 847.
[2] Thanh, P.N., Jin, W.-Y., Song, G.-Y., Bae, K., and Kang, S.S. (2004) *Arch. Pharm. Res.*, **27**, 1211.

P

Pacificins C and H

Physical data:
- Pacificin C: $C_{20}H_{34}O_4$, $[\alpha]_D^{25}$ $-46°$ (CHCl$_3$, c 0.1)
- Pacificin H: $C_{20}H_{36}O_5$, colorless amorphous solid, $[\alpha]_D^{25}$ $-18°$ (CHCl$_3$, c 0.1)

Structures:

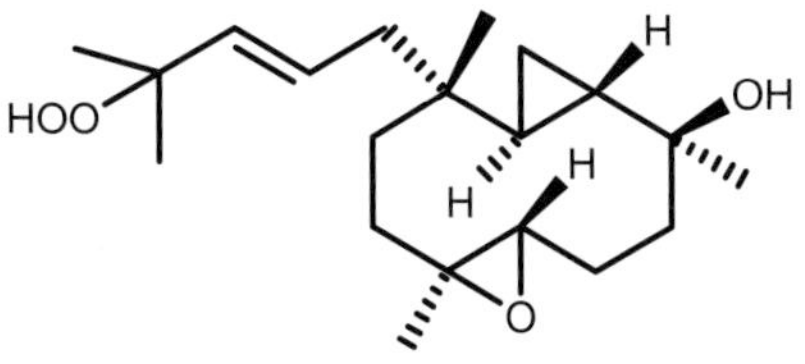

Pacificin C

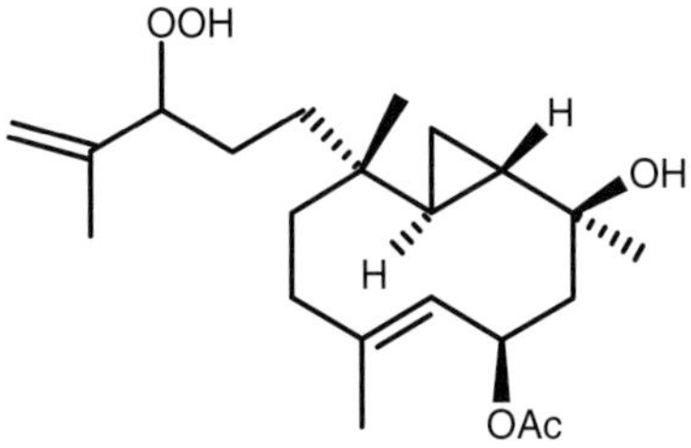

Pacificin H

Compound class: Prenylbicyclogermacrane-type diterpenoids

Source: Formosan soft coral *Nephthea pacifica* Kukenthal (family: Nephtheidae)

Pharmaceutical potential: *Cytotoxic*
Pacificins C and H showed significant cytotoxicity against P-388 (mouse lymphocytic leukemia) cells with ED$_{50}$ values of 1.44 and 2.01 μg/ml, respectively.

Reference

El-Gamal, A.A.H., Wang, S.-K., Dai, C.-F., Chen, I.-G., and Duh, C.-Y. (2005) *J. Nat. Prod.*, **68**, 74.

Pacificins K and L

Physical data:
- Pacificin K: $C_{23}H_{36}O_3$, $[\alpha]_D^{25}$ $-56.0°$ (CHCl$_3$, c 0.3)
- Pacificin L: $C_{20}H_{34}O_3$, $[\alpha]_D^{25}$ $-67.0°$ (CHCl$_3$, c 0.2)

Handbook of Pharmaceutical Natural Products, Volume 2.
Goutam Brahmachari
Copyright © 2010 WILEY-VCH Verlag GmbH & Co. KGaA, Weinheim
ISBN: 978-3-527-32148-3

Pacificin K

Structures:

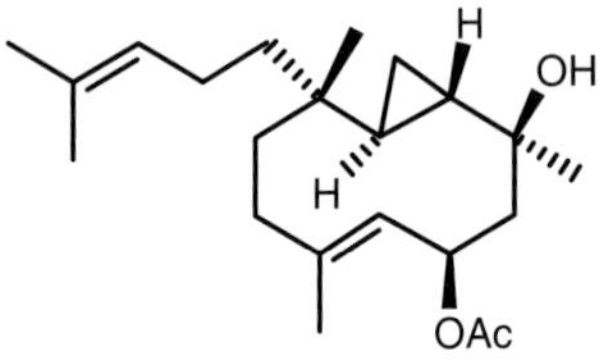

Pacificin K 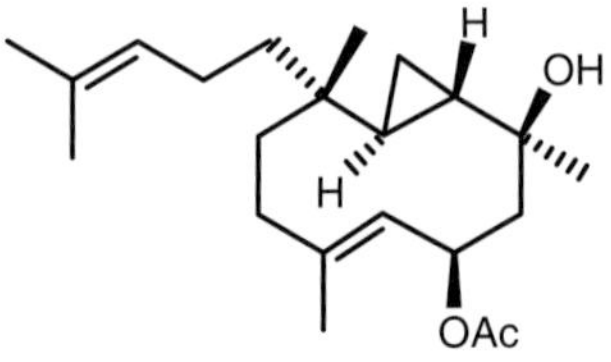   Pacificin K

Compound class: Diterpenoids (prenylbicyclogermacrane type)

Source: *Nephthea elongata* Kukenthal (Formosan soft coral; family: Nephtheidae)

Pharmaceutical potential: *Cytotoxic*
Pacificins K and L were reported to show potent cytotoxic activity against P-388 cells with ED_{50} values of 3.2 and 2.6 µg/ml, respectively.

Reference

El-Gamal, A.A.H., Wang, S.-K., and Duh, C.-Y. (2007) *Chem. Pharm. Bull.*, **55**, 890.

Paecilopeptin

Physical data: $C_{13}H_{24}N_2O_3$, colorless powder, mp 108–110 °C, $[\alpha]_D^{24}$ −106° (MeOH, *c* 0.2)

Structure:

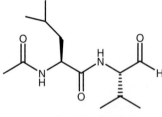

Paecilopeptin

Compound class: Antibiotic

Source: *Paecilomyces carneus* (fungal strain; culture broth)

Pharmaceutical potential: *Cathepsin S enzyme inhibitor*
The isolate was evaluated to have dose-dependant human cathepsin S enzyme (a cysteine protease) inhibitory potential with an IC_{50} value of 2.1 nM *in vitro*.

Reference

Shindo, K., Suzuki, H., and Okuda, T. (2002) *Biosci. Biotechnol. Biochem.*, **66**, 2444.

Paeonilide

Physical data: $C_{17}H_{18}O_6$, colorless needles (MeOH), mp 144–145 °C, $[\alpha]_D^{20}$ +54.3° (CHCl$_3$, c 0.44)

Structure:

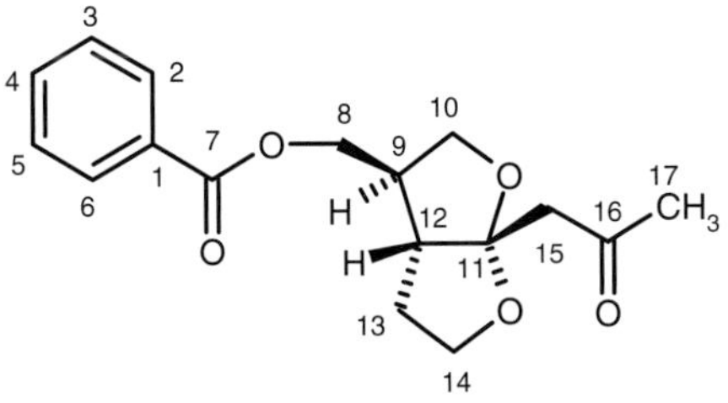

Paeonilide

Compound class: Monoterpenoid

Source: *Paeonia delavayi* (roots; family: Paeoniaceae)

Pharmaceutical potential: *Anti-PAF activity*
Paeonilide inhibited platelet aggregation induced by PAF (platelet activating factor) with an IC$_{50}$ value of 8.0 µg/ml; however, the compound showed no such inhibitory effect against ADP- or arachidonic acid (AA)-induced platelet aggregation.

Reference

Liu, J.K., Ma, Y.-B., Wu, D.-G., Lu, Y., Shen, Z.-Q., Zheng, Q.-T., and Chen, Z.-H. (2000) *Biosci. Biotechnol. Biochem.*, **64**, 1511.

Paeonins A and B

Physical data:
- Paeonin A: $C_{30}H_{32}O_{12}$, colorless gummy solid, $[\alpha]_D^{25}$ +10° (MeOH, c 0.012)
- Paeonin B: $C_{23}H_{28}O_{11}$, colorless gummy solid, $[\alpha]_D^{25}$ +2.5° (MeOH, c 0.016)

Structures:

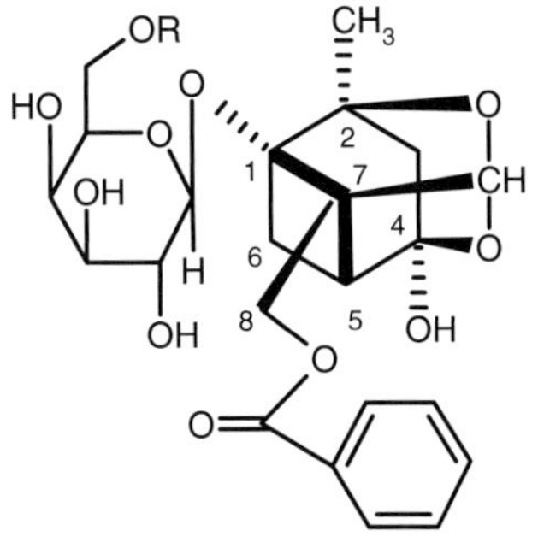

Paeonin A: R = COC$_6$H$_5$
Paeonin B: R = H

Compound class: Monoterpene galactosides

Source: *Paeonia emodi* (roots; family: Paeoniaceae)

Pharmaceutical potential: *Lipoxygenase inhibitors*
Both the compounds were found to be potent inhibitors of lipoxygenase; paeonin B showed greater inhibitory potential than the other. The IC_{50} values for paeonins A and B were determined to be 66.1 and 56.9 µM, respectively (baicalein was used as a positive control; $IC_{50} = 22.4$ µM).

Reference

Riaz, N., Anis, I., Malik, A., Ahmed, Z., Aziz-Ur-Rehman, Muhammad, P., Nawaz, S.A., and Choudhary, M.I. (2003) *Chem. Pharm. Bull.*, **51**, 252.

Palau'amide

Physical data: $C_{46}H_{69}N_5O_{10}$, colorless oil, $[\alpha]_D^{23}$ −22° (MeOH, *c* 0.4)

Structure:

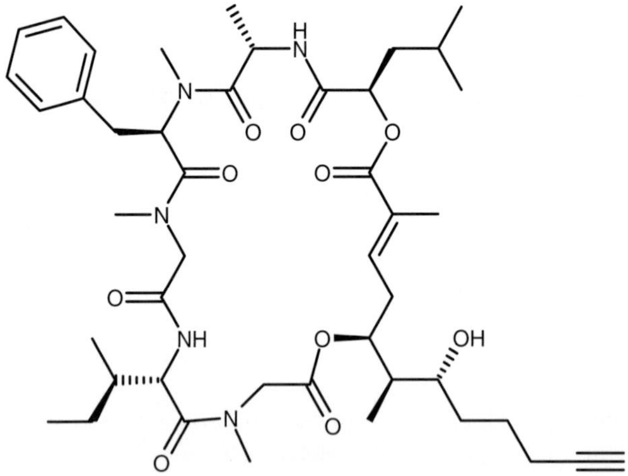

Palau'amide

Compound class: Cyclopeptide

Source: *Lyngbya* sp. (cyanobacterium; family: Oscillatoriaceae)

Pharmaceutical potential: *Cytotoxic (antitumor)*
Palau'amide was evaluated to display strong *in vitro* cytotoxicity against KB (human nasopharyngeal carcinoma) cells with an IC_{50} value of 13 nM.

Reference

Williams, P.G., Yoshida, W.Y., Quon, M.K., Moore, R.E., and Paul, V.J. (2003) *J. Nat. Prod.*, **66**, 1545.

Panaxynol and panaxydol

Physical data:
- Panaxynol: $C_{17}H_{24}O$, colorless oil, $[\alpha]_D$ $-33.0°$ (MeOH, c 2.1)
- Panaxydol: $C_{17}H_{24}O_2$, colorless oil, $[\alpha]_D$ $-79.3°$ (MeOH, c 1.01)

Structures:

Panaxynol

Panaxydol

Compound class: Polyacetylene derivatives

Sources: *Panax ginseng* (roots; family: Araliaceae) [1–7]; Panaxynol also from *Angelica japonica* (roots; family: Umbelliferae) [8]

Pharmaceutical potential: *Cytotoxic*
Both the polyacetylenic compounds displayed significant *in vitro* cytotoxic activities against a number of cancer cell lines. Matsunaga *et al.* [4] reported that panaxynol and panaxydol exhibit cytotoxicities against the MK-1, B-16, L-929, and MRC-5 cancer cell lines with respective ED_{50} values of 0.027 ± 0.004, 1.23 ± 0.03, 2.50 ± 0.28, and $17.10 \pm 1.3\,\mu g/ml$ for panaxynol and 0.016 ± 0.005, 1.5 ± 0.10, 2.6 ± 0.17, and $11.5 \pm 0.40\,\mu g/ml$ for panaxydol. Furthermore, Yang *et al.* [7] evaluated their cytotoxic activities *in vitro* against A549, SK-OV-3, SK-MEL-2, and HCT-15 cell lines – the respective ED_{50} values were determined as 6.04, 2.38, 4.06, and 4.01 µM for panaxynol and 10.4, 2.93, 4.05, and 3.94 µM for panaxydol.

Panaxynol isolated from *Angelica japonica* was also reported to display potent antiproliferative activity against the MK-1 cells with ED_{50} value of 0.3 µg/ml; besides, it exhibited moderate cytotoxicities against HeLa and B16F10 cells with respective ED_{50} values of 54.7 and 19.5 µg/ml [8].

References

[1] Takahashi, M. and Yoshikura, M. (1966) *Yakugaku Zasshi*, **86**, 1053.
[2] Poplawski, J., Wrobel, J.T., and Glinka, T. (1980) *Phytochemistry*, **19**, 1539.
[3] Kitagawa, I., Yoshikawa, M., Yoshihara, M., Hayashi, T., and Taniyama, T. (1983) *Yakugaku Zasshi*, **103**, 612.
[4] Matsunaga, H., Kitano, M., Yamamoto, H., Fujito, H., Mori, M., and Takata, K. (1990) *Chem. Pharm. Bull.*, **38**, 3480.

[5] Hirakura, K., Morita, M., Nakajima, K., Ikeya, Y., and Mitsuhashi, H. (1992) *Phytochemistry*, **31**, 899.
[6] Hirakura, K., Morita, M., Niitsu, K., Ikeya, Y., and Maruno, M. (1994) *Phytochemistry*, **35**, 963.
[7] Yang, M.C., Seo, D.S., Choi, S.U., Park, Y.H., and Lee, K.R. (2008) *Arch. Pharm. Res.*, **31**, 154.
[8] Fujioka, T., Furumi, K., Fujii, H., Kabe, H., Mihashi, K., Nakano, Y., Matsunaga, H., Katano, M., and Mori, M. (1999) *Chem. Pharm. Bull.*, **47**, 96.

Pancixanthone B

Physical data: $C_{18}H_{16}O_5$, yellow powder, $[\alpha]_D$ +3° (CHCl$_3$, c 0.117)

Structure:

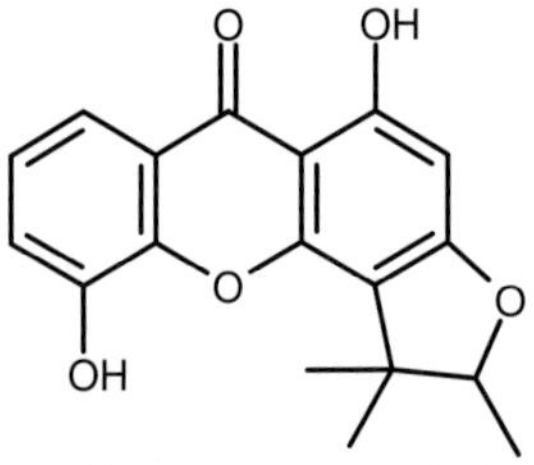

Pancixanthone B

Compound class: Xanthone

Sources: *Calophyllum pauciflorum* A.C. Smith (stem barks; family: Guttiferae) [1]; *Garcinia vieillardii* P. (stem barks; family: Guttiferae) [2]

Pharmaceutical potential: *Antileishmanial*
Hay *et al.* [2] evaluated the xanthone for its antileishmanial activity, and the test compound showed the activity against *Leishmania mexicana* and *L. infantum* with IC$_{50}$ values of 11.7 ± 3.6 and 3.2 ± 2.5 µg/ml, respectively; respective IC$_{50}$ values for amphotericin B and allopurinol were recorded as 0.03 ± 0.03 and 0.03 ± 0.01 and 21 ± 1 and 37 ± 2 µg/ml.

References

[1] Ito, C., Miyamoto, Y., Rao, K.S., and Furukawa, H. (1996) *Chem. Pharm. Bull.*, **44**, 441.
[2] Hay, A.-E., Merza, J., Landreau, A., Litaudon, M., Pagniez, F., Pape, P.L., and Richomme, P. (2008) *Fitoterapia*, **79**, 42.

Panduratin A

Systematic name: (1'RS, 2'SR, 6'RS) (2,6-Dihydroxyphenyl-4-methoxy)-[3'-methyl-2'-(3''-methylbut-2''-enyl)-6'-phenylcyclohex-3'-enyl] methanone

Physical data: $C_{26}H_{30}O_4$

Structure:

Panduratin A

Compound class: Cyclohexenyl chalcone derivative

Source: *Kaempferia pandurata* Roxb. (syn. *Boesenbergia pandurata*; rhizomes; family: Zingiberaceae) [1–9]

Pharmaceutical potentials: *Antioxidative; anti-inflammatory; cytotoxic; anticancerous; antimutagenic; antiaging; antibacterial*

Yun et al. [5] demonstrated that panduratin A effectively inhibits production of both nitric oxide (NO) and prostaglandin E_2 (PGE_2) induced by lipopolysaccharide (LPS) in RAW264.7 cells with respective IC_{50} values of 0.175 and 0.0195 µM; the test compound suppressed both inducible nitric oxide synthase (iNOS) and cyclooxygenase-2 (COX-2) enzyme expression without any appreciable cytotoxic effect on RAW264.7 cells in a dose-dependent manner. Panduratin A was also found to suppress the phosphorylation of inhibitor $\varkappa B\alpha$ ($I\varkappa B\alpha$) and degradation of $I\varkappa B\alpha$ associated with nuclear factor $\varkappa B$ (NF-$\varkappa B$) activation; hence, the compound was supposed to exert its strong inhibitory activity against LPS-induced production of NO and PGE_2 through the suppression of NF-$\varkappa B$ activation, thereby, indicating its potential for use as an anti-inflammatory chemotype [2, 5]. The same group of workers [6] investigated the effects of panduratin A on cytoplasmic levels of COX-2, as well as proliferation and apoptosis in human colon cancer cells HT-29 using MTT assay; it induced apoptosis in human colon cancer cells and displayed cytotoxicity with an IC_{50} value of 28 µM. The investigators also noted that treatment with an apoptosis-inducing concentration of panduratin A resulted in cleavage of poly (ADP-ribose) polymerase (PARP) with a concomitant decrease in procaspase-3 protein, thereby, suggesting its potential use as a cancer chemopreventive and therapeutic agent [6]. The chalcone derivative also evaluated for its protective effect against human androgen-independent prostate cancer cells, PC3 and DU145, and the compound was found to have potent antiproliferative effect against such cancer cells [7].

Panduratin A displayed significant protective effect on cytotoxicity induced by *tert*-Butyl-hydroperoxide (*t*-BHP) in human hepatoma (HepG2) cells as studied by Sohn et al. [8]; the protective effect of the compound was found to be associated with its antioxidative property [8, 11]. Panduratin A exhibited dose- and time-dependent protective effects against the overproduction of reactive oxygen species (ROS), lipid peroxidation, and disruption of intracellular antioxidant systems as stimulated by the *t*-BHP toxicity; at a dose of 10–15 µM, panduratin A showed a stronger reductive effect on ROS level than silybin, which was used as a positive control [8]. The chalcone derivative was also found to inhibit the growth of MCF-7 human breast cancer and HT-29 human colon adenocarcinoma cells with IC_{50} values of 3.75 and 6.56 µg/ml, respectively [9].

Very recently, Shim *et al.* [10] reported the effects of this compound on the expression of matrix metalloproteinase-1 (MMP-1) and type-1procollagen in UV-irradiated human skin fibroblasts; MMPs play a key role in the pathophysiological mechanism of phtoaging. Epigallocatechin 3-*O*-gallate (EGCG), which is well-known natural antiaging agent that effectively decreases the MMPs and increases the type-1 procollagen expression levels in the dermis due to its antioxidant effect [12, 13], was used as positive control in the experiment. Panduratin A was found to inhibit UV-induced MMP-1 expressions by 47% at 0.001 µM, 69% at 0.01 µM, and 87% at 0.1 µM compared to the UV-irradiated control (while EGCG inhibited MMP-1expression by 56% at 0.1 µM). Again, it enhanced the type-1 procollagen protein expression levels by 48% at 0.001 µM, 66% at 0.01 µM, and 74% at 0.1 µM compared with the UV-irradiated control, respectively, which was higher than 69% for EGCG at 0.1 µM. From their detailed experimental results, it was evident that the natural chalcone derivative significantly reduced the expression of MMP-1 and induced the expression of type-1 procollagen at the protein and also mRNA gene levels, thereby, suggesting it as a potential candidate for the prevention and treatment of skin aging [10].

Panduratin A was also reported to possess antibacterial activity against *Porphyromonas gingivalis* [4] including some antimicrobial activities also [14, 15].

References

[1] Tuntiwachwutiikul, P., Pancharoen, O., Reutrakul, U., and Byrne, L.T. (1984) *Aust. J. Chem.*, **37**, 449.
[2] Tuchinda, P., Reutrakul, V., Claeson, P., Ponprayoon, U., Sematong, T., Santosuk, T., and Taylor, W.C. (2002) *Phytochemistry*, **59**, 169.
[3] Pandji, C., Grimm, C., Wray, V., Witte, L., and Proksch, P. (1993) *Phytochemistry*, **34**, 415.
[4] Park, K.M., Choo, J.H., Sohn, J.H., Lee, S.H., and Hwang, J.K. (2005) *Food Sci. Biotechnol.*, **14**, 286.
[5] Yun, J.M., Kwon, H., and Hwang, J.K. (2003) *Planta Med.*, **69**, 1102.
[6] Yun, J.M., Kwon, H., Mukhtar, H., and Hwang, J.K. (2005) *Planta Med.*, **71**, 501.
[7] Yun, J.M., Kweon, M.H., Kwon, H.J., Hwang, J.K., and Mukhtar, H. (2006) *Carcinogenesis*, **27**, 1454.
[8] Sohn, J.H., Han, K.L., Lee, S.H., and Hwang, J.K. (2005) *Biol. Pharm. Bull.*, **28**, 1083.
[9] Chandra, K., Peter, J.G., Roland, R.I., and Howie, M.G. (2007) *J. Nat. Med.*, **61**, 131.
[10] Shim, J.-S., Kwon, Y.-Y., and Hwang, J.-K. (2008) *Planta Med.*, **74**, 239.
[11] Shindo, K., Kato, M., Kinoshita, A., Kobayashi, A., and Koike, Y. (2006) *Biosci. Biotechnol. Biochem.*, **70**, 2281.
[12] Demeule, M., Brossard, M., Page, M., Gingras, D., and Beliveau, R. (2000) *Biochem. Biophys. Acta*, **1478**, 51.
[13] Tobi, S.E., Gilbert, M., Paul, N., and McMillan, T.J. (2002) *Int. J. Cancer*, **102**, 439.
[14] Cheenpracha, S., Karalai, C., Ponglimanont, C., Subhadhirasakul, S., and Tewtrakul, S. (2006) *Bioorg. Med. Chem.*, **14**, 1710.

[15] Kiat, T.S., Pippen, R., Yusof, R., Ibrahim, H., Khalid, N., and Rahman, N.A. (2006) *Bioorg. Med. Chem. Lett.*, **12**, 3337.

Panicutine

Physical data: $C_{23}H_{29}NO_4$, colorless crystals

Structure:

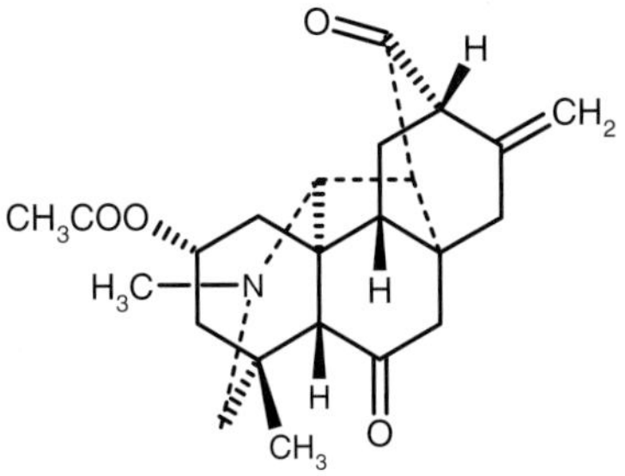

Panicutine

Compound class: Alkaloid (hetidine-type diterpenoid alkaloid)

Sources: *Aconitum paniculatum* Lam. *A. heterophylloides* (family: Ranunculaceae) [1–3]; *Delphinium denudatum* (roots; family: Ranunculaceae) [4]

Pharmaceutical potential: *Antifungal*
The alkaloid showed antifungal activities against some pathogenic fungi such as *Allescheria boydii*, *Aspergillus niger*, *Epidermophyton floccosum*, and *Pleurotus ostreatus* with respective MIC values of 75, 125, 200, and 125 μg/ml (using the agar diffusion tube method; nystatin and griseofulvin were used as positive controls) [4].

References

[1] Pelletier, S.W., Joshi, B.S., Desai, H.K., Al Panu, A., and Katz, A. (1986) *Heterocycles*, **24**, 1275.
[2] Katz, A. and Staehelin, E. (1982) *Helv. Chim. Acta*, **65**, 286.
[3] Pelletier, S.W., Mody, M.V., Finer-Moore, J., Desai, H.K., and Puri, H.S. (1981) *Tetrahedron Lett.*, **22**, 313.
[4] Atta-ur-Rahman, Nasreen, A., Akhtar, F., Shekhani, M.S., Clardy, J., Parvez, M., and Choudhary, M.I. (1997) *J. Nat. Prod.*, **60**, 472.

Papyriflavonol A

Systematic name: 3,5,7,3′,4′-Pentahydroxy-6,5′-di-(3-methyl-2-butenyl)flavone; alternatively, 5,7,3′,4′-tetrahydroxy-6,5′-di-(γ,γ-dimethylallyl)-flavonol

Physical data: $C_{25}H_{26}O_7$

Structure:

Papyriflavonol A

Compound class: Flavonoid (prenylated flavonol)

Source: *Broussonetia papyrifera* (root barks; family: Moraceae) [1, 2]

Pharmaceutical potentials: *Secretory phospholipase A_2s ($sPLA_2s$) inhibitor (anti-inflammatory); antimicrobial; cytotoxic; tyrosinase activity inhibitor*

Kwak et al. [2] evaluated that papyriflavonol A selectively inhibits recombinant human secretory phospholipase A_2s ($sPLA_2s$); the test compound displayed potent and irreversible inhibitory activity against human group IIA and V $sPLA_2s$ in a dose-dependent manner with respective IC_{50} values of 3.9 and 4.5 µM. Interestingly, the test compound exerted weak inhibitory effect toward bovine group IB ($IC_{50} = 76.9$ µM) and the human group X ($IC_{50} = 225$ µM) $sPLA_2s$ compared to the activity against human group IIA and V sPLA2s, while human group IIF $sPLA_2$ was not practically inhibited. Papyriflavonol A was also recently reported to show an inhibitory effect on 5-lipooxygenase, but not the cyclooxygenases (COXs), with an IC_{50} value of 7 µM. Papyriflavonol A was also found to inhibit the stimulus-induced production of leukotriene C_4 (LTC_4) with an IC_{50} value of about 0.64 µM in mouse bone marrow-derived mast cells [2]; in addition, the flavonoid derivative significantly retarded IgE-dependent passive cutaneous anaphylaxis reaction in rats. Hence, papyriflavonol A might provide a basis for novel types of anti-inflammatory drugs.

Papyriflavonol A exhibited a good antifungal activity with strong antibacterial activity against four bacterial and two fungal microorganisms (namely, *Candida albicans, Saccharomyces cerevisiae, Escherichia coli, Salmonella typhimurium, Staphylococcus epidermis*, and *S. aureus*); the test compound also showed moderate cytotoxicity against HepG2 cells with an IC_{50} value of 20.9 µg/ml [3]. Papyriflavonol A was also reported to inhibit tyrosinase enzyme activity by 16 and 39%, respectively, at 30 and 100 µM concentrations [4].

References

[1] Son, K.H., Kwon, S.J., Chang, H.W., Kim, H.P., and Kang, S.S. (2001) *Fitoterapia*, **72**, 456.
[2] Kwak, W.J., Moon, T.C., Lin, C.X., Rhyn, H.G., Jung, H., Lee, E., Kwon, D.Y., Son, K.H., Kim, H.P., Kang, S.S., Murakami, M., Kudo, I., and Chang, H.W. (2003) *Biol. Pharm. Bull.*, **26**, 299.
[3] Sohn, H.-Y., Son, K.H., Kwon, C.-S., Kwon, G.S., and Kang, S.S. (2004) *Phytomedicine*, **11**, 666.

[4] Lee, N.K., Son, K.H., Chang, H.W., Kang, S.S., Park, H., Heo, M.Y., and Kim, H.P. (2004) *Arch. Pharm. Res.*, **27**, 1332.

Pelagiomicin A

Physical data: $C_{20}H_{21}N_3O_6$, reddish orange needles, mp 130 °C, $[\alpha]_D^{20}$ +19.8° (CHCl$_3$, c 1.0)

Structure:

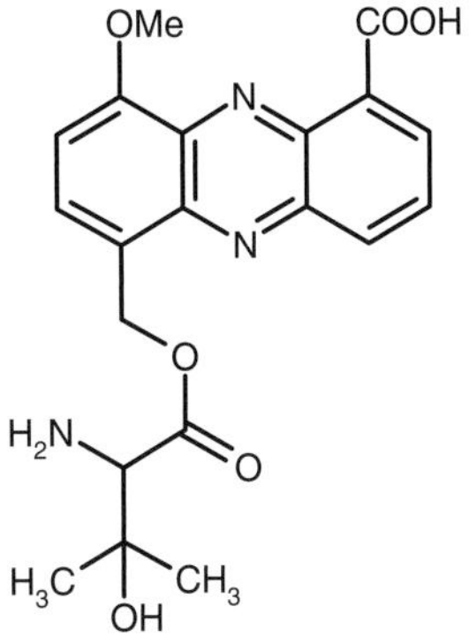

Pelagiomicin A

Compound class: Antibiotic

Source: *Pelagiobacter variabilis* (fermentation broth)

Pharmaceutical potentials: *Antimicrobial; cytotoxic*
The antibiotic exhibited strong activities against Gram-positive and Gram-negative bacteria but was not found to be active against yeast (*Candida albicans*) at a concentration of 87 µg/ml. The MIC values against the microbial strains *Staphylococcus aureus* ATCC 6538P, *Enterococcus hirae* ATCC 105f41, *Bacillus subtilis*, *Klebsiella pneumoniae* ATCC 1003, *Escherichia coli* ATCC26, *Pseudomonas aeruginosa*, *Salmonella choleraesuis* ATCC 9992, *Proteus vulgaris* ATCC 6897, and *Shigella sonnei* ATCC 9290 were determined to be 2.6, 0.16, 0.16, 0.16, 1.3, 5.2, 10, <0.04, and 1.3 µg/ml, respectively. The isolate also displayed cytotoxic activity against cultured cells such as HeLa, BALB3T3, and BALB3T3/H-ras with respective IC$_{50}$ values of 0.04, 0.2, and 0.07 µg/ml and weak antitumor activity against murine P388 leukemia *in vivo*.

Reference

Imamura, N., Nishijima, M., Takadera, T., Adachi, K., Sakai, M., and Sano, H. (1997) *J. Antibiot.*, **50**, 8.

Penicillide

See AS-186a

6-*n*-Pentyl-α-pyrone

Physical data: $C_{10}H_{14}O_2$

Structure:

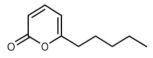

6-*n*-Pentyl-α-pyrone

Compound class: α-Pyrone derivative

Sources: *Myrothecium* sp. (marine-derived fungus) [1]; *Trichoderma koningii* (fungus; family: Hypocreaceae) [2]

Pharmaceutical potentials: *Tyrosinase inhibitory activity; antibiotic*
Li et al. [1] evaluated the α-pyrone derivative as a promising tyrosinase enzyme inhibitory agent; this cyclopentenone derivative was found to be more active ($ED_{50} = 0.8\,\mu M$) than kojic acid ($ED_{50} = 7.7\,\mu M$) currently being used as a functional personal care compound. The compound was also reported to have antibiotic property [2].

Tyrosinase, a key enzyme involved in the metabolism of melanin in melanocytes, is associated with skin coloring and local hyperpigmentation such as melasma, ephelis, and lentigo [3, 4]. Hence, tyrosinase inhibitory compounds find immense applications in formulating functional personal care products for skin-whitening effects and for preventive and therapeutic effects on the local hyperpigmentation diseases.

References

[1] Li, X., Kim, M.K., Lee, U., Kim, S.-K., Kang, J.S., Choi, H.D., and Son, B.W. (2005) *Chem. Pharm. Bull.*, **53**, 453.
[2] Simon, A., Dunlop, R.W., Ghisalberti, E.L., and Sivasithamparam, K. (1988) *Soil Biol. Biochem.*, **20**, 263.
[3] Shin, N.H., Lee, K.S., Kang, S.H., Min, K.R., Lee, S.H., and Kim, Y. (1997) *Nat. Prod. Sci.*, **3**, 111.
[4] Iwata, M., Corn, T., Iwata, S., Everett, M.A., and Fuller, B.B. (1990) *J. Invest. Dermatol.*, **95**, 9.

Perforamone B

Systematic name: 5-Hydroxy-7-methoxy-2-methyl-8-(1-hydroxy-3-methyl-3-butenyl)chromone

Physical data: $C_{16}H_{18}O_5$, colorless needles, mp 162–163 °C, $[\alpha]_D^{25}$ −18.2° (MeOH, *c* 0.12)

Structure:

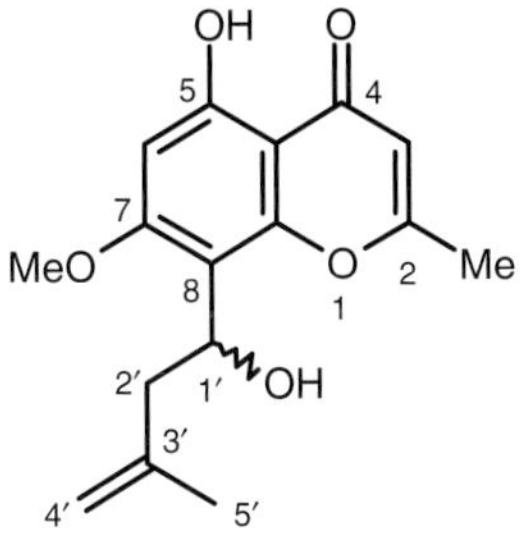

Perforamone B

Compound class: Chromone

Source: *Harrisonia perforata* (root barks; family: Simaroubaceae)

Pharmaceutical potentials: *Antiplasmodial; antimycobacterial*
The chromone isolate displayed significant *in vitro* antimalarial activity against *Plasmodium falciparum* (K1, multidrug resistant strain) with an EC_{50} value of 10.5 μg/ml; the test compound also exhibited weak antimycobacterial activity against *Mycobacterium tuberculosis* H37Ra with an EC_{50} value of 100 μg/ml.

Reference

Tuntiwachwuttikul, P., Phansa, P., Pootaeng-on, Y., and Taylor, W.C. (2006) *Chem. Pharm. Bull.*, **54**, 44.

Periglaucines A–D

Systematic names:
- Periglaucine A: (7α,8α,10β)-8,10-epoxy-7,8-dimethoxy-2,3-[methylenebis(oxy)]-17-methyl-6-oxohasubanan
- Periglaucine B: (7β,8β,10β)-8,10-epoxy-7,8-dimethoxy-2,3-[methylenebis(oxy)]-17-methyl-6-oxohasubanan
- Periglaucine C: (7α,8α,10β)-8,10-epoxy-7,8-dimethoxy-2,3-[methylenebis(oxy)]-17-methyl-6,16-dioxohasubanan
- Periglaucine D: (7β,8β,10β)-8,10-epoxy-7,8-dimethoxy-2,3-[methylenebis(oxy)]-16-propan-2-one-17-methyl-6-oxohasubanan

Physical data:
- Periglaucine A: $C_{20}H_{23}NO_6$, colorless crystals (acetone), mp 197–199 °C, $[\alpha]_D^{15}$ +78° (CHCl$_3$, *c* 0.55)
- Periglaucine B: $C_{20}H_{23}NO_6$, white powder, $[\alpha]_D^{22}$ +90.1° (CHCl$_3$, *c* 0.37)
- Periglaucine C: $C_{20}H_{21}NO_7$, white powder, $[\alpha]_D^{24}$ +131.7° (MeOH, *c* 0.12)
- Periglaucine D: $C_{23}H_{27}NO_7$, white powder, $[\alpha]_D^{24}$ +76.1° (MeOH, *c* 0.2)

Structures:

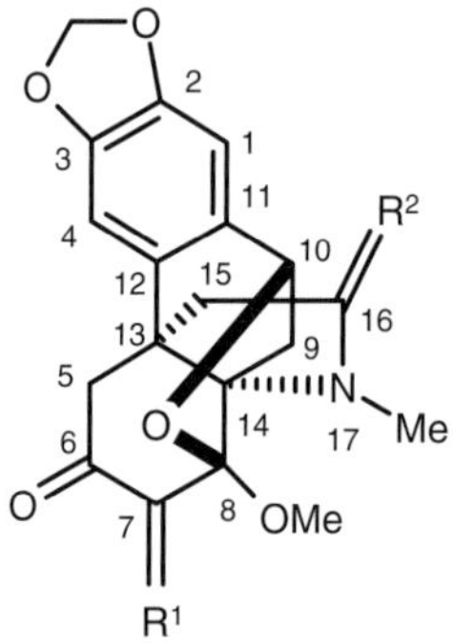

Periglaucine A: R¹ = H₂; R² = α-OCH₃ β-H
Periglaucine B: R¹ = H₂; R² = β-OCH₃ α-H
Periglaucine C: R¹ = O; R² = α-OCH₃ β-H

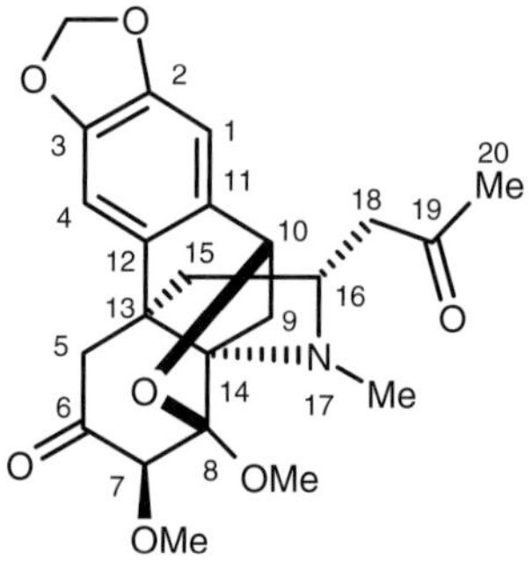

Periglaucine D

Compound class: Alkaloids (hasubanane type)

Source: *Pericampylus glaucus* (Lam.) Merr. (aerial parts; family: Menispermaceae)

Pharmaceutical potentials: *Antihepatitis B virus (HBV); anti-HIV*

Periglaucines A–D were evaluated for their antihepatitis B virus (HBV) activity in the HBV-transfected HepG2.2.15 cells; the alkaloids inhibited hepatitis B virus (HBV) surface antigen (HBsAg) secretion in the cell line. The respective IC_{50} values for periglaucines A–D were determined to be 1.04, 0.47, 1.72, 0.67, and 0.93 mM.

In addition, the hasubanane-type alkaloids displayed weak anti-HIV-1 activity in the syncytium assay; the respective EC_{50} values for periglaucines A–D were evaluated as 204.0, 388.6, 162.5, and 334.1 μM.

Reference

Yan, M.-H., Cheng, P., Jiang, Z.-Y., Ma, Y.-B., Zhang, X.-M., Zhang, F.-X., Yang, L.-M., Zheng, Y.-T., and Chen, J.-J. (2008) *J. Nat. Prod.*, **71**, 760.

Persenones A and B

Systematic names:
- Persenone A: (12Z,15Z)-2-hydroxy-4-oxoheneicosa-5,12,15-trien-1-yl acetate
- Persenone B: 2-hydroxy-4-oxononadeca-5-en-1-yl acetate

Physical data:
- Persenone A: $C_{23}H_{38}O_4$, colorless oil, $[\alpha]_D^{22}$ +17.6° (CHCl₃, *c* 0.48)
- Persenone B: $C_{21}H_{38}O_4$, colorless oil, $[\alpha]_D^{22}$ +8.3° (CHCl₃, *c* 0.33)

Structures:

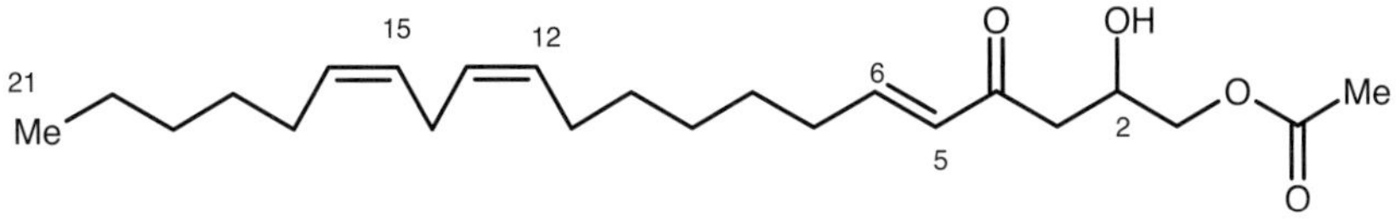

Persenone A

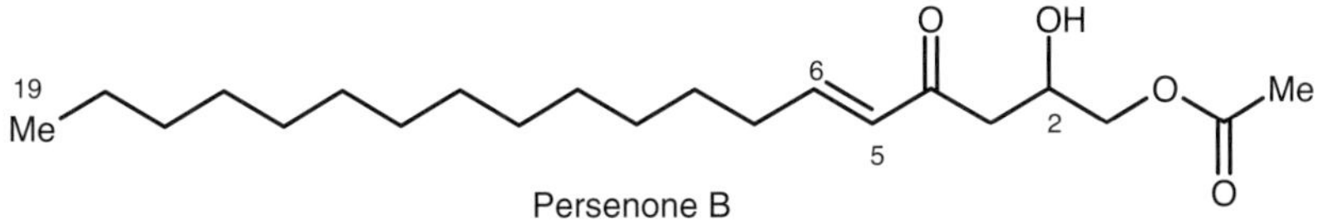

Persenone B

Compound class: Long-chain esters

Source: *Persea americana* P. Mill (fruits; family: Lauraceae)

Pharmaceutical potential: *Antioxidants (inhibitors to the production of NO and superoxide)*
Persenones A and B were evaluated as unique dual inhibitors of superoxide (O_2^-) and nitric oxide (NO) generation in inflammatory leukocytes. Their inhibitory potencies (IC_{50} values: persenone A, 1.2 µM; persenone B, 3.5 µM) against NO generation induced by lipopolysaccharide in combination with interferon-γ in mouse macrophage RAW 264.7 cells were measured to be higher than that of a natural NO generation inhibitor, docosahexaenoic acid (DHA; IC_{50} = 4.3 µM). The respective IC_{50} values for persenones A and B against TPA (12-O-tetradecanoylphorbol-13-acetate)-induced O_2^- production in differentiated human promyelocytic HL-60 cells were determined to be 1.4 and 1.8 µM. The present investigators pointed out that these compounds are the suppressors of both superoxide (O_2^-) and nitric oxide (NO)-generating biochemical pathways but are not the radical scavengers. Hence, these compounds are unique antioxidants and are expected to be notable cancer chemopreventive agents in inflammation-associated organs, including the stomach and colon.

Reference

Kim, O.K., Murakami, A., Nakamura, Y., Takeda, N., Yoshizumi, H., and Ohigashi, H. (2000) *J. Agric. Food Chem.*, **48**, 1557.

Pestalotheol C

Physical data: $C_{16}H_{24}O_4$, colorless powder, $[α]_D$ +230° (MeOH, *c* 0.6)

Structure:

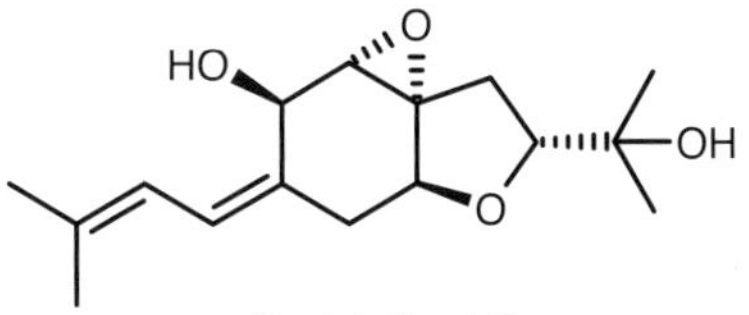

Pestalotheol C

Petasiformin A

Compound class: Chromenone type of metabolite

Source: *Pestalotiopsis theae* (plant endophytic fungus; culture broth)

Pharmaceutical potential: *Anti-HIV*
Pestalotheol C showed moderate *in vitro* inhibitory effect on HIV-1$_{LAI}$ replication in C8166 cells with an EC$_{50}$ value of 16.1 µM.

Reference

Li, E., Tian, R., Liu, S., Chen, X., Guo, L., and Che, Y. (2008) *J. Nat. Prod.*, **71**, 664.

Petasiformin A

Systematic name: 4-*O*-Sulfonyl-3-prenyl-*p*-coumaric acid

Physical data: C$_{14}$H$_{16}$ SO$_6$, white powder, mp >300 °C

Structure:

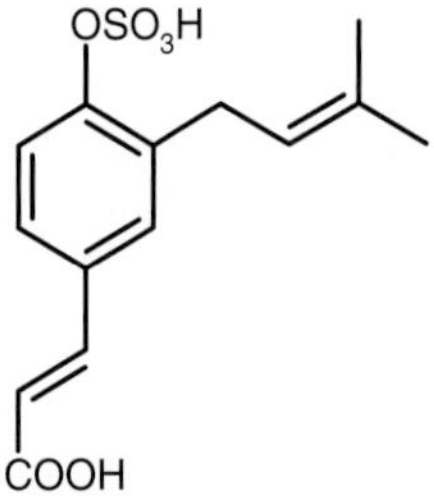

Petasiformin A

Compound class: Phenylpropenoyl sulfonic acid

Source: *Petasites formosanus* Kitamura (leaves; family: Compositae)

Pharmaceutical potential: *Antioxidant*
The test compound exhibited significant antioxidant activity as evaluated in the DPPH free radical scavenging assay; it showed free radical scavenging activity with an IC$_{50}$ value of 0.21 mg/ml (α-tocopherol was used as a positive control; IC$_{50}$ = 0.15 mg/ml).

Reference

Lin, C.-H., Li, C.-Y., and Wu, T.-S. (2004) *Chem. Pharm. Bull.*, **52**, 1151.

Peucedanone

Systematic name: 7-Hydroxy-6-(3-hydroxy-3-methyl-2-oxobutyl)coumarin

Physical data: C$_{14}$H$_{14}$O$_5$, pale yellow prisms (CHCl$_3$), [α]$_D$ +4.8° (CHCl$_3$–MeOH 1:4, *c* 0.5)

Structure:

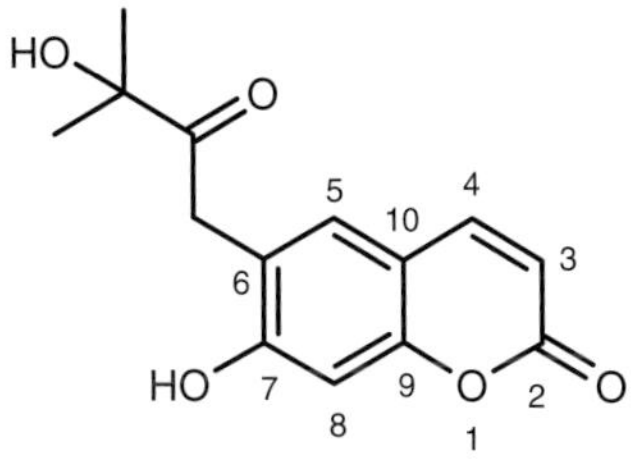

Peucedanone

Compound class: Coumarin

Source: *Angelica gigas* Nakai (roots; family: Umbelliferae)

Pharmaceutical potential: *Acetylcholinesterase (AChE) inhibitor*
The isolate was found to possess weak *in vitro* inhibitory activity against acetylcholinesterase (AChE) enzyme with an IC$_{50}$ value of 1.8×10^{-4} M.

Reference

Kang, S.Y., Lee, K.Y., Sung, S.H., Park, M.J., and Kim, Y.C. (2001) *J. Nat. Prod.*, **64**, 683.

PF 1163A and B

Physical data:
- PF 1163A: C$_{27}$H$_{45}$NO$_6$, colorless oil, $[\alpha]_D^{25}$ −91.84° (MeOH, *c* 1.0)
- PF 1163B: C$_{27}$H$_{43}$NO$_5$, colorless oil, $[\alpha]_D^{25}$ −111.59° (MeOH, *c* 1.2)

Structures:

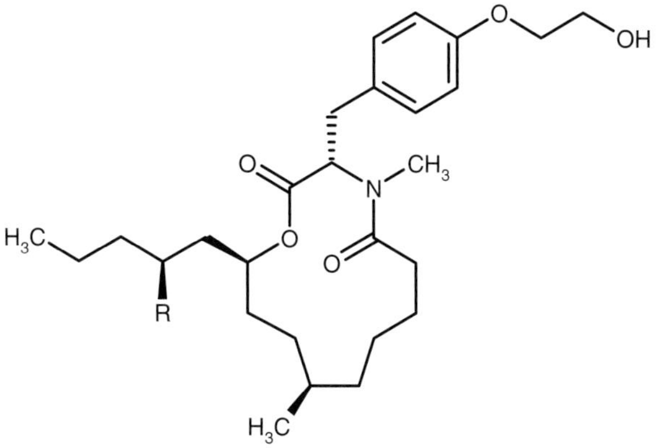

PF 1163A: R = OH
PF 1163B: R = H

Compound class: Antibiotics

Source: *Penicillium* sp. PF1163 (fermentation broth) [1, 2]

Pharmaceutical potential: *Antifungal*
PF 1163A and B were found to display potent growth inhibitory activity against the pathogenic fungal strain *Candida albicans* TIMM1768 with respective MIC values of 8 and 32 µg/ml; however, these compounds were inactive against *Aspergillus fumigatus*. Interestingly, both of them did not show cytotoxic activity against mammalian cells as evaluated against HepG2 human hepatoblastoma cell line – both the isolates did not show 50% growth inhibition against the cell line even at a concentration of 33.3 µg/ml. The investigators [1] established that the test compounds showed antifungal activity against *Candida albicans* by inhibiting the ergosterol biosynthesis in the organism; PF 1163A and B were proved to inhibit the biosynthetic pathway from lanosterol to ergosterol by the inhibitory assay of ergosterol synthesis. PF 1163A was found as potent as fluconazole (reference standard) in inhibiting the synthesis of ergosterol from [^{14}C]acetate – the IC$_{50}$ values for PF 1163A and B were evaluated as 12 and 34 ng/ml, respectively.

References

[1] Nose, H., Seki, A., Yaguchi, T., Hosoya, A., Sasaki, T., Hoshiko, S., and Shomura, T. (2000) *J. Antibioti.*, **53**, 33.
[2] Sasaki, T., Nose, H., Hosoya, A., Yoshida, S., Kawaguchi, M., Watanabe, T., Usui, T., Ohtsuka, Y., and Shomura, T. (2000) *J. Antibiot.*, **53**, 38.

Phakellistatin 14

Physical data: $C_{36}H_{53}N_7O_{10}S$, colorless solid, mp 189–191 °C, $[\alpha]_D^{27}$ −64.86° (MeOH, *c* 0.28)

Structure:

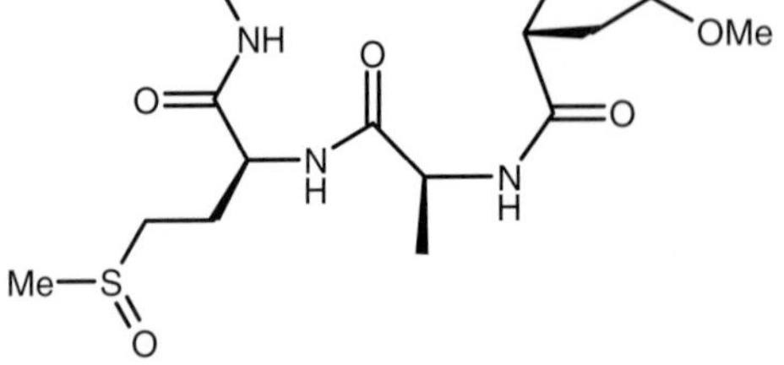

Phakellistatin **14**

Compound class: Cyclic peptide

Source: *Phakellia* sp. (marine sponge)

Pharmaceutical potential: *Cytotoxic*
Phakellistatin 14 displayed moderate *in vitro* cytotoxic activity against the murine P388 lymphocytic leukemia cells with an IC_{50} of 5.0 μg/ml; the isolate also showed moderate cancer cell growth inhibitory activity against a panel of human cancer cells ($GI_{50} = 0.75$–3.4 μg/ml).

Reference

Pettit, G.R. and Tan, R. (2005) *J. Nat. Prod.*, **68**, 60.

Phellamurin

Systematic name: 3,5,7,4′-Tetrahydroxy-8-(3-methylbut-2-enyl)flavanone-7-*O*-β-glucoside

Physical data: $C_{26}H_{30}O_{11}$, pale yellow amorphous powder

Structure:

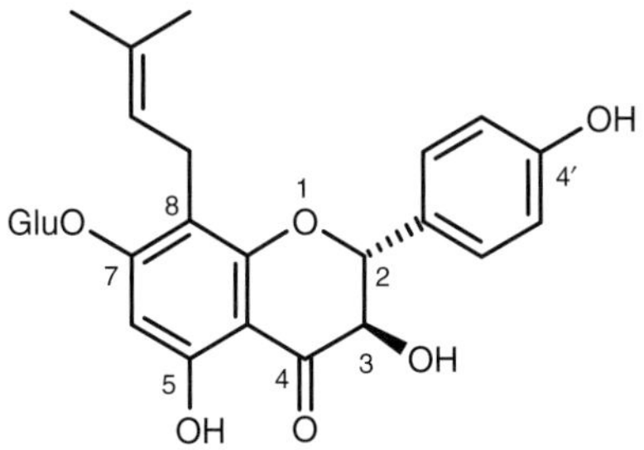

Glu = β-D-Glucopyranosyl
Phellamurin

Compound class: Flavonoid (prenylated dihydroflavanol glycoside)

Sources: *Commiphora africana* (stem wood; family: Burseraceae) [1]; *Phellodendron amurense* (family: Rutaceae) [2, 3]; *Phellodendron wilsonii* Hayata et Kanehira (leaves) [4]

Pharmaceutical potential: *DNA strand cleaving agent*
Ma *et al.* [1] evaluated that phellamurin induced *in vitro* DNA strand scission in the presence of 20 μM Cu^{2+} in a concentration-dependent manner; it was observed that ~62% conversion of DNA from supercoiled form to nicked circular form takes place at 50 μM concentration. The DNA cleaving activity of the compound was also detectable at a very low concentration of 1 μM, while at a very high concentration of 100 μM a little amount of linear duplex form of DNA was formed in addition to nicked circular form.

Chen et al. [4] demonstrated from their *in vitro* everted rat intestine study that phellamurin inhibits intestinal P-glycoprotein in a dose-dependent manner; it was also established from their *in vivo* study in the rat model that the flavonoid compound has a negative effect on cyclosporin absorption and disposition – coadministration of phellamurin with cyclosporin significantly decreased the absorption of the drug, arising out of a serious interaction between these two molecules; hence, to ensure the efficacy of cyclosporin, coadministration of phellamurin or *Phellodendron wilsonii* with cyclosporin should be avoided [4].

References

[1] Ma, J., Jones, S.H., and Hecht, S.M. (2005) *J. Nat. Prod.*, **68**, 115.
[2] Sakai, S. and Hasegawa, M. (1974) *Phytochemistry*, **13**, 303.
[3] Honda, K. and Hayashi, N. (1995) *J. Chem. Ecol.*, **21**, 1531.
[4] Chen, H.Y., Wu, T.S., Su, S.F., Kuo, S.C., Chao, P.D.L. (2002) *Planta Med.*, **68**, 138.

Phelligridimer A

Physical data: $C_{52}H_{32}O_{20}$, yellow amorphous powder, mp >205 °C, $[\alpha]_D^{20}$ −2° (DMSO, *c* 0.1)

Structure:

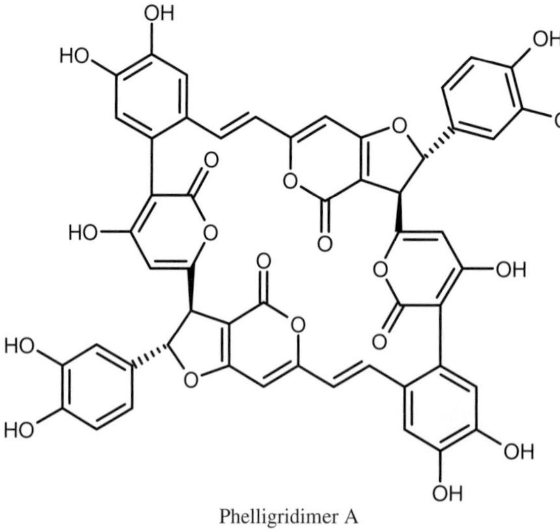

Phelligridimer A

Compound class: Pyranobenzopyranone derivative (dimeric form)

Source: *Phellinus igniarius* (DC. ex Fr.) Quel (fungus)

Pharmaceutical potential: *Antioxidant*

The dimeric metabolite displayed antioxidant activity inhibiting rat liver microsomal lipid peroxidation with an IC_{50} value of 10.2 µM, but was inactive to several human cancer cell lines tested.

Reference

Wang, Y., Wang, S.-J., Mo, S.-Y., Li, S., Yang, Y.-C., and Shi, J.-G. (2005) *Org. Lett.*, 7, 4733.

Phelligridin G

Systematic name: 8,9-Dihydroxy-3-{5′,6′-dihydroxy-5″-(*trans*-5″,6″-dihydroxystyryl)-3″-oxo-spiro [furan-2″ (3″*H*),1′-inden]-2′-yl}-1*H*,6*H*-pyrano[4,3-*c*][2]benzopyran-1,6-dione

Physical data: $C_{32}H_{18}O_{12}$, orange amorphous powder, mp >300 °C, $[\alpha]_D^{30}$ +75.9° ($CHCl_3$, *c* 1.40)

Structure:

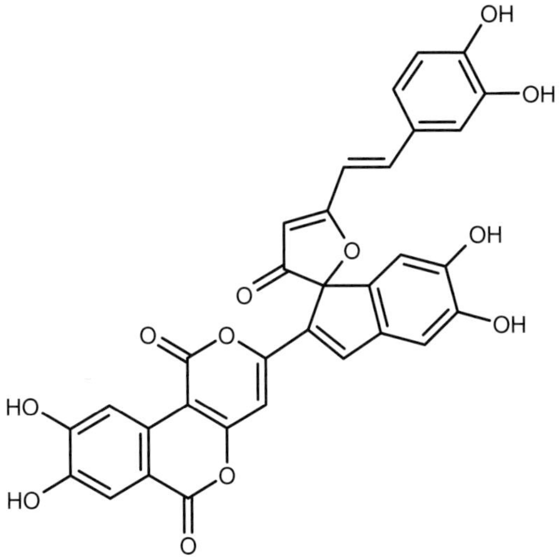

Phelligridin G

Compound class: Pyranobenzopyranone derivative

Source: *Phellinus igniarius* (DC. ex Fr.) Quel. (fungus; family: Polyporaceae)

Pharmaceutical potentials: ***Antioxidant; cytotoxic***
Phelligridin G exhibited antioxidant activity inhibiting rat liver microsomal lipid peroxidation with an IC_{50} value of 3.86 μM and moderate selective cytotoxic activities against human ovary cancer cell line (A2780) and human colon cancer cell line (HCT-8) with IC_{50} values of 20.4 and 30.2 μM, respectively.

Reference

Wang, Y., Mo, S.-Y., Wang, S.-J., Li, S., Yang, Y.-C., and Shi, J.-G. (2005) *Org. Lett.*, 7, 1675.

Phomacins A–C

Physical data:
- Phomacin A: $C_{25}H_{37}NO_5$, white powder, mp 130–133 °C, $[\alpha]$ −91° (CHCl$_3$, c 1.0)
- Phomacin B: $C_{25}H_{37}NO_5$, white powder, mp 98–100 °C, $[\alpha]$ −51.2° (CHCl$_3$, c 1.0)
- Phomacin C: $C_{25}H_{37}NO_4$, white powder, mp 112–114 °C, $[\alpha]$ −74.6° (CHCl$_3$, c 1.0)

Structures:

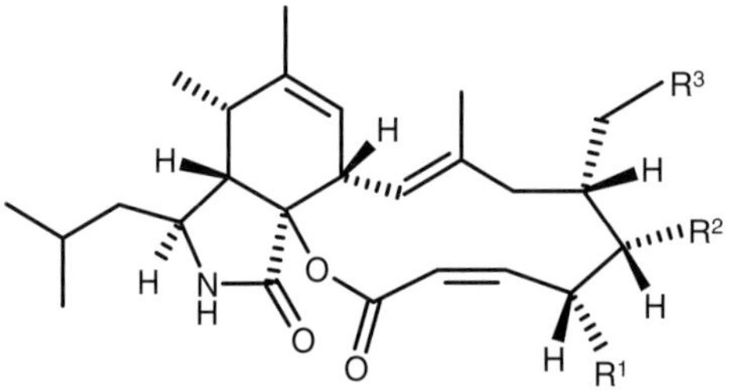

Phomacin A: R^1 = R^2 = OH; R^3 = H
Phomacin B: R^1 = R^3 = OH; R^2 = H

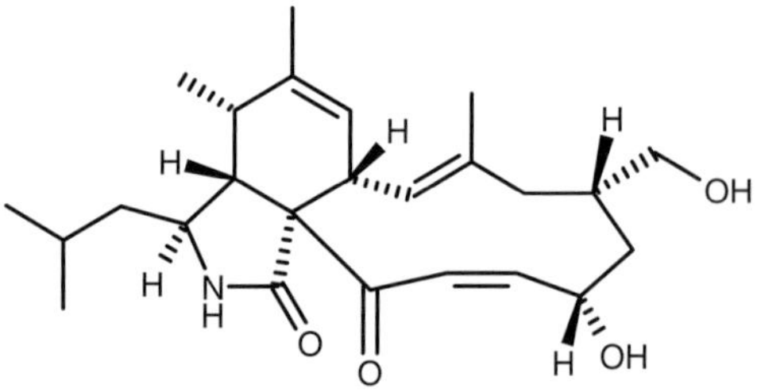

Phomacin C

Compound class: Antibiotics (cytochalasan derivatives)

Source: *Phoma* sp. (fungus; fermentation broth)

Pharmaceutical potential: *Antitumor*
All three isolates were evaluated to possess potent *in vitro* inhibitory activity against HT-29 colonic adenocarcinoma cells; phomacins A–C inhibited the growth of the adenocarcinoma cell line with respective IC$_{50}$ values of 0.6, 1.4, and 7.4 µg/ml, as determined in ^{3}H-thymidine incorporation (^{3}H-TdR) assay. In alamar blue assay, phomacins A and B were found to be cytototoxic (with respective IC$_{50}$ values of 17.4 and 10.1 µg/ml), while the other was nontoxic. Phomacin C, which exhibited potent inhibition of cellular proliferation as monitored by (^{3}H)-thymidine incorportion but remained nontoxic to the cells as determined by the alamar blue assay, might be an attractive candidate as an antitumor compound.

Reference

Alvi, K.A., Nair, B., Pu, H., Ursino, R., Gallo, C., and Mocek, U. (1997) *J. Org. Chem.*, **62**, 2148.

Phomoxanthones A and B

Physical data:
- Phomoxanthone A: $C_{38}H_{38}O_{16}$, yellow powder, mp 214–216 °C, $[\alpha]_D^{25}$ +99° (CHCl$_3$, c 0.4)

- Phomoxanthone B: $C_{38}H_{38}O_{16}$, yellow amorphous solid, mp 119–122 °C, $[\alpha]_D^{26}$ −120° (CHCl$_3$, c 0.15)

Structures:

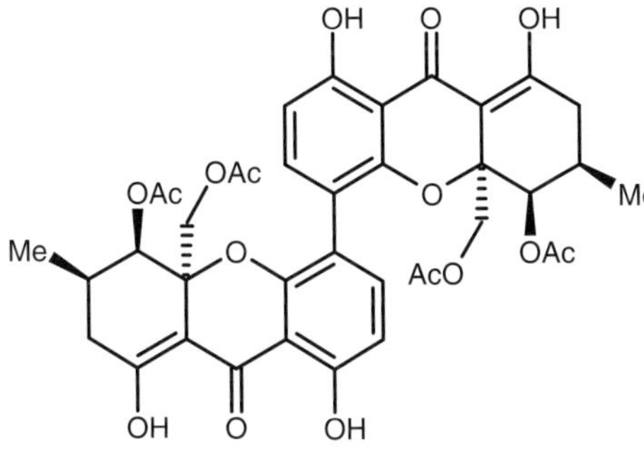

Phomoxanthone A

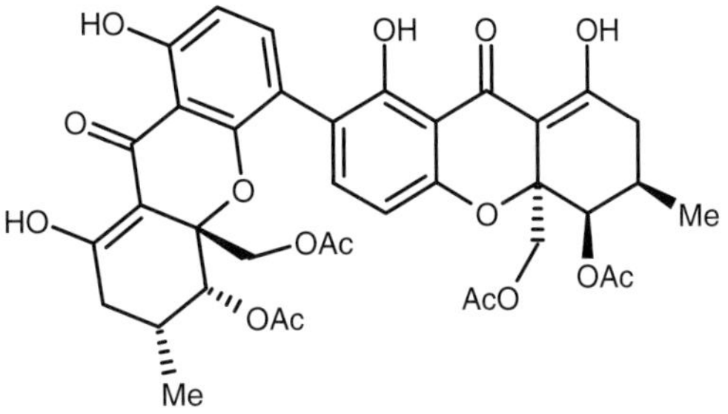

Phomoxanthone B

Compound class: Bixanthonolignoids

Source: *Phomopsis* sp. (endophytic fungus)

Pharmaceutical potentials: *Antimalarial; antitubercular; cytotoxic*

Phomoxanthones A and B, isolated from the endophytic fungus *Phomopsis* sp., exhibited significant *in vitro* antimalarial potential against *Plasmodium falciparum* (K1, multidrug-resistant strain) with IC$_{50}$ values of 0.11 and 0.33 μg/ml, respectively (IC$_{50}$ values for the reference standards: chloroquine diphosphate, 0.16 μg/ml; artemisinin, 0.0011 μg/ml). The compounds also displayed inhibitory activity against *Mycobacterium tuberculosis* H37Ra with respective MIC values of 0.50 and 6.25 mg/ml (MIC values for the reference standards: isoniazid, 0.05 μg/ml; kanamycin sulfate, 2.5 μg/ml). The bixanthonolignoids also showed moderate cytotoxicity against the KB cells, BC-1 cells, and Vero cells – the respective IC$_{50}$ values were determined to be 0.99, 0.51, and 1.4 μg/ml (phomoxanthone A); 4.1, 0.70, and 1.8 μg/ml (phomoxanthone B) (the IC$_{50}$ values measured for the reference standard ellipticine were 0.46 μg/ml against KB cells and 0.60 μg/ml against BC-1 cells).

Reference

Isaka, M., Jaturapat, A., Rukseree, K., Danwisetkanjana, K., Tanticharoen, M., and Thebtaranonth, Y. (2001) *J. Nat. Prod.*, **64**, 1015.

Phosphatoquinones A and B

Physical data:
- Phosphatoquinone A: $C_{21}H_{24}O_5$, brown oil, $[\alpha]_D^{25}$ −284° (CHCl$_3$, c 0.1)
- Phosphatoquinone B: $C_{21}H_{24}O_4$, reddish brown amorphous solid, $[\alpha]_D^{25}$ +4° (CHCl$_3$, c 0.1)

Phoyunbenes A–D

Structures:

Phosphatoquinone A

Phosphatoquinone B

Compound class: Antibiotics (naphthoquinone derivatives)

Source: *Streptomyces* sp. TC-0363 (culture broth) [1]

Pharmaceutical potential: *Protein tyrosine phosphatase (PTPase) inhibitors*
It has been demonstrated that protein tyrosine phosphatases (PTPase) and dual specific phosphatases (DSPase) are the key enzymes in signal transduction pathway for a wide range of cellular processes, usually responsible for various cancerous evolutions [2–5]. Phosphatoquinones A and B showed PTPase inhibitory activities with IC_{50} values of 28.0 and 2.9 µM, respectively, and appeared to be more potent than a well-known PTPase inhibitor, sodium vanadate; however, they did not inhibit serine/threonine protein phosphatases, PP1and PP2A [1].

References

[1] Kagamizono, T., Hamaguchi, T., Anado, T., Sugawara, K., Adachi, T., and Osada, H. (1999) *J. Antibiot.*, **52**, 75.
[2] Walton, K.M. and Dixson, J.E. (1993) *Ann. Rev. Biochem.*, **62**, 101.
[3] Honda, R., Ohba, Y., Nagata, A., Okayama, H., and Yasuda, H. (1993) *FEBS Lett.*, **318**, 331.
[4] Galaktionov, K., Lee, A.K., Eckstein, J., Draetta, G., Meckler, J., Lods, M., and Beach, D. (1995) *Science*, **29**, 1575.
[5] Gasparotto, D., Maestro, R., Piccinin, S., Vukosavljevic, T., Barzan, L., and Sulfaro, S. (1997) *Cancer Res.*, **57**, 2366.

Phoyunbenes A–D

Systematic names:
- Phoyunbene A: *trans*-3,3'-dihydroxy-2',4',5-trimethoxystilbene
- Phoyunbene B: *trans*-3,4'-dihydroxy-2',3',5-trimethoxystilbene
- Phoyunbene C: *trans*-3,3'-dihydroxy-2',5-dimethoxystilbene
- Phoyunbene D: *trans*-3-hydroxy-2',3',5-trimethoxystilbene

Physical data:
- Phoyunbene A: $C_{17}H_{18}O_5$, white needles, mp 170–172 °C
- Phoyunbene B: $C_{17}H_{18}O_5$, oil
- Phoyunbene C: $C_{16}H_{16}O_4$, oil
- Phoyunbene D: $C_{17}H_{18}O_4$, yellow prisms, mp 128–129 °C

Structures:

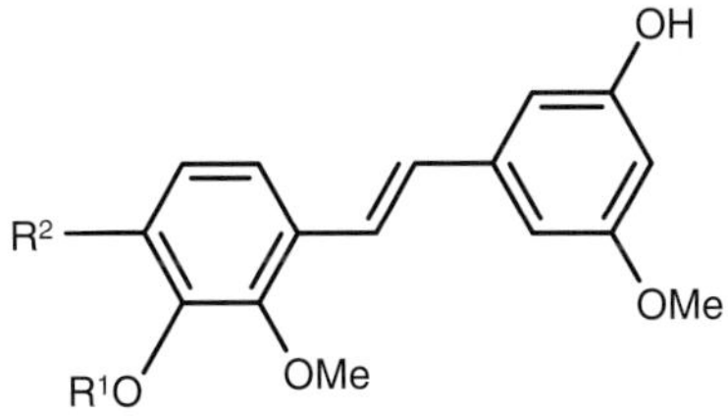

Phoyunbene A: $R^1 = H$; $R^2 = OMe$
Phoyunbene B: $R^1 = Me$; $R^2 = OH$
Phoyunbene C: $R^1 = R^2 = H$
Phoyunbene D: $R^1 = Me$; $R^2 = H$

Compound class: Stilbenes

Source: *Pholidota yunnanensis* Rolfe (whole plants; family: Orchidaceae)

Pharmaceutical potential: *Antioxidant (inhibitor to NO production)*
Phoyunbenes A–D were evaluated for their inhibitory effects on nitric oxide (NO) production in a murine macrophage-like cell line (RAW 264.7) activated by lipopolysaccharide (LPS) and interferon-γ (IFN-γ); all of them exhibited inhibitory effects on NO production without cytotoxicity with IC_{50} values of 32.9, 7.5, 49.0, and 87.3 μM, respectively. Interestingly, phoyunbene B showed much stronger efficacy than resveratrol ($IC_{50} = 29.8$ μM) used as a positive control in this assay.

Reference

Guo, X.-Y., Wang, J., Wang, N.-L., Kitanaka, S., Liu, H.-W., and Yao, X.-S. (2006) *Chem. Pharm. Bull.*, **54**, 21.

Pipercyclobutanamide A

Physical data: $C_{34}H_{38}N_2O_6$, colorless amorphous powder

Structure:

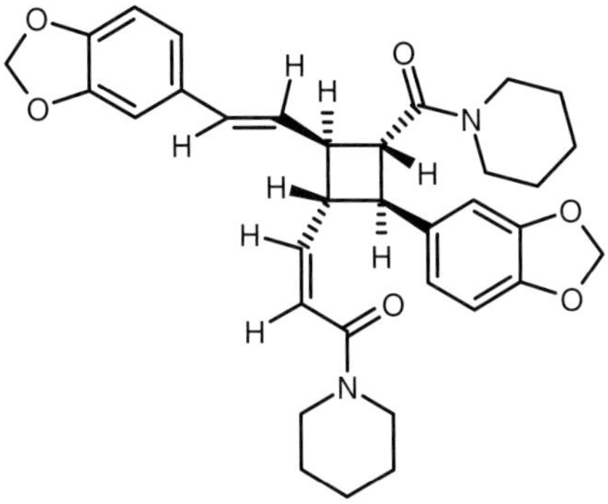

Pipercyclobutanamide A

Compound class: Alkaloid

Source: *Piper nigrum* (fruits [1]; MeOH extract of the plant [2]; family: Piperaceae)

Pharmaceutical potential: *CYP2D6 inhibitory*
The alkaloid was reported to show potent inhibition of human liver microsomal dextromethorphan O-demethylation activity, a selective marker of cytochrome P450 2D6 (CYP2D6); the IC_{50} value of the compound in the metabolism mediated by CYP2D6 was determined to be 0.34 µM (IC_{50} value for the positive control quinidine was observed to be 0.068 µM) [2].

References

[1] Fujiwara, Y., Naithou, K., Miyazaki, T., Hashimoto, K., Mori, K., and Yamamoto, Y. (2001) *Tetrahedron Lett.*, **42**, 2497.
[2] Subehan, Usia, T., Kadota, S., and Tezuka, Y. (2006) *Nat. Prod. Commun.*, **1**, 1.

Piptamine

Physical data: $C_{23}H_{41}N$, waxy solid

Structure:

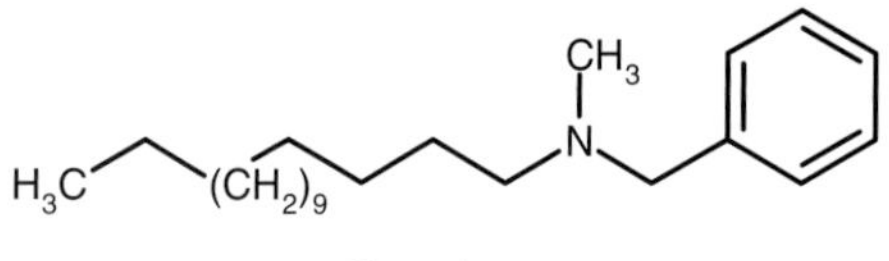

Piptamine

Compound class: Antibiotic

Source: *Piptoporus betulinus* Lu 9-1 (mushroom; malt agar slant culture)

Pharmaceutical potential: *Antimicrobial*
The isolate displayed antimicrobial activity against a number of microbial species such as *Staphylococcus aureus* 134/94, *S. aureus* SG 511, *Enterococcus faecalis* 1528, *Bacillus subtilis* ATCC6633, *Escherichia coli* SG 458, *Kluyveromyces marxianus* IMET 25148, *Candida albicans* BMSY212, and *Sporobolomyces salmonicolor* SBUG549 with respective MIC values of 6.25, 0.78, 1.56, 1.00, 12.5, 6.25, 6.25, and 6.25 µg/ml. The present investigators also determined hemolytic activity of the antibiotic at a dose of 10–50 µg/ml using heparinized blood of Beagle dogs.

Reference

Schlegel, B., Luhmann, U., Hart, A., and Grafe, U. (2000) *J. Antibiot.*, **53**, 973.

Plastoquinones 1 and 2

Physical data:
- Plastoquinone 1: $C_{27}H_{42}O_4$, pale yellowish oil, $[\alpha]_D^{25}$ +10.3° ($CHCl_3$, *c* 0.53)

- Plastoquinone **2**: $C_{27}H_{40}O_4$, pale yellowish oil, $[\alpha]_D^{25}$ +11.3° ($CHCl_3$, c 0.59)

Structures:

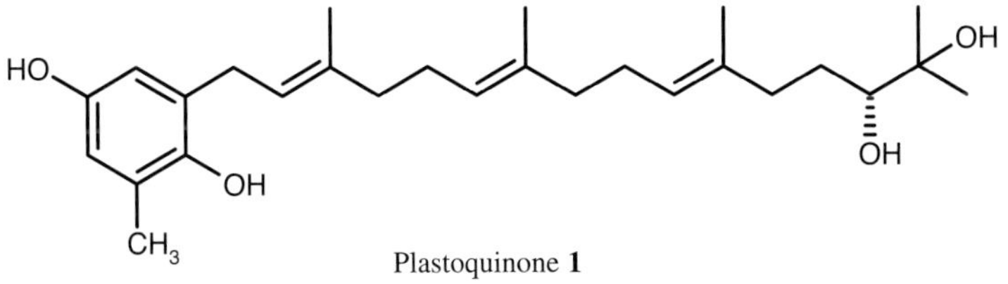

Plastoquinone **1**

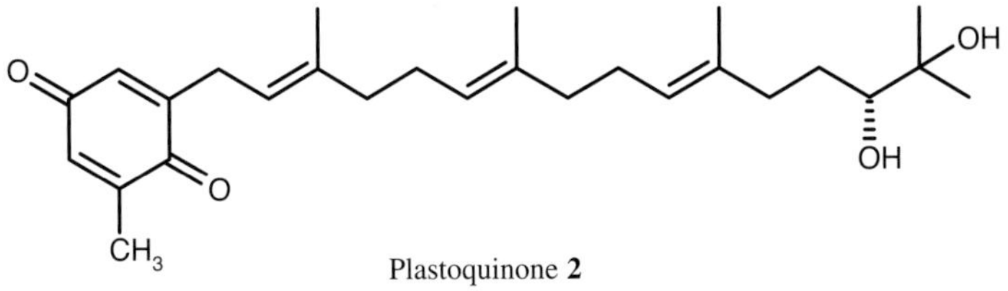

Plastoquinone **2**

Compound class: Plastoquinones (2-geranylgeranyl-6-methylbenzoquinone and its hydroquinone)

Source: *Sargassum micracanthum* (brown alga; family: Sargassaceae, Fucales) [1, 2]

Pharmaceutical potential: *Antioxidants*
Iwashima et al. [2] reported that both the compounds possess significant antioxidant and antiviral activities. Plastoquinones **1** and **2** showed inhibitory effect on NADPH-dependent lipid peroxidation in rat liver microsomes with IC_{50} values of 0.11 and 1.0 µg/ml, respectively (IC_{50} values of α-tocopherol and L-ascorbic acid (positive controls) = 34 and 28.0 µg/ml, respectively). The respective IC_{50} values for the test compounds in scavenging DPPH radicals were determined to be 11.0 and 400 µg/ml (IC_{50} values of α-tocopherol and L-ascorbic acid = 10.0 and 2.5 µg/ml, respectively). In addition, the plastoquinones were also reported to exert weak to moderate antiviral activities [2].

References

[1] Inaoka, K., Nishizawa, Y., Tone, H., and Kamiya, T. (1992) *Jpn. Kokai Tokkyo Koho* JP H04-49259.
[2] Iwashima, M., Mori, J., Ting, X., Matsunaga, T., Hayashi, K., Shinoda, D., Saito, H., Sankawa, U., and Hayashi, T. (2005) *Biol. Pharm. Bull.*, **28**, 374.

Platycodon saponins

Systematic names:
- Triterpenoid saponin **1**: 3-*O*-β-D-glucopyranosyl-2β,12α,16α,23,24-pentahydroxyoleanane-28(13)-lactone
- Triterpenoid saponin **2**: 3-*O*-β-D-glucopyranosyl-(1 → 3)-β-D-glucopyranosyl-2β,12α,16α,23α-tetrahydroxyoleanane-28(13)-lactone

Physical data:
- Triterpenoid saponin **1**: $C_{36}H_{58}O_{13}$, white powder, mp 222–223 °C, $[\alpha]_D^{20}$ +11.76° (MeOH, c 0.017)
- Triterpenoid saponin **2**: $C_{42}H_{68}O_{17}$, white powder, mp 212–213 °C, $[\alpha]_D^{20}$ +31.71° (MeOH, c 0.041)

Structures:

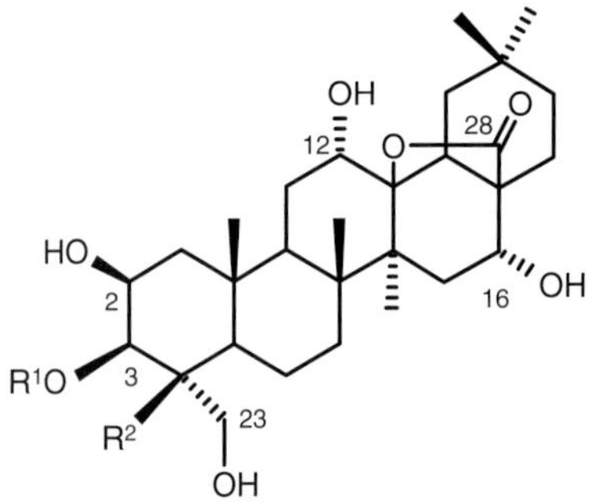

Triterpenoid saponin **1**: R^1 = Glu; R^2 = CH_2OH
Triterpenoid saponin **2**: R^1 = Glu (1→3) Glu; R^2 = CH_3

Compound class: Triterpenoid saponins

Source: *Platycodon grandiflorum* (roots; family: Campanulaceae)

Pharmaceutical potential: *Cytotoxic*
The triterpenoid saponins **1** and **2** were found to exhibit potent cytotoxicity against human Eca-109 cells with IC_{50} values of 0.649 and 0.503 μg/ml, respectively (topotecan was used as a positive control; IC_{50} = 0.032 μg/ml).

Reference

Zhang, L., Liu, Z.-H., and Tian, J.-K. (2007) *Molecules*, **12**, 832.

Plectranthols A and B

Systematic names:
- Plectranthol A: 19-*O*-(3,4-dihydroxybenzoyl)-11,12-dihydroxy-20(10 → 5)-*abeo*-abieta-1(10),6,8,11,13-tetraene
- Plectranthol B: 12-*O*-(3-methyl-2-butenoyl)-19-*O*-(3,4-dihydroxybenzoyl)-11-hydroxyabieta-8,11,13-triene

Physical data:
- Plectranthol A: $C_{27}H_{31}O_6$, brownish oil, $[\alpha]_D^{25}$ −154.2° (MeOH, c 0.22)
- Plectranthol B: $C_{32}H_{41}O_7$, brownish amorphous powder, $[\alpha]_D^{25}$ −20.6° (MeOH, c 0.20)

Structures:

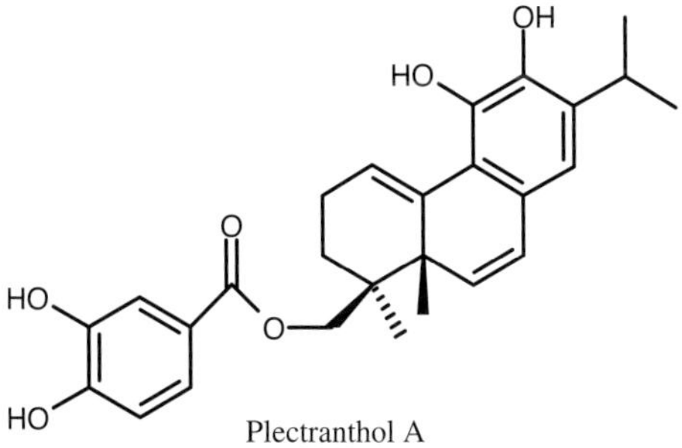

Plectranthol A

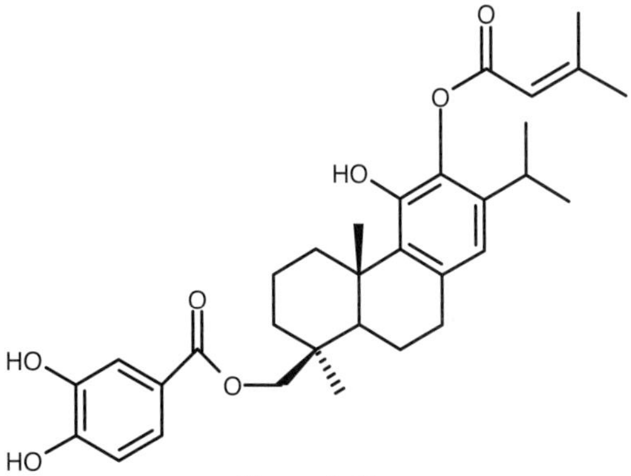

Plectranthol B

Compound class: Diterpenoids

Source: *Plectranthus nummularius* Briq. (fresh leaves; family: Labiatae)

Pharmaceutical potential: *Antioxidants*

Plectranthols A and B were found to possess potent DPPH free radical scavenging activity with EC_{50} values of 0.073 and 0.099 mM, respectively; their antioxidant efficacies were found to be greater than that of α-tocopherol used as the positive control ($EC_{50} = 0.134$ mM).

Reference

Narukawa, Y., Shimizu, N., Shimotohno, K., and Takeda, T. (2001) *Chem. Pharm. Bull.*, **49**, 1182.

Pleosporone

Physical data: $C_{15}H_{14}O_6$, yellow oil

Structure:

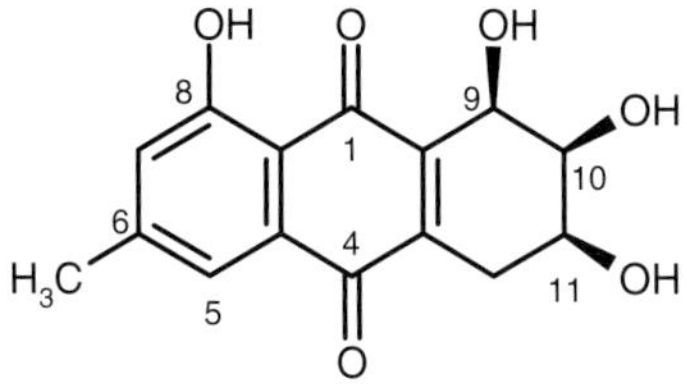

Pleosporone

Compound class: Naphthoquinone derivative

Source: Pleosporalean ascomycete (MF7028), an endophytic fungus from *Anthyllis vulneraria* L. (Fabaceae) collected in Madrid, Spain

Pharmaceutical potentials: *Antibacterial; cytotoxic*

The isolate showed modest antibacterial activities with MIC values ranging from 1 to 64 μg/ml against a number of bacterial strains; it displayed the highest sensitivity for *Streptococcus pneumoniae* and *Haemophilus influenzae* with MIC values of 4 and 1 μg/ml, respectively. Pleosporone showed modest selectivity for the inhibition of *S. aureus* RNA synthesis ($IC_{50} = 1.3$ μg/ml) compared to DNA ($IC_{50} = 8.4$ μg/ml) and protein synthesis ($IC_{50} = 15.4$ μg/ml). In addition, it exhibited cytotoxicity against HeLa cells with an IC_{50} value of 0.23 μg/ml.

Reference

Zhang, C., Ondeyka, J.G., Zink, D.L., Basilio, A., Vicente, F., Collado, J., Platas, G., Huber, J., Dorso, K., Motyl, M., Byrne, K., and Singh, S.B. (2009) *Bioorg. Med. Chem.*, **17**, 2162.

Plumbagin

Systematic name: 5-Hydroxy-2-methyl-1,4-naphthoquinone; alternatively, 5-hydroxy-2-methyl-naphthalene-1,4-dione)

Physical data: $C_{11}H_8O_3$, yellow pigment

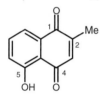

Plumbagin

Compound class: Naphthoquinone

Sources: It is widely spread in nature; few sources are cited such as *Juglans regia*, *J. cinerea*, and *J. nigra* (roots, leaves, barks, and wood; family: Juglandaceae) [1]; *Nepenthes thorelii* Lecomte (roots; family: Nepenthaceae) [2]; *Plumbago* sp. mainly, *P. europea*, *P. rosea*, *P. zeylanica*, *P. scandens* (roots; family: Plumbaginaceae) [3, 4, 6–8].

Pharmaceutical potentials: *Antimalarial; antimicrobial; anticancer; antitumor; cardiotonic; antifertile; anti-atherosclerotic; radiosensitizing*

Plumbagin was reported to exhibit a strong antimalarial activity against *Plasmodium falciparum* [2]; the respective IC_{50} value was found to be 0.27 μM, which is interesting when compared with conventional antimalarial agents such as chloroquine (0.09 μM) and pyrimethamine (11.29 μM). From the study, it was also evidenced that the quinone structure is essential for the activity, whereas the presence of a heteroatom such as oxygen or chlorine at position-3 of the naphthoquinone nucleus causes the weakening or loss of activity [2]. On the basis of their studies on NADH dehydrogenase of respiratory complex I, Krungkrai *et al.* [5] established that the malarial enzyme required both NADH and ubiquinone for maximal catalysis; at the same time plumbagin, a ubiquinone analog, showed strong inhibitory effect against the purified malarial enzymes. This result could explain the *in vitro* activity of plumbagin.

Plumbagin present in *Plumbago zeylanica* and *P. rosea* has been prescribed for cancer in the "Siddha" system of medicine [9]. Plumbagin has been shown to exert anticancer and antiproliferative activities in animal models as well as in cells in culture [10–14]. Sugie *et al.* [15] showed that plumbagin significantly inhibits azoxymethane-induced intestinal carcinogenesis in rats, suggesting its chemopreventive activity. Plumbagin has also been shown to induce S-G2/M cell cycle arrest through the induction of p21 (an inhibitor of cyclin-dependent kinase) [16]. A recent report showed that plumbagin has a chemotherapeutic potential as an anticancer agent in ovarian cancer cells with the mutated *BRCA1* gene [17]. The cytotoxic action of plumbagin in keratinocytes and cervical cancer cells was found to be due to a change in the redox status of the cell [5, 18]. It is evident that plumbagin may have potential as a chemotherapeutic agent; it induces reactive oxygen species, which mediate apoptosis in human cervical cancer cells, ME-180 [18]. The investigators concluded from their experimental results that the naphthoquinone inhibits the growth of ME-180 (human cervical cancer) cells in a concentration and time-dependent manner. The cytotoxic effect of plumbagin induced cell death is through the generation of reactive oxygen species (ROS) and subsequent induction of apoptosis as demonstrated by the investigators [18]. Moreover, plumbagin-induced apoptosis involved release of mitochondrial cytochrome c and apoptosis inducing factor (AIF), thus activation of caspase-dependent and -independent pathways, as shown by the plumbagin-mediated activation of caspase-3 and -9.

In embryonic kidney and brain tumor cells, plumbagin inhibited the enzyme NAD(P)H oxidase [19], linked with anticarcinogenic [10–14] and atherosclerotic effects [20]. NAD(P)H oxidase contributes to the pathogenesis of cancer and cardiovascular diseases such as hypertension, atherosclerosis, restenosis, cardiac hypertrophy and heart failure; Ding *et al.* [19] assumed that the beneficial effect of plumbagin is due to the inhibition of NAD(P)H oxidase. Human embryonic kidney 293 (HEK293) and brain tumor LN229 cells express mainly Nox-4, a renal NAD(P)H oxidase; the investigators examined the effect of plumbagin on Nox-4 activity in HEK293 and LN229 cells using lucigenin-dependent chemiluminescence assay. Plumbagin was found to inhibit the activity of Nox-4 in a time- and dose-dependent manner in HEK293 and LN229 cells, thereby, inhibiting the superoxide production in Nox-4 transfected COS-7 cells [19].

Hsu *et al.* [21] determined the cell growth inhibition activity of plumbagin by using *in vitro* and *in vivo* experimental models, and examined its effect on cell cycle distribution and apoptosis in human nonsmall cell lung cancer cells A549 as well as the anticancer mechanism of the drug. Human nonsmall cell lung cancer A549 cells were found to be highly sensitive to growth inhibition by plumbagin both in *in vitro* and *in vivo* experimental models. Plumbagin exhibited effective cell growth inhibition by inducing cancer cells to undergo G2/M phase arrest and apoptosis; plumbagin can inhibit cell cycle progression at the G2/M phase by increasing p21 expression in a p53-dependent manner, and by decreasing the expression of Cdc2, Cdc25C, and cyclinB1. The investigators observed that the plumbagin-induced cell growth inhibition in the A549 cells is mediated by activation of c-Jun N-terminal kinase (JNK), which stabilizes p53 by phosphorylation of p53 at Ser15 and decreasing the interaction of p53 and MDM2; c-Jun N-terminal kinase also phosphorylates Bcl-2, leading to alter function of Bcl-2 to apoptosis. Hence, plumbagin may be a promising chemopreventive agent against human nonsmall cell lung cancer [21].

Kavimani *et al.* [22] evaluated the antitumor activity of plumbagin against Dalton's lymphoma (DAL) in Swiss albino mice; a significant enhancement of mean survival time of plumbagin treated tumor bearing mice was noticed with respect to control group. Pulmbagin treatment was found to enhance peritoneal cell counts. On i.p. incoculation of the plumbagin-treated animals with DAL cells, tumor cell growth was found to be inhibited; besides, the plumbagin-treated groups at 14 days after transplantation were able to reverse the changes in the hematological parameters, protein and PVC consequent to tumor inoculation [22].

Plumbagin has also been proved to have anticancer effect against fibrosarcomas (ED_{50} 0.75 mg/kg body weight) and P388 lymphocytic leukemia (ED_{50} 4 mg/kg body weight) by Krishnaswamy and Purushothaman [14]. It was also reported to have modulatory effect on macrophage functions in BALB/c mice [23]. Kuo et al. [24] investigated the effect of plumbagin in human breast cancer cells in vitro, and established that the naphthoquinone effectively inhibits cell proliferation by inducing the cells to undergo G_2-M arrest and autophagic cell death by inhibiting the AKT/mammalian target of rapamycin pathway in the breast cancer cells; further, it inhibits tumor cell growth in nude mice also. The detailed experimental results are in good agreement with a critical role for AKT inhibition in plumbagin-induced G_2-M arrest and autophagy of human breast cancer cells. The investigators suggested that plumbagin may prove to be a valuable tool for inhibition of Cdc2/cyclin B1 and Cdc2/cyclin A complex in breast cancers because of the down-regulation of plumbagin on cyclin B1 and cyclin A expression, the induction of p21/WAF1 by plumbagin (which may subsequently inhibit the function of Cdc2 by forming Cdc2/p21/WAF complex), and also of the increase in activated phospho-Chk2 followed by an increase in inactivated phospho-Cdc25C, suggesting that an increase in Chk2 activation is followed by an increase in Cdc25C, which loses phosphatase function for dephosphorylating and activating Cdc2 [24].

Sankar et al. [25] studied lipid peroxidation in plumbagin administered rats; the drug was administered to rats at a concentration of 1, 2, 4, 8, and 16 mg/kg body weight and it was noted that after 24 h lipid peroxide levels decrease in subcellular fractions of liver. From their detailed study, it was evident that plumbagin inhibited ascorbate and nicotinamide adenine dinucleotide phosphate (reduced)-dependent lipid peroxidation, but it has no effect on cumene hydroperoxide-dependent lipid peroxidation. Injection of 16 mg of plumbagin per kg body weight was found to decrease liver total reduced glutathione and also fcrosomal glucose-6-phosphatase. Inbaraj and Chignell [1] studied the mechanistic pathway of cytotoxic action of plumbagin using HaCaT keratinocytes; exposure to plumbagin (at a range of 1–20 µM) resulted in a concentration-dependent decrease in cell viability. Redox cycling results in the generation of the corresponding semiquinone radical as detected by electron paramagnetic resonance; incubation of keratinocytes with the quinone was found to generate hydrogen peroxide (H_2O_2) resulting in the oxidation of glutathione (GSH) to GSSG. From their detailed study, the investigators concluded that the cytotoxicity of plumbagin is mainly the result of redox cycling, H_2O_2 production, and GSH oxidation; these findings suggest that topical preparations containing plumbagin should be used with care as their use may damage the skin. However, it is probable that the antifungal, antiviral, and antibacterial properties of the quinone are the result of redox cycling [1].

Parimala and Sachdanandam [12] evaluated the effect of the naphthoquinone on some glucose metabolizing enzymes studied in rats in experimental hepatoma. Its cardiotonic action was also studied on guinea-pig papillary muscle [26]. Bhargava, S.K. [27] also evaluated the effects of the compound on reproductive function of male dog. Besides such effects, plumbagin also exhibited radiosensitizing properties in experimental mouse tumors as well as in tumor cells in vitro [28–30]. Sandur et al. [8] suggested that plumbagin mediates its various activities through suppression of the transcription factor nuclear factor-κB (NF-κB); they found that the naphthoquinone inhibits NF-κB activation induced by a variety of agents and in a variety of cell lines. NF-κB activity was inhibited because plumbagin suppressed IKK activation, thus resulting in inhibition of IκBκ phosphorylation and degradation. Apart from this, plumbagin also inhibited the binding of the p65 subunit of NF-κB to the DNA. This resulted in suppression of NF-κB-regulated reporter gene transcription and gene products involved in cell proliferation (e.g., cyclin D1 and COX-2), antiapoptosis (e.g., survivin, IAP1, IAP2, Bcl-2, Bcl-xL, Bfl-1/A1, and cFLIP), angiogenesis (e.g., VEGF), and invasion (e.g., MMP-9). Suppression of NF-κB by plumbagin enhanced the apoptosis induced by TNF and paclitaxel. The detailed experimental observations by the present investigators [8] demonstrated that the antiproliferative, proapoptotic, anti-metastatic, chemosensitive, radiosensitive, and anti-inflammatory properties assigned to plumbagin are most likely mediated through the suppression of κF–κB activation as described here.

Plumbagin possesses potent antimicrobial activities [31, 32]; de Paiva et al. [7] reported that the naphthoquinone exhibits relatively specific activity against bacteria and yeast – it inhibited the growth of *Staphylococcus aureus* and *Candida albicans* with the respective MIC values of 1.56 and 0.78 μg/ml. From the study of Farr et al. [33] it was evident that actively growing *Escherichia coli* cells exposed to plumbagin, a redox cycling quinone that increases the flux of O_2^- radicals in the cell, are mutagenized or killed by this treatment; however, the toxicity of plumbagin was not found to be mediated by membrane damage. It was also observed that cells pretreated with plumbagin could partially reactivate lambda phage damaged by exposure to riboflavin plus light, a treatment that produces active oxygen species; the result suggested the induction of a DNA repair response. On the basis of their detailed observations, the investigators proposed that *E. coli* has an inducible DNA repair response specific for the type of oxidative damage generated during incubation with plumbagin, however, this response appears to be qualitatively distinct from the SOS response and the repair response induced by H_2O_2.

References

[1] Inbaraj, J.J. and Chignell, C.F. (2004) *Chem. Res. Taxicol.*, **17**, 55.

[2] Likhitwitayawuid, K., Kaewamatawong, R., Ruangrungsi, N., and Krungkrai, J. (1998) *Planta Med.*, **64**, 237.

[3] Bhattacharya, J. and De Carvalho, V.R. (1986) *Phytochemistry*, **25**, 764;Panichayupakaranant, P. and Tewtrakul, S. (2002) *Electronic J. Biotech.*, **5**, 228.

[4] Hsieh, Y.J., Lin, L.C., and Tsai, T.H. (2005) *J. Chromatogr. A*, **1083**, 141;Sankaram, A.V.B., Reddy, V.V.N., and Marthandamurthi, M. (1986) *Phytochemistry*, **25**, 2867;Yue, J., Lin, Z., Wang, D., Feng, Y., and Suu, H. (1994) *Phytochemistry*, **35**, 1023.

[5] Krungkrai, J., Kanchanarithisak, R., Krungkrai, S.R., and Rochanakij, S. (2002) *Exp. Parasitol.*, **100**, 54.

[6] Van der Vijver, L.M. (1972) *Phytochemistry*, **11**, 3247.

[7] de Paiva, S.R., Figueiredo, M.R., Aragão, T.V., and Kaplan, M.A.C. (2003) *Mem. Inst. Oswaldo. Cruz, Rio de Janeiro*, **98**, 959.

[8] Sandur, S.K., Ichikawa, H., Sethi, G., Ahn, K.S., and Aggarwal, B.B. (2006) *J. Biol. Chem.*, **281**, 17023.

[9] Mudaliar, M. (1969) Materia Medica, Published by Government of Tamil Nadu, India, 311.

[10] Hazra, B., Sarkar, R., Bhattacharyya, S., Ghosh, P.K., Chel, G., and Dinda, B. (2002) *Phytother. Res.*, **16**, 133.

[11] Naresh, R.A., Udupa, N., and Devi, P.U. (1996) *J. Pharm. Pharmacol.*, **48**, 1128.

[12] Parimala, R. and Sachdanandam, P. (1993) *Mol. Cell. Biochem.*, **12**, 59.

[13] Singh, U.V. and Udupa, N. (1997) *Indian J. Physiol. Pharmacol.*, **41**, 171.
[14] Krishnaswamy, M. and Purushothaman, K.K. (1980) *Indian J. Exp. Biol.*, **18**, 876.
[15] Sugie, S., Okamoto, K., Rahman, K.M., Tanaka, T., Kawai, K., Yamahara, J., and Mori, H. (1998) *Cancer Lett.*, **127**, 177.
[16] Jaiswal, A.S., Bloom, L.B., and Narayan, S. (2002) *Oncogene*, **21**, 5912.
[17] Srinivas, G., Annab, L.A., Gopinath, G., Banerji, A., and Srinivas, P. (2004) *Mol. Carcinog.*, **39**, 15.
[18] Srinivas, P., Gopinath, G., Banerji, A., Dinakar, A., and Srinivas, G. (2004) *Mol. Carcinog.*, **40**, 201.
[19] Ding, Y., Chen, Z.J., Liu, S., Che, D., Vetter, M., and Chang, C.H. (2005) *J. Pharm. Pharmacol.*, **57**, 111.
[20] Sharma, I., Gusain, D., and Dixit, V.P. (1991) *Indian J. Physiol. Pharmacol.*, **35**, 10.
[21] Hsu, Y.-L., Cho, C.-Y., Kuo, P.-L., Huang, Y.-T., and Lin, C.-C. (2006) *J. Pharmacol. Exp. Ther.*, **318**, 484.
[22] Kavimani, S., Ilango, R., Madheswaran, M., Jayakar, B., Gupta, M., and Majumdar, U.K. (1996) *Indian J. Pharm. Sci.*, **58**, 194.
[23] Abdul, K.M. and Ramchender, R.P. (1995) *Immunopharmacology*, **30**, 231.
[24] Kuo, P.-L., Hsu, Y.-L., and Cho, C.-Y. (2006) *Mol. Cancer Ther.*, **5**, 3209.
[25] Sankar, R., Devamanoharan, P.S., Raghupathi, G., Krishnasamy, M., and Shyamala Devi, C.S. (1987) *J. Biosci.*, **12**, 267.
[26] Itoigawa, M., Takeya, K., and Furukawa, H. (1991) *Planta Med.*, **57**, 317.
[27] Bhargava, S.K. (1984) *Indian J. Exp. Biol.*, **22**, 153.
[28] Devi, P.U., Rao, B.S., and Solomon, F.E. (1998) *Indian J. Exp. Biol.*, **36**, 891.
[29] Ganasoundari, A., Zare, S.M., and Devi, P.U. (1997) *Braz. J. Radiol.*, **70**, 599.
[30] Prasad, V.S., Devi, P.U., Rao, B.S., and Kamath, R. (1996) *Indian J. Exp. Biol.*, **34**, 857.
[31] Mossa, J.S., El-Feraly, F.S., and Muhammad, I. (2004) *Phytother. Res.*, **18**, 934.
[32] Didry, N., Dubrevil, L., and Pinkas, M. (1994) *Die Pharmazie*, **49**, 681.
[33] Farr, S.B., Natvig, D.O., and Kogoma, T. (1985) *J. Bacteriol.*, **164**, 1309.

PM-94128

Physical data: $C_{22}H_{34}N_2O_6$, solid, mp 172–173 °C, $[\alpha]_D^{25}$ −88.9° (CHCl$_3$, c 2.0)

Structure:

PM-94128

Compound class: Isocoumarin antibiotic

Source: *Bacillus* sp. PhM-PHD-090 (culture broth)

Pharmaceutical potential: *Antitumor*
The isolate exhibited potent antitumor activity against the tumor cell lines such as P-388, A-549, HT-29, and MEL-28 with an identical IC_{50} value of 0.05 µM.

References

Canedo, L.M., Puentes, J.L.F., Baz, J.P., Acebal, C., de la Calle, F., Gravalos, D.G., and de Quesada, T.G. (1997) *J. Antibiot.*, **50**, 175.

Pochonin G

Physical data: $C_{18}H_{17}ClO_6$, colorless oil, $[\alpha]_D^{20}$ −159° (Me_2CO, *c* 1.0)

Structure:

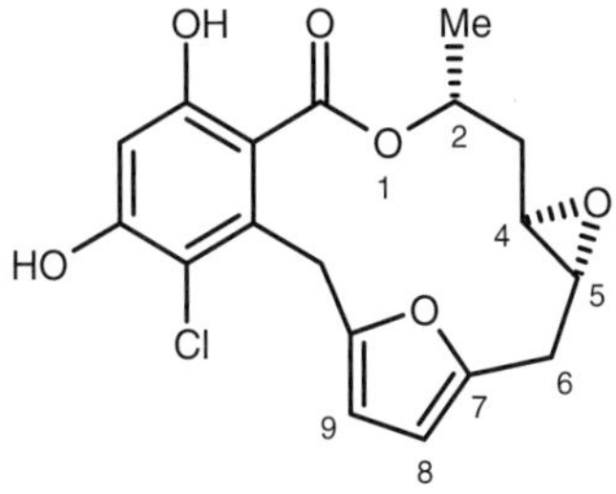

Pochonin G

Compound class: Macrocyclic antibiotic (resorcylic acid lactone)

Source: *Pochonia chlamydosporia* TF-0480 (fungus; culture broth) [1]

Pharmaceutical potential: *WNT-5A expression inhibitor (thus acts as hair growth stimulator)*
WNT-5A (wingless-type mouse mammary tumor virus integration site family, member 5A), a secretory glycoprotein, is known to be associated with the proliferation of dermal papilla cells that regulate the proliferation, differentiation, and apoptosis of these follicular keratinocytes, thereby controlling the hair cycle [2]. WNTs are regarded as important intercellular signaling molecules that regulate axis formation and organ formation during the fetal stage [3, 4]; thus, compounds having WNT inhibitory activity might be promising hair growth stimulators. Shinonaga et al. [1] evaluated pochonin G (a radicicol analog) as a significant WNT-5A expression inhibitor with an IC_{50} value of 8.15 µM; moreover, the test compound showed no cytotoxicity at concentrations above 100 µM.

References

[1] Shinonaga, H., Kawamura, Y., Ikeda, A., Aoki, M., Sakai, N., Fujimoto, N., and Kawashima, A. (2009) *Tetrahedron Lett.*, **50**, 108.

[2] Krus, S., Pytkowska, K., and Arct, J. (2007) *J. Appl. Cosmetol.*, **25**, 59.
[3] Cadigan, K.M. and Nusse, R. (1997) *Gene Dev.*, **11**, 3286.
[4] Wodarz, A. and Nusse, R. (1998) *Annu. Rev. Cell Dev. Biol.*, **14**, 59.

Polacandrin

Systematic name: 1β,3α,12β,25-Tetrahydroxy-20(S),24(S)-epoxydammarane

Physical data: $C_{30}H_{53}O_5$, colorless prisms, mp 234–238 °C, $[\alpha]_D$ −86.5° (CHCl$_3$, c 0.89)

Structure:

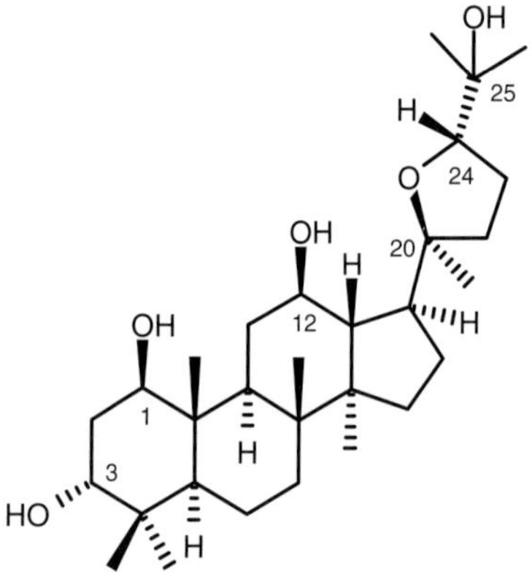

Polacandrin

Compound class: Dammarane triterpenoid

Source: *Polanisia dodecandra* (L.) DC (whole plants; family: Capparidaceae)

Pharmaceutical potential: *Cytotoxic (antitumor)*
The isolate showed potent cytotoxicity against KB (human nasopharynx carcinoma), P-388 (murine leukemia), and RPMI-7951 (melanoma) cell lines with ED$_{50}$ values of 0.60, 0.90, and 0.62 µg/ml, respectively; it also demonstrated marginal cytotoxic activity against HCT-8 (ED$_{50}$ 6.06 µg/ml) colon carcinoma cells.

Reference

Shi, Q., Chen, K., Fujioka, T., Kashiwada, Y., Chang, J.-J., Kozuka, M., Estes, J.R., McPhail, A.T., McPhail, D.R., and Lee, K.-H. (1992) *J. Nat. Prod.*, **55**, 1488.

Polpunonic acid (maytenonic acid)

Systematic name: 3-Oxofriedelan-29-oic acid

Physical data: $C_{30}H_{48}O_3$, solid, mp 274–275 °C [1], 261–262 °C [3], $[\alpha]_D$ −33.6° (CHCl$_3$, c 0.7) [1], −41.6° (CHCl$_3$, c 1.5) [3]

Structure:

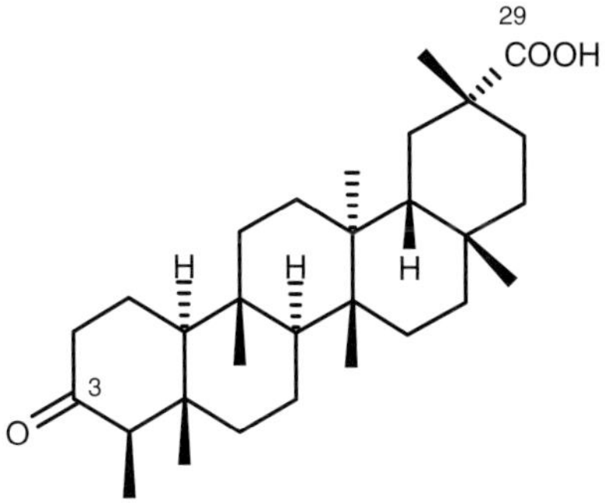

Polpunonic acid (maytenonic acid)

Compound class: Triterpenoid

Sources: *Maytenus diversifolia* (Gray) Hou (stems; family: Celastraceae) [1]; *Gymnosporia emarginata* (family: Celastraceae) [3]

Pharmaceutical potential: *Cytotoxic (anticancerous)*
The triterpenoid [1–3] was found to exhibit potent cytotoxicity against the A-549 lung carcinoma cells with an ED_{50} value of 0.21 µg/ml [1].

References

[1] Nozaki, H., Matsuura, Y., Hirono, S., Kasai, R., Chang, J.-J., and Lee, K.-H. (1990) *J. Nat. Prod.*, **53**, 1039.
[2] Monache, F.D., Mello, J.F., Marini-Bettolo, G.V., Lima, O.G., and Albuquerque, I.L. (1972) *Gazz. Chim. Ital.*, **102**, 636.
[3] Ramaiah, P.A., Devi, R.U., Frolow, F., and Lavie, D. (1984) *Phytochemistry*, **23**, 2251.

Polyhydroxylated cyclic sulfoxide

Physical data: $C_{12}H_{24}O_9S$, colorless amorphous solid, $[\alpha]_D^{20}$ +12.1° (H_2O, *c* 0.31)

Structure:

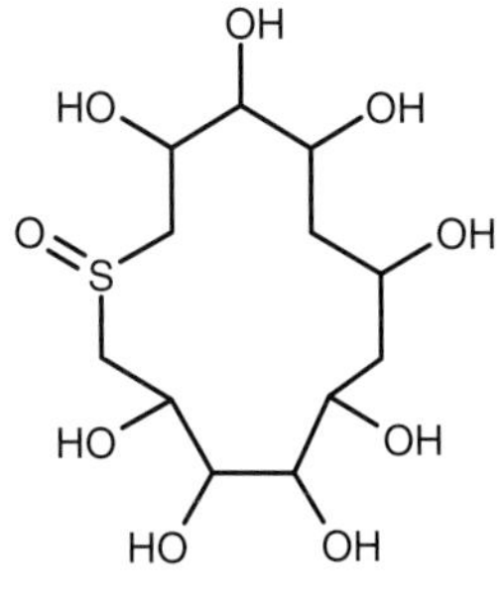

Polyhydroxylated cyclic sulfoxide

Compound class: Polyhydroxylated cyclic sulfoxide

Source: *Salacia reticulata* Wight (stems; family: Hippocrateaceae)

Pharmaceutical potential: *α-Glucosidase inhibitor*
The compound showed promising α-glucosidase inhibitory activity (IC_{50}: maltase, 0.227 μM; sucrase, 0.186 μM; isomaltase, 0.099 μM); however, it did not inhibit α-amylase from porcine pancreas. Such α-glucosidase inhibitory activity of the cyclic sulfoxide was found to be much greater than those of the previously identified compounds, salacinol and kotalano isolated from the same plant. The present investigators anticipated that the ring structure and the presence of a sulfoxide moiety along with orientation of the hydroxy groups in the test molecule might be important for such potent activity.

Reference

Ozaki, S., Oe, H., and Kitamura, S. (2008) *J. Nat. Prod.*, **71**, 981.

Polyketomycin

Physical data: $C_{44}H_{48}O_{18}$, orange powder, mp 194–196 °C, $[\alpha]_D^{24}$ −58.9° (MeOH, *c* 1.0)

Structure:

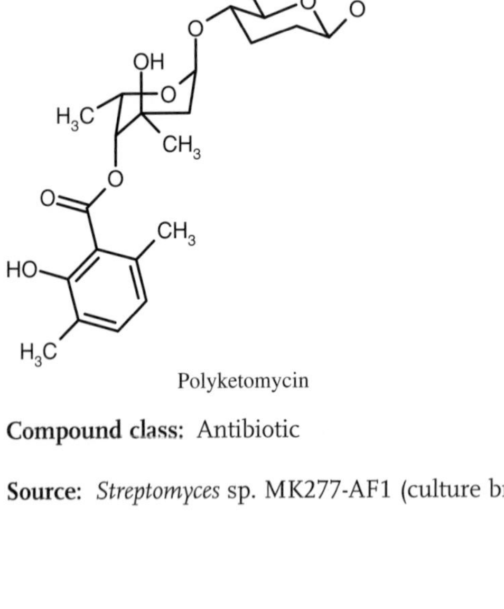

Polyketomycin

Compound class: Antibiotic

Source: *Streptomyces* sp. MK277-AF1 (culture broth) [1, 2]

Pharmaceutical potentials: *Antibacterial; cytotoxic*
Polyketomycin exhibited potent antibacterial activity against several Gram-positive bacterial strains such as *Staphylococcus aureus* FDA209P, *S. aureus* Smith, *S. aureus* MS9610, *S. aureus* MS16526 (MRSA), *S. aureus* TY-04282 (MRSA), *Micrococcus luteus* IFO3333, *Bacillus subtilis* PC1219, and *Corynebacterium bovis* 1810 with MIC values of 0.1, 0.1, 0.1, 0.025, 0.2, 0.1, <0.006, and 0.1 µg/ml, respectively [1]. In addition, the test compound was found to possess strong cytotoxicity against a number of tumor cell lines such as L1210 leukemia, EL-4 leukemia, P388 leukemia, Ehrlich carcinoma, IMC carcinoma, colon 26 adenocarcinoma, Meth A fibrosarcoma, FS-3 fibrosarcoma, and B16-BL10 melanoma with respective IC_{50} values of 3.3, 2.1, 5.2, 1.0, 0.9, 1.8, 2.4, 1.5, and 1.6 µg/ml [1].

References

[1] Momosae, I., Chen, W., Kinoshita, N., Iinuma, H., Hamada, M., and Takeuchi, T. (1998) *J. Antibiot.*, **51**, 21.
[2] Momosae, I., Chen, W., Nakamura, H., Naganawa, H., Iinuma, H., and Takeuchi, T. (1998) *J. Antibiot.*, **51**, 26.

Pomolic acid

Systematic name: 3β,19α-Dihydroxy-urs-12-en-28-oic acid

Physical data: $C_{30}H_{48}O_4$, colorless glossy crystalline powder, mp 271–273 °C [4], $[\alpha]_D^{20}$ +17.8° (MeOH, *c* 0.1) [4]

Structure:

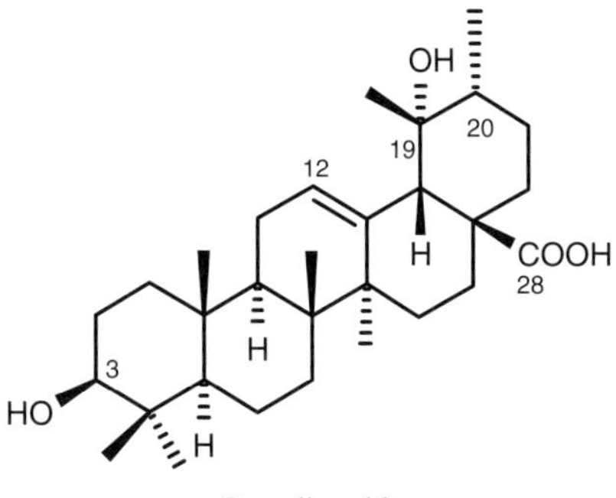

Pomolic acid

Compound class: Triterpenoid

Sources: *Euscaphis japonica* Pax. (pericarps of capsules; family: Staphyleaceae) [1]; *Rosa woodsii* Lindl. (leaves; family: Rosaceae), *Hyptis capitata* Jacq. (whole plants; family: Labiatae) [2]; *Cecropia pachystachya* Mart. (syn. *C. adenopus* Mart. leaves; family: Moraceae) [3]; *Weigela subsessilis* L.H. Bailley (leaves and stems; family: Caprifoliaceae) [4]

Pharmaceutical potentials: *Anti-HIV; anti-inflammatory; cytotoxic; anticancerous; anticomplementary*
Pomolic acid was reported to exhibit potent anti-HIV activity with an EC_{50} value of 1.4 µg/ml and also to inhibit uninfected cell growth with an IC_{50} value of 23.3 µg/ml, having a therapeutic index (TI) of 16.6 [2]. Fernandes *et al.* [5] showed that the triterpenoid can inhibit the growth of K562 cell line originated from chronic myeloid leukemia (CML) in blast crisis- and its vincristine-resistant derivative K562-Lucena1; induced apoptosis of leukemia cells (HL-60) was found to depend on the activation of caspase-3 and caspase-9 and dissipation of the mitochondrial transmembrane potential ($\Delta\psi_m$); hence, pomolic acid may find useful applications in overcoming apoptosis resistance within cancerous cells like leukemia as also evidenced from the *in vivo* experiment outcomes by Vasconcelos *et al.* [6].

Schinella *et al.* [3] reported that the triterpenoid inhibits the viability of human polymorphonuclear (PMN) cells through apoptosis in a time- and dose-dependent fashion; in annexin V-FITC binding assay, it was found to enhance the total of apoptotic cells by 42% at 100 µM and by 71% at 200 µM with respect to the control group. The investigators [3] suggested that the test compound can behave as an anti-inflammatory lead molecule since it regulates human neutrophil functions in this way.

Thuong *et al.* [4] evaluated pomolic acid (isolated from *Weigela subsessilis*) to have potent anticomplement activity against complement-induced hemolysis via the classical pathway, with an IC_{50} value of 4 ± 0.3 µM (rosmarinic acid was used as a positive control; $IC_{50} = 182 \pm 27.7$ µM). From detailed structure–activity relationships of such type of compounds, the present investigators suggested that a carboxylic group (at C-28) of ursane-type triterpenoids seems to play an important role in inhibiting the hemolytic activity of human serum against erythrocytes, and the hydroxy group at C-19 enhances the anticomplement activity of these compounds.

References

[1] Takahashi, K., Kawagushi, S., Nishura, K., Kubota, K., Tanabe, Y., and Takani, M. (1974) *Chem. Pharm. Bull.*, **22**, 650.

[2] Kashiwada, Y., Wang, H.-K., Nagao, T., Kitanaka, S., Yasuda, I., Fujioka, T., Yamagishi, T., Cosentino, O.L.M., Kozuka, M., Okabe, H., Ikeshiro, Y., Hu, C.-Q., Yeh, O.E., and Lee, K.-H. (1998) *J. Nat. Prod.*, **61**, 1090.

[3] Schinella, G., Aquila, S., Dade, M., Giner, R., del Carmen Recio, M., Spegazzini, E., de Buschiazzo, P., Tournier, H., and Ríos, J.L. (2008) *Planta Med.*, **74**, 215.

[4] Thuong, P.T., Min, B.-S., Jin, W.Y., Na, M.K., Lee, J.P., Seong, R.S., Lee, Y.-M., Song, K.S., Seong, Y.H., Lee, H.-K., Bae, K.H., and Kang, S.S. (2006) *Biol. Pharm. Bull.*, **29**, 830;Cheng D.L. and Cao X.P. (1992) *Phytochemistry*, **31**, 1317.

[5] Fernandes, J., Weinlich, R., Oliveira Castilho, R., Coelho Kaplan, M., Amarante-Mendes, G., and Gattass, C. (2005) *Cancer Lett.*, **219**, 49.

[6] Vasconcelos, F., Gattass, C., Rumjanek, V., and Maja, R. (2007) *Invest. New Drugs*, **25**, 525.

(2S)-Poncirin

Systematic name: (2*S*)-5-Hydroxy-4′-methoxyflavanone-7-*O*-(β-D-glucopyranosyl-(1 → 2)-α-L-rhamnopyranoside)

Physical data: $C_{28}H_{34}O_{14}$, white powder, $[\alpha]_D^{23}$ −81.6° (MeOH, c 0.18)

Structure:

R = β-D-Glu-(1→2)-α-L-Rha
(2S)-Poncirin

Compound class: Flavonoid (flavanone glycoside)

Source: *Poncirus trifoliata* (fruits; family: Rutaceae) [1–3]

Pharmaceutical potential: *Anti-inflammatory*
Han et al. [3] reported that the flavanone glycoside exhibited considerable inhibitory activity against lipopolysaccharide (LPS)-induced prostaglandin E_2 (PGE$_2$) and interleukin-6 (IL-6) production and mRNA expression in RAW 264.7 murine macrophage cells. It dose dependently inhibited the production of PGE$_2$ with an IC$_{50}$ value of 21.0 μM and reduced the production of IL-6 with an IC$_{50}$ value of 49.9 μM. The compound was found not to affect the cell viability of RAW 264.7 cells in either the presence or absence of LPS at 200 μM. The investigators also studied whether the inhibitory effect of the compound on such inflammatory mediators (PGE2 and IL-6) was related to modulation of COX-2 and IL-6 induction; for this purpose, they examined the mRNA expression levels by RT-PCR (reverse transcriptase polymerase chain reaction) – it was found that in response to LPS, COX-2, and IL-6, the corresponding mRNA expression levels were markedly upregulated and the compound significantly inhibited these COX-2 and IL-6 mRNA inductions. Hence, the compound may be a promising anti-inflammatory drug candidate, demanding for its further investigations including mechanism of action. It was supposed that the presence of 4′-methoxyl and 7-O-sugar moiety might be responsible for such pronounced effects [3].

References

[1] Hattori, S., Hasegawa, M., and Shimokoriyama, M. (1944) *Acta Phytochim.*, **14**, 1.
[2] Shimokoriyama, M. (1957) *J. Am. Chem. Soc.*, **79**, 4199.
[3] Han, A.-R., Kim, J.-B., Lee, J., Nam, J.-W., Lee, I.-S., Shim, C.-K., Lee, K.-T., and Seo, E.-K. (2007) *Chem. Pharm. Bull.*, **55**, 1270.

Pordamacrines A and B

Physical data:
- Pordamacrine A: $C_{23}H_{31}NO_5$, colorless solid, $[\alpha]_D^{27}$ −27° (MeOH, c 0.2)
- Pordamacrine B: $C_{23}H_{31}NO_4$, colorless solid, $[\alpha]_D^{27}$ +37° (MeOH, c 0.3)

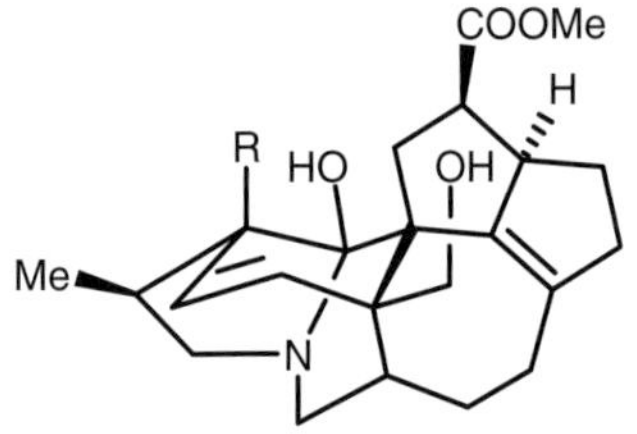

Pordamacrine A: R = OH
Pordamacrine B: R = H

Compound class: Alkaloids

Source: *Daphniphyllum macropodum* (leaves; family: Daphniphyllaceae)

Pharmaceutical potential: *Vasorelaxant*
Pordamacrines A and B were reported to possess moderate vasorelaxant activity; the *Daphniphyllum* alkaloids, when administered at a dose of 10^{-4} M, showed relaxation activity of 50 and 47.1%, respectively, against norepinephrine (3×10^{-7} M)-induced contractions of thoracic rat aortic rings with endothelium.

Reference

Matsuno, Y., Okamoto, M., Hirasawa, Y., Kawahara, N., Goda, Y., Shiro, M., and Morita, H. (2007) *J. Nat. Prod.*, **70**, 1516.

Porrigenic acid

Systematic name: (14*S*)-(10*E*,12*E*)-14-Hydroxy-9-oxo-10,12-octadecadienoic acid

Physical data: $C_{18}H_{31}O_4$, colorless solid, $[\alpha]_D^{28}$ $-10.5°$ (MeOH, *c* 0.088)

Structure:

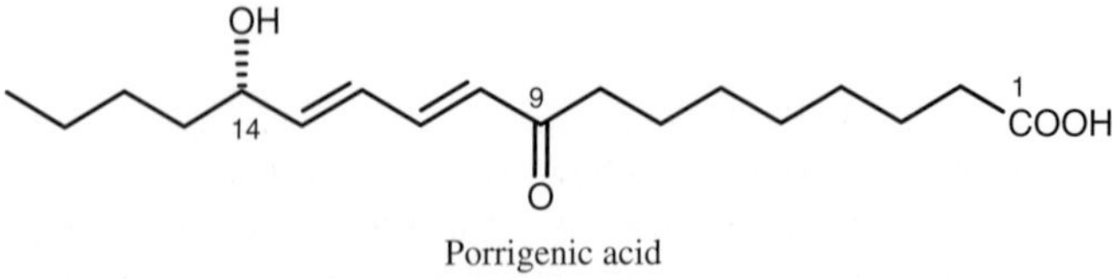

Porrigenic acid

Compound class: Conjugated ketonic fatty acid

Source: *Pleurocybella porrigens* (fungus; edible mushroom; family: Tricholomataceae)

Pharmaceutical potential: *Cytotoxic*
The investigators evaluated the fatty acid for its cytotoxic activity in both cultured human myeloma THP-1 cells and murine melanoma B16F1 cells; the compound was found to be active only against cultured THP-1 cells with an IC_{50} value of 46.5 µg/ml. Hence, the cytotoxic fatty acid (isolated from an edible mushroom) is supposed to have lineage-specific cell cytotoxicity.

Reference

Hasegawa, T., Ishibashi, M., Takata, T., Takano, F., and Ohta, T. (2007) *Chem. Bull. Pharm.*, **55**, 1708.

Primin

Systematic name: 2-Methoxy-6-*n*-pentyl-1,4-benzoquinone

Physical data: $C_{12}H_{16}O_3$, yellow crystals, mp 62–64 °C

Structure:

Primin

Compound class: Benzoquinone

Sources: *Primula obconica* (family: Primulaceae) [1–5]; *Miconia eriodonta* DC [5]; *Miconia* sp. [7]; *M. lepidota* [8]; *Botryosphaeria mamane* PSU-M76 (endophytic fungus) [9]

Pharmaceutical potentials: *Antimicrobial; antifeedant; antineoplastic*

Primin was reported to have significant antimicrobial and antitumor activity [6]. It showed antibacterial activity against *Staphylococcus aureus* ATCC 25923 and methicillin-resistant *S. aureus* SK1 with similar MIC value of 8 µg/ml [9]. Insect antifeedant [7], antimicrobial [6, 10], and antineoplastic [6, 10] activities of primin were also evaluated. Gunatilaka *et al.* [8] reported cytotoxic effects of the quinone studied in two cell lines – the IC_{50} values of the compound were determined to be 10 and 2.9 µg/ml, respectively, in the M109 and A2780 cell lines.

References

[1] Bloch, B. and Karrer, P. (1927) *Beiblatt Vierteljahr. Naturforsch. Ges. Zurich*, **13**, 1.
[2] Schildknecht, H., Bayer, I., and Schmidt, H. (1967) *Z. Naturforsch., Teil B*, **22**, 36.
[3] Schildknecht, H., Bayer, I., and Schmidt, H. (1967) *Z. Naturforsch., Teil B*, **22**, 287.
[4] Bieber, L.W., Chiappeta, A.D.A., Souza, M.A.D.M., and Generino R.M. (1990) *J. Nat. Prod.*, **53**, 706.
[5] Schlegel, R., Ritzau, M., Ihn, W., Stengel, C., and Grafe, U. (1995) *Nat. Prod. Lett.*, **6**, 171.
[6] de Lima, O.G., Marini-Bettolo, G.B., Delle Monache, F., Coelho, J.S.B., D'Albuqueque, I.L., Maciel, G.M., Lacerda, A.L., and Martins, D.G. (1970) *Rev. Inst. Antibiot., Univ. Fed. Pernambuco, Recife*, **10**, 29.
[7] Bernays, E., Lupi, A., Bettolo, R.M., Mastrofrancesco, C., and Tagliatesta, P. (1984) *Experientia*, **40**, 1010.

[8] Gunatilaka, A.A.L., Berger, J.M., Evans, R., Miller, J.S., Wisse, J.H., Neddermann, K.M., Bursuker, I., and Kingston, D.G.I. (2001) *J. Nat. Prod.*, **64**, 2.
[9] Pongcharoen, W., Rukachaisirikul, V., Phongpaichit, S., and Sakayaroj, J. (2007) *Chem. Pharm. Bull.*, **55**, 474.
[10] Marini-Bettolo, G.B., Delle Monache, F., Goncalves da Lima, O., and de Barros Coelho, S. (1971) *Gazz. Chim. Ital.*, **101**, 41.

Propindilactone L

Physical data: $C_{31}H_{42}O_{12}$, white powder, mp 170–171 °C, $[\alpha]_D^{24.9}$ +40.0° (MeOH, *c* 0.35)

Structure:

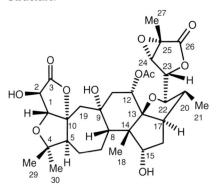

Propindilactone L

Compound class: Nortriterpenoid (18(13 → 14)-*abeo*-schiartane skeleton)

Source: *Schisandra propinqua* var. *propinqua* (stems; family: Schisandraceae)

Pharmaceutical potential: *Anti-hepatitis B virus (anti-HBV)*

Propindilactone L was evaluated to possess promising anti-HBV activity *in vitro*. It exhibited significant inhibitory potential against both HBsAg (hepatitis B virus surface antigen) and HBeAg (hepatitis B virus e antigen) with IC_{50} values of 0.488 and 1.309 mg/ml, respectively; the test compound was found to show low cytotoxicity and respective SI (selective index) values were calculated as 2.68 and 1.11.

Reference

Lei, C., Pu, J.-X., Huang, S.-X., Chen, J.-J., Liu, J.-P., Yang, L.-B., Ma, Y.-B., Xiao, W.-L., Li, X.-N., and Sun, H.-D. (2009) *Tetrahedron*, **65**, 164.

Propolis neoflavonoids 1 and 2

Physical data:
- Neoflavonoid **1**: $C_{15}H_{16}O_2$, brown oil, $[\alpha]_D^{22}$ +173.7° (CHCl$_3$, *c* 0.05)
- Neoflavonoid **2**: $C_{15}H_{16}O_2$, brown oil, $[\alpha]_D^{22}$ +214.0° (CHCl$_3$, *c* 0.09)

Protoxylocarpins A–E

Structures:

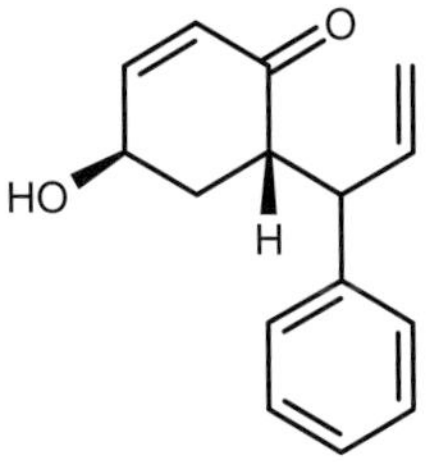

1, 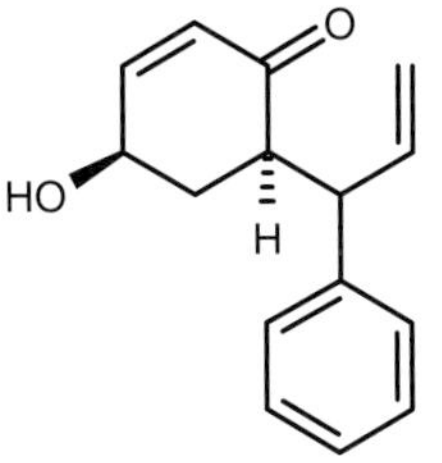 **2**

Compound class: Neoflavonoids

Source: Nepalese propolis (a resinous substance collected by bees from various plants)

Pharmaceutical potential: *Inhibitors of nitric oxide (NO) production*
The isolates displayed potent inhibitory activity on nitric oxide (NO) production in a lipopolysaccharide (LPS)-activated macrophage-like J774.1 cells assay with an equal IC_{50} value of 0.5 μM; the efficacy was found to be greater than that of the positive control, caffeic acid phenethyl ester (CAPE; $IC_{50} = 4.8$ μM).

Reference

Awale, S., Shrestha, S.P., Tezuka, Y., Ueda, J., Matsushige, K., and Kadota, S. (2005) *J. Nat. Prod.*, **68**, 858.

Protoxylocarpins A–E

Systematic names:
- Protoxylocarpin A: (5R,7R,8R,9R,10R,13S,17S)-17-{(2R,3S,5R)-5-[(1R)-1,2-dihydroxy-2-methylpropyl]-2-ethoxytetrahydrofuran-3-yl}-4,5,6,7,8,9,10,11,12,13,16,17-dodecahydro-7-hydroxy-4,4,8,10,13-pentamethyl-3H-cyclopenta[a]phenanthren-3-one
- Protoxylocarpin B: (5R,7R,8R,9R,10R,13S,17S)-17-{(2S,3S,5R)-5-[(1R)-1,2-dihydroxy-2-methylpropyl]-2-ethoxytetrahydrofuran-3-yl}-4,5,6,7,8,9,10,11,12,13,16,17-dodecahydro-7-hydroxy-4,4,8,10,13-pentamethyl-3H-cyclopenta[a]phenanthren-3-one
- Protoxylocarpin C: (5R,7R,8R,9R,10R,13S,17S)-17-{(2S,3S,5R)-2-ethoxy-5-[(1R)-2-ethoxy-1-hydroxy-2-methylpropyl]tetrahydrofuran-3-yl}-4,5,6,7,8,9,10,11,12,13,16,17-dodecahydro-7-hydroxy-4,4,8,10,13-pentamethyl-3H-cyclopenta[a]phenanthren-3-one
- Protoxylocarpin D: (5R,7R,8R,9R,10R,13S,17S)-17-{(2S,3S,5R)-5-[(1R)-1,2-dihydroxy-2-methylpropyl]-2-methoxytetrahydrofuran-3-yl}-7-hydroxy-4,4,8,10,13-pentamethyl-4,5,6,7,8,9,10,11,12,13,16,17-dodecahydro-3H-cyclopenta[a]phenanthren-3-one
- Protoxylocarpin E: (5R,7R,8R,9R,10R,13S,17S)-17-{(2S,3S,5R)-5-[(1R)-1,2-dihydroxy-2-methylpropyl]-2-methoxytetrahydrofuran-3-yl}-4,5,6,7,8,9,10,11,12,13,16,17-dodecahydro-4,4,8,10,13-pentamethyl-3-oxo-3H-cyclopenta[a]phenanthrene-7,11-diyl diacetate

Physical data:
- Protoxylocarpin A: $C_{32}H_{50}O_6$, white amorphous powder, $[\alpha]_D^{25}$ $-32°$ (MeOH, c 0.9)
- Protoxylocarpin B: $C_{32}H_{50}O_6$, white amorphous powder, $[\alpha]_D^{25}$ $-22°$ (MeOH, c 1.5)
- Protoxylocarpin C: $C_{34}H_{54}O_6$, white amorphous powder, $[\alpha]_D^{25}$ $-28°$ (MeOH, c 1.2)
- Protoxylocarpin D: $C_{31}H_{48}O_6$, white amorphous powder, $[\alpha]_D^{25}$ $-24°$ (MeOH, c 1.1)
- Protoxylocarpin E: $C_{35}H_{52}O_9$, white amorphous powder, $[\alpha]_D^{25}$ $-30°$ (MeOH, c 0.9)

Structures:

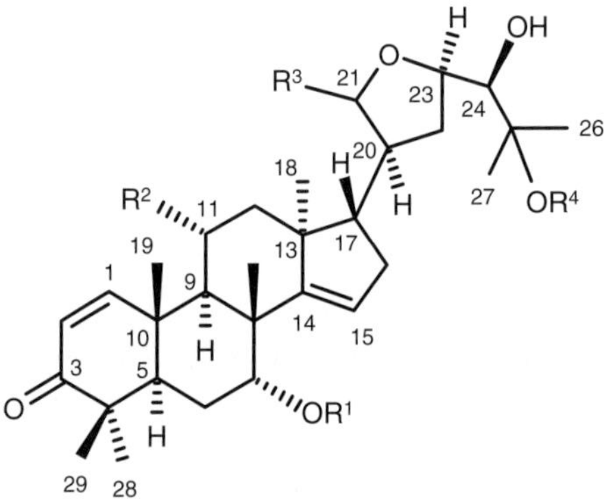

Protoxylocarpin A: R¹ = R² = R⁴ = H, R³ = α-EtO

Protoxylocarpin B: R¹ = R² = R⁴ = H, R³ = β-EtO

Protoxylocarpin C: R¹ = R² = H, R³ = α-EtO, R⁴ = Et

Protoxylocarpin D: R¹ = R² = R⁴ = H, R³ = β-MeO

Protoxylocarpin E: R¹ = Ac, R² = Aco, R³ = α-MeO, R⁴ = H

Compound class: Protolimonoids

Source: *Xylocarpus granatum* Koenig (dried fruit rinds; family: Meliaceae)

Pharmaceutical potential: *Cytotoxic (antitumor)*
All the isolates were evaluated for their *in vitro* antitumor activity and were found to exert potent cytotoxicities against the tumor cell lines HCT-8, Bel-7402, BGC-823, A-549, and A-2780; the respective IC$_{50}$ (μM) values were determined to be 2.83, 2.73, 3.35, 19.00, and 0.38 (protoxylocarpin A); 4.52, 5.52, 6.76, 13.52, and 0.97 (protoxylocarpin B); 4.21, 3.75, 5.87, 16.60, and 4.52 (protoxylocarpin C); 5.38, 6.21, 8.82, 9.62, and 3.35 (protoxylocarpin D); 3.87, 4.45, 4.62, 7.87, and 2.54 (protoxylocarpin E); 0.21, 0.02, 1.37, 2.76, and 0.03 (taxol, positive control).

Reference

Cui, J., Deng, Z., Xu, M., Proksch, P., Li, Q., and Lin, W. (2009) *Helv. Chim. Acta*, **92**, 139.

Przewalskin B

Physical data: $C_{20}H_{26}O_4$, colorless needles, $[\alpha]_D^{18}$ $-25.7°$ (CHCl$_3$, c 0.49)

Structure:

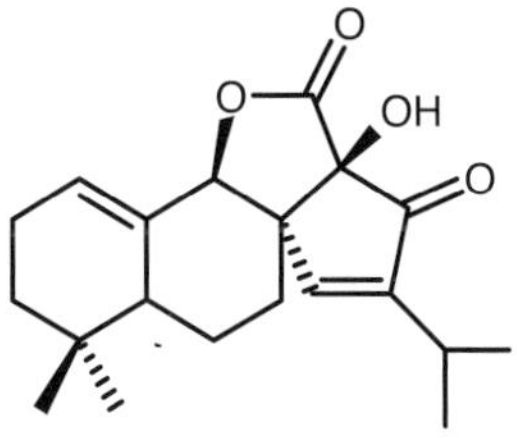

Przewalskin B

Compound class: Diterpenoid

Source: *Salvia przewalskii* Maxim. (whole plants; family: Labiatae)

Pharmaceutical potential: *Anti-HIV*
The isolate was found to show moderate cytopathic effects against HIV-1$_{IIIB}$ activity with an EC$_{50}$ value of 30.32 mg/ml and SI (selectivity index) of 3.32; however, it showed no significant cytotoxicity against a number of human cancer cell lines.

Reference

Xu, G., Hou, A.-J., Zheng, Y.-T., Zhao, Y., Li, X.-L., Peng, L.-Y., and Zhao, Q.-S. (2007) *Org. Lett.*, **9**, 291.

Pseudopteroxazole and *seco*-pseudopteroxazole

Physical data:
- Pseudopteroxazole: $C_{21}H_{27}NO$, yellowish oil, $[\alpha]_D^{25}$ +101° (CHCl$_3$, c 1.0)
- *seco*-Pseudopteroxazole: $C_{21}H_{29}NO$, yellowish oil, $[\alpha]_D^{25}$ +28.2° (CHCl$_3$, c 0.85)

Structures:

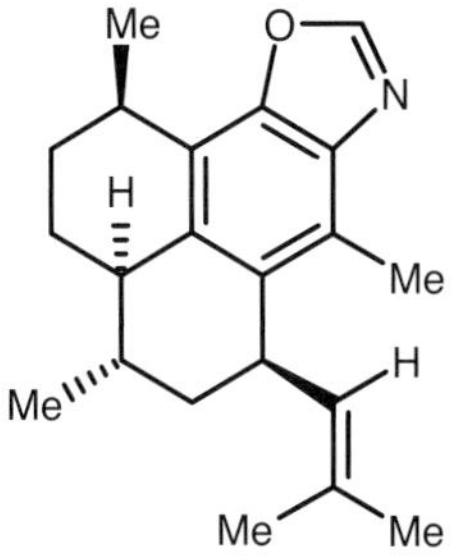

Pseudopteroxazole 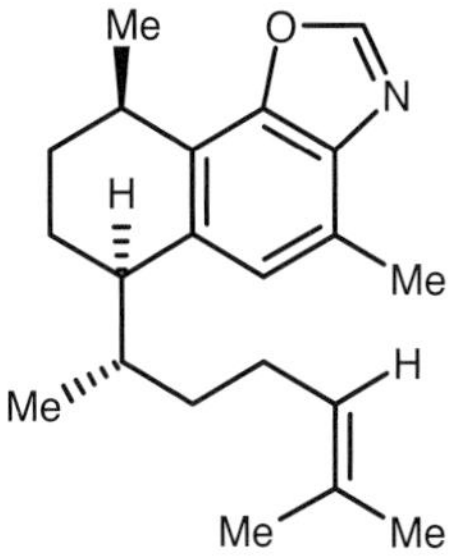   *seco*-Pseudopteroxazole

Compound class: Diterpenoid alkaloids

Source: *Pseudopterogorgia elisabethae* (Bayer) (West Indian gorgonian coral)

Pharmaceutical potential: *Antitubercular*

Pseudopteroxazole was evaluated as a potent antitubercular agent; the alkaloid inhibited the growth of *Mycobacterium tuberculosis* H37Rv *in vitro* by 97% at a concentration of 12.5 μg/ml, whereas *seco-* pseudopteroxazole showed only 66% inhibition against the mycobacterial growth on application at the same concentration. Hence, it appeared that the potent activity of the former might be attributed, at least in part, to the benzoxazole function. However, pseudopteroxazole was found not to have any significant *in vitro* cytotoxicity when tested in the National Cancer Institute's (NCI) 60-cell line tumor panel.

Reference

Rodrguez, A.D., Ramrez, C., Rodrguez, I.I., and Gonzlez, E. (1999) *Org. Lett.*, **1**, 527.

Pterocidin

Physical data: $C_{23}H_{34}O_6$, pale yellow oil, $[\alpha]_D^{20}$ −27.7° (CHCl$_3$, *c* 0.46)

Structure:

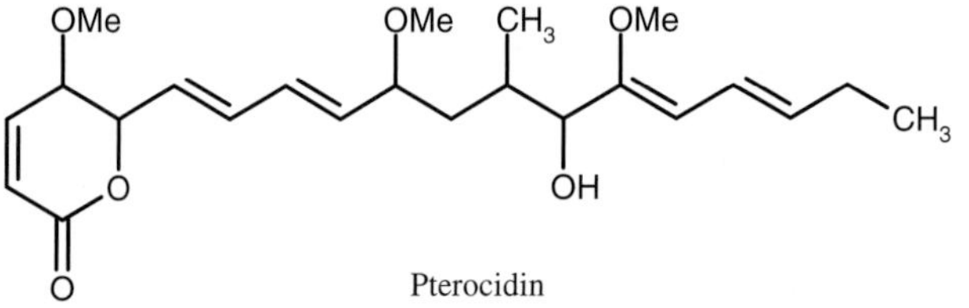

Pterocidin

Compound class: Antibiotic (polyketide δ-lactone)

Source: *Streptomyces hygroscopicus* TP-A0451 (fermentation broth)

Pharmaceutical potential: *Cytotoxic*

Pterocidin exhibited cytotoxicity against some cancer cell lines such as NCI-H522, OVCAR-3, SF539, and LOX-IMVI with IC$_{50}$ values of 2.9, 3.9, 5.0, and 7.1 μM, respectively.

Reference

Igarashi, Y., Miura, S.-S., Fujita, T., and Furumai, T. (2006) *J. Antibiot.*, **59**, 193.

Pterostilbene

Systematic name: 4-[(*E*)-2-(3,5-Dimethoxyphenyl)ethenyl]phenol [or 3,5-Dimethoxy-4′-hydroxy-*trans*-stilbene]

Physical data: $C_{16}H_{16}O_3$, mp 84 °C

Structure:

Pterostilbene

Compound class: Stilbenoid

Sources: *Pterocarpus marsupium* Roxb. (heartwoods; family: Leguminosae) [1, 2]; *Vaccinium* sp. (berries; family: Vacciniaceae) [8]; *Vitis vinifera* (leaves and berries; family: Vitaceae) [9–11].

Pharmaceutical potentials: *Antidiabetic; anticancerous, antioxidant; antihypercholesterolemic, antihypertriglyceridemic; anti-inflammatory*

Manickam et al. [3] evaluated antihyperglycemic activity of the phenolic compound against streptozotocin (STZ)-induced hyperglycemic rats; pterostibene was found to decrease significantly the plasma glucose level of STZ-induced diabetic rats, and the efficacy is comparable to that of the reference standard, metformin. The test compound at a dose of 20 mg/kg body weight (administered for 3 days) lowered 42% in plasma glucose level, while metformin lowered the plasma glucose level by 48% at a dose of 30 mg/kg body weight (administered for 3 days). Pterostibene did not alter the basal plasma glucose level in nondiabetic animals; furthermore, it significantly decreased the body weight in comparison to the vehicle-treated animals. Pterostibene might be useful in noninsulin-dependent *diabetes mellitus*. The investigators [3] supposed that the test compound may have insulin-like effects on several tissues as in the case of the oral hypoglycemic agents such as metformin [4–7]; however, detailed investigations are needed to elucidate the exact mode of action.

Pterostilbene was found to be effective in preventing carcinogen-induced preneoplastic lesions in a mouse mammary organ culture model [12]; besides, i.v. administration of this compound to mice inhibited metastatic growth of B16M-F10 melanoma cells in the liver, a common site for metastasis development [13]. In addition to the aforementioned activity in the mouse mammary organ culture model, pterostilbene was also found to exhibit strong antioxidant and antihypolipidemic activity [12, 14–16]. It was established by Suh et al. [17] that the compound is of great interest for the prevention of colon cancer; administration of the test compound at a dose of 40 ppm for 8 weeks significantly suppressed aberrant crypt foci (ACF) formation (57% inhibition, $p < 0.001$) as well as multiple clusters of aberrant crypts (29% inhibition, $p < 0.01$) in azoxymethane-induced colon carcinogenesis model in rats. The investigators [17] also observed that pterostilbene inhibited the induction of iNOS protein expression in the colon cancer cell line *in vitro* in a dose-dependent manner in HT-29 cells (14, 61, and 77% inhibition of iNOS expression, respectively at 1, 10, and 30 µmol/l concentrations). Hence, it might be suggested that the suppression of ACF formation is mediated through the inhibition of colonic cell proliferation and iNOS expression. Pan et al. [18] showed that pterostilbene was able to inhibit cell proliferation and induce apoptosis in human gastric carcinoma cells in a concentration- and time-dependent manner; from the results of their detailed investigations, the workers suggested that the Bcl-family of proteins, the mitochondrial pathway, and activation of the caspase cascade are responsible for pterostilbene-induced apoptosis. Pan and his group [19] showed that the observed response of pterostilbene is consistent with its known hypolipidemic properties, and the induction of mitochondrial genes is consistent with its demonstrated role in apoptosis in human cancer cell lines; it was also demonstrated that the stilbenoid has a significant effect on methionine metabolism [19].

Amarnath Satheesh and Pari [20] evaluated the antioxidant potential of pterostilbene on streptozotocin-nicotinamide-induced diabetic rats; at a dose of 40 mg/kg for 6 weeks it significantly decreased the activity of superoxide dismutase, catalase, glutathione peroxidase, glutathione-S-transferase and reduced glutathione in liver and kidney of diabetic animals when compared with normal control. The increased levels of lipid peroxidation measured as thiobarbituric acid reactive substances (TBARS) in liver and kidney of diabetic rats were also normalized by treatment with pterostilbene [20]. Remsberg et al. [21] also evaluated preclinical pharmacokinetics and pharmacodynamics, metabolism, anticancer, anti-inflammatory, antioxidant, and analgesic activities of this compound.

References

[1] Adinarayan, D. and Syamasundar, K.V. (1982) *Phytochemistry*, **21**, 1083.
[2] Maurya, R., Ray, A.B., Duah, F.K., Slatkin, D.J., and Schiff, P.L. Jr (1984) *J. Nat. Prod.*, **47**, 179.
[3] Manickam, M., Ramanathan, M., Farboodniay Jahromi, M.A., Chansouria, J.P.N., and Ray, A.B. (1997) *J. Nat. Prod.*, **60**, 609.
[4] Jackson, R.A., Hawa, M.I., Japan, J.B., Sim, B.M., Silvio, D., Featherbe, L., and Kurtz, D. (1987) *Diabetes*, **36**, 632.
[5] Hermann, L.S. (1979) *Diabetic Metab.*, **5**, 233.
[6] Lord, J.M., Atkins, T.W., and Bailey, C.J. (1983) *Diabetologia*, **25**, 108.
[7] Puah, J.A. and Bailey, C.J. (1984) *Diabetologia*, **27**, 322.
[8] Rimando, A.M., Kalt, W., Magee, J.B., Dewey, J., and Ballington, J.R. (2004) *J. Agric. Food Chem.*, **52**, 4713.
[9] Langcake, P., Cornford, C.A., and Pryce, R.J. (1979) *Phytochemistry*, **18**, 1025.
[10] Adrian, M., Jeandet, P., Douillet-Breuil, A.C., Tesson, L., and Bessis, R. (2000) *J. Agric. Food Chem.*, **48**, 6103.
[11] Pezet, R. and Pont, V. (1998) *Plant Physiol. Biochem. (Paris)*, **26**, 603.
[12] Rimando, A.M., Cuendet, M., Desmarchelier, C., Mehta, R.G., Pezzuto, J.M., and Duke, S.O. (2002) *J. Agric. Food Chem.*, **50**, 3453.
[13] Ferrer, P., Asensi, M., Segarra, R. et al. (2005) *Neoplasia*, **7**, 37.
[14] Stivala, L.A., Savio, M., Carafoli, F. et al. (2001) *J. Biol. Chem.*, **276**, 22586.
[15] Rimando, A.M., Nagmani, R., Feller, D.R., and Yokoyama, W. (2005) *J. Agric. Food Chem.*, **53**, 3403.
[16] Miura, D., Miura, Y., and Yagasaki, K. (2003) *Life Sci.*, **73**, 1393.
[17] Suh, N., Paul, S., Hao, X., Simi, B., Xiao, H., Rimando, A.M., and Reddy, B.S. (2007) *Clin. Cancer Res.*, **13**, 350.
[18] Pan, M.-H., Chang, Y.-H., Badmaev, V., Nagabhushanam, K., and Ho, C.-T. (2007) *J. Agric. Food Chem.*, **55**, 7777.

[19] Pan, Z., Agarwal, A.K., Xu, T., Feng, Q., Baersoni, S.R., Duke, S.O., and Rimando, A.M. (2008) *BMC Med. Genomics*, **1**, 7, http://www.biomedcentral.com/content/pdf/1755-8794-1-7.pdf. doi: 10.1186/1755-8794-1-7.
[20] Amarnath Satheesh, M. and Pari, L. (2006) *J. Pharm. Pharmacol.*, **58**, 1483.
[21] Remsberg, C.M., Yanez, J.A., Ohgami, Y., Vega-Villa, K.R., Rimando, A.M., and Davies, N.M. (2007) *Phytother. Res.*, **22**, 169.

Pulchellalactum

Physical data: $C_9H_{13}NO$, oily material

Structure:

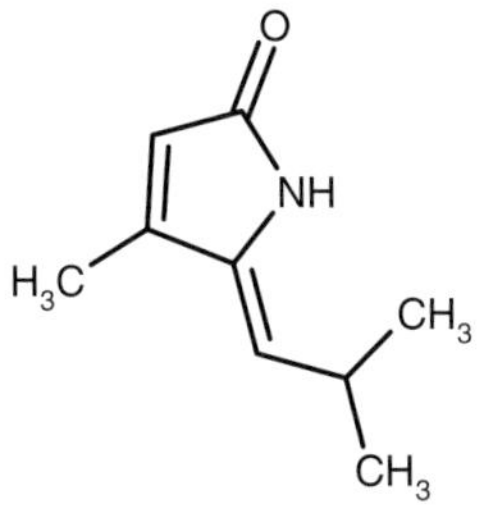

Pulchellalactum

Compound class: Antibiotic

Source: *Corollospora pulchella* ATCC 62554 (marine fungus; culture broth)

Pharmaceutical potential: *Tyrosine phosphatase inhibitor*
Pulchellalactum was evaluated as a moderate tyrosine phosphatase inhibitor; it showed dose-dependent inhibition of CD45 activity with an IC_{50} value of 124 μg/ml – the potency was almost comparable with that of sodium orthovanadate, a known tyrosine phosphatase inhibitor (IC_{50} = 91.9 μg/ml). In addition, the inhibition was found to be specific for CD45 as another protein tyrosine phosphatase, PTP1B, was not inhibited by the test compound.

Reference

Alvi, K.A., Casey, A., and Nair, B.G. (1998) *J. Antibiot.*, **51**, 515.

Pulsatilla saponin D

Systematic name: Hederagenin 3-*O*-α-L-rhamnopyranosyl (1 → 2)-[β-D-glucopyranosyl-(1 → 4)]-α-L-arabinopyranoside

Physical data: $C_{47}H_{76}O_{17}$, amorphous white, mp 239–241 °C, $[\alpha]_D^{22}$ +23.6° (MeOH, *c* 0.2)

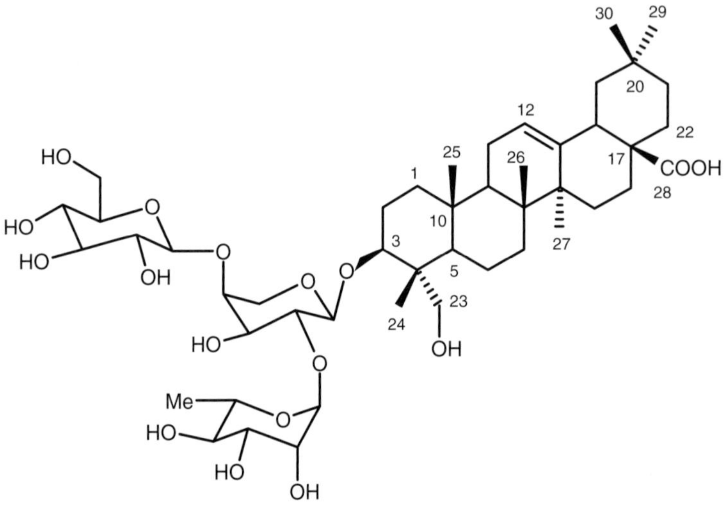

Pulsatilla saponin D

Compound class: Triterpenoid saponin

Sources: *Pulsatilla koreana* Nakai (roots; family: Ranunculaceae), *P. cernua* [2, 3], *P. chinensis* (roots) [4, 5]

Pharmaceutical potential: *Antitumor*
The triterpene saponin was evaluated for its *in vivo* antitumor as well as *in vitro* cytotoxic activity. It showed potent inhibition rate (IR) of tumor growth by 82% at a dose of 6.4 mg/kg body weight in the BDF1 mice bearing Lewis lung carcinoma cells; the inhibition rate of the growth of such solid tumor was found to be higher than that of adriamycin (IR = 64%) under the same conditions [1]. In addition, the compound showed moderate cytotoxic activity (ED_{50} value ranging from 6.3 to >10 µg/ml) against human lung (A-549), melanoma (SK-MEL-2), and breast (MCF-7) cancer cell lines [1]. Yoshihiro *et al.* [4] also reported that the compound displayed potent *in vitro* cytotoxicity against human leukemia cells (HL-60) with an IC_{50} value of 2.7 µg/ml.

References

[1] Kim, Y., Bang, S.-C., Lee, J.-H., and Ahn, B.-Z. (2004) *Arch. Pharm. Res.*, **27**, 915.
[2] Shimizu, M., Shingyouchi, K.-I., Morita, N., Kizu, H., and Tomimori, T. (1978) *Chem. Pharm. Bull.*, **26**, 1666.
[3] Kang, S.S. (1989) *Arch. Pharm. Res.*, **12**, 42.
[4] Yoshihiro, M., Akihito, Y., Minpei, K., Tomoki, A., and Yutaka, S. (1999) *J. Nat. Prod.*, **62**, 1279.
[5] Ekabo, O.A., Farnsworth, N.R., Henderson, T.O., Mao, G., and Mukherjee, R. (1996) *J. Nat. Prod.*, **59**, 431.

6-(9'-Purine-6',8'-diolyl)-2β-suberosanone

Physical data: $C_{20}H_{26}N_4O_3$, white powder, $[\alpha]_D^{20}$ +28° (CHCl$_3$, c 0.2)

Structure:

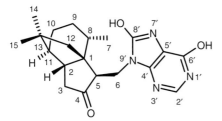

6-(9'-Purine-6',8'-diolyl)-2β-suberosanone

Compound class: Sesquiterpene alkaloid

Source: *Subergorgia suberosa* (South China Sea gorgonian coral)

Pharmaceutical potential: *Cytotoxic*
Qi et al. evaluated the cytotoxic efficacy of the alkaloid toward the human breast carcinoma MDA-MB-231 and MCF cancer cell lines; the test compound displayed moderate cytotoxicity against the MDA-MB-231 cell line with an IC$_{50}$ value of 8.87 μg/ml and potential cytotoxicity against the MCF cells at a concentration of 50 μM (MTT assay).

Reference

Qi, S.-H., Zhang, S., Li, X., and Li, Q.-X. (2005) *J. Nat. Prod.*, **68**, 1288.

Purpactin A

See AS-186b

Pyranojacaeubin

See Rheediaxanthone A

Pyranoxanthones 1 and 2

Systematic names:
- Pyranoxanthone **1**: 1,5,6-trihydroxy-6',6'-dimethyl-2*H*-pyrano(2',3':3,4)-2-(3-methylbut-2-enyl) xanthone
- Pyranoxanthone **2**: 1,6,7-trihydroxy-6',6'-dimethyl-2*H*-pyrano(2',3':3,2)-4-(3-methylbut-2-enyl) xanthone

Physical data: $C_{23}H_{22}O_6$, yellow amorphous powder (for both)

Structures:

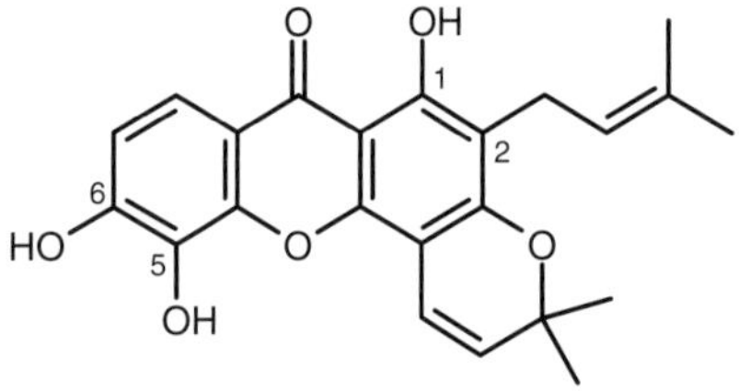

Compound class: Xanthones

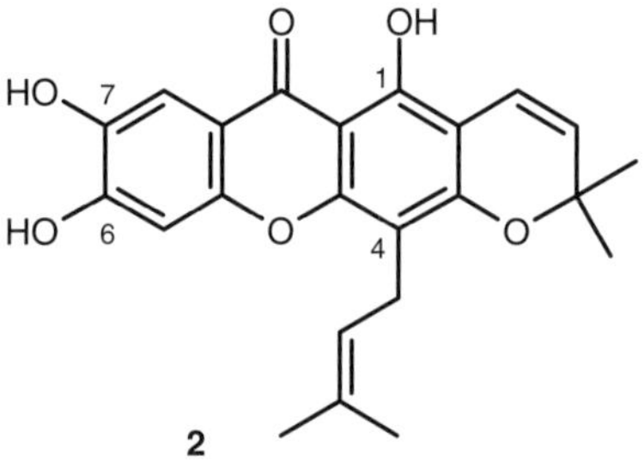

Source: *Garcinia lancilimba* (barks; family: Guttiferae)

Pharmaceutical potential: *Cytotoxic*
The present investigators evaluated both the xanthonoids for their cytotoxic potential against human breast cancer MDA-MB-435S cells; the compounds exhibited moderate activity against the target cells with respective IC_{50} values of 5.88 ± 0.49 and $6.05 \pm 0.21\,\mu g/ml$ (IC_{50} value of adriamycin used as a positive control $= 0.24 \pm 0.01\,\mu g/ml$).

Reference

Yang, N.-Y., Han, Q.-B., Cao, X.-W., Qiao, C.-F., Song, J.-Z., Chen, S.-L., Yang, D.-J., Yiu, H., and Xu, H.-X. (2007) *Chem. Pharm Bull.*, **55**, 950.

Pyripyropenes I–L

Physical data:
- Pyripyropene I: $C_{34}H_{43}NO_{10}$, colorless crystal, $[\alpha]_D^{28}$ +69° (MeOH, *c* 1.0)
- Pyripyropene J: $C_{33}H_{41}NO_{10}$, colorless powder, mp 248–250 °C, $[\alpha]_D^{28}$ +65° (MeOH, *c* 1.0)
- Pyripyropene K: $C_{33}H_{41}NO_{10}$, colorless crystal, $[\alpha]_D^{28}$ +58° (MeOH, *c* 1.0)
- Pyripyropene L: $C_{33}H_{41}NO_{10}$, colorless crystal, mp 148–150 °C, $[\alpha]_D^{28}$ +58° (MeOH, *c* 1.0)

Structures:

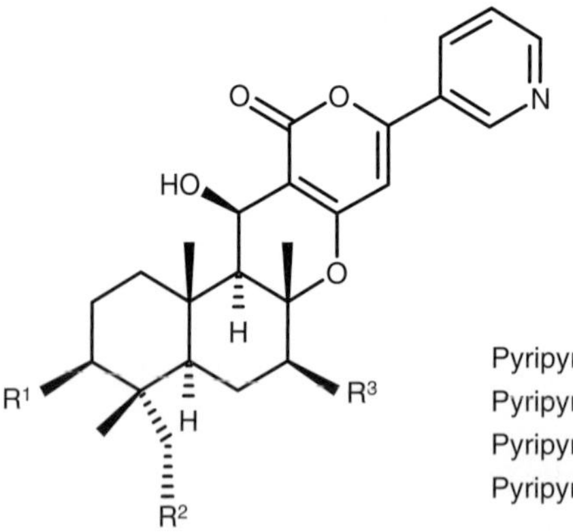

Pyripyropene I: $R^1 = R^2 = R^3 = OCOCH_2CH_3$
Pyripyropene J: $R^1 = R^3 = OCOCH_2CH_3$; $R^2 = OCOCH_3$
Pyripyropene K: $R^1 = R^2 = OCOCH_2CH_3$; $R^3 = OCOCH_3$
Pyripyropene L: $R^1 = OCOCH_3$; $R^2 = R^3 = OCOCH_2CH_3$

Compound class: Antibiotics

Source: *Aspergillus fumigatus* FO-1289-2501 (culture broth)

Pharmaceutical potential: *Acyl-CoA:cholesterol acyltransferase (ACAT) inhibitors*
All the isolates were found to have acyl-CoA:cholesterol acyltransferase (ACAT) inhibitory activity in rat liver microsomes; pyripyropene L was evaluated to be the most potent among them with an IC_{50} value of 0.27 µM, followed by pyripyropenes J ($IC_{50} = 0.85$ µM), I ($IC_{50} = 2.45$ µM), and K ($IC_{50} = 2.65$ µM).

Reference

Tomoda, H., Tabata, N., Yang, D.-J., Takayanagi, H., Nishida, H., and Omura, S. (1995) *J. Antibiot.*, **48**, 495.

Pyrolaside B

Systematic name: 5-5'-Dehydro-[(4-O-5'')-dehydro-di(3-methyl-4-hydroxyphenyl-1-O-β-D-glucopyranoside)]-(3-methyl-4-hydroxyphenyl-1-O-β-D-glucopyranoside

Physical data: $C_{39}H_{50}O_{21}$, white amorphous powder, $[\alpha]_D^{20}$ −47.4° (H_2O, *c* 0.1)

Structure:

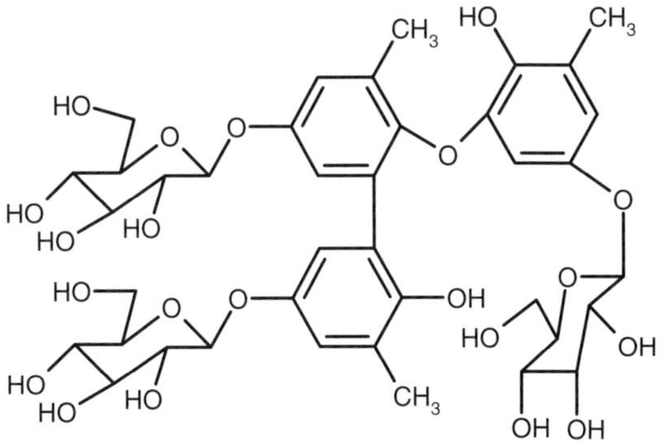

Pyrolaside B

Compound class: Phenolic glycoside trimer

Source: *Pyrola rotundifolia* (whole parts; family: Pyrolaceae)

Pharmaceutical potential: *Antibacterial*
Pyrolaside B displayed significant activity against two Gram-positive organisms, *Staphylococcus aureus* and *Micrococcus luteus*, with respective MIC values of 35.0 and 20.5 µM (bakuchiol was used as a positive control; respective MIC values were 20.0 and 10.0 µM).

Reference

Chang, J. and Inui, T. (2005) *Chem. Pharm. Bull.*, **53**, 1051.

6- and 8-(2-Pyrrolidinone-5-yl)-(−)-epicatechin

Physical data:
- 6-(2-Pyrrolidinone-5-yl)-(−)-epicatechin: $C_{19}H_{19}NO_7$, white powder, mp 195–197 °C, $[\alpha]_D^{20}$ −16.3° (MeOH, c 0.2)
- 8-(2-Pyrrolidinone-5-yl)-(−)-epicatechin: $C_{19}H_{19}NO_7$, white powder, mp 195–197 °C, $[\alpha]_D^{20}$ −1.0° (MeOH, c 0.2)

Structures:

6-Isomer: $R^1 =$ (2-pyrrolidinone-5-yl); $R^2 = H$

8-Isomer: $R^1 = H$; $R^2 =$ (2-pyrrolidinone-5-yl)

Compound class: Flavonoids (flavan-3-ols)

Source: *Actinidia arguta* (roots; family: Actinidiaceae) [1]

Pharmaceutical potential: *Inhibitors to the formation of advanced glycation end products (AGEs)*
Jang et al. [1] evaluated the flavonoid isolates for their potential to inhibit the formation of advanced glycation end products (AGEs); both of them were found to possess significant inhibitory efficacy against the formation of AGEs with an IC_{50} value of 13.5 µg/ml (36.0 µM) for 6-(2-pyrrolidinone-5-yl)-(−)-epicatechin and 17.9 µg/ml (47.8 µM) for the 8-isomer. The present investigators evaluated the potential also for (−)-catechin and (+)-catechin showing respective IC_{50} values of 36.1 µg/ml (125.2 µM) and 4.0 µg/ml (13.6 µM) (aminoguanidine was used as a positive control; $IC_{50} = 71.1$ µg/ml (961 µM)) [1].

AGEs not only are regarded as markers, but also are the important causative factors for the pathogenesis of diabetes, diabetic nephropathy, cataracts, atherosclerosis, and neurodegenerative diseases, including Alzheimer's disease [2–7]. Inhibitors of AGEs formation can thus offer a promising therapeutic approach in designing drugs for the prevention of diabetic or other pathogenic complications.

References

[1] Jang, D.S., Lee, G.Y., Lee, Y.M., Kim, Y.S., Sun, H., Kim, D.-H., and Kim, J.S. (2009) *Chem. Pharm. Bull.*, **57**, 397.
[2] Reddy, V.P. and Beyaz, A. (2006) *Drug Discov. Today*, **11**, 646.
[3] Yamamoto, Y., Doi, T., Kato, I., Shinohara, H., Sakurai, S., Yonekura, H., Watanabe, T., Myint, K.M., Harashima, A., Takeuchi, M., Takasawa, S., Okamoto, H., Hashimoto, N., Asano, M., and Yamamoto H. (2005) *Ann. N.Y. Acad. Sci.*, **1043**, 562.
[4] Stitt, A.W. (2005) *Ann. N.Y. Acad. Sci.*, **1043**, 582.
[5] Ahmed, N. (2005) *Diabetes Res. Clin. Pract.*, **67**, 3.

[6] Forbes, J.M., Yee, L.T., Thallas, V., Lassila, M., Candido, R., Jandeleit-Dahm, K.A., Thomas, M.C., Burns, W.C., Deemer, E.K., Thorpe, S.R., Cooper, M.E., Allen, T.J. (2004) *Diabetes*, **53**, 1813.
[7] Jono, T., Kimura, T., Takamatsu, J., Nagai, R., Miyazaki, K., Yuzuriha, T., Kitamura, T., Horiuchi, S. (2002) *Pathol. Int.*, **52**, 563.

Q

Quassimarin

Physical data: $C_{27}H_{36}O_{11}$, needles, mp 237.5–238.5 °C, $[\alpha]_D^{26}$ +22.4° ($CHCl_3$, *c* 0.29)

Structure:

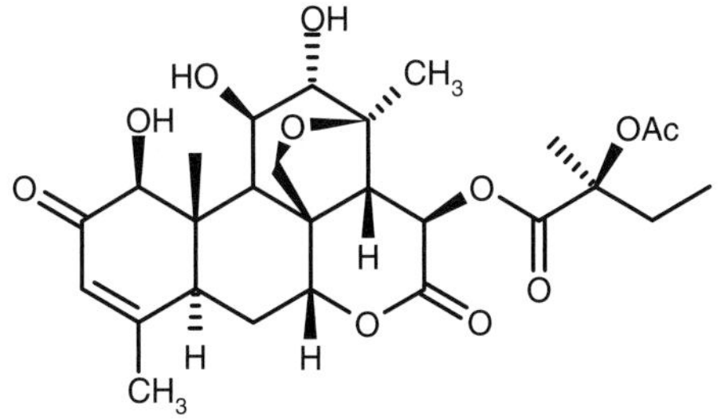

Quassimarin

Compound class: Quassinoid (simaroubolide)

Sources: *Quassia amara* L. (sap; family: Simaroubaceae) [1]; *Leitneria floridana* Chapman (aerial parts; family: Leitneriaceae) [2]

Pharmaceutical potential: ***Antitumor***
Xu *et al.* [2] evaluated that the isolate possesses potent *in vitro* antitumor efficacy by inhibiting the growth of a panel of human tumor cell lines such as KB, A-549, HCT-8, CAKI-1, MCF-7, and SK-MEL-2 with respective ED_{50} values of 0.06, 0.03, 0.012, 0.05, 0.006, and 0.05 μg/ml. Antileukemic potential of the compound was also reported by Kupchan *et al.* [1].

References

[1] Kupchan, S.M. and Streelman, D.R. (1976) *J. Org. Chem.*, **41**, 3481.
[2] Xu, Z., Chang, F.-R., Wang, H.-K., Kashiwada, Y., McPhail, A.T., Bastow, K.F., Tachibana, Y., Cosentino, M., and Lee, K.-H. (2000) *J. Nat. Prod.*, **63**, 1712.

Quercetin-3-*O*-α-(6″-caffeoylglucosyl-β-1,2-rhamnoside)

Physical data: $C_{36}H_{36}O_{19}$, solid, $[\alpha]_D^{25}$ −147.6° (MeOH, *c* 0.21)

Handbook of Pharmaceutical Natural Products, Volume 2.
Goutam Brahmachari

ISBN: 978-3-527-32148-3

Quercetin 3-O-α-L-[6″-p-coumaroyl-β-D-glucopyranosyl-(1→2)-rhamnopyranoside]

Structure:

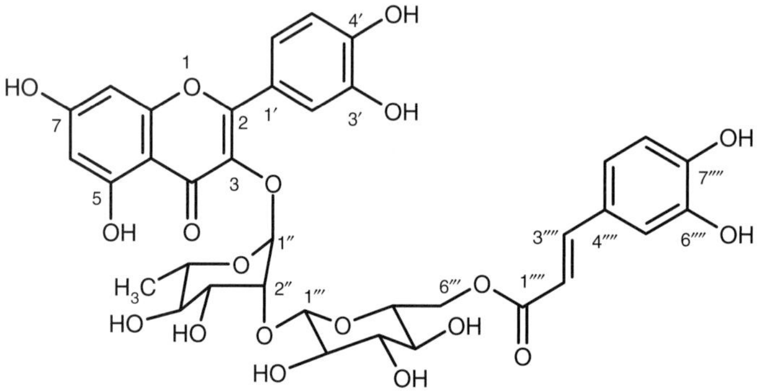

Quercetin-3-O-α-(6‴-caffeoylglucosyl-β-1,2-rhamnoside)

Compound class: Flavonoid glycoside

Source: *Sedum sarmentosum* Bunge (aerial parts; family: Crassulaceae) [1]

Pharmaceutical potential: *Angiotensin converting enzyme (ACE) inhibitory*

The isolate inhibited angiotensin converting enzyme (ACE) activity *in vitro* in a concentration-dependent manner with an IC_{50} value of $158.9 \pm 11.1\ \mu M$ (captopril was used as a positive control; $IC_{50} = 0.02\ \mu M$). Angiotensin II, a potent vasoconstrictor, is generated from angiotensin I by direct involvement of ACE (a zinc-dipeptidyl dipeptidase); angiotensin II is known to stimulate both the synthesis and release of aldosterone from the adrenal cortex, thereby increasing blood pressure via sodium retention [2]. Therefore, the inhibition of ACE has been considered to be one of the effective therapeutic approaches for the treatment of cardiovascular diseases such as hypertension.

References

[1] Oh, H., Kang, D.-G., Kwon, J.-W., Kwon, T.-O., Lee, S.-Y., Lee, D.-B., and Lee, H.-S. (2004) *Biol. Pharm. Bull.*, **27**, 2035.
[2] Lacaille-Dubois, M.A., Franck, U., and Wagner, H. (2001) *Phytomedicine*, **8**, 47.

Quercetin 3-O-α-L-[6″-p-coumaroyl-β-D-glucopyranosyl-(1→2)-rhamnopyranoside]

Physical data: $C_{36}H_{36}O_{18}$

Structure:

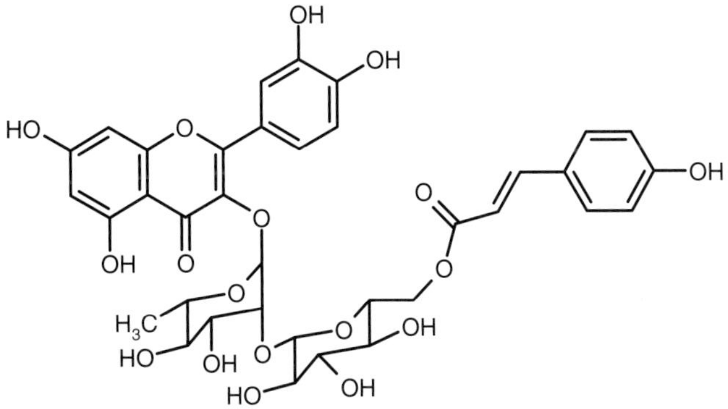

Quercetin 3-*O*-α-L-[6‴-*p*-coumaroyl-β-D-glucopyranosyl-(1→2)-rhamnopyranoside]

Compound class: Flavonoid glycoside (flavonol glycoside)

Source: *Ginkgo biloba* L. (leaves; family: Ginkgoaceae) [1–3]

Pharmaceutical potential: *Antioxidant*
Tang et al. [1] evaluated the flavonoid glycoside as an important antioxidant; it inhibited *in vitro* TPA-induced free radical generation in the cytochrome *c* reduction assay using the HL-60 cell culture system as well as scavenged free radical in the DPPH assay; the IC_{50} values were determined to be 13.2 μg/ml (17.5 μM) and 14.0 μg/ml (18.5 μM), respectively (gallic acid was used as a positive control; IC_{50} = 3.0 μg/ml (17.6 μM) in the cytochrome *c* reduction assay; 3.6 μg/ml (21.2 μM) in the DPPH assay) [1].

References

[1] Tang, Y., Lou, F., Wang, J., Li, Y., and Zhuang, S. (2001) *Phytochemistry*, **58**, 1251.
[2] Nasr, C., Lobstein-Guth, A., Haag-Berrurier, M., and Anton, R. (1987) *Phytochemistry*, **26**, 2869.
[3] Markham, K.R., Geiger, H., and Jaggy, H. (1992) *Phytochemistry*, **31**, 1009.

Quercetin 3-*O*-α-L-[6″-*p*-coumaroyl-β-D-glucopyranosyl-(1→2)-rhamnopyranoside]-7-*O*-β-D-glucopyranoside

Physical data: $C_{42}H_{47}O_{23}$, yellow amorphous powder

Structure:

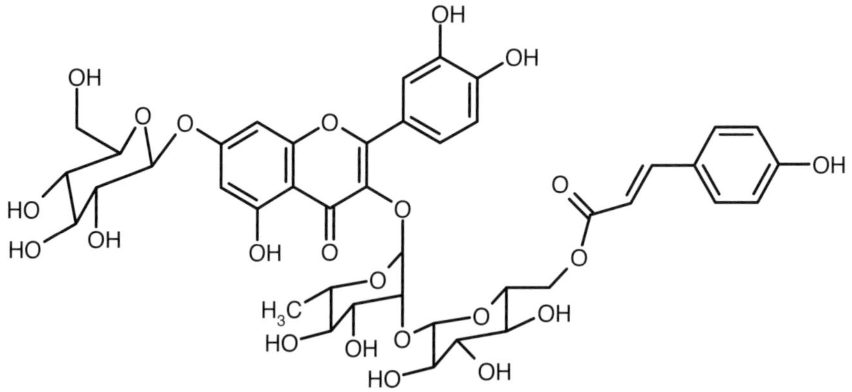

Quercetin 3-O-α-L-[6'''-p-coumaroyl-β-D-glucopyranosyl-(1→2)-rhamnopyranoside]-7-O-β-D-glucopyranoside

Compound class: Flavonoid glycoside (flavonol glycoside)

Source: *Ginkgo biloba* L. (leaves; family: Ginkgoaceae) [1, 2]

Pharmaceutical potential: *Antioxidant*

Tang et al. [1] reported that the flavonoid glycoside possesses significant *in vitro* antioxidant activity; it was noted to inhibit TPA-induced free radical generation in the cytochrome *c* reduction assay using the HL-60 cell culture system as well as to scavenge free radical in the DPPH assay; the IC$_{50}$ values were determined to be 13.5 µg/ml (14.7 µM) and 14.5 µg/ml (15.8 µM), respectively (gallic acid was used as a positive control; IC$_{50}$ = 3.0 µg/ml (17.6 µM) in the cytochrome *c* reduction assay; 3.6 µg/ml (21.2 µM) in the DPPH assay) [1].

References

[1] Tang, Y., Lou, F., Wang, J., Li, Y., and Zhuang, S. (2001) *Phytochemistry*, **58**, 1251.
[2] Hasler, A., Gross, G.A., Meier, B., and Sticher, O. (1992) *Phytochemistry*, **31**, 1391.

Quercetin 3-O-(2'',3''-digalloyl)-β-D-galactopyranoside

Physical data: $C_{35}H_{28}O_{20}$, yellow amorphous powder, $[\alpha]_D^{23}$ −28.7° (MeOH, *c* 0.30)

Structure:

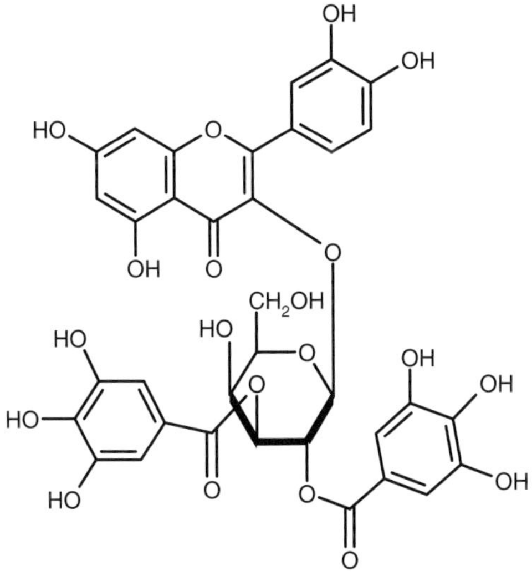

Quercetin 3-O-(2″,3″-digalloyl)-β-D-galactopyranoside

Compound class: Flavonoid

Source: *Euphorbia lunulata* (whole plants; family: Euphorbiaceae) [1]

Pharmaceutical potential: *Insulin-like activity*
In continuation to their search for identifying low molecular compounds that mimic human ligands [2], the investigators [1] established the quercetin galactoside as a promising lead for the development of a nonpeptidyl insulin substitutional medicine. The test compound exhibited insulin-like activity; it was found to stimulate *in vitro* proliferation activity of the insulin-dependent cells BAF/InsR in a dose-dependent manner showing a maximal efficiency (51.1 mU/ml) at 10 μg/ml [1].

References

[1] Nishimura, T., Wang, L.-Y., Kusano, K., and Kitanaka, S. (2005) *Chem. Pharm. Bull.*, **53**, 305.
[2] Kusano, K., Ebara, S., Tachibana, K., Nishimura, T., Sato, S., Kuwaki, T., and Taniyama, T. (2004) *Blood*, **103**, 836.

Quercetin 3-O-β-D-glucopyranosyl-(1→2)-rhamnopyranoside

Physical data: $C_{27}H_{30}O_{16}$

Structure:

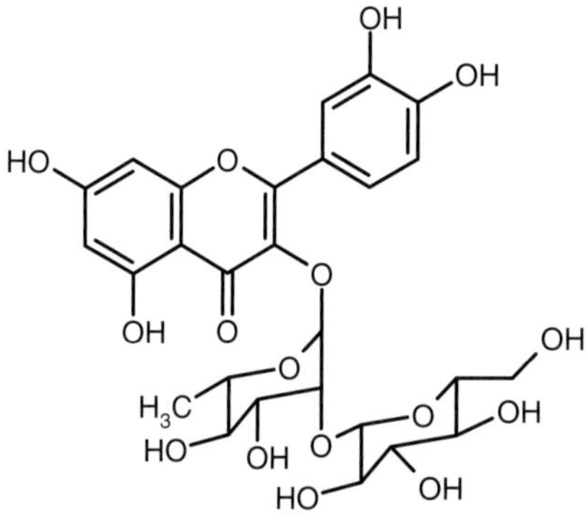

Quercetin 3-*O*-β-D-glucopyranosyl-(1→2)-rhamnopyranoside

Compound class: Flavonoid glycoside (flavonol glycoside)

Source: *Ginkgo biloba* L. (leaves; family: Ginkgoaceae) [1, 2]

Pharmaceutical potential: *Antioxidant*

The quercetin glycoside was reported to inhibit TPA-induced free radical generation in the cytochrome *c* reduction assay using the HL-60 cell culture system and also to scavenge free radical in the DPPH assay with IC_{50} values of 14.8 µg/ml (23.3 µM) and 16.1 µg/ml (26.4 µM), respectively (gallic acid was used as a positive control; IC_{50} = 3.0 µg/ml (17.6 µM) in the cytochrome *c* reduction assay; 3.6 µg/ml (21.2 µM) in the DPPH assay) [1].

References

[1] Tang, Y., Lou, F., Wang, J., Li, Y., and Zhuang, S. (2001) *Phytochemistry*, **58**, 1251.
[2] Hasler, A., Gross, G.A., Meier, B., and Sticher, O. (1992) *Phytochemistry*, **31**, 1391.

Quercetin 4′-*O*-rhamnopyranosyl-3-*O*-β-D-allopyranoside

Physical data: $C_{27}H_{30}O_{16}$, yellow amorphous powder, $[\alpha]_D^{20}$ −25.1° (MeOH, *c* 0.19)

Structure:

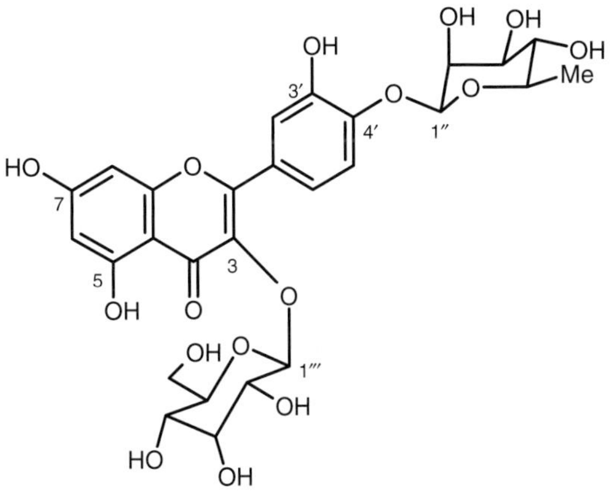

Quercetin 4'-O-rhamnopyranosyl-3-O-β-D-allopyranoside

Compound class: Flavonoid

Source: *Acacia pennata* Willd. (leaves; family: Mimosaceae)

Pharmaceutical potential: *Cyclooxygenase (COX) inhibitor*

The flavonoid glycoside was evaluated to possess potent inhibitory activity against cyclooxygenase-1 (COX-1) *in vitro* with an IC_{50} value of 11.6 µg/ml; it showed 80.4% inhibition at a dose of 1×10^{-4} g/ml. However, its efficacy against COX-2 is relatively lower (only 12.6% inhibition at the same concentration). Commercial aspirin was used as a positive control, which showed percent inhibition (IC_{50} value) of 54.2% (70 µg/ml) against COX-1 and 27.5% (238 µg/ml) against COX-2 at the same concentration.

Reference

Dongmo, A.B., Miyamoto, T., Yoshikawa, K., Arihara, S., and Lacaille-Dubois, M.-A. (2007) *Planta Med.*, **73**, 1202.

Quillaic acid glycosidic ester

Systematic name: Quillaic acid, α-L-arabinopyranosyl-(1→4)-α-L-arabinopyranosyl-(1→3)-β-D-xylopyranosyl-(1→4)-α-L-rhamnopyranosyl-(1→2)-β-D-fucopyranosyl ester

Physical data: $C_{57}H_{90}O_{25}$, white amorphous powder, $[\alpha]_D^{25}$ +8.1° (MeOH, *c* 0.1)

Structure:

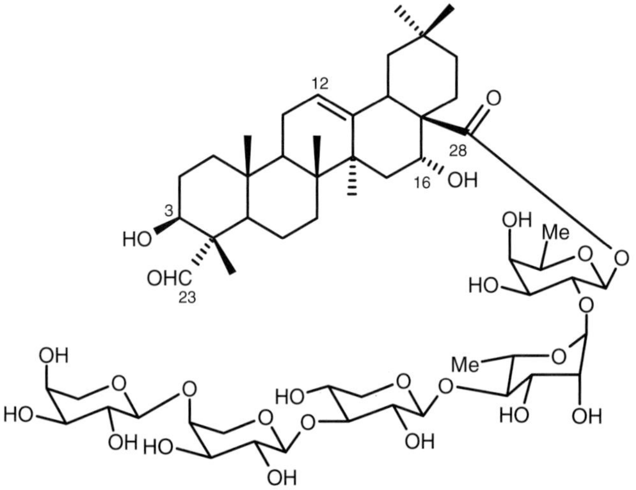

Quillaic acid glycosidic ester

Compound class: Monodesmosidic triterpene saponin

Source: *Gypsophila oldhamiana* Miq. (roots; family: Caryophyllaceae)

Pharmaceutical potential: *Immunomodulatory*

The triterpene saponin was evaluated for its immunomodulatory effect in an *in vitro* lymphocyte proliferation test system and also in the *in vitro* T-cell activation assay; the isolate was found to have an immunomodulatory effect in a concentration-dependent manner. In the concentration range of 10–100 µg/ml, it exhibited a significant enhancement of granulocyte phagocytosis (40–75%), and in the concentration range of 100 ng/ml to 1 pg/ml, it exerted an immunosuppressive effect (65–22%) in the T-cell activation assay.

Reference

Luo, J.-G., Kong, L.-Y., Takaya, Y., and Niwa, M. (2006) *Chem. Pharm. Bull.*, **54**, 1200.

R

Ravenic acid

Physical data: $C_{15}H_{16}NO_3$, orange–yellow amorphous solid

Structure:

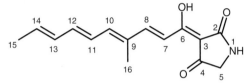

Ravenic acid

Compound class: Antibiotic (polyene tetramic acid)

Source: *Penicillium* sp. (fungus; culture broth)

Pharmaceutical potential: ***Antibacterial***
Ravenic acid was found to inhibit the growth of a methicillin-resistant *Staphylococcus aureus* (MRSA) strain down to 25 μg/ml.

Reference

Michael, A.P., Grace, E.J., Kotiw, M., and Barrow, R.A. (2002) *J. Nat. Prod.*, **65**, 1360.

Rediocides A, C, E, F, and G

Physical data:
- Rediocide A: $C_{44}H_{58}O_{13}$, white powder, mp 213–215 °C [1], 193–195 °C [3], $[\alpha]_D^{22}$ +80° (MeOH, *c* 0.7) [1], $[\alpha]_D^{30}$ −124° (MeOH, *c* 0.1) [3]
- Rediocide C: $C_{46}H_{54}O_{13}$, pale yellow gum [2], white powder [3], mp 191–193 °C [3], $[\alpha]_D^{22}$ + 130.7° (MeOH, *c* 1.75) [2], $[\alpha]_D^{30}$ −46° (MeOH, *c* 0.1) [3]
- Rediocide E: $C_{43}H_{56}O_{13}$, pale yellow gum [2], white powder [3], mp 197–198 °C [3], $[\alpha]_D^{23}$ + 131.3° (MeOH, *c* 0.53) [2], $[\alpha]_D^{30}$ −14° (MeOH, *c* 0.1) [3]
- Rediocide F: $C_{45}H_{52}O_{13}$, white powder, mp 199–200 °C, $[\alpha]_D^{30}$ −292° (MeOH, *c* 0.1) [3]
- Rediocide G: $C_{46}H_{54}O_{13}$, white powder, mp >230 °C (decomp.) [4]

Handbook of Pharmaceutical Natural Products, Volume 2.
Goutam Brahmachari

ISBN: 978-3-527-32148-3

Structures:

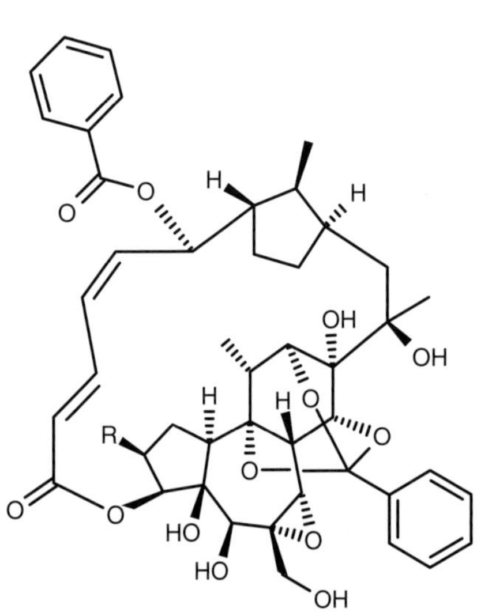

Rediocide C: R = CH₃
Rediocide F: R = H

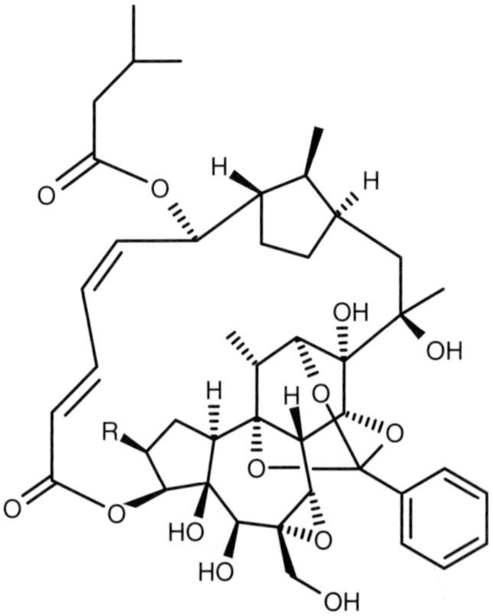

Rediocide A: R = CH₃
Rediocide E: R = H

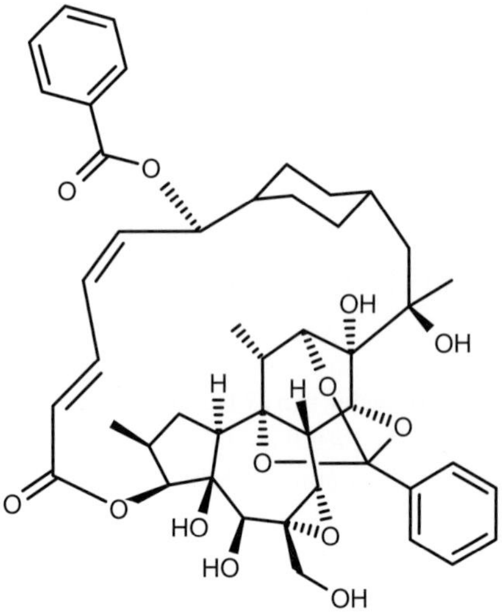

Rediocide G

Compound class: Daphnane diterpenoids

Source: *Trigonostemon reidioides* (roots; family: Euphorbiaceae) [1–4]

Pharmaceutical potentials: *Acaricidal (antiallergic); insecticidal*

House dust mite (HDM) is considered as one of the most important causative agents in the etiology of allergy [5–8]; Soonthornchareonnon et al. [3] evaluated such acaricidal potential of the isolated diterpenoids on *Dermatophagoides pteronyssinus* (HDM allergen). Rediocides A, C, E, and F exhibited promising killing effect against HDM with respective EC_{50} values (observed at the seventh day of incubation) of 2.53, 0.78, 5.59, and 0.92 μg/cm^2, which are found to be more potent than the reference standard, benzyl benzoate (LC$_{50}$ = 6.6 μg/cm^2). Rediocide A also exhibited cytotoxic activity against HeLa (ED$_{50}$ = 5.0 μg/ml) and HepG2 (ED$_{50}$ = 6.7 μg/ml) [4]. Temperam et al. [4] showed that rediocide G exhibits cytotoxicity against various cell lines such as GepGII, HeLa, HuCCA-1, and KB with ED$_{50}$ values of 6.4, 4.8, 5.0, and 5.0 μg/ml, respectively.

Rediocides are known to exhibit insecticidal activities against mosquito larvae (*Aedes aegypti*) and ectoparasitic flea (*Ctenocephalides felis*) [1, 2].

References

[1] Jayasuriya, H., Zink, D.L., Singh, S.B., Borris, R.P., Nanakorn, W., Beck, H.T., Balick, M.J., Goetz, M.A., Slayton, L., Gregorn, L., Zakson, M.Z., and Shoop, W. (2000) *J. Am. Chem. Soc.*, **122**, 4998.
[2] Jayasuriya, H., Zink, D.L., Borris, R.P., Nonakorn, W., Beck, H.T., Balick, M.J., Goetz, M.A., Gregom, L., Shoop, W.L., Singh, S.B. (2004) *J. Nat. Prod.*, **67**, 228.
[3] Soonthornchareonnon, N., Sakayarojkul, M., Isaka, M., Mahakittikun, V., Chuakul, W., and Wongsinkongman, P. (2005) *Chem. Pharm. Bull.*, **53**, 241.
[4] Temperam, A., Thasana, N., Pavaro, C., Chuakul, W., Siripong, P., and Ruchirawat, S. (2005) *Chem. Pharm. Bull.*, **53**, 1321.
[5] Mitchell, E.B., Wilkins, S., McCallum, D.J., and Platts-Mills, T.A.E. (1985) *Clin. Allergy*, **15**, 235.
[6] Platts-Mills, T.A.E. (1992) *J. Allergy Clin. Immunol.*, **89**, 1046.
[7] Ehnert, B., Lau-Schadendorf, S., Weber, A., Buettner, P., Schou, C., and Wahn, U. (1992) *J. Allergy Clin. Immunol.*, **90**, 135.
[8] McDonald, L.G. and Tovey, E. (1992) *J. Allergy Clin. Immunol.*, **90**, 599.

Remangiflavanones A and B

Systematic names:
- Remangiflavanone A: (2S)-5,7,4′-trihydroxy-8-lavandulylflavanone
- Remangiflavanone B: (2S)-5,7,2′,4′-tetrahydroxy-8-lavandulylflavanone

Physical data:
- Remangiflavanone A: $C_{25}H_{28}O_5$, yellow crystals, mp 147–149 °C, $[\alpha]_D^{24}$ −83.3° (MeOH, c 0.18)
- Remangiflavanone B: $C_{25}H_{28}O_6$, yellow crystals, mp 168–170 °C, $[\alpha]_D^{24}$ −46.4° (MeOH, c 0.14)

Structures:

Remangiflavanone A: R = H
Remangiflavanone B: R = OH

Compound class: Flavonoids (flavanones)

Source: *Physena madagascariensis* Noronha ex Thouars (leaves; family: Capparaceae)

Pharmaceutical potential: *Antibacterial*
Remangiflavanones A and B were found to have significant antibacterial potential (bacteriocidal as well as growth inhibitory) against a number of bacteria, such as *Staphylococcus aureus, Staph. epidermidis, Salmonella sonnei, Escherichia coli, Enterobacter aerogenes, Listeria monocytogenes*, and *Enterococcus* sp., at concentrations as low as 4 µM.

Reference

Deng, Y., Lee, J.P., Tianasoa-Ramamonjy, M., Snyder, J.K., Etages, A.D.S., Kanada, D., Snyder, M. P., and Turner, C.J. (2000) *J. Nat. Prod.*, **63**, 1082.

Respirantin and a related cyclodepsipeptide

Physical data:
- Cyclodepsipeptide **1** (respirantin): $C_{37}H_{53}N_3O_{13}$, colorless amorphous powder, mp 118–120 °C
- Cyclodepsipeptide **2**: $C_{36}H_{52}N_3O_{13}$, colorless amorphous powder, mp 117–120 °C

Structures:

Cyclodepsipeptide **1** (respirantin): R = CH$_2$CH(CH$_3$)$_2$
Cyclodepsipeptide **2**: R = CH(CH$_3$)$_2$

Compound class: Cyclodepsipeptides

Sources: Cyclodepsipeptide **1**: *Streptomyces* sp. [1], *Kitasatospora* sp. [2]; cyclodepsipeptide **2**: *Kitasatospora* sp. [2]

Pharmaceutical potential: *Anticancerous*
Pettit et al. [2] evaluated both the isolates as powerful inhibitors of cancer cell growth; cyclodepsipeptides **1** and **2** were found to inhibit the growth of murine P388 lymphocytic leukemia with respective ED$_{50}$ values of 0.0037 and 0.033 µg/ml. They strongly inhibited the growth of a variety of cancer cell lines such as BXPC-3 (pancreas), MCF-7 (breast), SF268 (CNS), KM20L2 (colon), and DU-145 (prostrate) with GI$_{50}$ values of 0.47 and 1.2; 0.0006 and 0.00064; 0.0016 and 0.00063; 0.0006 and 0.00058; 0.00018 and <0.0001 µg/ml, respectively. In addition, cyclodepsipeptide **1**

was also reported to show good activity against the pathogenic fungus *Cryptococcus neoformans* with an MIC value of 2.0 μg/ml, whereas cyclodepsipeptide **2** showed marginal activity against the bacterium *Micrococcus luteus* with an MIC value of 64 μg/ml [2].

References

[1] Urushibata, I., Isogai, A., Matsumoto, S., and Suzuki, A. (1993) *J. Antibiot.*, **46**, 701.
[2] Pettit, G.R., Tan, R., Pettit, R.K., Smith, T.H., Feng, S., Doubek, D.L., Richert, L., Hamblin, J., Weber, C., and Jean-Charles Chapuis, J.-C. (2007) *J. Nat. Prod.*, **70**, 1069.

Reynosin

Physical data: $C_{15}H_{20}O_3$, $[\alpha]_D^{25}$ +122° (EtOH, *c* 0.10)

Structure:

Reynosin

Compound class: Sesquiterpene lactone

Sources: *Ambrosia confertiflora* DC (family: Asteraceae) [1]; *Laurus nobilis* (bay leaf) [2]; *Magnolia grandiflora* L. (family: Magnoliaceae) [1]; *Saussurea lappa* Clarke (Saussureae Radix; roots; family: Compositae) [1, 3, 4]

Pharmaceutical potentials: *Melanogenesis inhibitor; cytotoxic; antitumor; inhibitor to TNF-α and CINC-1/IL-8*
Choi *et al.* [3] evaluated the isolate as a potent antimelanogenesis agent; the sesquiterpene lactone displayed promising inhibitory effect on the 3-isobutyl-1-methylxanthine (IBMX)-induced production of melanin in B-16 melanoma cells in a dose-dependant manner with an IC_{50} value of 2.5 μg/ml (arbutin was used as a positive control; IC_{50} = 29.0 μg/ml). Such potent inhibitor of the cellular production of melanin could find useful applications as active materials for skin whitening cosmetics [3].

Reynosin inhibited the LPS-induced production of tumor necrosis factor-α (TNF-α), a proinflammatory cytokine, in murine macrophage-like cell (RAW264.7 cells) in a dose-dependant manner with an IC_{50} value of 21.7 μg/ml (i.e., 87.4 μM) [4]. The isolate was also found to be a potent inhibitor to CINC-1 (cytokine-induced neutrophil chemoattractant-1) induction in LPS-stimulated rat kidney epithelioid NRK-52E cells; it exhibited the inhibitory effect in a dose-dependant manner with an IC_{50} value of 1.0 μM [1]. In addition, the sesquiterpene lactone was reported to induce apoptosis of leukemia cells [5].

References

[1] Jung, J.H., Ha, J.Y., Min, K.R., Shibata, F., Nakagawa, H., Kang, S.S., Chang, I.-M., and Kim, Y. (1998) *Planta Med.*, **64**, 454.
[2] Fang, F., Sang, S., Chen, K.Y., Gosslau, A., Ho, C.T., and Rosen, R.T. (2005) *Food Chem.*, **93**, 497.
[3] Choi, J.Y., Choi, E.H., Jung, H.W., Oh, J.S., Lee, W.-H., Lee, J.G., Son, J.-K., Kim, Y., and Lee, S.H. (2008) *Arch. Pharm. Res.*, **31**, 294.
[4] Cho, J.Y., Park, J., Yoo, E.S., Baik, K.U., Jung, J.H., Lee, J., and Park, M.H. (1998) *Planta Med.*, **64**, 594.
[5] Rivero, A., Qintana, J., Eiroa, J.L., Lopez, M., Triana, J., Bermejo, J., and Estevez, F. (2003) *Eur. J. Pharmacol.*, **482**, 77.

Rheediaxanthone A (pyranojacaeubin)

Systematic name: 4,8-Dihydroxy-6',6'-dimethyl-2H-pyrano(2',3':3,2)-6'',6''-dimethyl-2H-pyrano(2'',3'':6,7)xanthone

Physical data: $C_{23}H_{20}O_6$

Structure:

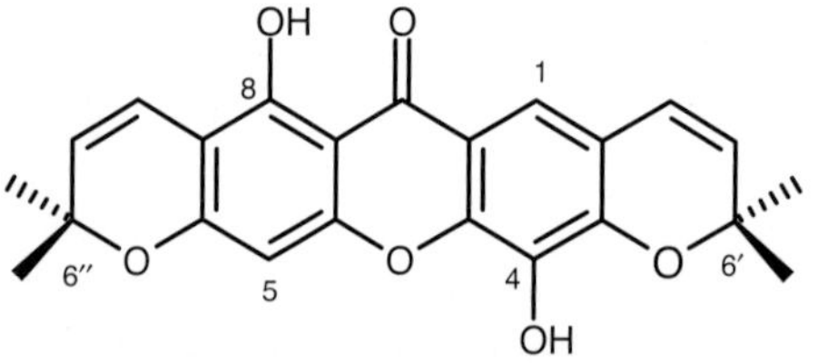

Rheediaxanthone A (pyranojacaeubin)

Compound class: Xanthone (pyranoxanthone)

Sources: *Garcinia densivenia* (stem barks; family: Clusiaceae/Guttiferae) [1]; *Rheedia gardneriana* (roots; family: Clusiaceae/Guttiferae) [2]; *Calophyllum blancoi* (roots; family: Guttiferae) [3]

Pharmaceutical potential: *Antiviral*
The pyranoxanthone exhibited *in vitro* antiviral activity against coronavirus by inhibiting virus-induced cytopathic effects with an EC_{50} value of 15.0 µg/ml.

References

[1] Waterman, P.G. and Crichton, E.G. (1980) *Phytochemistry*, **19**, 2723.
[2] Monache, G.D., Monache, F.D., Waterman, P.G., Clichton, E.G., and Lima, R.A. (1984) *Phytochemistry*, **23**, 1757.
[3] Shen, Y.-C., Wang, L.-T., Khalil, A.T., Chiang, L.C., and Cheng, P.-W. (2005) *Chem. Pharm. Bull.*, **53**, 244.

Rhinacanthins C and D

Physical data:
- Rhinacanthin C: $C_{25}H_{30}O_5$, yellow oil
- Rhinacanthin D: $C_{23}H_{20}O_7$, yellow powder

Structures:

Rhinacanthin C: R =

Rhinacanthin D: R =

Compound class: Naphthoquinones

Source: *Rhinacanthus nasutus* (L.) Kurz (family: Acanthaceae) [1–3]

Pharmaceutical potentials: *Antiviral; antiproliferative (antitumor)*

Both rhinacanthins C and D were reported to display potent inhibitory activity against human cytomegalovirus (hCMV) *in vitro* with EC_{50} values of 0.02 and 0.22 µg/ml, respectively [1].

Both the naphthoquinone esters were found to possess antiproliferative activities [2, 3]. Akinobu *et al.* [2] assessed the antiproliferative activity of the ethanol extract of root and aqueous extract of leaves of *R. nasutus*, a traditional Thai medicine for cancer treatment, and the supposed active moiety rhinacanthin C *in vitro* using the human cervical carcinoma cell line HeLa, its MDR1 overexpressing subline Hvr100-6, human prostate carcinoma PC-3 cells, and human bladder carcinoma T24 cells. The antiproliferative activity of the *R. nasutus* extracts was also assessed *in vivo* using sarcoma 180-bearing mice. On the basis of their detailed studies, the investigators observed that rhinacanthin C showed potent *in vitro* antiproliferative activity for MDR1 overexpressing Hvr100-6 cells, similarly to parent HeLa cells, and the *in vitro* antiproliferative activity of the ethanol extract of root *R. nasutus* was due to rhinacanthin C, whereas that of the aqueous extract of leaves of *R. nasutus* was due to constituents other than rhinacanthin C. Both of the *R. nasutus* extracts showed *in vivo* antiproliferative activity after oral administration once daily for 14 days [2]. Pongpun *et al.* [3] also studied the *in vitro* antiproliferative activity against a panel of 10 kinds of cancer cells (e.g., KB, Hep-2, MCF-7, HepG2, HeLa, SiHa, C-32, LLC, Colon-26, and P388) and nontumorigenic Vero cells using the MTT assay. Antitumor activity of aqueous extracts of the roots and stems of the plant, the chloroform extract, and rhinacanthin C in Meth-A sarcoma-bearing BALB/c mice was also evaluated. The naphthoquinones showed apparent antiproliferative activity against cancer cells with the IC_{50} values of 0.29–54.4 µM, whereas they showed moderate activity against Vero cells. Regarding *in vivo* antitumor activity, the chloroform extract and rhinacanthin C (25 mg/kg/day) as well as aqueous extracts of the roots and stems (500 mg/kg/day) significantly suppressed the growth of Meth-A sarcoma-bearing mice [3]; the investigators also suggested that that the presence of phenolic group(s) in quinine and quinolate parts of these compounds may be important for increasing their antiproliferative activity.

References

[1] Sendl, A., Chen, J.L., Jolad, S.D., Stoddart, C., Rozhon, E., and Kernan, M. (1996) *J. Nat. Prod.*, **59**, 808.
[2] Akinobu, G., Toshiyuki, S., Takashi, K., Toshiro, S., Yoshitaka, W., Atsuchi, W., Tetsutaro, K., Yoshiji, T., Akira, I., Seigo, I., Midori, H., Hisako, T., and Katsuhiko, O. (2004) *Biol. Pharm. Bull.*, **27**, 1070.
[3] Pongpun, S., Kwanjai, K., Suratsawadee, P., Jantana, Y., Rittichai, C., Somsak, R., and Naoto, O. (2006) *J. Tradit. Med.*, **23**, 166.

Rhodiocyanoside A

Physical data: $C_{11}H_{17}NO_6$, white powder, $[\alpha]_D^{25}$ $-16.1°$ (MeOH, c 0.4)

Structure:

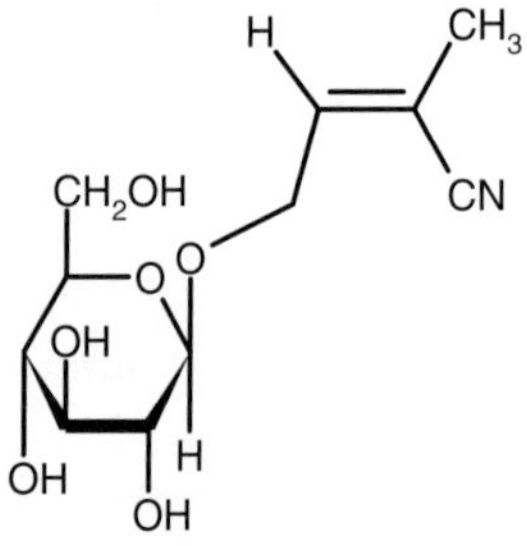

Rhodiocyanoside A

Compound class: Cyanoglycoside

Source: *Rhodiola quadrifida* (Pall.) Fisch. et Mey. (underground parts; family: Crassulaceae)

Pharmaceutical potential: *Antiallergic*
The isolate was evaluated to possess inhibitory potential against histamine release from rat peritoneal exudate mass cells induced by antigen–antibody reaction; it exhibited $(60.7 \pm 3.4)\%$ inhibition at a concentration of 10^{-4} M.

Reference

Yoshikawa, M., Shimada, H., Shimada, H., Murakami, N., Yamahara, J., and Matsuda, H. (1996) *Chem. Pharm. Bull.*, **44**, 2086.

Rhuscholide A

Systematic name: 5-Hydroxy-3-(propan-2-ylidene)-7-3,7,11,15-tetramethylhexadeca-2,6,10,11--teraenyl)-2(3*H*)-benzofuranone

Physical data: $C_{31}H_{42}O_3$, pale yellow amorphous powder

Structure:

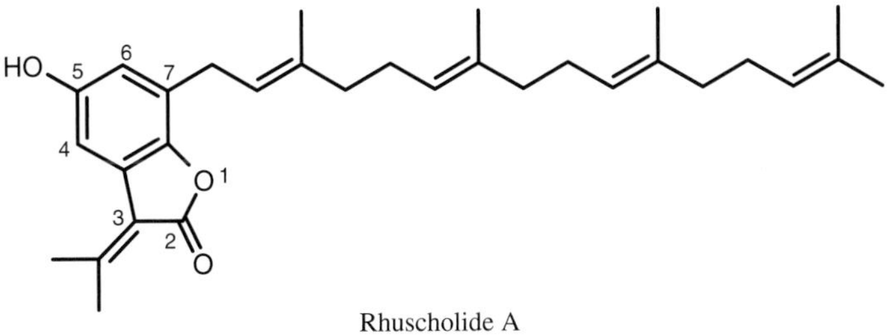

Rhuscholide A

Compound class: Benzofuran lactone

Source: *Rhus chinensis* (stem; family: Anacardiaceae)

Pharmaceutical potential: *Anti-HIV-1*

Anti-HIV-1 bioassay studies revealed that the benzofuran lactone possesses significant anti-HIV-1 activity *in vitro* with an EC_{50} value of 1.62 μM and a therapeutic index (TI) of 42.40; Gu *et al.* [1] further noted that the substitution of propan-2-ylidene at C-3 within rhuscholide A molecule is quite important for its anti-HIV activity since the C-3 unsubstituted benzofuranone isolated from the same plant [1] as well as from *Iryanthera grandis* [2] was found to be not so active (EC_{50} = 3.7 μM with TI of only 3.28).

References

[1] Gu, Q., Wang, R.-R., Zhang, X.-M., Wang, Y.-H., Zheng, Y.-T., Zhou, J., and Chen, J.-J. (2007) *Planta Med.*, **73**, 279.
[2] Vieira, P.C., Gottlieb, O.R., and Gottlieb, H.E. (1983) *Phytochemistry*, **22**, 2281.

Robustaflavone and its 7,4′,7″-trimethyl ether

Physical data:
- Robustaflavone: $C_{30}H_{18}O_{10}$, yellowish powder
- Robustaflavone 7,4′,7″-trimethyl ether: $C_{33}H_{24}O_{10}$, yellow powder

Structures:

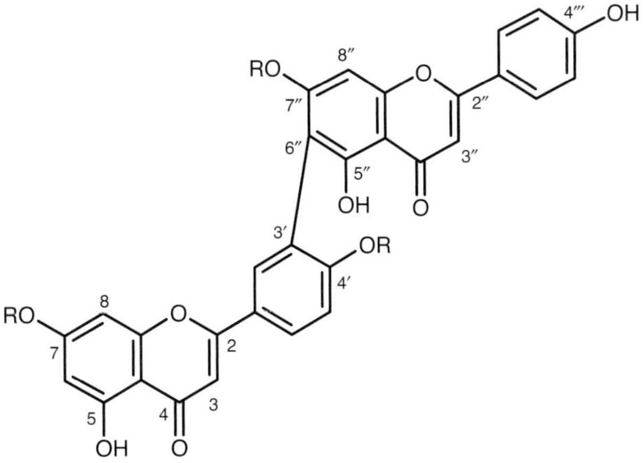

Robustaflavone: R = H
Trimethyl ether: R = Me

Compound class: Flavonoid (biflavonoid)

Sources: Robustaflavone: *Rhus succedanea* (seed kernels; family: Anacardiaceae) [1–3]; *Selaginella delicatula* (whole plants; family: Selaginellaceae) [4]; *Thuja orientalis* L. (syn. *Biota orientalis* (L.) Endl., *Platycladus orientalis* (L.) Franco) (fruits; family: Cupressaceae) [5, 6]; Robustaflavone 7,4′,7″-trimethyl ether: *Selaginella doederleinii* Hiron. (whole plants; family: Selaginellaceae) [7, 8]

Pharmaceutical potentials: *Anti-hepatitis B virus (anti-HBV); human neutrophil elastase (HNE) inhibitor*

Robustaflavone was reported to exhibit promising anti-hepatitis B activity [2, 3, 9]; Lin et al. [2] evaluated the compound as a potent *in vitro* inhibitor of hepatitis B with an EC_{50} value of 0.25 µM and an *in vitro* selectivity index (SI = IC_{50}/EC_{90}) of 153. The investigators suggested that inhibition of HBV DNA polymerase is the mechanism of action. Zembower et al. [3] also demonstrated such anti-hepatitis B activity of the biflavonoid and its hexaacetate derivative was observed to inhibit HBV replication with an EC_{50} value of 0.73 µM, but exhibited no cytotoxicity at concentrations up to 1000 µM. Combinations of robustaflavone with penciclovir and lamivudine displayed synergistic anti-HBV activity, having the most pronounced effects when the combination ratios were similar to the ratio of EC_{50} potencies. Thus, a 1:1 combination of robustaflavone and penciclovir exhibited an EC_{50} value of 0.11 µM and an SI of 684, while a 10:1 combination of robustaflavone and lamivudine exhibited an EC_{50} value of 0.054 µM and an SI of 894 [3]. The biflavone was also reported to exhibit significant human neutrophil elastase (HNE) inhibitory activity with an IC_{50} value of 1.33 µM [6].

Robustaflavone 7,4′,7″-trimethyl ether was reported to be cytotoxic against human cancer cell lines such as HCT (colorectal carcinoma), NCI-H358 (bronchioalveolar carcinoma), and K562 (chronic myelogenous leukemia) with respective EC_{50} values of 15.6, 20.1, and 22.5 µM [8].

References

[1] Lin, Y.-M. and Chen, F.-C. (1974) *Phytochemistry*, **13**, 1617.
[2] Lin, Y.-M., Zembower, D.E., Flavin, M.T., Schure, R.M., Anderson, H.M., Korba, B.E., and Chen, F.-C. (1997) *Bioorg. Med. Chem. Lett.*, **7**, 2325.
[3] Zembower, D.E., Lin, Y.-M., Flavin, M.T., Chen, F.-C., and Korba, B.E. (1998) *Antiviral Res.*, **39**, 81.
[4] Lin, L.-C., Kuo, Y.-C., and Chou, C.-J. (2000) *J. Nat. Prod.*, **63**, 627.
[5] Khabir, M., Khatoon, F., and Ansari, W.H. (1985) *Curr. Sci.*, **54**, 1180.
[6] Xu, G.-H., Ryoo, I.-J., Kim, Y.-H., Choo, S.-J., and Yoo, I.-D. (2009) *Arch. Pharm. Res.*, **32**, 275.
[7] Gu, Y., Xu, Y., Fang, S., and He, Q. (1990) *Zhiwu Xuebao*, **32**, 631.
[8] Lee, N.-Y., Min, H.-Y., Lee, J., Nam, J.-W., Lee, Y.-J., Han, A.-R., Wiryawan, A., Suprapto, W., Lee, S.K., and Seo, E.-K. (2008) *Chem. Pharm. Bull.*, **56**, 1360.
[9] Zembower, D.E. and Zhang, H. (1998) *J. Org. Chem.*, **63**, 9300.

Rocaglamide

Physical data: $C_{29}H_{31}NO_7$, crystalline (MeOH), mp 118–119 °C, $[\alpha]_D^{25}$ −96° (CHCl$_3$, c 1.0) [1]; mp 117–118 °C (*iso*-Pr$_2$O–Et$_2$O), $[\alpha]_D^{20}$ −93° (CHCl$_3$, c 1.88) [2]

Structure:

[Chemical structure of Rocaglamide]

Rocaglamide

Compound class: Benzofuran (1H-2,3,3a,8b-tetrahydrocyclopenta[b]benzofuran)

Sources: *Aglaia elliptifolia* Merr. (roots and stems; family: Meliaceae) [1]; *A. odorata* (leaves) [2]; *A. duperreana* (roots) [3]

Pharmaceutical potentials: *Antileukemic; cytotoxic; antitumor; inhibitor of platelet aggregation; insecticidal*

This compound showed an optimal T/C value of about 156% at a dose of 1.0 mg/kg against the P-388 murine lymphocytic leukemia *in vivo* test system [1].

Rocaglamide exhibited potent cytotoxic activity against six cancer cell lines such as KB (nasal pharyngeal carcinoma), A-549 (human lung carcinoma), HCT-8 (human colon carcinoma), P-388 (murine leukemia), RPMI-7951 (human melanoma), and TE-671 (human medulloblastoma) with respective IC_{50} values of 0.006, 0.006, 0.007, 0.005, 0.002, and 0.006 μM [4]. The antitumor activity of the isolate was favorably comparable with those of standard antitumor drugs (drug IC_{50} values) [5]: etoposide 0.12 (KB) and 2.62 (P-388); vinblastine sulfate 0.002 (A-549), 0.005 (HLT-8); doxorubicin hydrochloride 0.15 (A-549) and 0.3 (HCT-8). The compound demands further investigation as a lead molecule for potentially useful antitumor and antiproliferative agent.

The isolate was reported to possess potent inhibition activity toward TNF-α and NF-ϰB in T cells, thereby being anti-inflammatory [6]. Besides these pharmaceutical potentials, rocaglamide also showed potent insecticidal properties as tested against *Spodoptera littoralis*; the EC_{50} and LC_{50} values were found to be 0.08 and 0.9 ppm, respectively, determined in a chronic feeding assay (7 days) with the neonate larvae (positive control, azadirachtin: $EC_{50} = 0.06$ ppm; $LC_{50} = 0.7$ ppm) [3].

References

[1] King, M.L., Chiang, C.C., Ling, H.C., Fujita, E., Ochiai, M., and McPhail, A.T. (1982) *J. Chem. Soc., Chem. Commun.*, 1150.
[2] Ishibashi, F., Satasook, C., Isman, M.B., and Towers, G.H. (1993) *Phytochemistry*, **32**, 307.
[3] Hiort, J., Chaidir, Bohnenstengel, F.I., Nugroho, B.W., Schneider, C., Wray, V., Witte, L., Hung, P.D., Kiet, L.C., and Proksch, P. (1999) *J. Nat. Prod.*, **62**, 1632.
[4] Wu, T.-S., Liou, M.-J., Kuoh, C.-S., Teng, C.-M., Nagao, T., and Lee, K.-H. (1997) *J. Nat. Prod.*, **60**, 606–608.
[5] Lee, K.H., Lin, Y.M., Wu, T.S., Zhang, D.C., Yamagishi, T., Hayashi, T., Hall, I.H., Chang, J.J., Wu, R.Y., and Yang, T.H. (1988) *Planta Med.*, **54**, 308–311.

Rostratins A–D

[6] Baumann, B., Bohnenstengel, F., Siegmund, D., Wajant, H., Weber, C., Herr, I., Debatin, K.-M., Proksch, P., and Wirtha, T. (2002) *J. Biol. Chem.*, **277**, 44791.

Rostratins A–D

Physical data:
- Rostratin A: $C_{18}H_{24}N_2O_6S_2$, colorless gum, $[\alpha]_D^{20}$ −185° (MeOH–CH$_2$Cl$_2$ 1: 1, *c* 0.0037)
- Rostratin B: $C_{18}H_{20}N_2O_6S_2$, colorless gum, $[\alpha]_D^{20}$ −210° (MeOH–CH$_2$Cl$_2$ 1: 1, *c* 0.0004)
- Rostratin C: $C_{18}H_{20}N_2O_8S_2$, colorless gum, $[\alpha]_D^{20}$ −167° (MeOH–CH$_2$Cl$_2$ 1: 1, *c* 0.0022)
- Rostratin D: $C_{18}H_{20}N_2O_6S_4$, colorless gum, $[\alpha]_D^{20}$ +108° (MeOH–CH$_2$Cl$_2$ 1: 1, *c* 0.0069)

Structures:

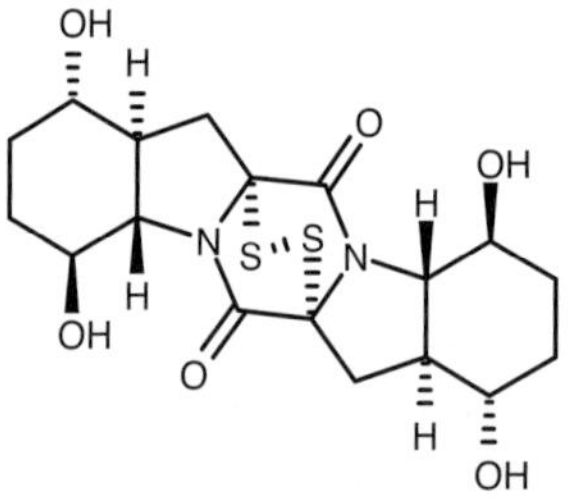

Rostratin A

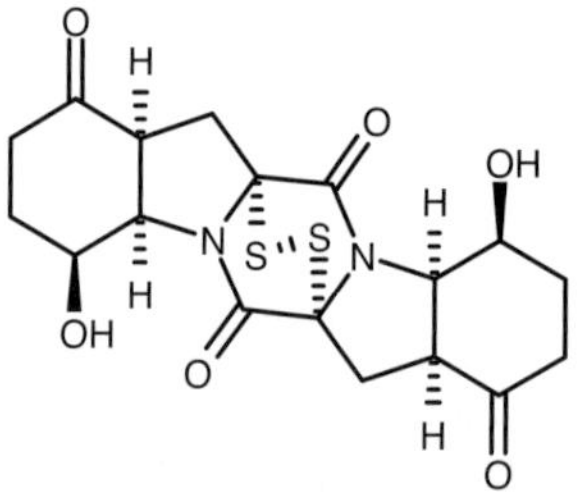

Rostratin B

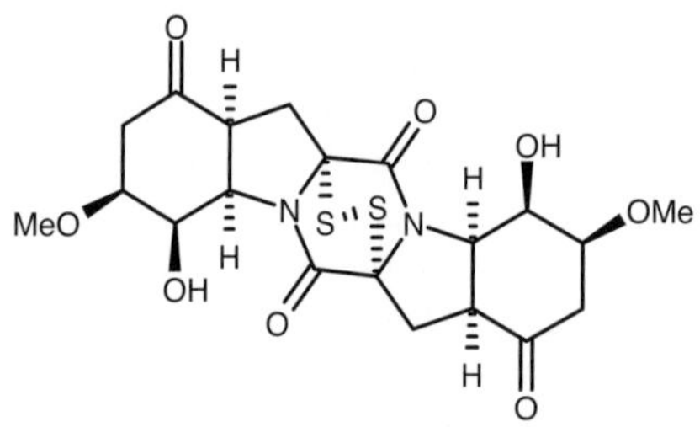

Rostratin C

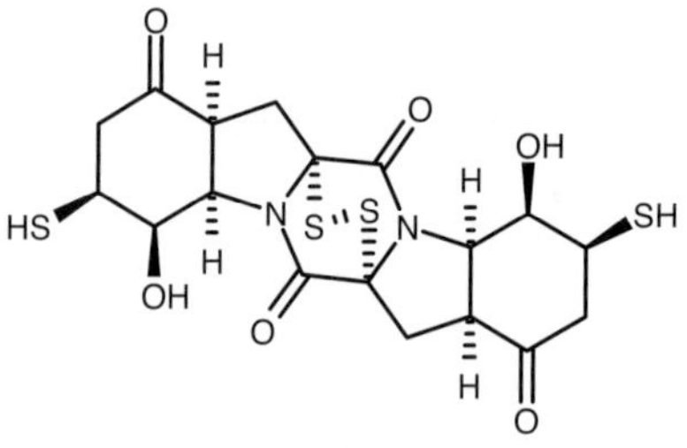

Rostratin D

Compound class: Antibiotics (dithioalkaloids)

Source: *Exserohilum rostratum* (marine-derived fungus; culture broth)

Pharmaceutical potential: *Cytotoxic*

Rostratins A–D displayed *in vitro* cytotoxicity against human colon carcinoma (HCT-116) with IC$_{50}$ values of 8.5, 1.9, 0.76, and 16.5 μg/ml, respectively.

Reference

Tan, R.X., Jensen, P.R., Williams, P.G., and Fenical, W. (2004) *J. Nat. Prod.*, **67**, 1374.

Rotenone

Systematic name: 1,2,12,12a-Tetrahydro-8,9-dimethoxy-2-(1-methylethenyl)-[1]-benzopyrano[3,4-b]furo[2,3-h]-[1]-benzopyan-6(6H)-one

Physical data: $C_{23}H_{22}O_6$, crystalline solid, mp 162–163 °C, $[\alpha]_D^{25}$ −226−228° (benzene)

Structure:

Rotenone

Compound class: Flavonoid

Sources: *Derris elliptica* Benth. (family: Leguminosae) [1–6]; *Mundulea chapelieri* (Baill.) R. Viguier ex Du Puy & Labat (bark, leaves and flowers; family: Fabaceae) [7]; *Erycibe expansa* (stems; family: Convolvulaceae) [8]

Pharmaceutical potentials: *Cytotoxic; antitumor; NO production inhibitor; antiproliferative; insecticidal and piscicidal*

Rotenone was found to possess potent antitumor and growth inhibitory activities as studied in both cultured cells and experimental tumors [9–13]. Blasko *et al.* [14] noted intense, but nonspecific, broad cytotoxic activity of the test compound against P-388 lymphocytic leukemia, KB carcinoma of the nasopharynx, and a number of human cancer cell types such as HT-1080 human fibrosarcoma, LU-1 lung cancer, COL-2 colon cancer, MEL-2 melanoma, and BC-1 breast cancer cell lines *in vitro*; the respective ED_{50} values were determined to be 0.005, 0.067, 0.047, 0.044, 0.15, 0.092, and 0.039 µg/ml. It exerted potent cytotoxicity *in vitro* against the A2780 human ovarian cancer cells with an IC_{50} value of 0.7 µg/ml [7]. In addition, rotenone exhibited significant inhibitory activity against lipopolysaccharide-activated nitric oxide (NO) production in mouse peritoneal macrophages with an IC_{50} of 27 µM [8]. Rotenone is widely used as an agricultural and horticultural insecticide and piscicide [14, 15].

References

[1] Biichi, G., Crombie, L., Godin, P.J., Kaltenbronn, J.S., Siddalingaiah, K.S., and Whiting, D.A. (1961) *J. Chem. Soc.*, 2843.
[2] Crombie, L. and Godin, P.J. (1961) *J. Chem. Soc.*, 2861.
[3] Crombie, L. and Lown, J.W. (1962) *J. Chem. Soc.*, 775.
[4] Adam, D.J., Crombie, L., and Whiting, D.A. (1966) *J. Chem. Soc. C*, 542.
[5] Carlson, D.G., Weisleder, D., and Tallent, W.H. (1973) *Tetrahedron*, **29**, 2731.
[6] Nishizawa, Y. and Casida, J.E. (1965) *J. Agric. Food Chem.*, **13**, 522.

[7] Cao, S., Schilling, J.K., Miller, J.S., Andriantsiferana, R., Rasamison, V.E., and Kingston, D.G. I. (2004) *J. Nat. Prod.*, **67**, 454.
[8] Morikawa, T., Xu, F., Matsuda, H., and Yoshikawa, M. (2006) *Chem. Pharm. Bull.*, **54**, 1530.
[9] Sinha, S.N., Ray, B.K., Goswami, B.B., Bhattacharrya, P., and Kipak, K.D. (1975) *Indian J. Med. Res.*, **63**, 1737.
[10] Figueras, M.J. and Gosalvez, M. (1973) *Eur. J. Cancer*, **9**, 529.
[11] Lee, K.H., Anuforo, D.C., Hung, D.-C., and Piantadosi, C. (1972) *J. Pharm. Sci.*, **61**, 626.
[12] Terranova, T., Galeotti, T., Baldi, S., and Negri, G. (1967) *Biochem. Z.*, **346**, 439.
[13] Loffler, M. and Schneider, F. (1982) *Mol. Cell Biochem.*, **48**, 77.
[14] Blasko, G., Shieh, H.-L., Pezzuto, J.M., and Cordell, G.A. (1989) *J. Nat. Prod.*, **52**, 1363.
[15] Ingham, J.L. (1983) *Fortschr. Chem. Org. Naturst.*, **43**, 1.

Rotundifoliosides A, H, I, and J

Systematic names:
- Rotundifolioside A: 13β,28-epoxy-16α-hydroxyurs-11-en-3β-yl β-D-xylopyranosyl-(1→2)-β-D-glucopyranosyl-(1→2)-β-D-glucopyranoside
- Rotundifolioside H: 13β,28-epoxy-16α,23-dihydroxyurs-11-en-3β-yl β-D-xylopyranosyl-(1→2)-β-D-glucopyranosyl-(1→2)-β-D-fucopyranoside
- Rotundifolioside I: 13β,28-epoxy-16α-hydroxyurs-11-en-3β-yl β-D-xylopyranosyl-(1→2)-β-D-glucopyranosyl-(1→2)-β-D-fucopyranoside
- Rotundifolioside J: 3β,28-epoxy-16α-hydroxyurs-11-en-3β-yl α-L-rhamnopyranosyl-(1→2)-β-D-glucopyranosyl-(1→2)-β-D-fucopyranoside

Physical data:
- Rotundifolioside A: $C_{47}H_{76}O_{17}$, white powder, $[\alpha]_D^{24}$ −62° (pyridine, *c* 0.18)
- Rotundifolioside H: $C_{47}H_{76}O_{17}$, white powder, $[\alpha]_D^{24}$ −9.4° (pyridine, *c* 0.64)
- Rotundifolioside I: $C_{47}H_{76}O_{16}$, white powder, $[\alpha]_D^{24}$ −10.1° (pyridine, *c* 0.99)
- Rotundifolioside J: $C_{48}H_{78}O_{16}$, white powder, $[\alpha]_D^{24}$ +31.3° (pyridine, *c* 1.0)

Structures:

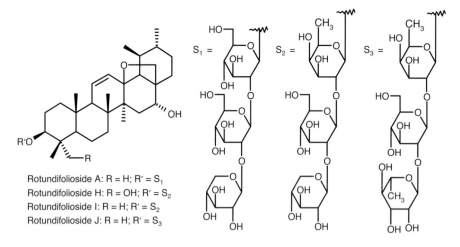

Rotundifolioside A: R = H; R′ = S₁
Rotundifolioside H: R = OH; R′ = S₂
Rotundifolioside I: R = H; R′ = S₂
Rotundifolioside J: R = H; R′ = S₃

Compound class: Triterpenoid saponins (ursane-type saikosaponin analogs)

Source: *Bupleurum rotundifolium* (fruits; family: Umbelliferae)

Pharmaceutical potential: *Antiproliferatives*
All the triterpenoid glycosides were evaluated for their *in vitro* antiproliferative potentials against MK-1, human uterus carcinoma (HeLa), and murine melanoma (B16F10) cells; the respective 50% growth inhibition (GI_{50}) values were determined to be 48, 71, and 31 µM (rotundifolioside A); 18, 31, and 18 µM (rotundifolioside H); 20, 37, and 18 µM (rotundifolioside I); 16, 21, and 11 µM (rotundifolioside J).

Reference

Fujioka, T., Yoshida, K., Fujii, H., Nagao, T., Okabe, H., and Mihashi, K. (2003) *Chem. Pharm. Bull.*, **51**, 365.

Rubraxanthone

Systematic name: 1,3,6-Trihydroxy-7-methoxy-8-[(2*E*)-3,7-diemethyloct-2,6-dienyl]xanthone

Physical data: $C_{24}H_{26}O_6$, yellow powder, mp 205–206 °C [1], 207–209 °C [8]

Structure:

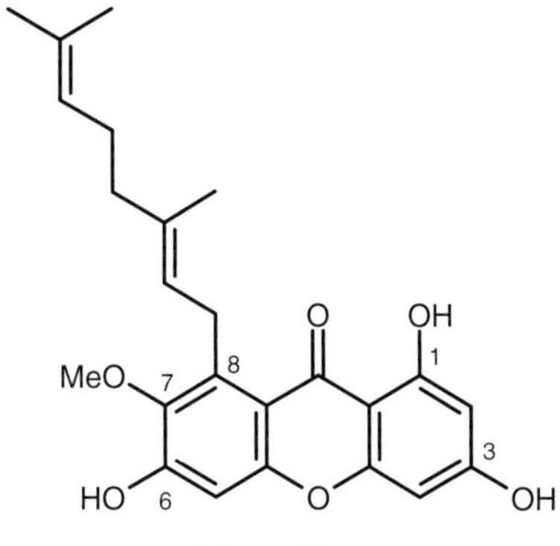

Rubraxanthone

Compound class: Prenylated xanthone

Sources: *Garcinia merguensis* (woods; family: Guttiferae) [1, 2], *G. dioica* [3], *G. cowa* [4], *G. parvifolia* Miq. [5, 6]; *Allanblackia monticola* (stem barks; family: Clusiaceae) [7]; *Mesua corneri* (stem barks; family: Guttiferae) [8]; *Allanblackia monticola* (stem barks; family: Guttiferae) [9]

Pharmaceutical potentials: *Cytotoxic; anticancerous; platelet activating factor (PAF) inhibitor; antimicrobial*
Rubraxanthone was found to possess promising anticancerous potential; Kijjoa *et al.* [2] showed that the compound displayed prominent cytotoxicity against a panel of cancer cell lines such as MCF-7, MDA-MB-231, NCI-H460, and SF-268 with respective GI_{50} values of 9.0, 16.8 ± 1.4, 18.5 ± 1.4, and 17.0 ± 2.5 µM (doxorubicin was used as a positive control with respective GI_{50}

values of 42.8 ± 8.2, 10.86 ± 1.28, 94.0 ± 8.7, and 94.0 ± 7.0 nM). The geranylated xanthone molecule also displayed significant cytotoxicity against the CEM-SS cell line (T-lymphoblastic leukemia) with an IC_{50} value of 5.0 µg/ml [8].

Rubraxanthone exhibited a strong inhibitory effect on platelet activating factor (PAF) binding to rabbit platelets with an IC_{50} value of 18.2 µM; from the structure–activity relationships of xanthones, the present investigators [6] suggested that a geranyl group substituted at C-8 becomes beneficial to the binding, while a hydroxylated prenyl group at C-4 resulted in a significant loss in binding to the PAF receptor.

Rubraxanthone also reported to exhibit growth inhibitory activity against *Staphylococcus aureus*, *Pseudomonas aeruginosa*, *Bacillus subtilis*, and *Escherichia coli* [10]. Iinuma *et al.* [3] showed that the test compound displayed potent inhibitory activity against *Staphylococcus* strains with MIC values of 0.31–1.25 µg/ml, an activity that was greater than that of the antibiotic vancomycin (MIC 3.13–6.25 µg/ml); this strong *in vitro* antibacterial activity of the xanthone derivative against both methicillin-resistant and methicillin-sensitive *Staphylococcus aureus* suggests that the compound might find wide pharmaceutical uses [3].

References

[1] Ampofo, S.A. and Waterman, P.G. (1986) *Phytochemistry*, **25**, 2351.
[2] Kijjoa, A., Gonzalez, M.J., Pinto, M.M., Nascimento, M.S.J., Campos, N., Mondranondra, I.-O., Silva, A.M.S., Eaton, G., and Herz, W. (2008) *Planta Med.*, **74**, 864.
[3] Iinuma, M., Tosa, H., Tanaka, T., Asai, F., Kobayashi, Y., Shimano, R., and Miyauchi, K. (1996) *J. Pharm. Pharmacol.*, **48**, 861.
[4] Wahyuni, F.S., Byrne, L.T., Dachriyanus, Dianita, R., Jubahar, J., Lajis, N.H., and Sargent, M.V. (2004) *Aust. J. Chem.*, **57**, 223.
[5] Rukachaisirikul, V., Naklue, W., Phongpaichit, S., Towatana, N.H., and Maneenoon, K. (2006) *Tetrahedron*, **62**, 8578.
[6] Jantan, I., Pisar, M.M., Idris, M.S., Taher, M., and Ali, R.M. (2002) *Planta Med.*, **68**, 1133.
[7] Azebaze, A.G.B., Meyer, M., Valentin, A., Nguemfo, E.L., Fomum, Z.T., and Nkengfack, A.E. (2006) *Chem. Pharm. Bull.*, **54**, 111.
[8] Ee, G.C.L., Izzaddin, S.A., Rahmani, M., Sukari, M.A., and Lee, H.L. (2006) *Nat. Prod. Sci.*, **12**, 138.
[9] Azebaze, A.G.B., Meyer, M., Bodo, B., and Nkengfack, A.E. (2004) *Phytochemistry*, **65**, 2561.
[10] Taher, M., Idris, M.S., and Arbain, D. (1999) *Bull. Kimia*, **14**, 51.

Rubriflordilactone B

Physical data: $C_{22}H_{28}O_4$, colorless prisms, mp 201–202 °C, $[\alpha]_D^{27.9}$ −119.84° (MeOH, *c* 0.514)

Structure:

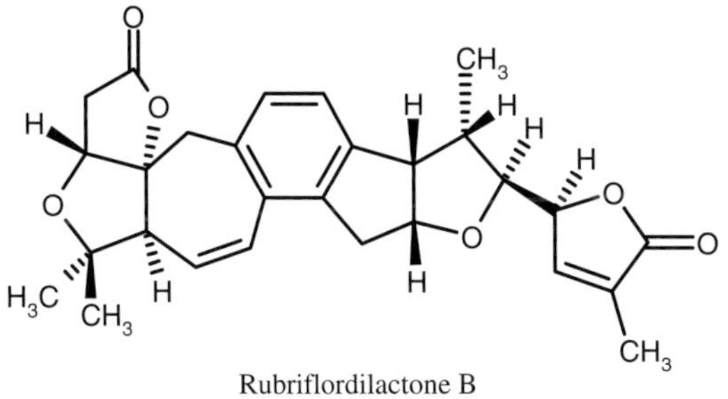

Rubriflordilactone B

Compound class: Bisnortriterpenoid

Source: *Schisandra rubriflora* (leaves and stems; family: Schisandraceae)

Pharmaceutical potential: *Anti-HIV*
The isolate was found to exert inhibitory activity against HIV-1 replication with low cytotoxicity; it inhibited HIV-1$_{IIIB}$-induced syncytium formation in C8166 cells with an EC$_{50}$ value of 9.75 μg/ml and with a selectivity index (SI) of 12.39. The present investigators also noted that the test compound exerted its obvious protection against lytic effects of HIV-1$_{IIIB}$-induced MT-4 host cells with a selectivity index of 6.09.

Reference

Xiao, W.-L., Yang, L.-M., Gong, N.-B., Wu, L., Wang, R.-R., Pu, J.-X., Li, L., Huang, S.-X., Zheng, Y.-T., Li, R.-T., Lu, Y., Zheng, Q.-T., and Sun, H.-D. (2006) *Org. Lett.*, **8**, 991.

Rubrisandrin A

Physical data: C$_{22}$H$_{28}$O$_6$, obtained as an inseparable mixture of **1** and **2** (at a ratio of about 2 : 1), amorphous solid, $[\alpha]_D^{22}$ +18.5° (MeOH, *c* 0.17)

Structure:

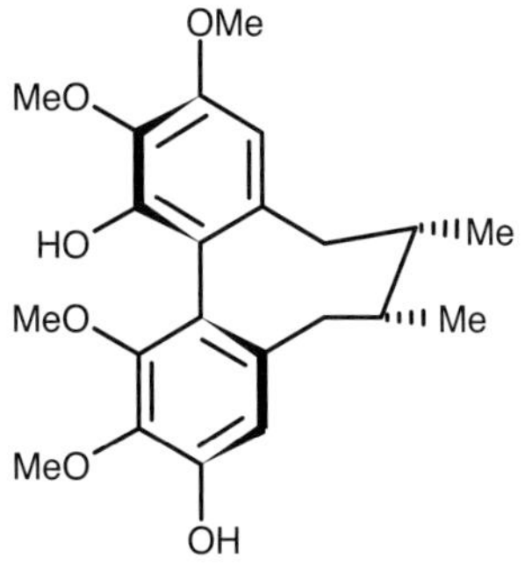

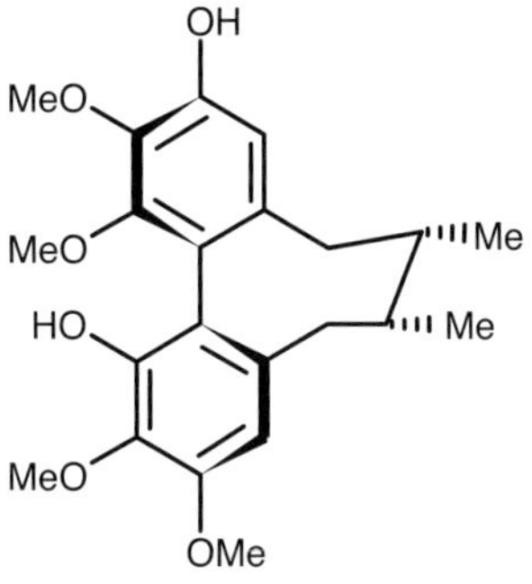

Rubrisandrin A (inseparable mixture of **1** and **2**)

Rubrisandrin A

Compound class: Dibenzocyclooctadiene lignan

Source: *Schisandra rubriflora* (fruits; family: Schisandraceae)

Pharmaceutical potential: *Anti-HIV*

Rubrisandrin A exhibited *in vitro* inhibitory effect on HIV replication in H9 lymphocytes with an EC_{50} value of 11.3 µM (IC_{50} >64 µM).

Reference

Chen, M., Kilgore, N., Lee, K.-H., and Chen, D.-F. (2006) *J. Nat. Prod.*, **69**, 1697.

S

Salaspermic acid

Physical data: $C_{30}H_{48}O_4$, colorless prisms (acetone), mp >320 °C

Structure:

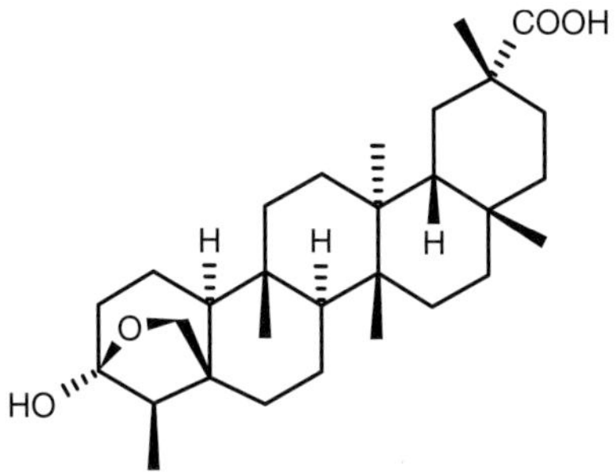

Salaspermic acid

Compound class: Triterpenoid (friedelane type)

Sources: *Salacia macrosperma* (family: Hippocrateaceae) [1, 2]; *Tripterygium wilfordii* Hook. (roots; family: Celastraceae) [3]

Pharmaceutical potential: ***Anti-HIV***
Salaspermic acid was found to have moderate inhibitory activity against HIV replication in H9 lymphocyte cells with an ED_{50} value of 5 µg/ml (10 µM); it inhibited uninfected H9 cell growth with an IC_{50} value of 53 µM [3]. The compound also showed an inhibitory effect against HIV-1 recombinant reverse transcriptase-associated reverse transcriptase activity with an EC_{50} value of ~16 µg/ml [3].

References

[1] Viswanathan, N.I. (1979) *J. Chem. Soc. Perkin Trans. I*, **2**, 349.
[2] Zhang, W.J., Pan, D.J., Zhang, L.X., and Shao, Y.D. (1986) *Acta Pharm. Sin.*, **21**, 592.
[3] Chen, K., Shi, Q., Kashiwada, Y., Zhang, D.-C., Hu, C.-Q., Jin, J.-Q., Nozaki, H., Kilkuskie, R. E., Tramontano, E., Cheng, Y.-C., McPhail, D.R., McPhail, A.T., and Lee, K.-H. (1992) *J. Nat. Prod.*, **55**, 340.

(−)-Salutaridine

Systematic name: 4-Hydroxy-3,6-dimethoxy-17-methyl-5,6,8,14-tetradehydromorphinan-7-one

Physical data: $C_{19}H_{21}NO_4$

Handbook of Pharmaceutical Natural Products, Volume 2.
Goutam Brahmachari

ISBN: 978-3-527-32148-3

Structure:

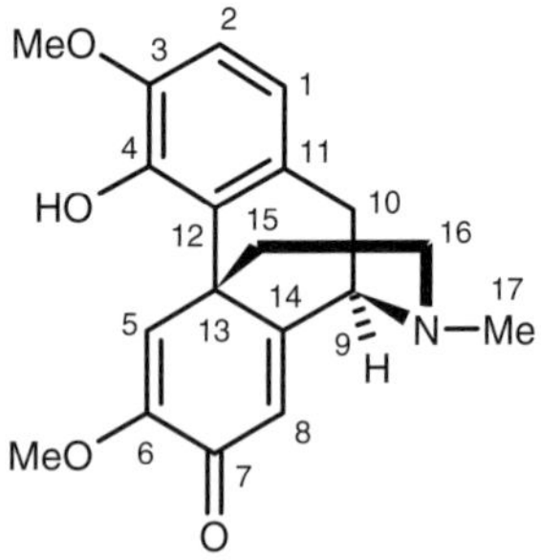

(−)-Salutaridine

Compound class: Alkaloid (morphinane alkaloid)

Sources: *Papaver bracteatum* (family: Papaveraceae) [1]; *Corydalis saxicola* Bunting (roots; family: Fumariodeae) [2]

Pharmaceutical potential: *Anti-hepatitis B virus*
Wu et al. [2] evaluated that the alkaloid possesses potent inhibitory potential against the secretions of HBsAg (hepatitis B virus surface antigen) and HBeAg (hepatitis B virus e antigen) in the HepG2.2.15 cell line with IC_{50} values of 0.27 µM (SI = 2.1) and 0.45 µM (SI = 1.3), respectively; lamivudine was used as a positive control (IC_{50} value against HBsAg: 15.37 µM (SI = 2.9); IC_{50} value against HBeAg: 44.85 µM (SI = 1.0)).

References

[1] Sariyar, G., Gulgeze, H.B., and Gozler, B. (1992) *Planta Med.*, **58**, 368.
[2] Wu, Y.-R., Ma, Y.-B., Zhao, Y.-X., Yao, S.-Y., Zhou, J., Zhou, Y., and Chen, J.J. (2007) *Planta Med.*, **73**, 787.

Salvileucalin B

Physical data: $C_{20}H_{15}O_5$, colorless amorphous solid, $[\alpha]_D$ +44.7° (CHCl$_3$, *c* 0.10)

Structure:

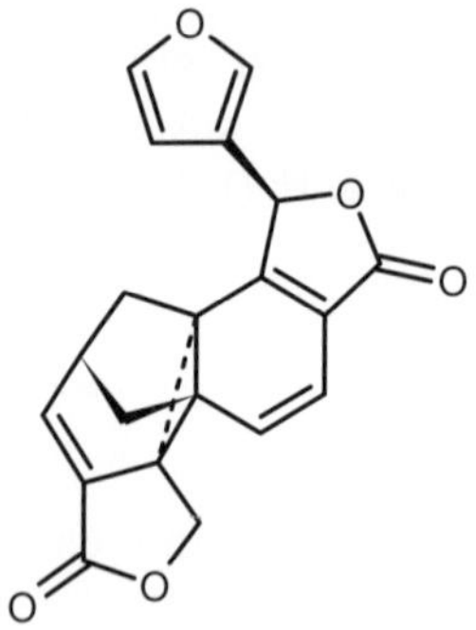

Salvileucalin B

Compound class: Diterpenoid (neoclerodane type)

Source: *Salvia leucantha* Cav. (aerial parts; family: Labiatae)

Pharmaceutical potential: *Cytotoxic*
The compound displayed cytotoxicity against A549 (human lung adenocarcinoma) and HT-29 (human colon adenocarcinoma) cells with IC$_{50}$ values of 5.23 and 1.88 µg/ml, respectively.

Reference

Aoyagi, Y., Yamazaki, A., Nakatsugawa, C., Fukaya, H., Takeya, K., Kawauchi, S., and Izum, H. (2008) *Org. Lett.*, **10**, 4429.

Salvins A and B

Systematic names:
- Salvin A: 2α,3β,5α-trihydroxyurs-12-en-28-oic acid
- Salvin B: 3α,6α,24-trihydroxyolean-12-en-28-oic acid

Physical data:
- Salvin A: C$_{30}$H$_{48}$O$_5$, colorless amorphous solid, $[\alpha]_D^{26}$ −74.4° (MeOH, *c* 0.062)
- Salvin A: C$_{30}$H$_{48}$O$_5$, amorphous powder, $[\alpha]_D^{26}$ −53.8° (MeOH, *c* 0.013)

Structures:

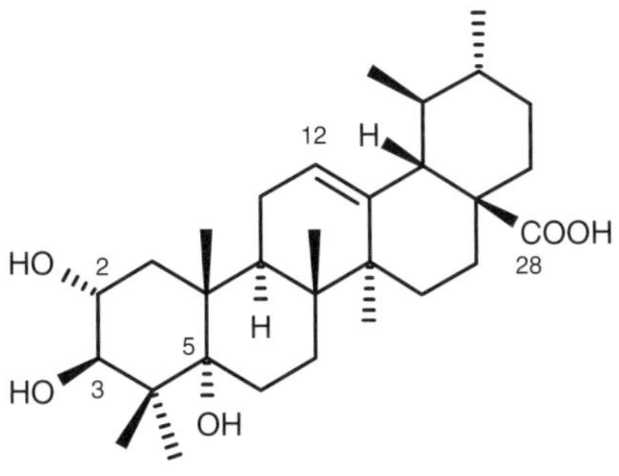

Salvin A

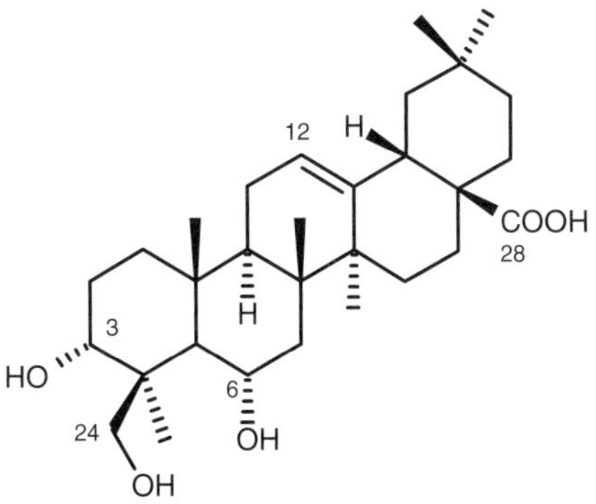

Salvin B

Compound class: Triterpenoids

Source: *Salvia santolinifolia* (whole plants; family: Labiatae)

Pharmaceutical potential: *Butyrylcholinesterase (BChE) inhibitors*
Both the isolates exerted significant inhibitory potential against butyrylcholinesterase (BChE) enzyme with IC$_{50}$ values of 12.5 ± 0.08 and 65.5 ± 0.01 µM (galantamine was used as a positive control; IC$_{50}$ = 8.7 ± 0.01 µM).

Reference

Mehmood, S., Riaz, N., Nawaz, S.A., Afza, N., Malik, A., and Choudhary, M.I. (2006) *Arch. Pharm. Res.*, **29**, 195.

Salzmannianosides A and B

Systematic names:
- Salzmannianoside A: 3-O-{[β-D-glucopyranosyl-(1→4)]-[α-L-rhamnopyranosyl-(1→2)]-α-L-arabinopyranosyl}gypsogenin
- Salzmannianoside B: 3-O-{[β-D-glucopyranosyl-(1→4)]-[α-L-arabinopyranosyl-(1→3)-α-L-rhamnopyranosyl-(1→2)]-α-L-arabinopyranosyl}hederagenin

Physical data:
- Salzmannianoside A: $C_{47}H_{74}O_7$, $R_f = 0.11$ ($CHCl_3$–MeOH–H_2O (7:3:1))
- Salzmannianosides B: $C_{52}H_{85}O_{21}$, $R_f = 0.06$ ($CHCl_3$–MeOH–H_2O (7:3:1))

Structures:

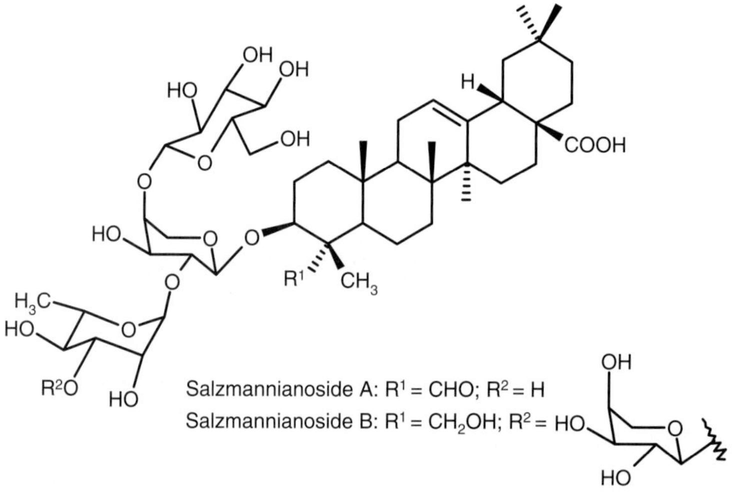

Salzmannianoside A: R^1 = CHO; R^2 = H
Salzmannianoside B: R^1 = CH_2OH; R^2 = HO-

Compound class: Saponins (triterpenoid glycosides)

Source: *Serjania salzmanniana* Schlecht. (stems; family: Sapindaceae)

Pharmaceutical potential: *Antifungal*
Salzmannianosides A and B showed antifungal activities against certain pathogenic fungi such as *Cryptococcus neoformans*, *Candida albicans*, and *Aspergillus fumigatus* with respective MIC values of 8, 16, and 125 and 8, 125, and 250 µg/ml (amphotericin B was used as a positive control).

Reference

Ekabo, O.A., Farnsworth, N.R., Henderson, T.O., Mao, G., and Mukherjee, R. (1996) *J. Nat. Prod.*, **59**, 431.

Samaderines B, C, E, X, Y, and Z

Physical data:
- Samaderine B: $C_{19}H_{23}O_7$
- Samaderine C: $C_{19}H_{25}O_7$
- Samaderine E: $C_{20}H_{26}O_8$, solid, mp 202–207 °C, $[\alpha]_D^{25}$ −11.7° (pyridine, c 0.23)
- Samaderine X: $C_{22}H_{28}O_9$, colorless prisms (n-hexane–CHCl$_3$), mp 171–172 °C, $[\alpha]_D^{24}$ −5.1° (pyridine, c 0.7)
- Samaderine Y: $C_{20}H_{26}O_7$, colorless prisms (n-hexane–CHCl$_3$), mp 182–183 °C, $[\alpha]_D^{26}$ −29.8° (pyridine, c 0.5)
- Samaderine Z: $C_{20}H_{26}O_8$, white amorphous solid, $[\alpha]_D^{25}$ +46.1° (MeOH, c 0.3)

Structures:

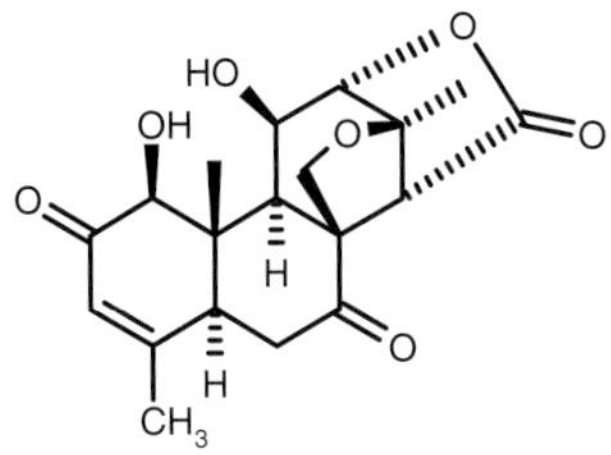

(Samaderine B) 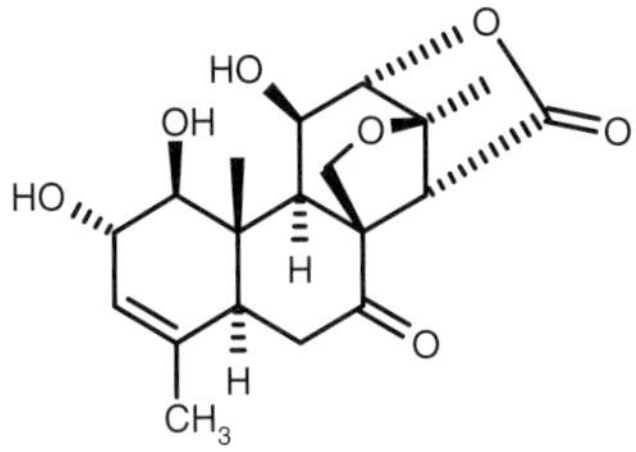 (Samaderine C) 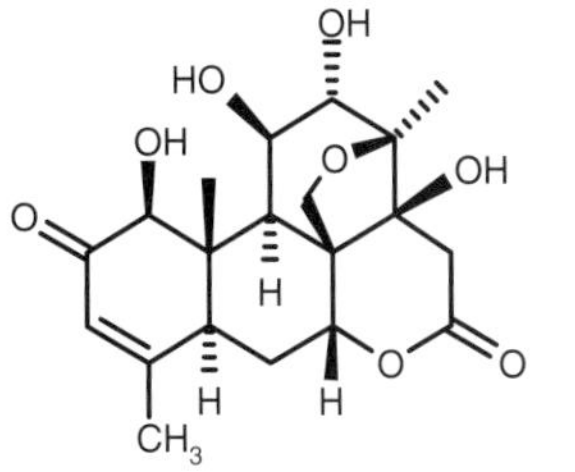 (Samaderine E)

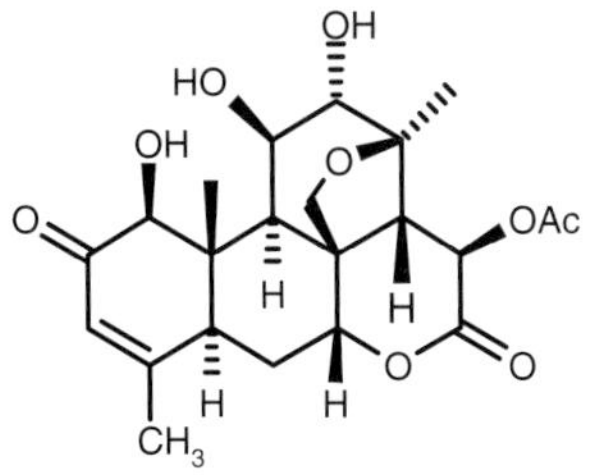

(Samaderine X) 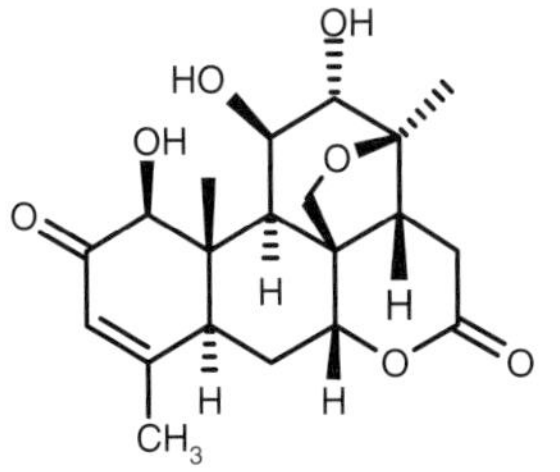 (Samaderine Y) 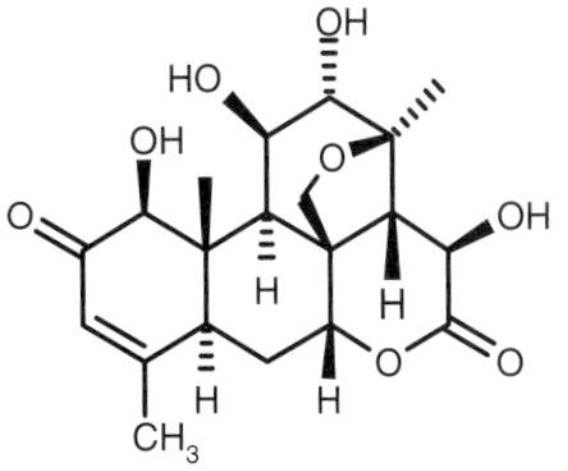 (Samaderine Z)

Compound class: Quassinoids (modified triterpene)

Sources: Samaderine B: *Quassia indica* (stems; family: Simaroubaceae) [1, 2], *Samadera madagascariensis* (leaves; family: Simaroubaceae) [3]; Samaderine C: *Quassia indica* (stems; family: Simaroubaceae) [1, 2]; Samaderine E: *Quassia indica* (stems) [1], *Samadera indica* [4]; Samaderines X and Z: *Quassia indica* (stems; family: Simaroubaceae) [1]; Samaderines Y: *Ailanthus malabarica* (family: Simaroubaceae) [5], *Quassia indica* (stems; family: Simaroubaceae) [1]

Pharmaceutical potentials: *Cytotoxic; antimalarial; anti-inflammatory*

All the isolates were found to exhibit potent *in vitro* cytotoxic activity against KB cells; among them samaderines B, E, and X showed greater efficacies. The individual IC$_{50}$ values for the quassinoids (samaderines B, C, E, X, Y, and Z) were determined to be 0.07, 0.40, 0.04, 0.02, 0.10, and 0.20 μg/ml, respectively [1].

Samaderines B, E, X, and Z were reported to exhibit significant *in vitro* antimalarial potential tested against the parasite *Plasmodium falciparum* (a chloroquine-resistant K1 strain in human erythrocytes); the respective IC_{50} and IC_{90} values (μM) were determined to be 0.21 and 0.69 μM for samaderine B, 0.056 and 0.093 μM for samaderine E, 0.014 and 0.069 μM for samaderine X, and 0.071 and 0.19 μM for samaderine Z [1, 6].

Furthermore, samaderines B and X were observed to show anti-inflammatory activity in Sprague–Dawley rats; Kitagawa et al. [1] measured the exudate volume and the number of leukocytes in the pleural cavity 4 h after carrageenin injection; samaderines B and X, administered 0 and 60 min after carrageenin injection at a dose of 1.0 mg/kg body weight, inhibited the exudate volume by 78 and 79%, respectively, and the number of leukocytes by 94 and 95%, respectively [1].

References

[1] Kitagawa, I., Mahmud, T., Yokota, K., Nakagawa, S., Mayumi, T., Kobayashi, M., and Shibuya, H. (1996) *Chem. Pharm. Bull.*, **44**, 2009.
[2] Zylber, J. and Polonsky, J. (1964) *Bull. Soc. Chim. Fr.*, 2016.
[3] Coombes, P.H., Naidoo, D., Mulholland, D.A., and Randrianarivelojosia, M. (2005) *Phytochemistry*, **66**, 2734.
[4] Wani, M.C., Taylor, H.L., Wall, M.E., McPhail, A.T., and Onan, K.D. (1977) *J. Chem. Soc., Chem. Commun.*, 295.
[5] Aono, H., Koike, K., Kanako, J., and Ohmoto, T. (1994) *Phytochemistry*, **37**, 579.
[6] Tanabe, K., Izumo, A., Kato, M., Miki, A., and Doi, S. (1989) *J. Protozool.*, **36**, 139.

Sandwicensin

Systematic name: 3-Hydroxy-9-methoxy-10-prenylpterocarpan

Physical data: $C_{21}H_{22}O_4$, clear glass, $[α]_D$ −116° ($CHCl_3$, *c* 0.24)

Structure:

Sandwicensin

Compound class: Pterocarpan (isoflavonoid)

Sources: *Erythrina mildbraedii* (family: Leguminosae) [1]; *E. glauca* Willd. (bark) [2]; *E. lysistemon* (aerial parts including flowers) [3]; *E. x bidwillii* [4]; *E. poeppigiana* (roots) [5]; *E. stricta* (roots) [6]

Pharmaceutical potentials: *Anti-HIV; antimycobacterial*

Sandwicensin was reported to exert moderate *in vitro* inhibition to the cytopathic effects of HIV-1 infection in a human T-lymphoblastoid cell line (CEM-SS) with EC_{50} and IC_{50} values of 2.0 and 7.0 μg/ml, respectively, showing a maximum of 50–60% protection [2]. It also showed weak antimycobacterial activity with an MIC value of 50 μg/ml [6].

References

[1] Mitscher, L.A., Okwute, S.K., Gollapudi, S.R., Drake, S., and Avona, E. (1988) *Phytochemistry*, **27**, 3449–3452.
[2] McKee, T.C., Bokesch, H.R., McCormick, J.L., Rashid, M.A., Spielvogel, D., Kirk, R., Gustafson, K.R., Alavanja, M.M., Cardellina, J.H., II, and Boyd, M.R. (1997) *J. Nat. Prod.*, **60**, 431–438.
[3] Juma, B.F. and Majinda, R.R.T. (2005) Proceedings of the 11th NAPRECA Symposium, August 9–12, Antananarivo, Madagascar, pp. 97–109.
[4] Tanaka, H., Tanaka, T., Hosoya, A., Kitade, Y., and Etoh, H. (1998) *Phytochemistry*, **47**, 1397.
[5] Sato, M., Tanaka, H., Yamaguchi, R., Oh-Uchi, T., and Etoh, H. (2003) *Lett. Appl. Microbiol.*, **37**, 81.
[6] Rukachaisirikul, T., Saekee, A., Tharibun, C., Watkuolham, S., and Suksamrarn, A. (2007) *Arch. Pharm. Res.*, **30**, 1398.

β-Sanshool and γ-sanshool

Physical data:
- β-Sanshool: $C_{16}H_{25}NO$, colorless oil
- γ-Sanshool: $C_{18}H_{27}NO$, colorless oil

Structures:

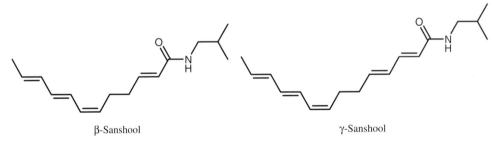

β-Sanshool   γ-Sanshool

Compound class: Aliphatic acid amides

Source: *Zanthoxylum piperitum* DC (stems; family: Rutaceae)

Pharmaceutical potential: *Inhibitors of cholesterol acyltransferase (hACAT) activity*

Cholesterol acyltransferase (ACAT) offers a vital role in the esterification process of cholesterol with its substrates, cholesterol and fatty acyl coenzyme A, to facilitate both intracellular storage and intercellular transport. ACAT-1 is usually involved in macrophage foam cell formation, while ACAT-2 facilitates the process of cholesterol absorption in intestinal enterocytes. β-Sanshool and γ-sanshool were evaluated to inhibit human ACAT-1 and -2 activities with IC_{50} values of 39.0 and 79.7 μM for the former and 12.0 and 82.6 μM for the latter, respectively. The compound may find useful applications against atherosclerosis.

Reference

Park, Y.-D., Lee, W.S., An, S., and Jeong, T.-S. (2007) *Biol. Pharm. Bull.*, **30**, 205.

Santolina sesquiterpene

Systematic name: (1R,2R,4S,5S,6R,7S)-4,5-Epoxygermacra-9Z-en-1,2,6-triol

Physical data: $C_{15}H_{26}O_4$, colorless amorphous solid, $[\alpha]_D^{25}$ +4.0° (CHCl$_3$, c 0.08)

Structure:

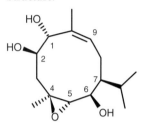

Santolina sesquiterpene

Compound class: Germacrane sesquiterpene

Source: *Santolina insularis* (aerial parts; family: Asteraceae)

Pharmaceutical potential: *Cytotoxic*

The isolated germacrane sesquiterpene was reported to display potent cytotoxic activity against human colon carcinoma cells (Caco-2) with an IC$_{50}$ value of 0.3 µg/ml (i.e., 1.1 µM) (an MTT assay was used).

Reference

Appendino, G., Aviello, G., Ballero, M., Borrelli, F., Fattorusso, E., Petrucci, F., Santelia, F.U., and Taglialatela-Scafati, O. (2005) *J. Nat. Prod.*, **68**, 853.

Sapinmusaponins Q and R

Systemic names: Sapinmusaponin Q: 21α-methoxy-3β,21(R),23(S)-epoxytirucall-7,24-diene-3-O-β-D-glucopyranosyl-(1→2)-β-D-glucopyranoside
Sapinmusaponin R: 21α-methoxy-3β,21(R),23(S)-epoxytirucall-7,24-diene-3-O-α-L-rhamnopyranosyl-(1→6)-β-D-glucopyranosyl-(1→2)-β-D-glucopyranoside

Physical data:
- Sapinmusaponin Q: $C_{43}H_{70}O_{13}$, white amorphous powder, $[\alpha]_D^{24}$ −24.6° (MeOH, c 0.4)
- Sapinmusaponin R: $C_{49}H_{80}O_{17}$, white amorphous powder, $[\alpha]_D^{24}$ −21.2° (MeOH, c 0.2)

Structures:

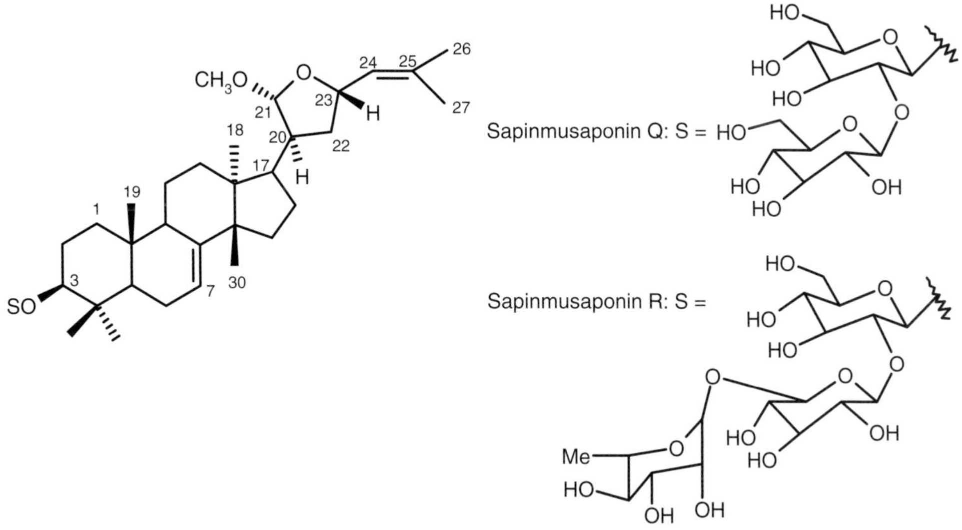

Sapinmusaponin Q: S = [trisaccharide structure shown]

Sapinmusaponin R: S = [trisaccharide structure shown with Me group]

Compound class: Triterpene (tirucallane-type tritrpenoids) saponins

Source: *Sapindus mukorossi* (galls; family: Sapindaceae)

Pharmaceutical potential: *Antiplatelet aggregation*

The investigators showed that both sapinmusaponins Q and R demonstrate more potent antiplatelet aggregation activity than aspirin. The IC_{50} values (μM) for sapinmusaponin Q in inhibiting platelet aggregation on washed rabbit platelets induced by platelet activity factor (PAF), TXA_2 (U46619, a thromboxane A2 agonist), thrombin (THB), and arachidonic acid (AA) were determined, respectively, as 7.7 ± 0.2, 3.4 ± 0.7, 8.4 ± 0.2, and 6.7 ± 0.4, while those for sapinmusaponin R were assessed, respectively, as 13.5 ± 0.3, 5.4 ± 0.2, 8.9 ± 0.4, and 7.7 ± 0.2.

Reference

Huang, H.-C., Tsai, W.-J., Liaw, C.-C., Wu, S.-H., Wu, Y.-C., and Kuo, Y.-H. (2007) *Chem. Pharm. Bull.*, **55**, 1412.

Saponaceol A

Physical data: $C_{46}H_{69}NO_{11}$, amorphous powder, $[\alpha]_D^{25}$ $-10.4°$ (MeOH, *c* 0.6)

Sappanchalcone

Structure:

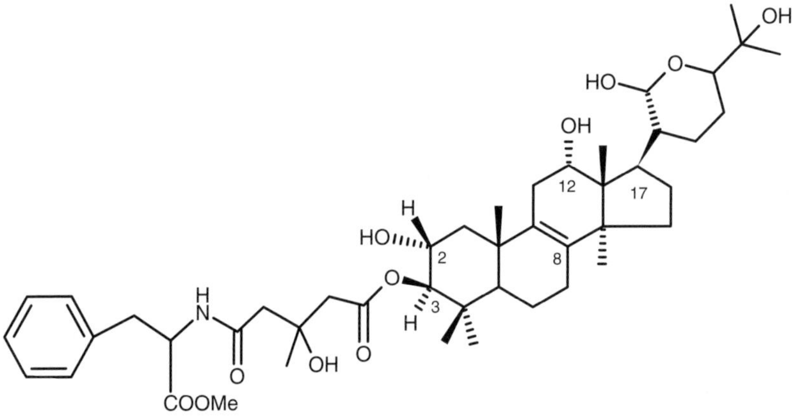

Saponaceol A

Compound class: Triterpenoid ester

Source: *Tricholoma saponaceum* (fresh fruit bodies; family: Tricholomataceae)

Pharmaceutical potential: *Cytotoxic*
The triterpenoid ester exhibited *in vitro* growth inhibitory activity against HL-60 human leukemia cells with an IC_{50} value of 8.9 μM.

Reference

Yoshikawa, K., Kuroboshi, M., Ahagon, S., and Arihara, S. (2004) *Chem. Pharm. Bull.*, **52**, 886.

Sappanchalcone

Systematic name: 3,4,4′-Trihydroxy-2′-methoxychalcone

Physical data: $C_{16}H_{14}O_5$, yellow needles (EtOH), mp 199–200 °C

Structure:

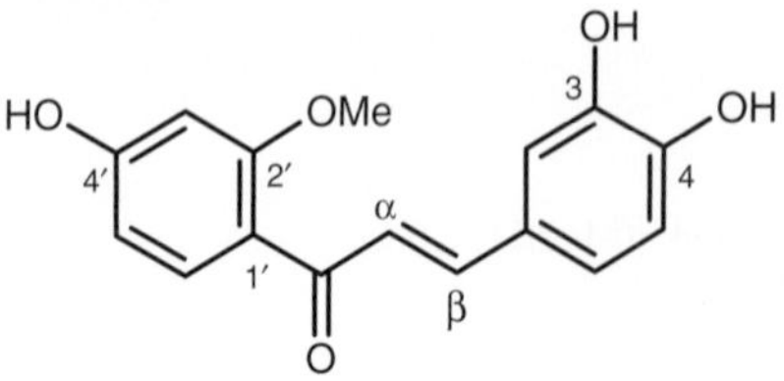

Sappanchalcone

Compound class: Flavonoid (chalcone derivative)

Source: *Caesalpinia sappan* L. (heartwoods; family: Leguminosae) [1–4]

Pharmaceutical potentials: *Anticonvulsant; anti-inflammatory; cytoprotective (antioxidative)*

Sappanchalcone was found to possess significant *in vitro* anticonvulsant activity by inhibiting the activities of the SSAR (succinic semialdehyde reductase) enzyme in a dose-dependent manner; inhibition of normal functioning of this enzyme is very much important in the elevation of neurotransmitter GABA (γ-aminobutyric acid) levels in CNS [2]. In addition, the chalcone derivative was evaluated to inhibit both nitric oxide (NO) and prostaglandin E_2 (PGE_2) production and also to exert suppressive effects on tumor necrosis factor-α (TNF-α), interleukin-6 (IL-6), cyclooxygenase-2 (COX-2), and inducible nitric oxide synthase (iNOS) mRNA expression with the IC_{50} values of 11.2, 7.7, 47.8, 17.4, 75.1, and 16.6 μM, respectively; hence, this flavonoid might be useful as an anti-inflammatory drug candidate [3].

In addition, Jeong *et al.* [4] found that at concentrations of 20–40 μM, the test compound showed significant cytoprotective effects against glutamate-induced oxidative damage in mouse hippocampal HT22 cells; the investigators reported that the compound enhanced cellular resistance to oxidative injury caused by glutamate-induced cytotoxicity in the HT22 cells through Nrf2-dependent HO-1 expression and via activation of the p38 MAPK pathway [4].

References

[1] Nagai, M., Gagumo, S., Eguchi, I., Lee, S.-M., and Suzuki, T. (1984) *Yakugaku Zasshi*, **104**, 935.
[2] Baek, N.-I., Jeon, S.G., Ahn, E.-M., Hahn, J.-T., Bahn, J.H., Jang, J.S., Cho, S.-W., Park, J.K., and Choi, S.Y. (2000) *Arch. Pharm. Res.*, **23**, 344.
[3] Washiyama, M., Sasaki, Y., Homokazu, T., and Nagumo, S. (2009) *Biol. Pharm. Bull.*, **32**, 941.
[4] Jeong, G.-S., Lee, D.-S., Kwon, T.-O., Lee, H.-S., An, R.-B., and Kim, Y.-C. (2009) *Biol. Pharm. Bull.*, **32**, 945.

Sarasinoside J

Systematic name: 3β-O-[β-D-Glucopyranosyl(1→2)-β-D-glucopyranosyl(1→6)-β-D-N-acetyl-2-amino-glucopyranosyl(1→2)-β-D-xylopyranosyl(4→1)-β-D-N-acetyl-2-aminogalactopyranosyl]-25-hydroxy-30-norlanosta-8(9)-en-23-one

Physical data: $C_{62}H_{102}N_2O_{27}$, yellow amorphous solid, $[α]_D^{20}$ −9.8° (MeOH, *c* 0.5)

Structure:

Sarasinoside J

Compound class: Norlanostane triterpenoidal saponin

Source: *Melophlus sarassinorum* (Indonesian sponge)

Pharmaceutical potential: *Antimicrobial*
Sarasinoside J was reported to exhibit strong activity against the yeast *Saccharomyces cerevisiae* and moderate antibacterial activity toward *Bacillus subtilis*. In an agar diffusion assay, using a loading concentration of 10 µg (in MeOH), the test compound exhibited zones of inhibition of 13 and 9 mm toward *S. cerevisiae* and *B. subtilis*, respectively.

Reference

Dai, H.-F., Edrada, R.A., Ebel, R., Nimtz, M., Wray, V., and Proksch, P. (2005) *J. Nat. Prod.*, **68**, 1231.

Sarcodonin-δ

Physical data: $C_{35}H_{38}N_2O_{13}$

Structure:

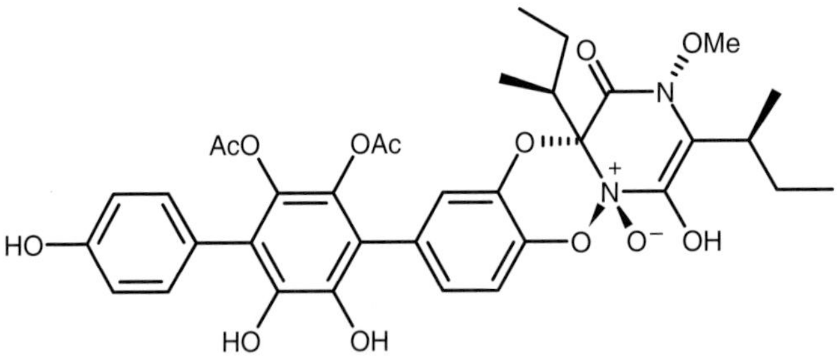

Sarcodonin-δ

Compound class: Nitrogen-containing terphenyl

Sources: *Sarcodon scabrosus* (inedible mushroom; family: Thelephoraceae) [1]; *Hydnellum suaveolens* (inedible mushroom, fresh fruit bodies, family: Thelephoraceae) [2]; *H. geogerirum* (inedible mushroom, fresh fruit bodies) [2]

Pharmaceutical potential: *Antioxidant*
The isolate showed moderate antioxidant activity by scavenging free radicals in the DPPH assay with an IC_{50} value of 25.0 µM (IC_{50} value of α-tocopherol (positive control) = 22.8 µM) [2].

References

[1] Ma, B.J. and Liu, K.J. (2005) *Z. Naturforsch.*, **60b**, 565.
[2] Hashimoto, T., Quang, D.N., Kuratsune, M., and Asakawa, Y. (2006) *Chem. Pharm. Bull.*, **54**, 912.

Sarcophyton polyhydroxysterol

Systematic name: 23,24-Dimethylcholest-16(17)-*E*-en-3β,5α,6β,20(*S*)-tetraol

Physical data: $C_{29}H_{50}O_4$, colorless amorphous powder, $[\alpha]_D^{20}$ −14.1° (EtOH, *c* 0.16)

Structures:

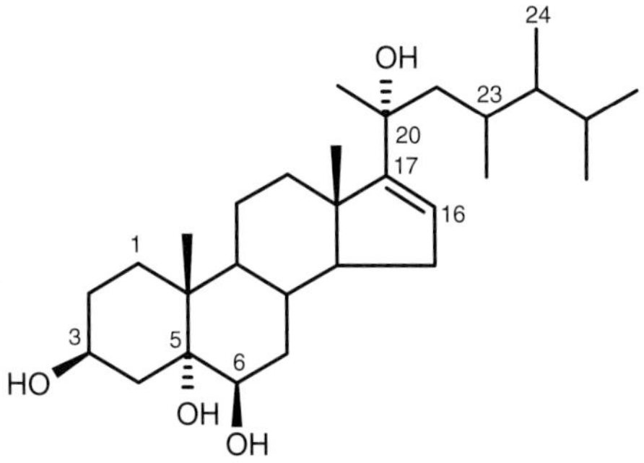

Sarcophyton polyhydroxysterol

Compound class: Polyhydroxysterol

Source: *Sarcophyton trocheliophorum* (soft coral)

Pharmaceutical potential: *Cytotoxic*
The polyhydroxysterol exhibited potent cytotoxic activity against the growth of human HL-60 leukemia, M-14 skin melanoma, and MCF-7 breast carcinoma cells with EC_{50} values of 2.8, 4.3, and 4.9 µg/ml, respectively.

Reference

Dong, H., Gou, Y.-L., Kini, R.M., Xu, H.-X., Chen, S.-X., Teo, S.L.M., and But, P.P.-H. (2000) *Chem. Pharm. Bull.*, **48**, 1087.

Saxicolaline A

Physical data: $C_{20}H_{22}NO_5$, yellow crystalline solid (MeOH), mp 223–225 °C, $[\alpha]_D^{27}$ +718.8° (C_5H_5N, *c* 0.16)

Structure:

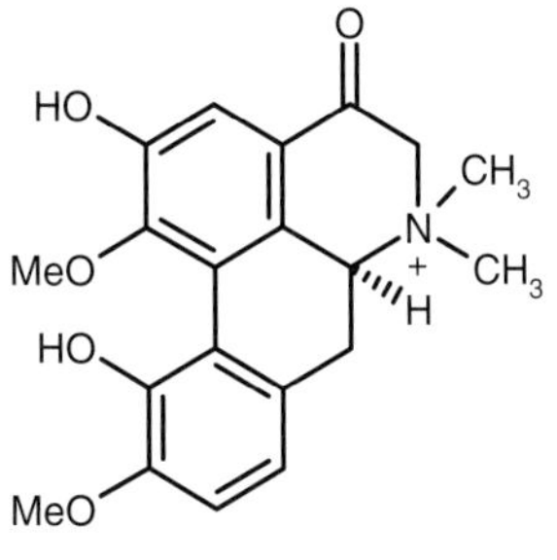

Saxicolaline A

Scalusamide A

Compound class: Quaternary alkaloid

Source: *Corydalis saxicola* Bunting (roots; family: Fumariodeae)

Pharmaceutical potential: *Antihepatitis B virus*
The alkaloid showed inhibitory effect against secretions of HBsAg (hepatitis B virus surface antigen) and HBeAg (hepatitis B virus e antigen) in the HepG2.2.15 cell line with IC_{50} values of 2.19 μM (SI > 1.3) and >2.81 μM (SI = 1.0), respectively; lamivudine was used as a positive control (IC_{50} against HBsAg: 15.37 μM (SI = 2.9); IC_{50} against HBeAg: 44.85 μM (SI = 1.0)).

Reference

Wu, Y.-R., Ma, Y.-B., Zhao, Y.-X., Yao, S.-Y., Zhou, J., Zhou, Y., and Chen, J.J. (2007) *Planta Med.*, **73**, 787.

Scalusamide A

Physical data: $C_{16}H_{27}NO_3$, colorless amorphous solid, $[\alpha]_D^{22}$ −28° (CHCl$_3$, c 1.0)

Structure:

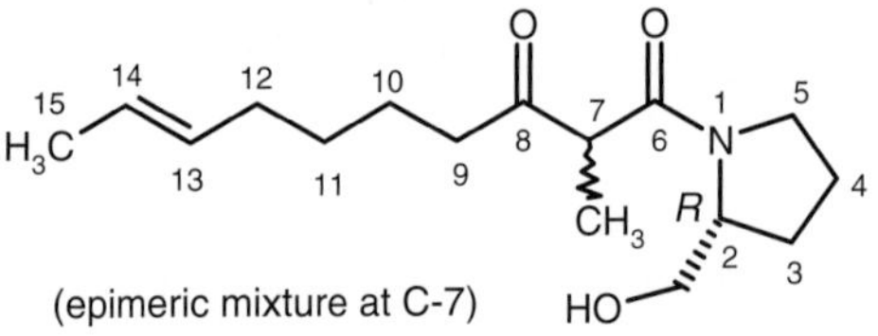

(epimeric mixture at C-7)

Scalusamide A

Compound class: Pyrrolidine alkaloid

Source: *Penicillium citrinum* strain N055 (marine-derived fungus; culture broth)

Pharmaceutical potentials: *Antifungal; antibacterial*
Scalusamide A exhibited moderate antifungal activity against *Cryptococcus neoformans* (MIC = 16.7 μg/ml) as well as moderate antibacterial activity against *Micrococcus luteus* (MIC = 33.3 μg/ml).

Reference

Tsuda, M., Sasaki, M., Mugishima, T., Komatsu, K., Sone, T., Tanaka, M., Mikami, Y., and Kobayashi, J. (2005) *J. Nat. Prod.*, **68**, 273.

Sch 213766

Physical data: $C_{26}H_{37}NO_6$, gummy substance, $[\alpha]_D^{25}$ +33.3° (Me$_2$CO, c 0.1)

Structure:

Sch 213766

Compound class: Antibiotic

Source: *Chaetomium globosum* (fungal fermentation broth) [1]

Pharmaceutical potential: *Chemokine receptor CCR-5 inhibitor (anti-HIV)*
It has been reported that functional inhibition of chemokine receptor CCR-5 may result in blocking viral entry at an initial stage of HIV-1 infection; hence, potential CCR-5 antagonists are expected to play a promising role in developing anti-HIV drugs [2–11]. Yang et al. [1] demonstrated that the fungal metabolite (Sch 213766) possesses significant inhibitory activity with an IC_{50} value of 8.6 µM in the CCR-5 membrane binding assay.

References

[1] Yang, S.-W., Mierzwa, R., Terracciano, J., Patel, M., Gullo, V., Wagner, N., Baroudy, B., Puar, M., Chan, T.-M., and Chu, M. (2007) *J. Antibiot.*, **60**, 524.

[2] Cocchi, F., DeVico, A.L., Grazino-Demo, A., Arya, S.K., Gallo, R.C., and Lusso, P. (1995) *Science*, **270**, 1811.

[3] Samson, M., Liber, F., Doranz, B.J., Rucker, J., Liesnard, C., Farber, C.-M., Saragosti, S., Lapoumeroulle, C., Cognanx, J., Forceille, C., Muyldermans, G., Vehofstede, C., Butonboy, G., Georges, M., Imai, T., Rana, S., Yi, Y., Smyth, R.J., Collman, R.G., Doms, R.W., Vassart, G., and Parmentier, M. (1996) *Nature*, **382**, 722.

[4] Feng, Y., Broder, C.C., Kennedy, P.E., and Berger, E.A. (1996) *Science*, **272**, 872.

[5] Allkhatib, G., Combadiere, C., Broder, C.C., Feng, Y., Kennedy, P.E., Murphy, P.M., and Berger, E.A. (1996) *Science*, **272**, 1955.

[6] Berger, E.A., Murphy, P.M., and Farber, J.M. (1999) *Annu. Rev. Immunol.*, **17**, 657.

[7] Strizki, J.M., Xu, S., Wagner, N.E., Wojcik, L., Liu, J., Hou, Y., Endres, M., Palani, A., Shapiro, S., Clader, J.W., Greenlee, W.J., Tagat, J.R., McCombie, S., Cox, K., Fawzi, A.B., Chou, C.-C., Pugliese-Sivo, C., Davies, L., Moreno, M.E., Ho, D.D., Trkola, A., Stoddart, C.A., Moore, J.P., Reyes, G.R., and Baroudy, B.M. (2001) *Proc. Natl. Acad. Sci. USA*, **98**, 12718.

[8] Shiraishi, M., Aramaki, Y., Seto, M., Imoto, H., Nishikawa, Y., Kanzaki, N., Okamoto, M., Sawada, H., Nishimura, O., Baba, M., and Fujino, M. (2000) *J. Med. Chem.*, **43**, 2049.

[9] Baba, M., Takashima, K., Miyake, H., Kanzaki, N., Teshima, K., Wang, X., Shirashi, M., and Iizawa, Y. (2005) *Antimicrob. Agents Chemother.*, **49**, 4584.
[10] Strizki, J.M., Tremblay, C., Xu, S., Wojcki, L., Wagner, N., Gonsiorek, W., Hipkin, W., Chou, C.C., Pugliesie-Sivo, C., Xiao Y., Tagat, J.R., Cox, K., Priestley, T., Sorota, S., Huang, W., Hirsch, M., Reyes, G.R., and Baroudy, B.M. (2005) *Antimicrob. Agents Chemother.*, **49**, 4911.
[11] Palani, A. and Tagat, J.R. (2006) *J. Med. Chem.*, **49**, 2851.

Sch 725432

Physical data: $C_{15}H_{23}O_3$, colorless gum

Structure:

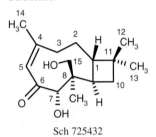

Sch 725432

Compound class: Sesquiterpenoid (caryophyllene type)

Source: *Chrysosporium pilosum* (fungus; fermentation broth)

Pharmaceutical potential: *Antifungal*
The fungal metabolite showed moderate antifungal activity *in vitro* against *Saccharomyces cerevisiae* (PM503) with an MIC value of 60 μg/ml.

Reference

Yang, S.-W., Chan, T.-M., Terracciano, J., Boehm, E., Patel, R., Chen, G., Loebenberg, D., Patel, M., Gullo, V., Pramanik, B., and Chu, M. (2009) *J. Nat. Prod.*, **72**, 484.

Sch 54445

Physical data: $C_{30}H_{29}N_2O_9Cl$, yellow–brown needles, mp 201–203 °C, $[\alpha]_D^{23}$ −558° (CHCl$_3$, *c* 0.2)

Structure:

Sch 54445

Compound class: Polycyclic xanthone

Source: Fermentation culture broth of an *Actinoplanes* sp. (SCC 2314, ATCC 55600) [1]

Pharmaceutical potential: *Antifungal*

Sch 54445, a polycyclic xanthone related to the albofungin family of compounds, was established as a highly potent broad-spectrum antifungal agent [1]. The geometric mean MIC values (μg/ml) for the antifungal activity of the compound against various yeasts and dermatophytes were observed to be ~0.00038 when tested against 12 strains of yeasts in Sabouraud dextrose broth (SDB medium, pH 5.7; six strains of *Candida albicans* and two strains each of *C. tropicalis, C. stellatoidea*, and *C. parapsilosis*); 0.00764 when tested against seven strains of dermatophytes in SDB medium (two strains each of *Trichophyton mentagrophytes, T. rubrum*, and *T. tonsurans* and one strain of *Microsporum canis*); 0.025 when tested against four strains of *Aspergillus* in SDB medium (two strains of *Aspergillus flavus* and one strain each of *A. niger* and *A. fumigatus*); and 0.5–6.4 when tested against nine strains of yeasts in Eagle's minimum essential medium (EMEM, pH 7.0; six strains of *C. albicans*, two strains of *C. tropicalis*, and one strain of *C. stellatoidea*).

The potency of Sch 54445 was found to be greater than that of its related compounds, Sch 42137 [2] and albofungin, against most yeasts and dermatophytes. It is promising to note that the compound exhibited better activity than both Sch 42137 and albofungin against *Aspergillus*, which is a clinically significant pathogen of fungal infections. Toxicity study revealed that Sch 54445, like its related compounds, showed *in vivo* toxicity with an ED_{50} value of 1 mg/kg in mice [1].

References

[1] Chu, M., Pruunees, I., Mierzwa, R., Terracciano, J., Patel, M., Loebenberg, D., Kaminski, J.J., Das, P., and Puar, M.S. (1997) *J. Nat. Prod.*, **60**, 525.
[2] Cooper, R., Truumees, I., Gunnarsson, I., Loebenberg, D., Horan, A., Marquez, J., Patel, M., Gullo, V., Das, P., and Mittelman, S. (1992), *J. Antibiot.*, **45**, 444.

Sch 642305

Physical data: $C_{14}H_{20}O_4$, white solid, mp 143–145 °C, $[\alpha]_D$ +67.44° (MeOH, *c* 0.5)

Schiprolactone A

Structure:

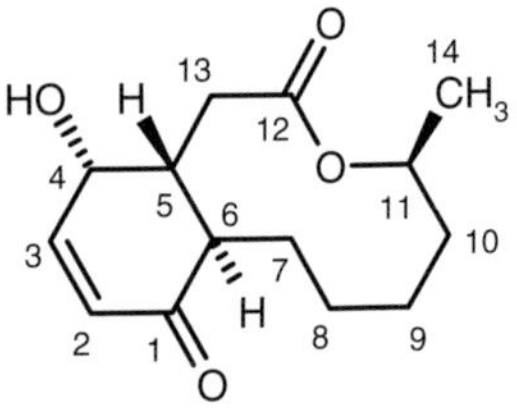

Sch 642305

Compound class: Antibiotic

Source: *Penicillium verrucosum* (fermentation broth)

Pharmaceutical potential: *Inhibitor to bacterial DNA primase*
The fungal metabolite exhibited inhibitory activity against *Escherichia coli* bacterial DNA primase with an EC_{50} value of 70 μM. In a cell-based assay, the test compound showed antibacterial activity (with an MIC value of 40 μg/ml) against an *E. coli* strain with a defective lipopolysaccharide layer and disruption of the efflux pump *acrAB*.

Inhibition of bacterial primase enzyme causes a catastrophic event in bacterial chromosome replication; hence, the primase enzyme is an attractive target for antibacterial drug discovery process.

Reference

Chu, M., Mierzwa, R., Xu, L., He, L., Terracciano, J., Patel, M., Gullo, V., Black, T., Zhao, W., Chan, T.-M., and McPhail, A.T. (2003) *J. Nat. Prod.*, **66**, 1527.

Schiprolactone A

Physical data: $C_{32}H_{42}O_6$, amorphous powder, mp 215–217 °C

Structure:

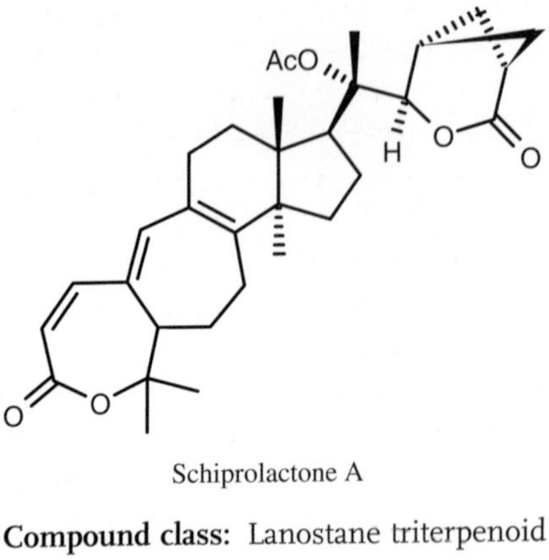

Schiprolactone A

Compound class: Lanostane triterpenoid

Sources: *Schisandra propinqua* (Wall.) Hook. f. et Thoms. (family: Schisandraceae) [1]; *S. henryi* (stems) [2]

Pharmaceutical potential: *Cytotoxic (anticancerous)*
Chen et al. [2] evaluated that the isolate possesses moderate cytotoxicity *in vitro* against leukemia and HeLa cells with respective IC$_{50}$ values of 5.06 µg/ml (or 0.0097 µmol/ml) and 50.6 µg/ml (or 0.097 µmol/ml).

References

[1] Chen, Y.G., Qin, G.W., and Xie, Y.Y. (2001) *Chin. J. Chem.*, **19**, 304.
[2] Chen, Y.-G., Wu, Z.-C., Lv, Y.-P., Gui, S.-H., Wen, J., Liao, X.-R., Yuan, L.-M., and Halaweish, F. (2003) *Arch. Pharm. Res.*, **26**, 912.

Schisanlactone A

Physical data: $C_{30}H_{40}O_4$, crystalline solid, mp 227–229 °C, $[\alpha]_D^{23}$ +365° (CHCl$_3$, *c* 0.2)

Structure:

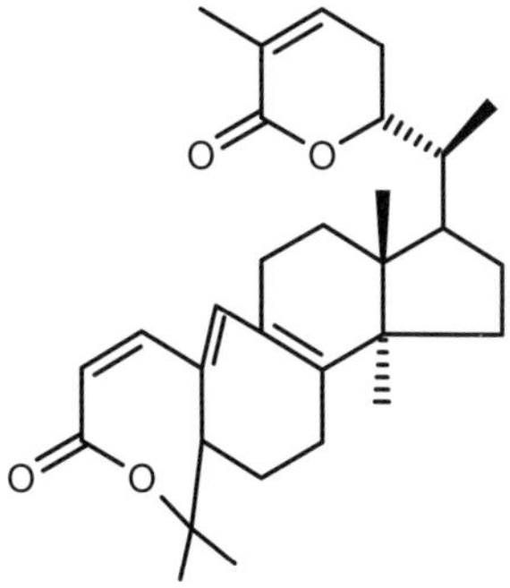

Schisanlactone A

Compound class: Triterpenoid

Sources: *Schisandra* sp. (family: Schisandraceae) [1]; *Kadsura longipedunculata* Finet. et Gagnep. (roots and stems; family: Schisandraceae) [2]

Pharmaceutical potential: *HIV-1 protease inhibitor*
The isolate was evaluated for its HIV-1 protease inhibitory potential, and the test compound exerted appreciable inhibition to the target by 94.8 ± 0.1% at 100 µg/ml concentration, thus having an IC$_{50}$ value of 20 µg/ml [2].

References

[1] Liu, J.-S., Huang, M.-F., Arnold, G.F., Arnold, E., Clardy, J., and Ayer, W.A. (1983) *Tetrahedron Lett.*, **24**, 2351.
[2] Sun, Q.-Z., Chen, D.-F., Diang, P.-L., Ma, C.-M., Kakuda, H., Nakamura, N., and Hattori, M. (2006) *Chem. Pharm. Bull.*, **54**, 129.

Schisanwilsonene A

Physical data: $C_{15}H_{26}O_2$, colorless crystals (acetone), mp 165–168 °C; $[\alpha]_D^{25}$ +52.3° (MeOH, c 0.02)

Structure:

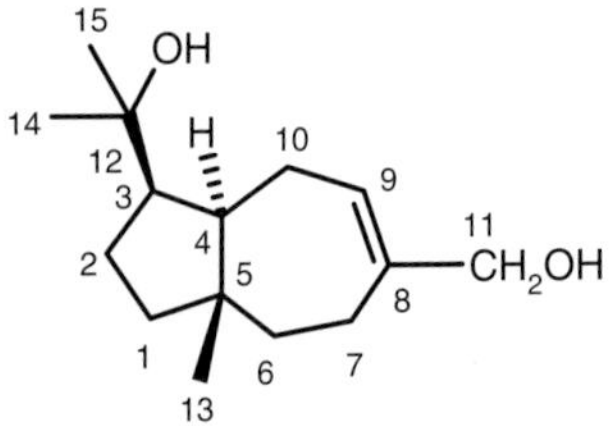

Schisanwilsonene A

Compound class: Sesquiterpenoid (carotene-type)

Source: *Schisandra wilsoniana* A.C. Smith (fruits; family: Schisandraceae)

Pharmaceutical potential: *Antihepatitis B virus (anti-HBV)*
The isolate was evaluated to possess antihepatitis B virus activity; it displayed the antiviral efficacy at a concentration of 50 mg/ml by inhibiting HBsAg and HBeAg secretion by 76.5 and 28.9%, respectively.

Reference

Ma, W.-H., Huang, H., Zhou, P., and Chen, D.-F. (2009) *J. Nat. Prod.*, **72**, 676.

Schizolaenones A and B

Systematic names:
- Schizolaenone A: (2S)-5,7,4′-trihydroxy-6-(3-methylbut-2-enyl)-3′-(geranyl)flavanone
- Schizolaenone B: (2S)-5,7, 3′,4′-tetrahydroxy-6-(3-methylbut-2-enyl)-5′-(geranyl)flavanone

Physical data:
- Schizolaenone A: $C_{30}H_{36}O_5$, yellow amorphous solid, $[\alpha]_D$ +2.7° (MeOH, c 0.44)
- Schizolaenone B: $C_{30}H_{36}O_6$, yellow amorphous solid, $[\alpha]_D$ −3.9° (MeOH, c 0.53)

Structures:

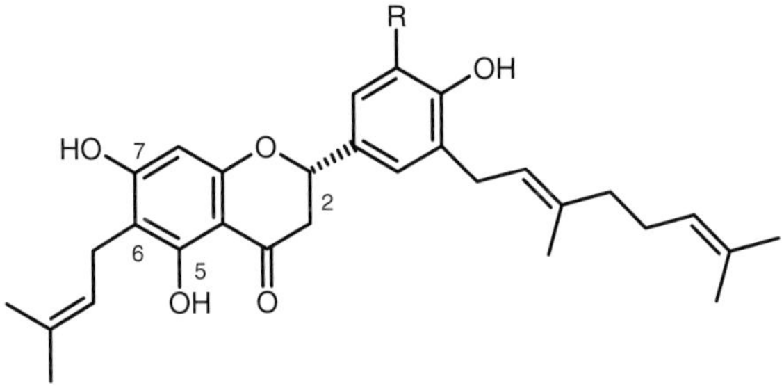

Schizolaenone A: R = H
Schizolaenone B: R = OH

Compound class: Flavonoids (prenylated flavanones)

Source: *Schizolaena hystrix* (fruits; family: Sarcolaenaceae)

Pharmaceutical potential: *Cytotoxic*
Schizolaenones A and B exhibited moderate cytotoxicity against A2780 human ovarian cancer cells with IC$_{50}$ values of 10 and 11 µg/ml, respectively.

Reference

Murphy, B.T., Cao, S., Norris, A., Miller, J.S., Ratovoson, F., Andriantsiferana, R., Rasamison, V.E., and Kingston, D.G.I. (2005) *J. Nat. Prod.*, **68**, 417.

Schweinfurthins E–H

Physical data:
- Schweinfurthin E: $C_{30}H_{38}O_6$, pale yellow solid, $[\alpha]_D^{22}$ +49.2° (MeOH, *c* 0.13)
- Schweinfurthin F: $C_{30}H_{38}O_5$, pale yellow solid, $[\alpha]_D^{22}$ +50.8° (MeOH, *c* 0.06)
- Schweinfurthin G: $C_{29}H_{36}O_5$, pale yellow solid, $[\alpha]_D^{22}$ +33.3° (MeOH, *c* 0.03)
- Schweinfurthin H: $C_{30}H_{38}O_7$, pale yellow solid, $[\alpha]_D^{22}$ +32.4° (MeOH, *c* 0.04)

Structures:

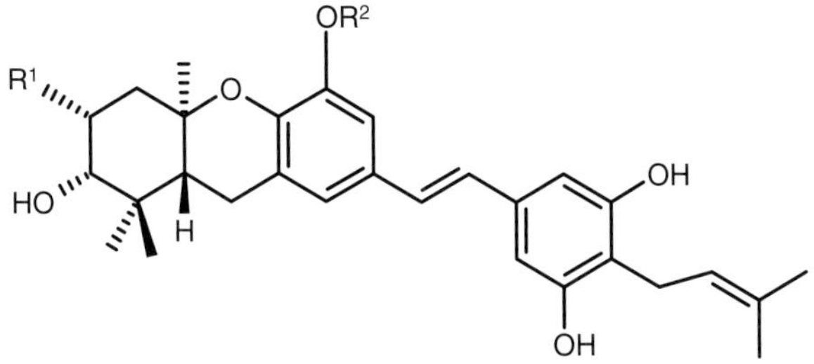

Schweinfurthin E: R^1 = OH; R^2 = Me
Schweinfurthin F: R^1 = OH; R^2 = Me
Schweinfurthin G: R^1 = R^2 = H

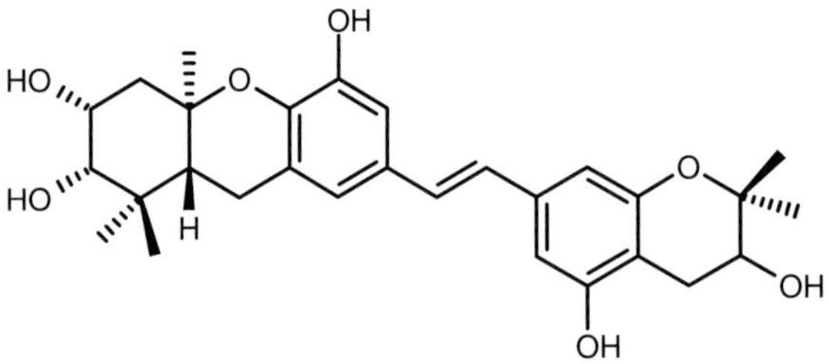

Schweinfurthin H

Compound class: Prenylated stilbenes

Source: *Macaranga alnifolia* (fruits; family: Euphorbiaceae)

Pharmaceutical potential: *Antiproliferative*

All the isolates, schweinfurthins E–H, were found to possess antiproliferative activity against the A2780 human ovarian cancer cell line with IC_{50} values of 0.26, 5.0, 0.39, and 4.5 µM, respectively. Schweinfurthin E ($IC_{50} = 0.26$ µM) was the most potent among the compounds.

A huge difference in the activity between schweinfurthin F ($IC_{50} = 5.0$ µM) and schweinfurthin G ($IC_{50} = 0.39$ µM) is indicative of the fact that the combination of the loss of the C-3 hydroxyl group with methylation of the C-5 hydroxyl group leads to a greater loss of activity; from this comparative study, the investigators suggested that such antiproliferative activity in these compounds increases with increase in polarity of the molecules, and hence, it is possible that the observed activity is limited in some way by aqueous solubility, but further experiments are needed to confirm this view.

Reference

Yoder, B.J., Cao, S., Norris, A., Miller, J.S., Ratovoson, F., Razafitsalama, J., Andriantsiferana, R., Rasamison, V.E., and Kingston, D.G.I. (2007) *J. Nat. Prod.*, **70**, 342.

Sciscllascilloside E-1

Systematic name: 15-Deoxoeucosterol 3-*O*-α-L-rhamnopyranosyl(1→2)-[β-D-glucopyranosyl(1→3)]-β-D-glucopyranosyl(1→2)-α-L-arabinopyranosyl(1→6)-β-D-glucopyranoside

Physical data: $C_{58}H_{94}O_{27}$, white needles (aq. MeOH), mp 221–223 °C, $[\alpha]_D^{25}$ −57.1° (MeOH, *c* 0.08)

Structure:

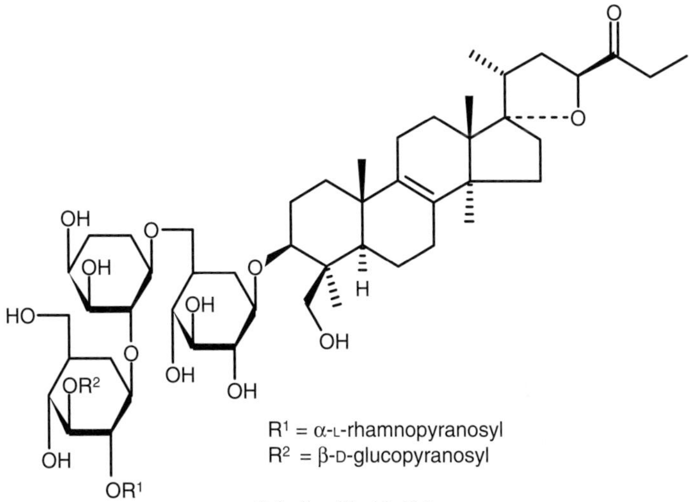

R¹ = α-L-rhamnopyranosyl
R² = β-D-glucopyranosyl

Sciscllascilloside E-1

Compound class: Nortriterpenoid oligoglycoside

Source: *Scilla scilloides* (Lind.) Druce (fresh bulbs; family: Liliaceae) [1, 2]

Pharmaceutical potential: *Cytotoxic (antitumor)*

The nortriterpenoid oligoglycoside showed potent cytotoxic activity against a variety of tumor cell lines such as HT1080, B16 (F-10), 3LL, MCF7, PC-3, HT29, LOX-IMVI, and A549 with ED_{50} values ranging from 1.53 to 3.06 nM [2]. The present investigators [2] evaluated the compound as a significant antitumor agent in both *in vitro* and *in vivo* experiments with B16 (F-10)- and Sarcoma 180-bearing mice – it was observed that the compound prolonged the life span of B16 (F-10)-bearing mice and Sarcoma 180-bearing mice at doses of 1.5 and 3 mg/kg/day, respectively (administered intraperitoneally once daily for 20 days with vehicle (0.9% saline), used as a negative control).

References

[1] Sholichin, M., Miyahara, K., and Kawasaki, T. (1985) *Chem. Pharm. Bull.*, **33**, 1756.

[2] Lee, S.-M., Chun, H.-K., Lee, C.-H., Min, B.-S., Lee, E.-S., and Kho, Y.-H. (2002) *Chem. Pharm. Bull.*, **50**, 1245.

Scleropyric acid

Physical data: $C_{17}H_{28}O_2$, colorless sticky solid

Scoparic acid A

Structure:

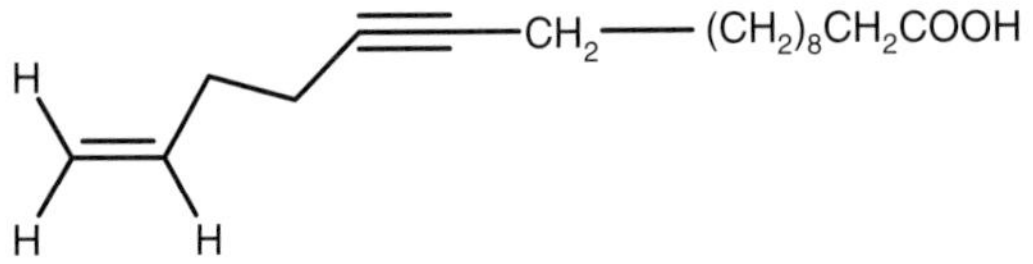

Scleropyric acid

Compound class: Unsaturated aliphatic carboxylic acid

Source: *Scleropyrum wallichianum* (Wight & Arn.) Aarn. (syn. *Scleropyrum maingayi* Hook. f. (twigs; family: Santalaceae)

Pharmaceutical potentials: *Antimycobacterial; antiplasmodial*
The unsaturated carboxylic acid displayed moderate antimycobacterial activity against *Mycobacterium tuberculosis* H37Ra with an MIC value of 25 µg/ml (positive controls rifampicin, isoniazid, and kanamycin sulfate showed MIC values of 0.004, 0.06, and 2.5 µg/ml, respectively) and also showed antiplasmodial activity against *Plasmodium falciparum* (K1, multidrug-resistant strain) with an IC_{50} value of 7.2 µg/ml (positive control artemisinin showed an IC_{50} value of 1 ng/ml).

Reference

Suksamrarn, A., Buaprom, M., Udtip, S., Nuntawong, N., Haritakun, R., and Kanokmedhakul, S. (2005) *Chem. Pharm. Bull.*, **53**, 1327.

Scoparic acid A

Physical data: $C_{27}H_{36}O_5$, colorless amorphous powder, $[\alpha]_D^{26}$ −38.3° (CHCl$_3$, *c* 1.0)

Structure:

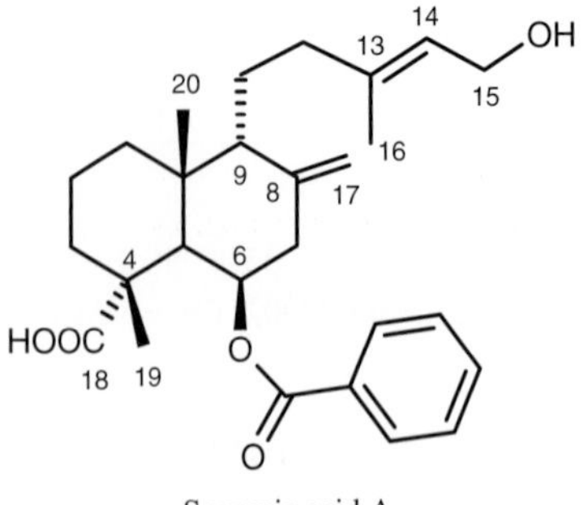

Scoparic acid A

Compound class: Labdane-type diterpenoid acid

Source: *Scoparia dulcis* L. (whole plants; family: Scrophulariaceae)

Pharmaceutical potential: *β-Glucuronidase inhibitor*
The diterpenoid derivative was found to possess significant inhibitory activity against bovine liver β-glucuronidase with an IC_{50} value of 6.8×10^{-6} M (glucosacchro-1:4-lactone was used as a

positive control; $IC_{50} = 1.8 \times 10^{-6}$ M). The present investigators showed that inhibition of β-glucuronidase by the compound was noncompetitive with *p*-nitrophenyl-β-D-glucuronide as a substrate; it was also investigated that scoparic acid A inhibits the activity of β-glucuronidase by binding specifically with the enzyme.

Reference

Hayashi, T., Kawasaki, M., Okamura, K., Tamada, Y., Morita, N., Tezuka, Y., Kikuchi, T., Miwa, Y., and Taga, T. (1992) *J. Nat. Prod.*, **55**, 748.

Scroside D

Systematic name: 2-(3,4-Dihydroxyphenyl)-ethyl-*O*-β-D-glucopyranosyl(1→3)-β-D-glycopyranoside

Physical data: $C_{20}H_{30}O_{13}$, white amorphous powder, mp 240–242 °C, $[\alpha]_D^{20}$ −42.6° (MeOH, *c* 0.8)

Structure:

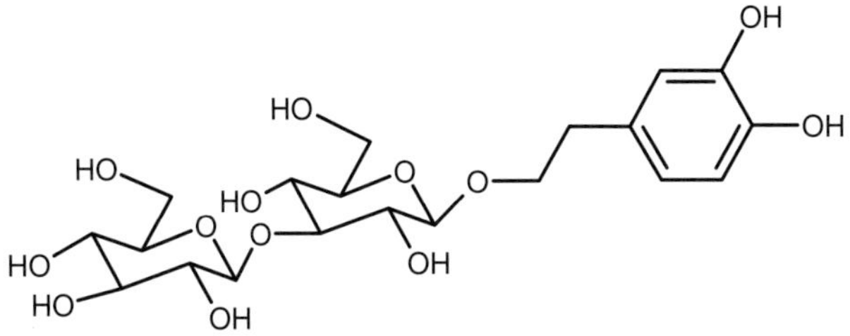

Scroside D

Compound class: Phenylethanoid glycoside

Source: *Picrorhiza scrophulariiflora* (roots; family: Scrophulariaceae)

Pharmaceutical potential: *Antioxidant (radical scavenger)*
The glycoside showed antioxidant activity by scavenging hydroxyl radical and superoxide anion with respective IC_{50} values of 48.7 and 84.5 μM (ascorbic acid was used as a positive control; respective IC_{50} values were recorded as 51.8 and 86.2 μM).

Reference

Wang, H., Sun, Y., Ye, W.-C., Xiong, F., Wu, J.-J., Yang, C.-H., and Zhao, S.-X. (2004) *Chem. Pharm. Bull.*, **52**, 615.

Scutebarbatines C–F

Physical data:
- Scutebarbatine C: $C_{33}H_{35}NO_8$, white needles, mp 156–158 °C, $[\alpha]_D^{29}$ −109.6° (MeOH, *c* 0.13)
- Scutebarbatine D: $C_{33}H_{35}NO_8$, white needles, mp 151–153 °C, $[\alpha]_D^{29}$ −98.4° (MeOH, *c* 0.12)

- Scutebarbatine E: $C_{33}H_{33}NO_8$, white needles, mp 154–156 °C, $[\alpha]_D^{29}$ −108.4° (MeOH, c 0.13)
- Scutebarbatine F: $C_{33}H_{37}NO_9$, white needles, mp 159–160 °C, $[\alpha]_D^{29}$ −65.9° (MeOH, c 0.14)

Structures:

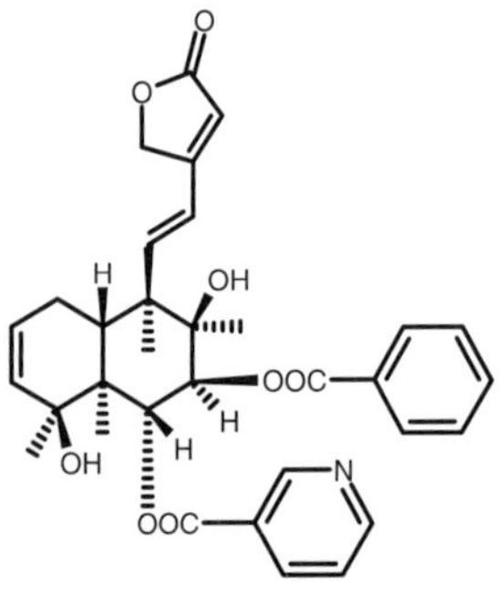

Scutebarbatine C

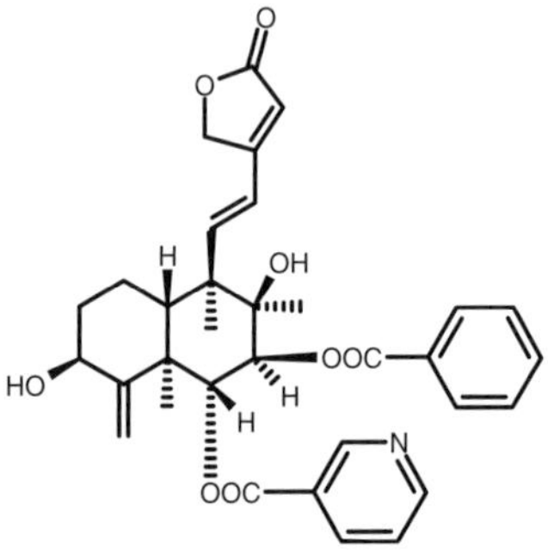

Scutebarbatine D

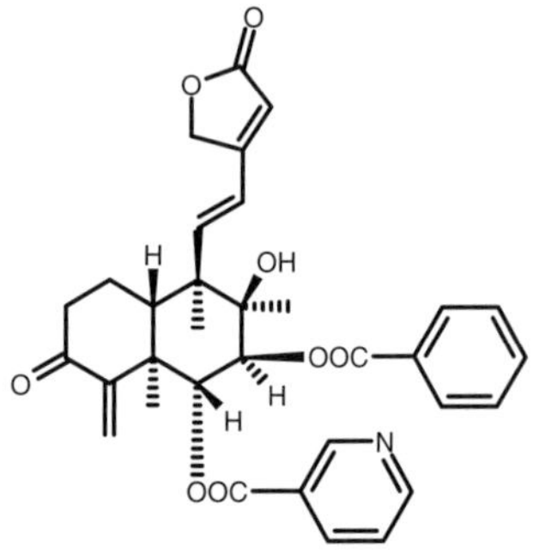

Scutebarbatine E

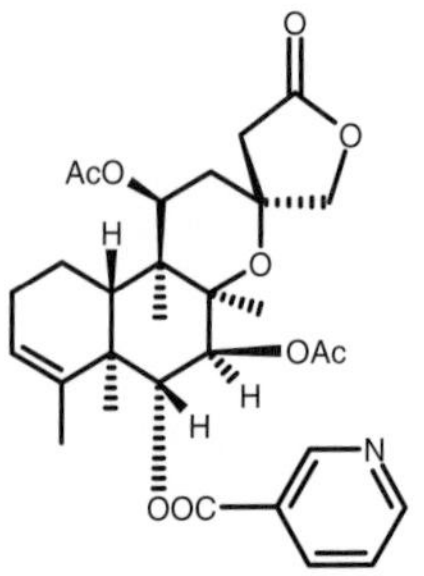

Scutebarbatine F

Compound class: *neo*-Clerodane diterpenoid alkaloids

Source: *Scutellaria barbata* D. Don (whole plants; family: Labiatae)

Pharmaceutical potential: *Cytotoxic*

Scutebarbatines C–F showed significant *in vitro* cytotoxic activities against three human cancer cell lines such as HONE-1 nasopharyngeal, KB oral epidermoid carcinoma, and HT29 colorectal carcinoma cells; their respective IC_{50} values (in µM) were determined to be 4.7 ± 2.0, 5.0 ± 2.1, 4.1 ± 1.5, and 4.4 ± 1.9 against HONE-1; 7.1 ± 2.6, 7.8 ± 1.8, 7.6 ± 2.2, and 6.1 ± 2.7 against KB; and 3.9 ± 2.1, 4.6 ± 1.5, 5.3 ± 2.0, and 4.6 ± 1.9 against HT29 cells.

Reference

Dai, S.-J., Chen, M., Liu, K., Jiang, Y.-T., and Shen, L. (2006) *Chem. Pharm. Bull.*, **54**, 869.

Scutebarbatines G and H

Physical data:
- Scutebarbatine G (1): $C_{26}H_{33}NO_7$, white needles, mp 156–158 °C, $[\alpha]_D^{29}$ −55.7° (MeOH, c 0.14)

- Scutebarbatine H (4): $C_{26}H_{31}NO_7$, white needles, mp 157–159 °C, $[\alpha]_D^{29}$ −69.798° (MeOH, c 0.14)

Other natural derivatives: 6,7-Di-O-nicitinoylscutebarbatine G (2): $C_{38}H_{39}N_3O_9$, white needles, mp 150–152 °C, $[\alpha]_D^{29}$ −57.9° (MeOH, c 0.13)
6-O-Nicitinoyl-7-O-acetylscutebarbatine G (3): $C_{34}H_{38}N_2O_9$, white needles, mp 149–151 °C, $[\alpha]_D^{29}$ −60.3° (MeOH, c 0.12)
7-O-Nicitinoyl scutebarbatine H (5): $C_{32}H_{34}N_2O_8$, white needles, mp 153–154 °C, $[\alpha]_D^{29}$ −73.5° (MeOH, c 0.13)

Structures:

1: R¹ = nicotinoyl; R² = R³ = H
2: R¹ = R² = R³ = nicotinoyl
3: R¹ = R² = nicotinoyl; R³ = acetyl

4: R¹ = nicotinoyl; R² = H
5: R¹ = R² = nicotinoyl

Nicotinoyl =

Compound class: *neo*-Clerodane diterpenoid alkaloids

Source: *Scutellaria barbata* (aerial parts; family: Labiatae)

Pharmaceutical potential: Cytotoxic
The present investigators examined the diterpenoid alkaloids (1–5) for their cytotoxic efficacies against three tumor cell lines (HONE-1 nasopharyngeal, KB oral epidermoid carcinoma, and HT29 colorectal carcinoma cells) and found that all of them showed significant cytotoxic activity with IC_{50} values in the range 3.4–8.5 μM. The IC_{50} (μM) values of each compound, respectively, against HONE-1, KB, and HT-29 cell lines were determined to be 3.7 ± 2.0, 7.1 ± 2.2, and 6.9 ± 1.3 (1); 5.0 ± 3.2, 8.5 ± 1.7, and 6.6 ± 2.4 (2); 4.5 ± 1.5, 6.1 ± 2.6, and 5.3 ± 2.0 (3); 4.4 ± 2.7, 6.1 ± 2.2, and 3.5 ± 3.4 (4); 3.7 ± 2.0, 6.7 ± 2.6, and 3.4 ± 2.1 (5); 0.5 ± 0.4, 0.8 ± 0.3, and 2.0 ± 0.7 (etoposide, positive control); and 3.6 ± 0.3, 4.1 ± 0.7, and 5.2 ± 1.5 (cisplatin, positive control).

Reference

Dai, S.-J., Wang, G.-F., Chen, M., Liu, K., and Shen, L. (2007) *Chem. Pharm. Bull.*, **55**, 1218.

Scutebarbatines I–L

Physical data:
- Scutebarbatine I: $C_{30}H_{41}NO_8$, white needles, mp 150–151 °C, $[\alpha]_D^{29}$ −13.9° (CHCl₃, c 0.12)
- Scutebarbatine J: $C_{30}H_{41}NO_8$, white needles, mp 149–150 °C, $[\alpha]_D^{29}$ −7.7° (CHCl₃, c 0.13)
- Scutebarbatine K: $C_{28}H_{33}NO_7$, white needles, mp 155–156 °C, $[\alpha]_D^{29}$ −110.8° (MeOH, c 0.14)
- Scutebarbatine L: $C_{33}H_{41}NO_9$, white needles, mp 153–155 °C, $[\alpha]_D^{29}$ −103.7° (MeOH, c 0.13)

Structures:

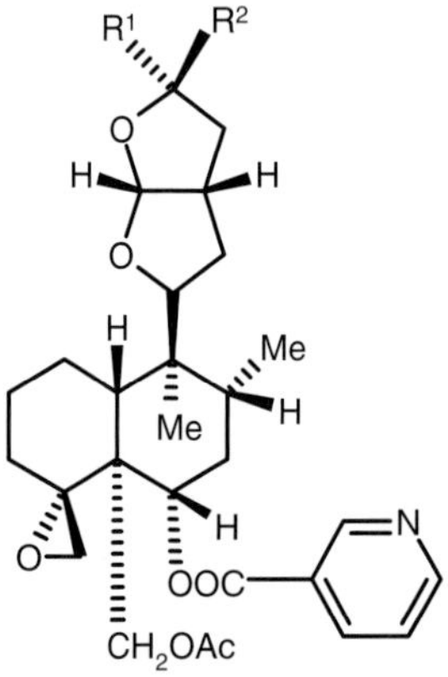

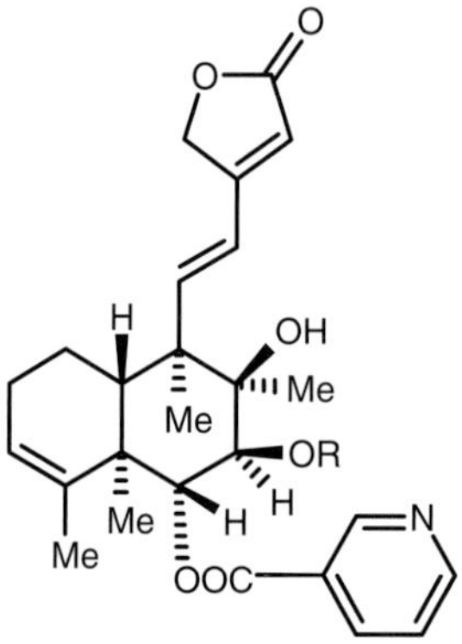

Scutebarbatine I: R¹ = H, R² = OEt
Scutebarbatine J: R¹ = OEt, R² = H

Scutebarbatine K: R = Ac
Scutebarbatine L: R = OCCH(OAc)CHMe$_2$

Compound class: *neo*-Clerodane diterpenoid alkaloids

Source: *Scutellaria barbata* D. Don (aerial parts; family: Labiatae)

Pharmaceutical potential: *Cytotoxic*

All the isolates were found to exhibit significant *in vitro* cytotoxic activities against three human cancer lines, HONE-1 nasopharyngeal, KB oral epidermoid carcinoma, and HT29 colorectal carcinoma cells, with IC$_{50}$ values in the range of 3.2–8.3 µM. The detailed IC$_{50}$ values for the test compounds against the respective cell lines (HONE-1, KB, and HT-29) were evaluated, respectively, as 4.2 ± 2.2, 4.7 ± 2.7, and 7.5 ± 2.6 µM (scutebarbatine I); 4.4 ± 1.9, 5.1 ± 1.8, and 8.3 ± 1.1 µM (scutebarbatine J); 3.9 ± 2.2, 5.5 ± 2.0, and 5.9 ± 2.7 µM (scutebarbatine K); and 3.2 ± 2.3, 5.6 ± 1.3, and 6.0 ± 1.5 µM (scutebarbatine L).

Reference

Dai, S.-J., Liang, D.-D., Ren, Y., Liu, K., and Shen, L. (2008) *Chem. Pharm. Bull.*, **56**, 207.

Scutianthraquinones A–C

Physical data:
- Scutianthraquinone A: $C_{39}H_{32}O_{13}$, light brown amorphous solid, $[\alpha]_D^{25}$ +60.8° (CHCl$_3$, *c* 0.06)
- Scutianthraquinone B: $C_{38}H_{30}O_{13}$, light brown amorphous solid, $[\alpha]_D^{25}$ +134.8° (CHCl$_3$, *c* 0.04)
- Scutianthraquinone C: $C_{34}H_{24}O_{12}$, light brown amorphous solid, $[\alpha]_D^{25}$ +122.2° (MeOH, *c* 0.04)

Structures:

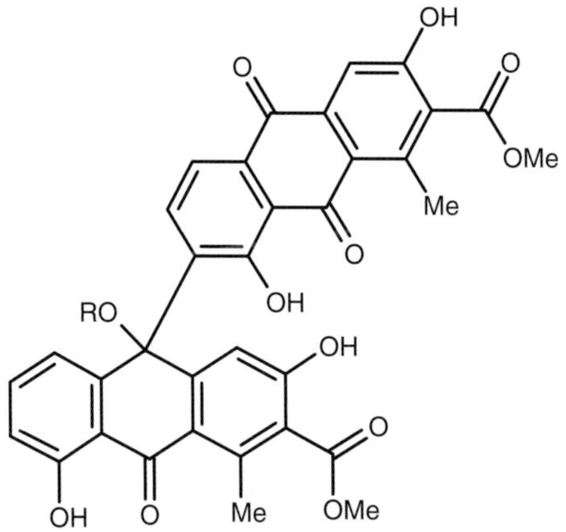

Scutianthraquinone A: R = COCH(CH₃)CH₂CH₃
Scutianthraquinone B: R = COCH(CH₃)₂
Scutianthraquinone C: R = H

Compound class: Anthraquinones

Source: *Scutia myrtina* (Burm. f.) Kurz (barks; family: Rhamnaceae)

Pharmaceutical potentials: *Antiproliferative; antimalarial*

The anthraquinones, scutianthraquinones A–C, exhibited moderate antiplasmodial activities against the chloroquine-resistant strains of *Plasmodium falciparum* Dd2 and FCM29; the respective IC$_{50}$ values were recorded as 1.23, 1.14, and 3.14 μM against *P. falciparum* Dd2 and 1.2, 5.4, and 15.4 μM against *P. falciparum* FCM29. In addition, scutianthraquinones A–C displayed weak antiproliferative activities against the A2780 human ovarian cancer cell line with respective IC$_{50}$ values of 7.6 ± 0.8, 5.8 ± 2.5, and >16 μM.

Reference

Hou, Y., Cao, S., Brodie, P.J., Callmander, M.W., Ratovoson, F., Rakotobe, E.A., Rasamison, V.E., Ratsimbason, M., Alumasa, J.N., Roepe, P.D., and Kingston, D.G.I. (2009) *Bioorg. Med. Chem.*, **17**, 2871.

Scytalidamides A and B

Physical data:
- Scytalidamide A: C$_{50}$H$_{67}$N$_7$O$_7$, white crystals, mp 147–150 °C, $[\alpha]_D^{25}$ −151.2° (MeOH, *c* 0.6)
- Scytalidamide B: C$_{51}$H$_{69}$N$_7$O$_7$, white crystals, mp 141-143 °C, $[\alpha]_D^{25}$ −156.9° (MeOH, *c* 0.6)

Structures:

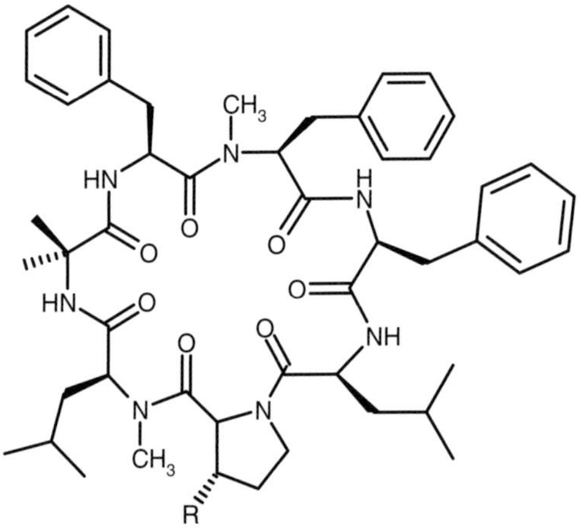

Scytalidamide A: R = H
Scytalidamide B: R = CH3

Compound class: Cyclic peptides

Source: *Scytalidium* sp. (marine fungus)

Pharmaceutical potential: *Cytotoxic (antitumor)*
Scytalidamides A and B exhibited significant *in vitro* cytotoxicity against the human colon carcinoma tumor cell line HCT-116 with IC$_{50}$ values of 2.7 and 11.0 µM, respectively. The present investigators also studied their cytotoxicity against the NCI 60 cell line panel; both of the test compounds showed moderate cytotoxicity in such a panel with respective mean GI$_{50}$ values of 7.9 and 4.1 µM. The most sensitive cell lines were reported to be MOLT-4 leukemia (3.0 µM) for scytalidamide A and Uacc-257 melanoma (1.2 µM) for scytalidamide B.

Reference

Tan, L.T., Cheng, X.C., Jensen, P.R., and Fenical, W. (2003) *J. Org. Chem.*, **68**, 8767.

Secoaggregatalactone A

Systematic name: (2*E*)-2-[(1*S*)-1-Hydroxy-2-oxopropyl]-2-tridecenoic acid methyl ester

Physical data: C$_{17}$H$_{30}$O$_4$, colorless oil, $[\alpha]_D^{25}$ +73.75° (CHCl$_3$, *c* 0.295)

Structure:

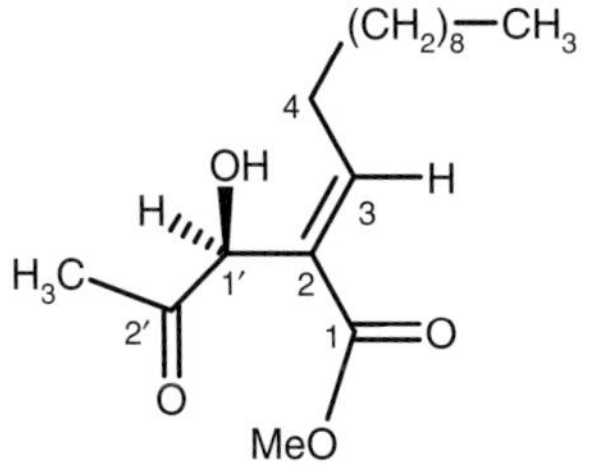

Secoaggregatalactone A

Compound class: Secobutanolide derivative

Source: *Lindera aggregata* (Sims) Kostern (*L. strychnifolia* Vill (leaves; family: Lauraceae)

Pharmaceutical potential: Cytotoxic

The isolate exhibited significant cytotoxicity against human hepatoma cells (HepG2 cell line) with an EC_{50} value of 6.61 µg/ml (22.1 µM). Detailed experimental observations revealed that the test compound induced apoptotic cell death in HepG2 cells by means of activation of caspase-8, Bid, and caspase-3, leading to cleavage of poly(ADP ribose) polymerase (PARP) and causing DNA fragmentation resulting in apoptosis.

Reference

Lin, C.-T., Chu, F.-H., Chang, S.-T., Chueh, P.-J., Su, Y.-C., Wu, K.-T., and Wang, S.-Y. (2007) *Planta Med.*, **73**, 1548.

3,4-Secoisopimara-4(18),7,15-triene-3-oic acid

Physical data: $C_{20}H_{30}O_2$, crystalline, mp 51–52 °C, $[\alpha]_D^{25}$ +50.3° (MeOH, *c* 0.94)

Structure:

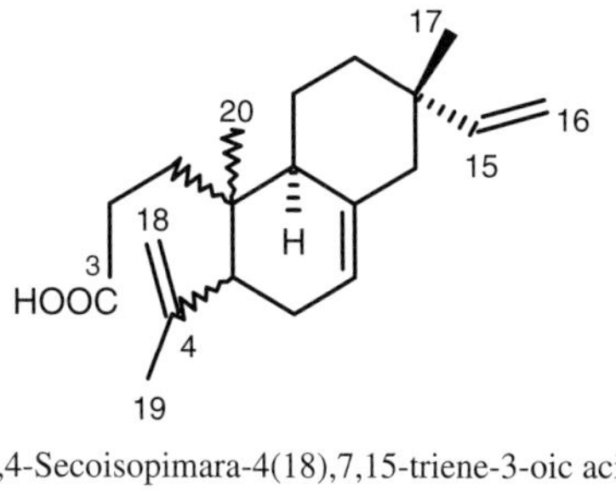

3,4-Secoisopimara-4(18),7,15-triene-3-oic acid

Compound class: Secoisopimarane diterpene

Source: *Salvia cinnabarina* M. Martens et Galeotti (leaves; family: Labiatae/Lamiaceae)

Pharmaceutical potential: *Antispasmodic*

The diterpenoid was found to have significant antispasmodic activity on histamine-, acetylcholine-, and barium chloride-induced contractions in the isolated guinea pig ileum; the IC_{50} values were estimated to be 2.7, 5.3, and 1.5 µg/ml, respectively, which are comparable with those obtained for the reference standard, papaverine. The inhibitory effect of the compound is, however, nonspecific as it blocks nonselectively the contractions induced by all the agonists tested. The investigators suggested that this compound could be treated in relieving gastrointestinal colic, diarrhea, and/or other gastrointestinal disorders; however, extent of efficacy is a subject of *in vivo* investigation.

Reference

Romussi, G., Ciarallo, G., Bisio, A., Fontana, N., De Simone, F., De Tommasi, N., Mascolo, N., and Pinto, L. (2001) *Planta Med.*, **67**, 153.

Selinone

Systematic name: 2,3-Dihydro-5,7-dihydroxy-2-{4-[3-methylbut-2-enyl)oxy]phenyl}-4*H*-1-benzopyran-4-one

Physical data: $C_{20}H_{20}O_5$, isolated as racemate; $[\alpha]_D^{23}$ +0° (MeOH, *c* 0.29)

Structure:

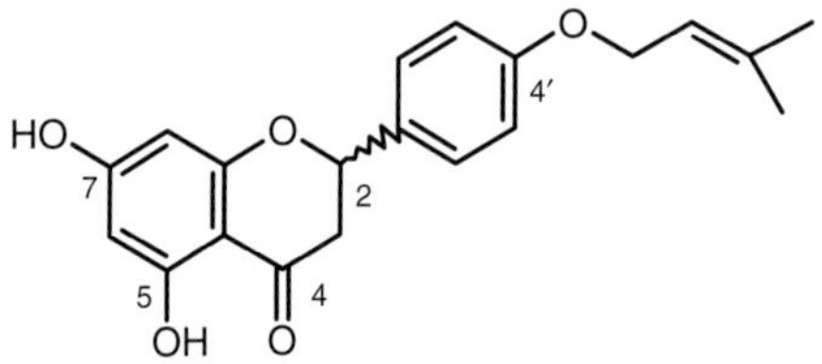

Selinone

Compound class: Flavonoid (flavanone)

Sources: *Selinum vaginatum* (family: Apiaceae) [1]; *Monotes engleri* (leaves; family: Dipterocarpaceae)

Pharmaceutical potential: *Antifungal*

The *O*-prenylated flavanone was found to possess antifungal activity against *Candida albicans* with an MIC value of 10 µg/ml [2].

References

[1] Seshadri, T.R. and Sood, M.S. (1967) *Tetrahedron Lett.*, **9**, 853.
[2] Garo, E., Wolfender, J.-L., Hostettmann, K., Hiller, W., Antus, S., and Mavi, S. (1998) *Helv. Chim. Acta*, **81**, 754.

Semicochliodinols A and B

Physical data:
- Semicochliodinol A: $C_{27}H_{23}N_2O_4$, solid, mp 195–197 °C
- Semicochliodinol B: $C_{27}H_{23}N_2O_4$, solid, mp >240 °C

Structures:

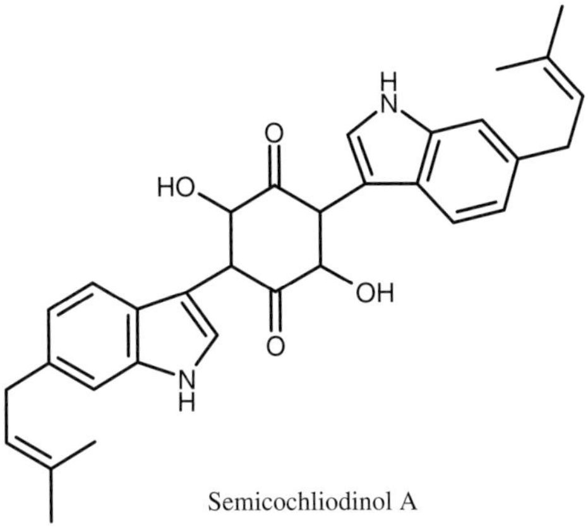

Semicochliodinol A

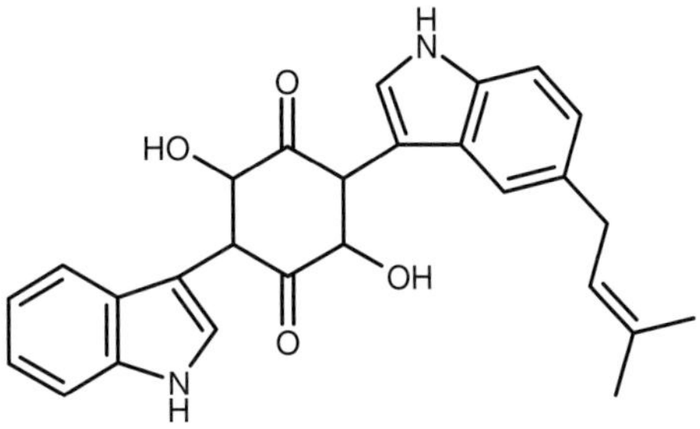

Semicochliodinol B

Compound class: Antibiotics (indole-quinone derivatives)

Source: *Chrysosporium mevdarium* P-5656 (fungus; fermentation broth)

Pharmaceutical potentials: *HIV-1 protease inhibitor; epidermal growth factor receptor protein tyrosine kinase (EGF-R PTK)*

Semicochliodinol A was found to exhibit greater inhibitory activity ($IC_{50} = 0.37 \mu M$) in comparison to semicochliodinol B ($IC_{50} \geq 0.5 \mu M$) against HIV-1 protease enzyme. Both the compounds also inhibited epidermal growth factor receptor protein tyrosine kinase with respective IC_{50} values of 20 and 60 µM.

Reference

Fredenhagen, A., Petersen, F., Tintelnot-Blomley, M., Rosel1, J., Mett, H., and Hug, P. (1997) *J. Antibiot.*, **50**, 395.

4-Senecioyloxymethyl-6,7-dimethoxycoumarin

Physical data: $C_{17}H_{18}O_6$, white needles, mp 141–142.5 °C

Structure:

4-Senecioyloxymethyl-6,7-dimethoxycoumarin

Compound class: Flavonoid (coumarin)

Source: *Crinum latifolium* L. (leaves; family: Amaryllidaceae)

Pharmaceutical potential: *Antiangiogenic*
The coumarin derivative was found to exhibit strong inhibitory activity against the *in vitro* tube-like formation of human umbilical venous endothelial cells (HUVECs); the inhibition percentage (IP) was measured to be 76.6 ± 3.9 at a dose of 3 µg/ml, comparable to that of suramin (positive control; IP $= 75.3 \pm 4.6$ at 30 µg/ml). The compound also showed quite significant activity (IP $= 53.5 \pm 3.8$ at 1 µg/ml). Although the test compound appears as a selective and potent inhibitor of *in vitro* HUVEC tube formation, it was found not to possess significant cytotoxicity against two tumor cell lines such as B16F10 (murine melanoma) and HCT116 (human colon carcinoma). The coumarin derivative might be a potential lead compound for the development of antiangiogenic anticancer agents.

Reference

Nam, N.-H., Kim, Y., You, Y.-J., Hong, D.-H., Kim, H.-M., and Ahn, B.-Z. (2004) *Nat. Prod. Res.*, **18**, 485.

E- and *Z*-Senegasaponins a and b

Systematic names:
- *E*- and *Z*-Senegasaponin a: 3-*O*-β-D-glucopyranosylpresenegenin 28-*O*-{[β-D-apiofuranosyl(1→3)]-β-D-galactopyranosyl(1→4)-β-D-xylopyranosyl(1→4)-α-L-rhamnopyranosyl(1→3)}{4-*O*-(*E*/*Z*)-4″-methoxycinnamoyl}-β-D-fucopyranoside
- *E*- and *Z*-Senegasaponin b: 3-*O*-β-D-glucopyranosylpresenegenin 28-*O*-{β-D-galactopyranosyl(1→4)-β-D-xylopyranosyl(1→4)-α-L-rhamnopyranosyl(1→2)}{4-*O*-(*E*/*Z*)-4″-methoxycinnamoyl}-β-D-fucopyranoside

Physical data:
- *E*-Senegasaponin a: $C_{74}H_{110}O_{35}$, colorless fine crystals, mp 228–231 °C, $[\alpha]_D^{28}$ $-12.9°$ (MeOH, *c* 1.0)
- *Z*-Senegasaponin a: $C_{74}H_{110}O_{35}$, colorless fine crystals, mp 237–240 °C, $[\alpha]_D^{27}$ $-22.0°$ (MeOH, *c* 1.0)
- *E*-Senegasaponin b: $C_{69}H_{102}O_{31}$, colorless fine crystals, mp 251–254 °C, $[\alpha]_D^{27}$ $+7.4°$ (MeOH, *c* 1.0)

- Z-Senegasaponin b: $C_{69}H_{102}O_{31}$, colorless fine crystals, mp 252–255 °C, $[\alpha]_D^{28}$ −13.2° (MeOH, c 1.0)

Structures:

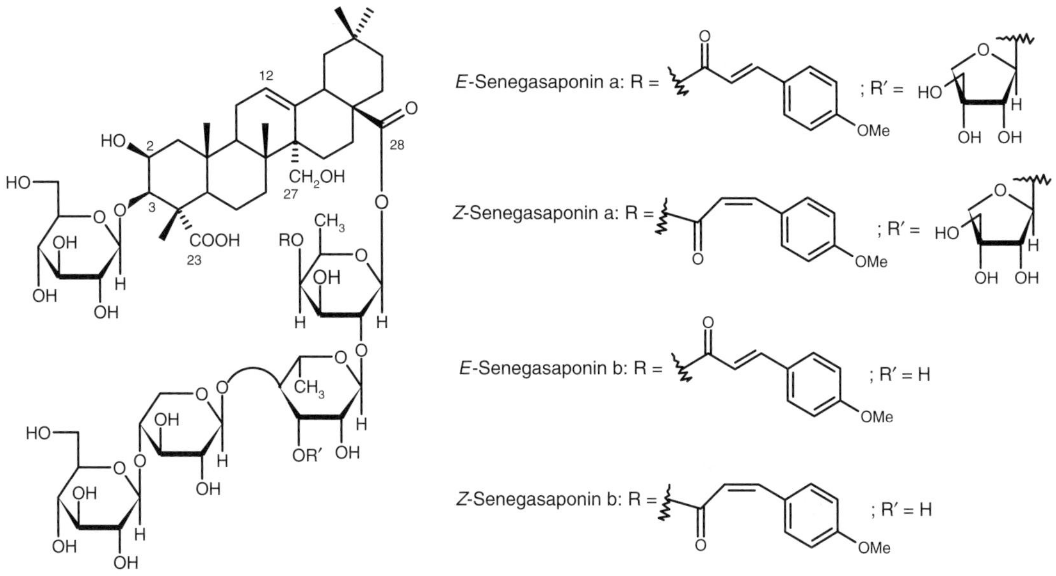

E-Senegasaponin a: R = (3,4-dimethoxycinnamoyl, E) ; R' = (arabinofuranosyl with OH, OH)

Z-Senegasaponin a: R = (3,4-dimethoxycinnamoyl, Z) ; R' = (arabinofuranosyl with OH, OH)

E-Senegasaponin b: R = (4-methoxycinnamoyl, E) ; R' = H

Z-Senegasaponin b: R = (4-methoxycinnamoyl, Z) ; R' = H

Compound class: Triterpenoid saponins

Source: *Polygala senega* Linn. var. *latifolia* Torrey et Gray (roots; family: Polygalaceae)

Pharmaceutical potentials: *Hypoglycemic; inhibitors of alcohol absorption*
The triterpenoid glycosides were reported to exert potent inhibitory effect on ethanol absorption and also to display hypoglycemic activity in the oral D-glucose tolerance test in rats. The saponins were found to display inhibitory effect on the elevation of plasma glucose level in the oral D-glucose tolerance test – at a dose of 100 mg/kg (p.o.), the differences in plasma glucose concentration between the normal control and each sample treatment after 0.5, 1.0, and 2.0 h were determined to be 22 ± 7.0, 23.4 ± 5.7, and 18.4 ± 6.2 mg/ml for E- and Z-senegasaponin a and 37.6 ± 4.1, 38.9 ± 5.3, and 27.7 ± 4.4 mg/ml for E- and Z-senegasaponin b, respectively.

Reference

Yoshikawa, M., Murakami, T., Ueno, T., Kadoya, M., Matsuda, H., Yamahara, J., and Murakami, N. (1995) *Chem. Pharm. Bull.*, 43, 2115.

Senegins II and III

Systematic names:
- Senegin II: 3-O-β-D-glucopyranosylpresenegenin 28-O-[β-D-galactopyranosyl(1→4)-β-D-xylopyranosyl(1→4)-α-L-rhamnopyranosyl(1→2)-4-(3′,4′-dimethoxycinnamoyl)-β-D-fucopyranoside]

- Senegin III: 3-O-β-D-glucopyranosylpresenegenin 28-O-{2-[β-D-galactopyranosyl(1→4)-β-D-xylopyranosyl(1→4)-α-L-rhamnopyranosyl]-3-(α-L-rhamnopyranosyl)-4-(4'-methoxycinnamoyl)-β-D-fucopyranoside}

Physical data:
- Senegin II: $C_{70}H_{104}O_{32} \cdot 4H_2O$, colorless needles, mp 247–248 °C, $[\alpha]_D^{20}$ −6.2° (MeOH)
- Senegin III: $C_{75}H_{112}O_{35} \cdot 3H_2O$, white powder, $[\alpha]_D^{20}$ −6.6° (MeOH)

Structures:

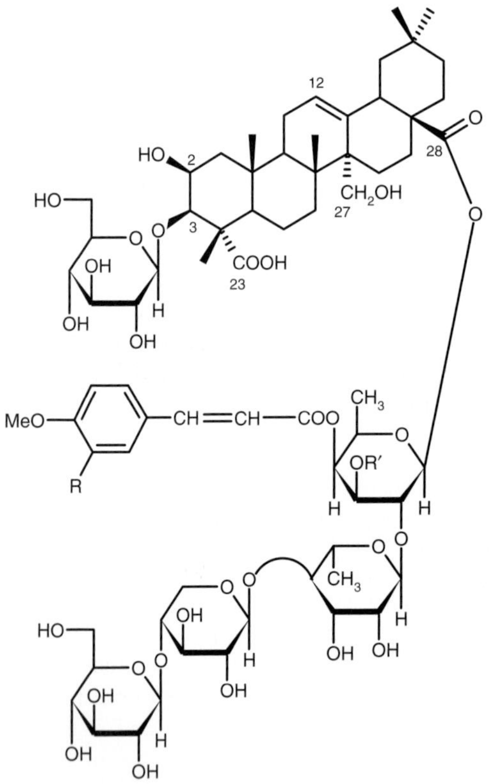

Senegin II: R = OCH₃; R' = H
Senegin III: R = H; R' = α-L-rhamnopyranosyl

Compound class: Triterpenoid saponins

Source: *Polygala senega* Linn. var. *latifolia* Torrey et Gray (roots/rhizomes; family: Polygalaceae) [1–4]

Pharmaceutical potential: *Hypoglycemic*
Both the glycosides were reported to reduce the blood glucose of normal mice as well as of KK-Ay mice, an animal model of obese non-insulin-dependent diabetes mellitus (NIDDM) with hyper-

insulinemia; the saponins showed the hypoglycemic activity at doses of 1 and 5 mg/kg, 4 h after intraperitoneal administration in both the systems [3]. The hypoglycemic effects exerted by the triterpenoid glycosides in normal mice were found to be dose dependent. Such hypoglycemic efficacies of the test compounds are summarized below.

Table Hypoglycemic effects of senegins II and III in normal and KK-Ay mice [3].

Test compounds	Dose (mg/kg)	Blood glucose level in normal mice (mg/100 ml)		Blood glucose level in KK-Ay mice (mg/100 ml)	
		0 h	4 h	0 h	4 h
Senegin II	1	211 ± 12	139 ± 8	395 ± 52	226 ± 11
	5	213 ± 11	137 ± 13	446 ± 41	217 ± 19
Senegin III	1	205 ± 4	167 ± 7	418 ± 61	195 ± 18
	5	210 ± 8	151 ± 2	438 ± 43	204 ± 16
Tolbutamide (a positive control)	50	200 ± 7	160 ± 9	537 ± 44	495 ± 66

The hypoglycemic effects of the compounds were potent, and glucose levels were similar to the basal level of normal mice, indicating that they may affect insulin resistance of peripheral tissues. Interestingly, these triterpenoid saponins were evaluated to possess much more potent hypoglycemic activity than that of the positive control, tolbutamide. Hence, these triterpenoid glycosides may be useful for treating NIDDM [3, 4].

References

[1] Tsukitani, Y., Kawanishi, S., and Shoji, J. (1973) *Chem. Pharm. Bull.*, **21**, 791.
[2] Tsukitani, Y. and Shoji, J. (1973) *Chem. Pharm. Bull.*, **21**, 1564.
[3] Kato, M., Miura, T., Nishiyama, Y., Ichimaru, M., Moriyasu, M., and Kato, A. (1997) *J. Nat. Prod.*, **60**, 604.
[4] Yoshikawa, M., Murakami, T., Matsuda, H., Ueno, T., Kadoya, M., Yamahara, J., and Murakami, N. (1996) *Chem. Pharm. Bull.*, **44**, 1305.

(+)-Sesamin

Physical data: $C_{20}H_{18}O_6$, colorless needles (EtOH), mp 119–121 °C, $[\alpha]_D^{25}$ +50° (CHCl$_3$, c 0.5)

Structure:

(+)-Sesamin

Compound class: Lignan

Source: *Wikstroemia lanceolata* (stems and roots; family: Thymelaeaceae) [1]; *Cinnamomum camphora* (stems; family: Lauraceae) [2]

Pharmaceutical potential: *Cytotoxic*
Hsieh *et al.* [2] demonstrated that (+)-sesamin exerts a reduction in the percentage of S phase of the cell cycle of HepG2 cells after incubation for 24 h in a dose-dependent manner; the percentage of cells in the S phase was noted to be decreased to 30% after incubation with 200 M of the test compound for 24 h. The results indicated that the lignan inhibits the DNA synthetic event in HepG2 cells [2].

References

[1] Lin, R.-W., Tsai, I.-L., Duh, C.-Y., Lee, K.-H., and Chen, I.-S. (2004) *Planta Med.*, **70**, 234.
[2] Hsieh, T.J., Chen, C.H., Lo, W.L., and Chen, C.Y. (2006) *Nat. Prod. Commun.*, **1**, 21.

Seselidiol

Systematic name: Heptadeca-1,8(Z)-diene-4,6-diyne-3,10-diol

Physical data: $C_{17}H_{24}O_2$, yellowish oil, $[\alpha]_D^{23}$ +192° (Et$_2$O, *c* 0.53)

Structure:

H$_2$C=CHCH(OH)—C≡C—C≡C—CH=CH—CH(OH)—(CH$_2$)$_6$CH$_3$

Seselidiol

Compound class: Polyacetylene derivative

Source: *Seseli marirei* Wolff (roots; family: Umbelliferae)

Pharmaceutical potential: *Cytotoxic (antitumor)*
The isolate displayed *in vitro* cytotoxicity against the tumor cell lines KB, HCT-8, P-388, and L-1210 with ED$_{50}$ values of 1.0, 10.0, 4.9, and 3.3 g/ml, respectively.

Reference

Hu, C.-Q., Chang, J.-J., and Lee, K.-H. (1990) *J. Nat. Prod.*, **53**, 932.

β-Sesquiphellandrene

Systematic name: (R)-3-Methylene-6-[(S)-6-methylhept-5-en-2-yl]-cyclohex-1-ene; alternatively, 2-methyl-6-(4-methylidene-1-cyclohex-2-enyl)hept-2-ene

Physical data: $C_{15}H_{24}$, colorless oil

Structure:

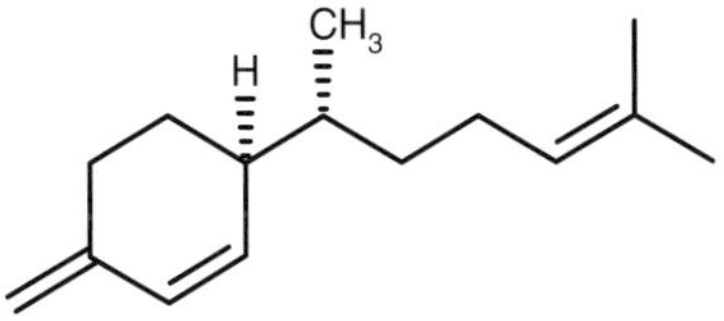

β-Sesquiphellandrene

Compound class: Sesquiterpenoid

Source: *Zingiber officinale* (dried rhizomes; family: Zingiberaceae)

Pharmaceutical potential: *Antirhinoviral*
The sesquiterpenoid compound was evaluated as a potent inhibitor of rhinovirus IB (BW333C85) *in vitro* with an ED_{50} value of 0.90 ± 0.10 µg/plate.

Reference

Denyer, C.V., Jackson, P., Loakes, D.M., Ellis, M.R., and Young, D.A.B. (1994) *J. Nat. Prod.*, **57**, 658.

Sesquiterpene hydroperoxides 1–3

Systematic names:
- **1**: 10α-hydroperoxy-guaia-1,11-diene
- **2**: 1α-hydroperoxy-guaia-10(15),11-diene
- **3**: 15α-hydroperoxy-guaia-1(10),11-diene

Physical data:
- **1**: $C_{15}H_{24}O_2$, colorless oil, $[\alpha]_D$ +24.5° (EtOH, *c* 0.35)
- **2**: $C_{15}H_{24}O_2$, colorless oil, $[\alpha]_D$ +71.8° (EtOH, *c* 0.58)
- **3**: $C_{15}H_{24}O_2$, colorless oil, $[\alpha]_D$ +7.3° (EtOH, *c* 0.23)

Structures:

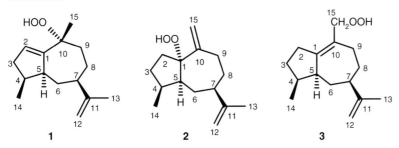

Compound class: Sesquiterpenoids (sesquiterpene hydroperoxides)

Source: *Pogostemon cablin* Benth. (whole plants; family: Labiatae)

Pharmaceutical potential: *Trypanocidal*

The sesquiterpene hydroperoxides **1–3** were reported to exhibit *in vitro* trypanocidal activity against epimastigotes of *Trypanosoma cruzi*; compound **1** showed a minimum lethal concentration (MLC) value of 0.84 µM, while the other two showed equal MLC values of 1.7 µM (gentian violet was used as a positive control; MLC = 6.3 µM).

Reference

Kiuchi, F., Matsuo, K., Ito, M., Qui, T.K., and Honda, G. (2004) *Chem. Pharm. Bull.*, **52**, 1495.

Sesterstatins 1–3

Physical data:
- Sesterstatin 1: $C_{25}H_{38}O_4$, colorless amorphous powder, mp 297–298 °C, $[\alpha]_D^{22}$ +16.3° (CHCl$_3$, c 0.12)
- Sesterstatin 2: $C_{25}H_{38}O_4$, colorless amorphous powder, mp 295–296 °C, $[\alpha]_D^{22}$ +13.8° (CHCl$_3$, c 0.09)
- Sesterstatin 3: $C_{25}H_{38}O_4$, colorless amorphous powder, mp 293–294 °C, $[\alpha]_D^{22}$ +27.2° (CHCl$_3$, c 0.22)

Structures:

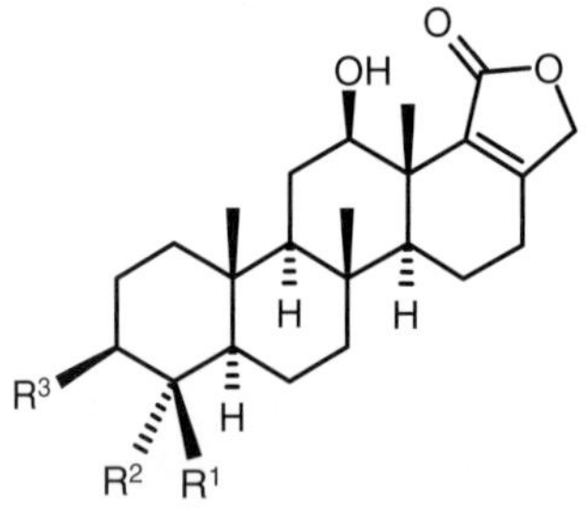

Sesterstatin 1: R^1 = R^2 = CH$_3$; R^3 = OH
Sesterstatin 2: R^1 = CH$_3$; R^2 = CH$_2$OH; R^3 = H
Sesterstatin 3: R^1 = CH$_2$OH; R^2 = CH$_3$; R^3 = H

Compound class: Sesterpenoids

Source: *Hyrtios erecta* (marine sponge; order: Dictyoceratida, family: Thorectidae) [1]

Pharmaceutical potentials: *Antineoplastic; antibacterial*

Sesterstatins **1–3** displayed growth inhibitory activity against the P-388 leukemia cells with ED$_{50}$ values of 0.46, 4.2, and 4.3 µg/ml, respectively. The cancer cell growth inhibitory activity of these sesterstatins is suggested to be quite related to the presence of butenolide-type lactone groups in these molecules.

Sesterstatin 2 inhibited growth of the Gram-positive bacterium *Staphylococcus aureus* with a minimum inhibitor concentration of 50–100 µg/disk.

Reference

Pettit, G.R., Cichacz, Z.A., Tan, R., Hoard, M.S., Melody, N., and Pettit, R.K. (1998) *J. Nat. Prod.*, **61**, 13.

Sg17-1-4

Physical data: $C_{24}H_{34}N_2O_9$, amorphous white powder, mp 147–1149 °C, $[\alpha]_D^{20}$ −110° (MeOH, *c* 0.1)

Structure:

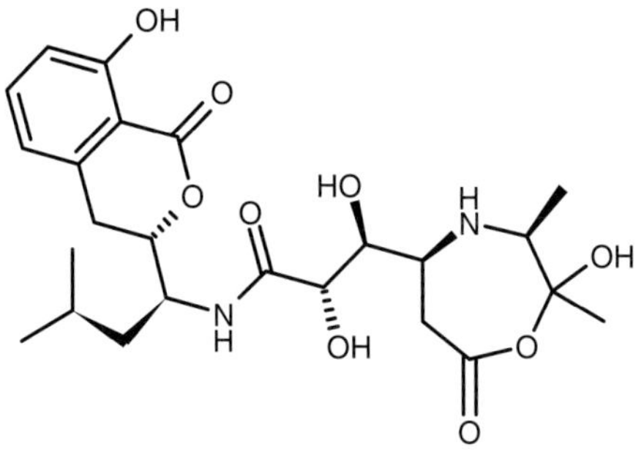

Sg17-1-4

Compound class: Antibiotic (isocoumarin)

Source: *Alternaria tenuis* Sg17-1 (marine fungus)

Pharmaceutical potential: *Cytotoxic*

The isolate was evaluated to possess moderate *in vitro* growth inhibitory activities against human malignant A375-S2 and human cervical cancer HeLa cells with IC_{50} values of 0.3 and 0.05 mM, respectively.

Reference

Huang, Y.-F., Li, L.-H., Tian, L., Qiao, L., Hua, H.-M., and Pei, Y.-H. (2006) *J. Antibiot.*, **59**, 355.

Siamenol

Physical data: $C_{18}H_{19}NO$, pale yellow solid

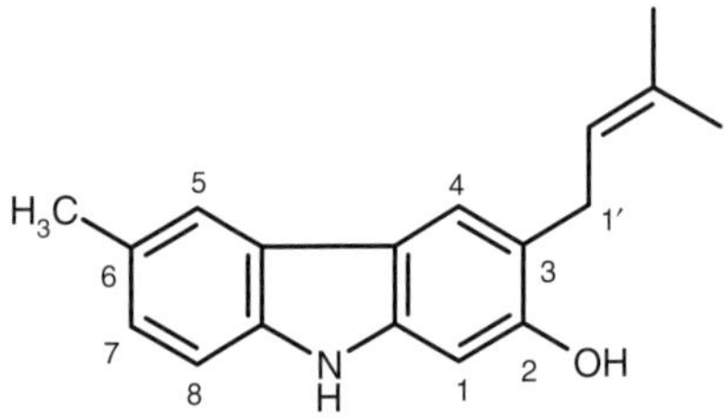

Siamenol

Compound class: Alkaloid

Source: *Murraya siamensis* (aerial parts (flowers, leaves, and twigs); family: Rutaceae)

Pharmaceutical potential: *Anti-HIV*
Siamenol was evaluated to display *in vitro* anti-HIV activity ($EC_{50} = 2.6\,\mu g/ml$), reaching 50–60% maximum protection in the XTT-tetrazolium assay.

Reference

Meragelman, K.M., McKee, T.C., and Boyd, M.R. (2000) *J. Nat. Prod.*, **63**, 427.

Sibiskoside

Systematic name: 1-*O*-β-D-Glucopyranosyl-geraniol-5,10-olide

Physical data: $C_{16}H_{25}O_8$, colorless amorphous powder, $[\alpha]_D^{22}$ $-20°$ (MeOH, *c* 1.0)

Structure:

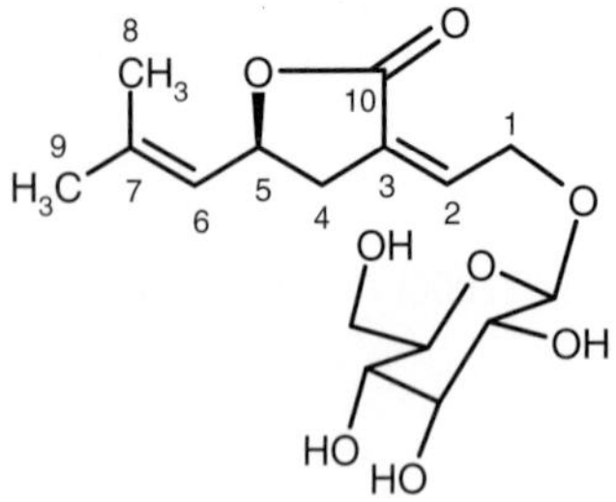

Compound class: Monoterpene glycoside

Source: *Sibiraea angustata* Rchd. (aerial parts; family: Rosaceae)

Pharmaceutical potential: *Antiobestic*
The present group of investigators evaluated the isolate as a promising natural agent having suppressive effect on the increase in body weight as studied in the rat model. Its oral administration to experimental mice reared with high-fat diet resulted in significant weight loss, which was

also reflected in serum triglyceride, cholesterol, and sugar levels and the weight of abdominal fat. Moreover, the compound was extremely less toxic; acute toxicity experiments revealed that oral administration of the test compound at a dose up to 2500 mg/kg body weight to mice resulted in no death and no evidence of abnormalities in internal organs.

Reference

Ito, Y., Kamo, S., Sadhu, S.K., Ohtsuki, T., Ishibashi, M., and Kano, Y. (2009) *Chem. Pharm. Bull.*, **57**, 294.

Sieboldine A

Physical data: $C_{16}H_{23}NO_4$, colorless needles (MeOH), mp 160 °C, $[\alpha]_D^{17}$ +139° (MeOH, *c* 0.3)

Structure:

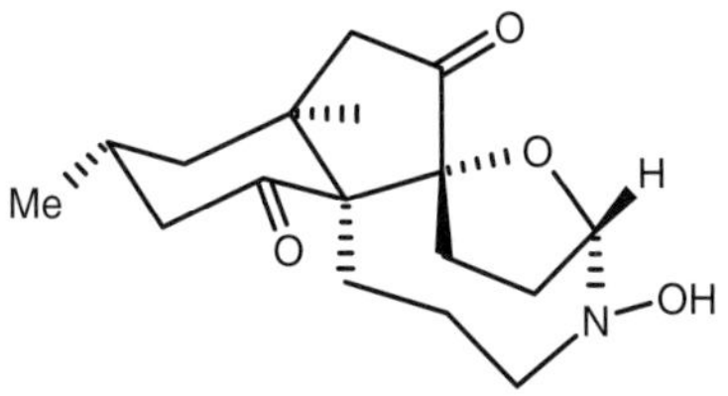

Sieboldine A

Compound class: Alkaloid (fused tetracyclic ring system consisting of an azacyclononane ring)

Source: *Lycopodium* species (club moss; family: Lycopodiaceae)

Pharmaceutical potentials: *Acetylcholinesterase (AChE) inhibitor; cytotoxic*
The tetracyclic alkaloid was evaluated to possess potent inhibitory activity against acetylcholinesterase (AChE) enzyme with an IC_{50} value of 2.0 μM, which was found to be comparable to that of (±)-huperzine A (IC_{50} = 1.6 μM). The isolate also exhibited *in vitro* cytotoxicity against murine lymphoma L1210 cells with an IC_{50} value of 5.1 μg/ml.

Reference

Hirasawa, Y., Morita, H., Shiro, M., and Kobayashi, J. (2003) *Org. Lett.*, **5**, 3991.

Sigmoidins A and B

Systematic names:
- Sigmoidin A: 5,7,3',4'-tetrahydroxy-2',5'-diprenylflavanone
- Sigmoidin B: 5,7,3',4'-tetrahydroxy-2'-prenylflavanone

Physical data:
- Sigmoidin A: $C_{25}H_{28}O_6$, needles, mp 181–182 °C, $[\alpha]_D^{28}$ −82° (MeOH, *c* 2.0)
- Sigmoidin B: $C_{20}H_{20}O_6$, granules, mp 217–218 °C, $[\alpha]_D^{28}$ −54° (MeOH, *c* 3.0)

Structures:

Sigmoidin A: R = CH$_2$—CH=C(CH$_3$)$_2$ (prenyl)
Sigmoidin B: R = H

Compound class: Flavonoids (flavanones)

Source: *Erythrina sigmoidea* Hua (barks; family: Leguminosae/Fabaceae) [1–3]

Pharmaceutical potentials: *Antioxidant; anti-inflammatory; antibacterial*
Both sigmoidins A and B showed effective antioxidant activity studied in the DPPH assay; they scavenged the radical formed in the assay by 93 and 86%, respectively, and the values are in the same range as that of quercetin-3-O-glucoside (92%), which was used as the reference standard [3]. The same investigators [3] evaluated both the prenylated flavanones for their *in vitro* as well as *in vivo* anti-inflammatory potential; during their study of the inhibition of arachidonic acid metabolism, the investigators demonstrated that both the flavanones were selective inhibitors of 5-lipoxygenase (5-LOX), with no effect on cyclooxygenase-1 (COX-1). At a concentration of 100 μM, neither of them reduced the production of 12-HHTrE (12-hydroxyheptadecatrienoic acid), and none of the tested compounds exhibited cytotoxicity in the MTT (3-(4,5-dimethylthiazol-2-yl)-2,5-diphenyltetrazolium bromide) test. Sigmoidin A at 100 μM was reported to reduce the production of leukotriene B$_4$ (LTB$_4$) from rat peritoneal leukocytes by 95% inhibition to give an IC$_{50}$ value of 31 μM. At the same concentration, sigmoidin B showed only a moderate effect (44%). *In vivo* anti-inflammatory activity of the compounds was examined on phospholipase A$_2$ (PLA$_2$)-induced paw edema as well as on 12-O-tetradecanoylphorbol 13-acetate (TPA)-induced ear edema in mice [3]. In the PLA$_2$-induced paw edema test, administration of sigmoidin B (5 mg/kg, i.p.) inhibited the edema by 59% 1 h after the PLA-2, while sigmoidin A showed only 20% inhibition 30 min after the injection. In TPA-induced ear edema, sigmoidins A and B were found to be effective at a dose of 0.25 mg/ear, decreasing the edema by 89 and 83%, respectively; for the reference drug indomethacin, the edema inhibition was found to be 83% at 0.5 mg/ear [3].
    Sigmoidins A and B exhibited significant antibiotic activity against Gram-positive bacteria [1]. Their antibacterial activity was also reported by Biyiti *et al.* [4]; the investigators studied the effect of the flavanones on the intracellular calcium compartment involved in the first cleavage of sea urchin eggs. When added after insemination, both compounds inhibited egg division with a half maximal dose of 7.5 μM for sigmoidin A and 12 μM for sigmoidin B. The first Ca^{2+} signal

following fertilization was not modified by the test compounds; however, the intracellular storage of calcium in isolated nonmitochondrial compartments was reduced by them in a dose-dependent manner. In *in vivo* test, the flavanones were found to dramatically reduce the capacity of storage of nonmitochondrial intracellular calcium compartments necessary for the cyclical elevation of cytosolic-free calcium during the cell cycle [4].

References

[1] Fomum, Z.T., Ayafor, J.F., Mbafor, J.T., and Mbi, C.M. (1986) *J. Chem. Soc., Perkin Trans. 1*, 33.
[2] Ndom, J.C., Mbafor, J.T., Formum, Z.T., Martin, M.T., and Bodo, M. (1993) *Magn. Res. Chem.*, **31**, 210.
[3] Njamen, D., Mbafor, J.T., Fomum, Z.T., Kamanyl, A., Mbanya, J.C., Reclo, M.C., Giner, R.M., Manez, S., and Rios, J.L. (2004) *Planta Med.*, **70**, 104.
[4] Biyiti, L., Pesando, D., Puiseux-Dao, S., Girad, J.P., and Payan, P. (1990) *Toxicon*, **28**, 275.

Silvestrol and *epi*-silvestrol

Physical data:
- Silvestrol: $C_{34}H_{38}N_3O_{13}$, white amorphous powder, mp 119–123 °C, $[\alpha]_D^{20}$ −137.0° (MeOH, *c* 0.2)
- *epi*-Silvestrol: $C_{34}H_{38}N_3O_{13}$, yellowish gum, $[\alpha]_D^{20}$ −94.5° (CHCl$_3$, *c* 0.43)

Structures:

Silvestrol: 5'''R
epi-Silvestrol: 5'''S

Compound class: Rocaglate derivatives

Source: *Aglaia silvestris* (M. Roemer) Merrill (syn. *A. pyramidata* Hance) (fruits and twigs; family: Meliaceae)

Pharmaceutical potential: *Cytotoxic*
Silvestrol and its epimer were found to show remarkable cytotoxicity against human lung cancer (Lu1), hormone-dependent human prostate cancer (LNCaP), human breast cancer (MCF-7), and human umbilical vein endothelial cells (HUVEC). Silvestrol was found to be three times more active than *epi*-silvestrol. The respective ED$_{50}$ values as determined against the aforementioned cell lines are 1.2, 1.5, 1.5, and 4.6 nM (silvestrol); 3.8, 3.8, 5.5, and 15.3 nM (*epi*-silvestrol). The

respective ED_{50} values for the positive controls used were determined to be 2.3, 4.7, 0.7, and 105.5 nM (paclitaxel); 28.7, 28.7, 28.7, and 1258.6 nM (camptothecin).

Reference

Hwang, B.Y., Su, B.-N., Chai, H., Mi, Q., Kardono, L.B.S., Afriastini, J.J., Riswan, S., Santarsiero, B.D., Mesecar, A.D., Wild, R., Fairchild, C.R., Vite, G.D., Rose, W.C., Farnsworth, N.R., Cordell, G.A., Pezzuto, J.M., Swanson, S.M., and Kinghorn, A.D. (2004) *J. Org. Chem.*, **69**, 3350.

Simalikalactone D

Physical data: $C_{25}H_{34}O_9$, white platelets (acetone), mp 230 °C

Structure:

Simalikalactone D

Compound class: Quassinoid

Sources: *Simaba multiflora* A. Juss. (family: Simaroubaceae) [1], *S. guianensis* Aubl. (barks) [2]; *Quassia afrikana* Baill. (ground root barks; family: Simaroubaceae) [3, 4], *Q. amara* L. [3, 5, 6]; *Leitneria floridana* Chapman (aerial parts; family: Leitneriaceae) [7]

Pharmaceutical potentials: *Antimalarial; antiviral; antitumor*
The isolate was reported to have potent *in vitro* antimalarial activity [2, 5, 8]. The isolate exhibited such activity against two different strains of *Plasmodium falciparum*, namely, W-2 Indochina (a chloroquine-resistant strain) and the D-6 Sierra Leone (a mefloquine-resistant strain), with IC_{50} values of 1.6 and 1.5 ng/ml, respectively [2]; the respective IC_{50} values for the known antimalarials used as standards were determined to be 63.2 and 2.9 ng/ml (chloroquine); 1.5 and 8.7 ng/ml (mefloquine); 1.1 and 2.1 ng/ml (artemisinin); and 68.6 and 12.0 ng/ml (quinine) [2]. Bertani et al. [5] observed that the quassinoid shows activity *in vitro* against FcB1 *Plasmodium falciparum* chloroquine-resistant strain with an IC_{50} value of 10 nM; the investigators also observed that the test compound inhibits 50% of *Plasmodium yoelii yoelii* rodent malaria parasite at 3.7 mg/kg/day *in vivo* by oral route. In addition, the compound was reported to possess antiviral potential, and it was suggested that the ester group at C-15 and the epoxymethano bridge between C-8 and C-13 in the molecule might be responsible for its antiviral activity [4].

Xu et al. [7] evaluated that the isolate possesses potent *in vitro* antitumor efficacy by inhibiting the growth of a panel of human tumor cell lines such as KB, A-549, HCT-8, CAKI-1, MCF-7, and

SK-MEL-2 with respective ED_{50} values of 0.018, 0.04, 0.013, 0.26, 0.014, and 0.13 µg/ml. Antileukemic potential of the compound was also reported by Kupchan et al. [6].

References

[1] Arisawa, M., Fujita, A., Morita, N., Kinghorn, A.D., Cordell, G.A., and Farnsworth, N.R. (1985) *Planta Med.*, **51**, 348.
[2] Cabral, J.A., McChesney, J.D., and Milhous, W.K. (1993) *J. Nat. Prod.*, **56**, 1954.
[3] Tresca, J.P., Alais, L., and Polonsky, J. (1971) *C. R. Acad. Sci. Paris*, **273**, 601.
[4] Apers, S., Cimanga, K., Vanden, B.D., van Meenen, E., Longanga, A.O., Foriers, A., Vlietinck, A., and Pieters, L. (2002) *Planta Med.*, **68**, 20.
[5] Bertani, S., Houël, E., Stien, D., Chevolot, L., Jullian, V., Garavito, G., Bourdy, G., and Deharo, E. (2006) *J. Ethnopharmacol.*, **108**, 155.
[6] Kupchan, S.M. and Streelman, D.R. (1976) *J. Org. Chem.*, **41**, 3481.
[7] Xu, Z., Chang, F.-R., Wang, H.-K., Kashiwada, Y., McPhail, A.T., Bastow, K.F., Tachibana, Y., Cosentino, M., and Lee, K.-H. (2000) *J. Nat. Prod.*, **63**, 1712.
[8] O'Neill, M.J., Bray, D.H., Boardman, P., Phillipson, J.D., Warhurst, D.C., Peters, W., and Suffness, M. (1986) *Antimicrob. Agents Chemother.*, **30**, 101.

Sinularia glycoside

Physical data: $C_{36}H_{58}O_8$, white needles (MeOH), mp 172.5–173 °C, $[\alpha]_D^{28}$ −6.3° (CHCl$_3$, c 5.26)

Structure:

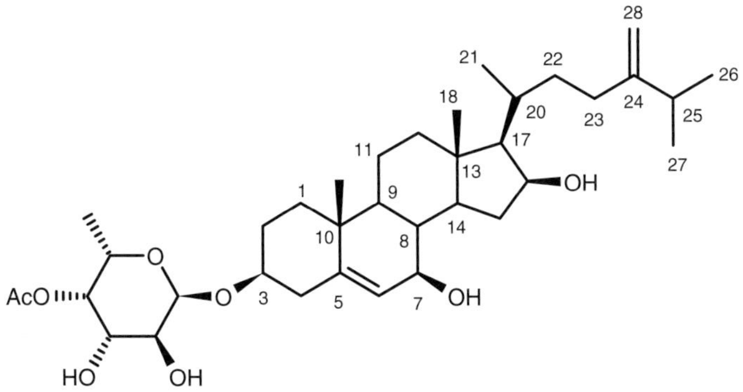

Sinularia glycoside

Compound class: Steroidal glycoside

Source: *Sinularia crispa* (Sri Lankan soft coral)

Pharmaceutical potential: *Spermatostatic*
The steroidal glycoside was evaluated to exhibit spermatostatic activity on rat cauda epididymal spermatozoa at 0.5 mg/ml.

Sinularolides B–E

Reference

Tillekeratne, L.M.V., Liyanage, G.K., Ratnasooriya, W.D., Ksebati, M.B., and Schmitz, F.J. (1989) *J. Nat. Prod.*, **52**, 1143.

Sinularolides B–E

Physical data:
- Sinularolide B: $C_{20}H_{28}O_5$, colorless crystal, mp 137–138 °C, $[\alpha]_D^{25}$ −134.3° (CHCl$_3$, c 0.05)
- Sinularolide C: $C_{20}H_{28}O_5$, colorless oil, $[\alpha]_D^{25}$ −56.3° (CHCl$_3$, c 0.07)
- Sinularolide D: $C_{20}H_{28}O_4$, colorless oil, $[\alpha]_D^{25}$ −31.6° (CHCl$_3$, c 0.04)
- Sinularolide E: $C_{20}H_{28}O_4$, colorless oil, $[\alpha]_D^{25}$ −29.7° (CHCl$_3$, c 0.08)

Structures:

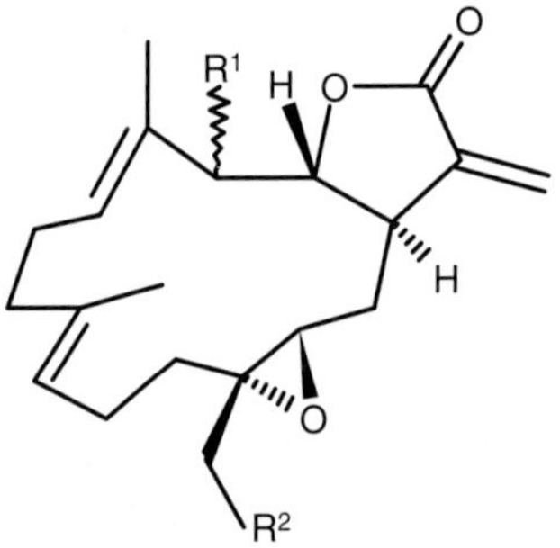

Sinularolide B: R^1 = β-OH, R^2 = OH
Sinularolide C: R^1 = α-OH, R^2 = OH
Sinularolide D: R^1 = H, R^2 = OH
Sinularolide E: R^1 = β-OH, R^2 = H

Compound class: Cembranoid diterpenoids

Source: Soft coral *Sinularia gibberosa* (Tixier-Durivault)

Pharmaceutical potential: *Antitumor*

Sinularolides B–E were reported to display moderate cytotoxic activity against cultured human tumor cell lines, such as HL-60 (human leukemic cancer cell), BGC-823 (human gastric cancer cell), and MDA-MB-435 (human breast cancer cell), with respective IC$_{50}$ values of 5.2, 6.3, and 8.0 μM (sinularolide B); 5.1, 5.2, and 5.7 μM (sinularolide C); 2.3, 6.1, and >10 μM (sinularolide D); 6.0, 8.6, and 2.1 μM (sinularolide E). The activity was studied using the MTT method for the HL-60 assay and the SRB method for the other two assays.

On the basis of bioassay results, the investigators suggested that the epoxide unit at C-12/C-13 is essential for the observed activity against the tested cell lines.

Reference

Li, G., Zhang, Y., Deng, Z., van Ofwegen, L., Proksch, P., and Lin, W. (2005) *J. Nat. Prod.*, **68**, 649.

β-Sitosterol glucoside-3′-O-hexacosannoicate

Physical data: $C_{61}H_{110}O_7$, white amorphous powder, $[\alpha]_D^{25}$ $-20°$ (MeOH, c 0.2)

Structure:

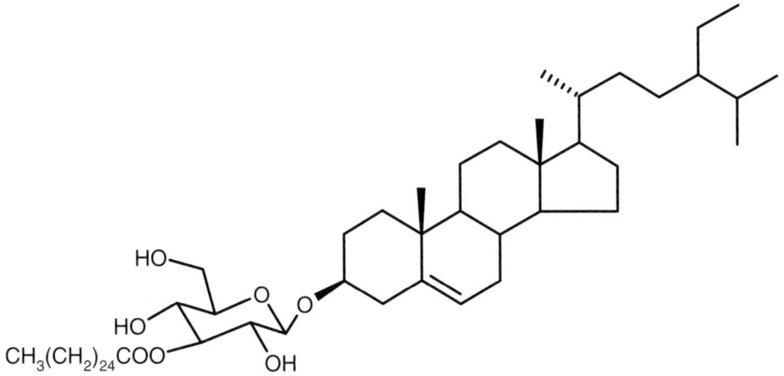

β-Sitosterol glucoside-3′-O-hexacosannoicate

Compound class: Steroidal glycoside

Source: *Acer okamotoanum* Nakai (leaves; family: Aceraceae)

Pharmaceutical potential: *Anticomplement activity*
The steroidal glycoside was evaluated to possess significant anticomplement activity *in vitro* – the compound exhibited potent inhibitory activity against complement activity induced by the classical pathway (CP) with an IC_{50} value of $0.2 \pm 0.08\,\mu M$; it was found to be more efficacious than tiliroside used as a positive control ($IC_{50} = 78.12 \pm 2.4\,\mu M$). The steroidal derivative may find useful applications in inhibition of the unwanted and excessive activation of the complement system that plays a key role in host defense.

Reference

Jun, W.Y., Min, B.-S., Lee, J.P., Thuong, P.T., Lee, H.-K., Song, K., Seong, Y.H., and Bae, K. (2007) *Arch. Pharm. Res.*, **30**, 172.

Somocystinamide A

Physical data: $C_{42}H_{70}N_4O_4S_2$, white amorphous solid, $[\alpha]_D^{22}$ $+13.5°$ (CHCl$_3$, c 0.75)

Sophoraflavanone G

Structure:

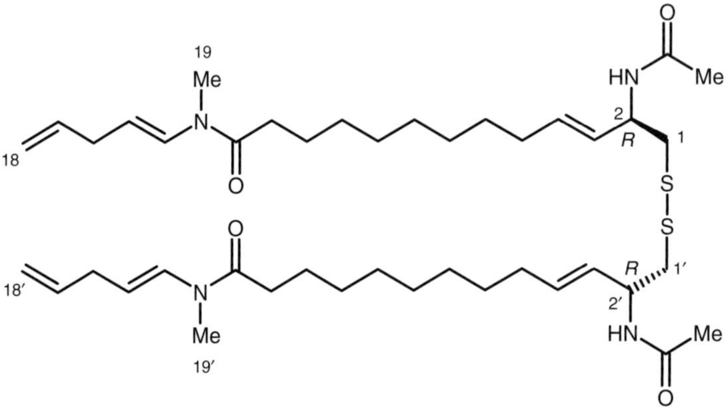

Somocystinamide A

Compound class: Disulfide dimer

Source: *Lyngbya majuscula/Schizothrix* sp. (mixed assemblage of marine cyanobacteria)

Pharmaceutical potential: *Cytotoxic*
The disulfide dimer exhibited significant cytotoxicity against mouse neuro-2a neuroblastoma cells with an IC_{50} value of 1.4 µg/ml.

Reference

Nogle, L.M. and Gerwick, W.H. (2002) *Org. Lett.*, **4**, 1095.

Sophoraflavanone G

Systematic name: (2S)-5,7,2′,4′-Tetrahydroxy-8-lavandulylflavanone

Physical data: $C_{25}H_{28}O_6$, colorless needles (benzene), mp 173–175 °C, $[\alpha]_D^{25}$ −49.0° (MeOH, *c* 1.0)

Structure:

Sophoraflavanone G

Compound class: Flavonoid (flavanone)

Sources: *Sophora moorcroftiana* Benth. ex Baker (roots; family: Leguminoase) [1]; *S. exigua* [2]; *S. flavescens* (roots) [3]; *S. leachiano* (roots) [4]; *S. pachycarpa* (roots) [5]

Pharmaceutical potentials: *Cytotoxic; antioxidant; antimicrobial; antimalarial*
Sophoraflavanone G showed moderate cytotoxic activity against human myeloid leukemia HL-60 cells with an IC_{50} value of 12.5 µM (IC_{50} value of cisplatin (positive control) = 2.3 µM) [3]. The compound also displayed significant antioxidant activity as studied in the DPPH assay; it was found to scavenge the DPPH free radical with an IC_{50} value of 5.26 ± 0.97 µg/ml [6].

Fakhimi et al. [5] showed that an ultralow concentration of sophoraflavanone G potentiates the antimicrobial action of gentamicin – it was observed that the MIC value of gentamicin decreases from 32 to 8 µg/ml (a fourfold decrease) in the presence of 0.03 µg/ml of sophoraflavanone G; hence, the flavanone may find possible application in combination therapy against *S. aureus*. Pronounced synergistic effects of the combinations of sophoraflavanone G separately with ampicillin and gentamicin were noticed against a number of bacterial and fungal strains [7]. Sakagami et al. [8] evaluated anti-MRSA (anti-methicillin-resistant *Staphylococcus aureus*) of this flavonoid against 27 strains – the MIC values were determined within the range of 3.13–6.25 µg/ml. They also studied synergism between sophoraflavanone G and a number of antibacterial agents such as vancomycin hydrochloride, fosfomycin, methicillin, cefzonam, gentamicin, minocycline, and levofloxacin; the experimental findings support the fruitful use of the test compound in combination with the established antimicrobial drugs [8]. Antimicrobial potentiality of this flavanone was also evaluated by Sohn et al. [9]. The flavanone is considered to exert an antibacterial effect by reducing the fluidity of cellular membranes [2]. In addition, sophoraflavanone G showed moderate *in vitro* antimalarial activity against *Plasmodium falciparum* with an EC_{50} value of 2.6 µM [10].

References

[1] Shirataki, Y., Yokoe, I., Noguchi, M., Tomimori, T., and Komatsu, M. (1988) *Chem. Pharm. Bull.*, **36**, 2220.
[2] Tsuchiya, H. and Iinuma, M. (2000) *Phytomedicine*, **7**, 161.

[3] Kang, T.-H., Jeong, S.-J., Ko, W.-G., Kim, N.-Y., Lee, B.-H., Inagak, M., Miyamoto, T., Ryuichi, R., and Kim, Y.-C. (2000) *J. Nat. Prod.*, **63**, 680.
[4] Iinuma, M., Tanaka, T., Mizuno, M., Shirataki, Y., Yokoe, I., Komatsu, M., and Lang, F.A. (1990) *Phytochemistry*, **29**, 2667.
[5] Fakhimi, A., Iranshahi, M., Emami, S.A., Amin-Ar-Ramimeh, E., Zarrini, G., and Reza, A. (2006) *Z. Naturforsch.*, **61c**, 769.
[6] Piao, X.-L., Piao, X.S., Kim, S.W., Park, J.H., Kim, H.Y., and Cai, S.-Q. (2006) *Biol. Pharm. Bull.*, **29**, 1911.
[7] Cha, J.D., Jeong, M.R., Jeong, S.I., and Lee, K.Y. (2007) *J. Microbiol. Biotechnol.*, **17**, 858.
[8] Sakagami, Y., Minura, M., Kajimura, K., Yokoyama, H., Iinuma, K., Tanaka, T., and Ohyama, O. (1998) *Lett. Appl. Microbiol.*, **27**, 98.
[9] Sohn, H.-Y., Son, K.H., Kwon, C.-S., Kwon, G.-S., and Kang, S.S. (2004) *Phytomedicine*, **11**, 666.
[10] Kim, Y.C., Kim, H.-S., Wataya, Y., Sohn, D.H., Kang, T.H., Kim, M.S., Kim, Y.M., Lee, G.-M., Chang, J.-D., and Park, H. (2004) *Biol. Pharm. Bull.*, **27**, 748.

Sophoraflavanone L

Systematic name: 5,2′,4′-Trihydroxy-7-(γ,γ-dimethylallyloxy)-8-(γ,γ-dimethylallyl) flavanone

Physical data: $C_{25}H_{28}O_6$, colorless prisms (MeOH), mp 125–127 °C, $[\alpha]_D^{25} \pm 0°$ (MeOH, *c* 0.5)

Structure:

Sophoraflavanone L

Compound class: Prenylated flavonoid (flavanone)

Source: *Sophora flavescens* Ait. (roots; family: Leguminosae)

Pharmaceutical potential: *Cytotoxic*
Sophoraflavanone L displayed moderate cytotoxic activity against the KB epidermoid carcinoma cell line with an IC_{50} value of 7.1 μg/ml, while the value for the positive control, camptothecin, was evaluated as 2.24 μg/ml under the same conditions.

Reference

Shen, C.-C., Lin, T.-W., Yu-Ling Huang, Y.-L., Shr-Ting Wan, S.-T., Shien, B.-J., and Chen, C.-C. (2006) *J. Nat. Prod.*, **69**, 1237.

Sorocenols G and H

Physical data:
- Sorocenol G: $C_{39}H_{32}O_8$, red amorphous solid, $[\alpha]_D^{25}$ +44° (MeOH, c 0.2)
- Sorocenol H: $C_{39}H_{36}O_9$, red amorphous solid, $[\alpha]_D^{25}$ +46° (MeOH, c 0.2)

Structures:

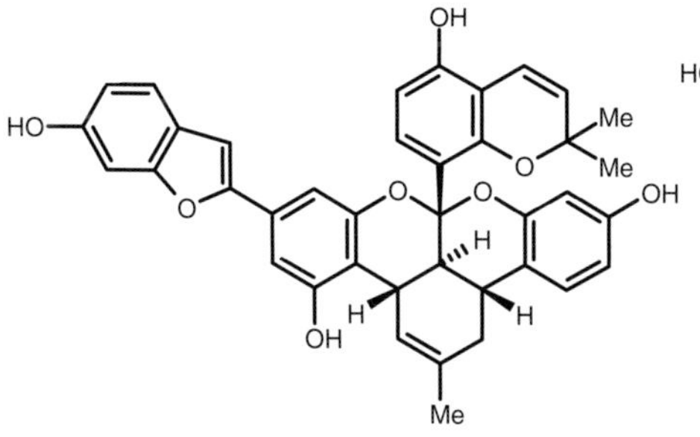

Sorocenol G

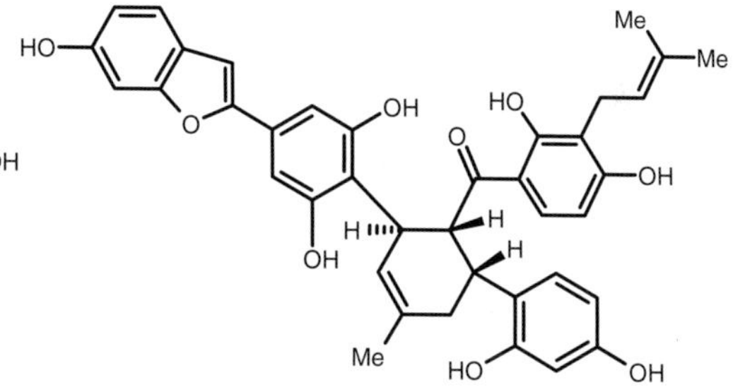

Sorocenol H

Compound class: Oxygen heterocyclic Diels–Alder-type adducts

Source: *Sorocea bonplandii* Baillon (roots; family: Moraceae)

Pharmaceutical potentials: *Antibacterial; antifungal*

Sorocenols G and H were reported to display significant and selective activity against methicillin-resistant *Staphylococcus aureus* with respective IC$_{50}$ values of 1.5 and 0.5 μM. Sorocenol H also exhibited antifungal activity against *Candida albicans*, *Cryptococcus neoformans*, and *Aspergillus fumigatus* with IC$_{50}$ values of 5.4, 5.4, and 10.0 μM, respectively.

Reference

Ross, S.A., Rodriguez-Guzman, R., Radwan, M.M., Jacob, M., Ding, Y., Li, X.-C., Ferreira, D., and Manly, S.P. (2008) *J. Nat. Prod.*, **71**, 1764.

Soulattrolide

Systematic name: (−)-[10*S*,11*R*,12*S*]-10,11-*trans*-Dihydro-12-hydroxy-6,6,10,11-tetramethyl-4-phenyl-2*H*,6*H*-benzo[1,2-*b*:3,4-*b'*:5,6-*b''*]tripyran-2-one

Physical data: $C_{25}H_{24}O_5$, light yellow needles, mp 177–181 °C, $[\alpha]_D^{25}$ −14° (CHCl$_3$, c 0.25)

Structure:

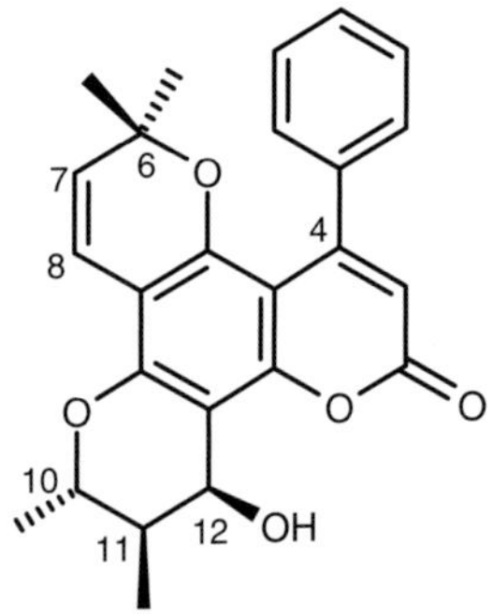

Soulattrolide

Compound class: Dipyranocoumarin

Sources: *Calophyllum soulattri* Burm f. (barks; family: Clusiaceae/Guttiferae) [1]; *C. brasiliense* (leaves) [2]; *C. teysmannii* (latex) [3]

Pharmaceutical potential: *Anti-HIV*

Huerta-Reyes *et al.* [2] evaluated that the isolate shows approximately 78% inhibition against the HIV-1 reverse transcriptase with an IC_{50} value of 0.665 μM. From their detailed study, Pengsuparp *et al.* [3] also reported that soulattrolide is a potent inhibitor of HIV-1 RT with an IC_{50} value of 0.34 μM; inhibition was remarkably specific, with no appreciable activity being observed toward HIV-2 RT, AMV RT, RNA polymerase, or DNA polymerases α or β.

References

[1] Gunasekera, S.P., Jayatilake, G.S., Selliah, S.S., and Sultanbawa, M.U.S. (1977) *J. Chem. Soc., Perkin Trans. I*, 1505; Shi, X., Attygalle, A.B., Liwo, A., Hao, M.-H., and Meinwald, J. (1998) *J. Org. Chem.*, **63**, 1233.

[2] Huerta-Reyes, M., del Carmen Basualdo, M., Abe, F., Jimenez-Estrada, M., Soler, C., and Reyes-Chilpa, R. (2004) *Biol. Pharm. Bull.*, **27**, 1471.

[3] Pengsuparp, T., Serit, M., Hughes, S.H., Soejarto, D.D., and Pezzuto, J.M. (1996) *J. Nat. Prod.*, **59**, 839.

SPF-32629A and B

Physical data:
- SPF-32629A: $C_{17}H_{19}NO_4$, pale yellow powder, $[\alpha]_D^{25}$ +26.4° (MeOH, *c* 0.58)
- SPF-32629B: $C_{18}H_{19}NO_6$, pale yellow powder, $[\alpha]_D^{25}$ +24.5° (MeOH, *c* 0.40)

Structures:

SPF-32629A: R = H
SPF-32629B: R = COOH

Compound class: Antibiotics

Source: *Penicillium* sp. SPF-32629 (culture broth) [1]

Pharmaceutical potential: *Human chymase inhibitors*

Chymase inhibitors are considered useful as therapeutic agents for various ailments including cardiovascular diseases, arteriosclerosis, and skin inflammation [2–7]. SPF-32629A and B were evaluated as specific human chymase inhibitors having IC_{50} values of 0.25 and 0.42 μg/ml, respectively [1]. The present investigators [1] observed that SPF-32629A exhibits inhibition against both human cathepsin G and human elastase with an IC_{50} value of 4.9 μg/ml and also against human chymotrypsin by only 30% at a concentration of 10 μg/ml, while SPF-32629B inhibited human elastase with an IC_{50} value of 4.5 μg/ml but did not inhibit human chymotrypsin or human cathepsin G at a concentration of 10 μg/ml. In addition, both of them did not display cytotoxicity against HL60 human leukemia cells at a concentration of 10 μg/ml [1].

References

[1] Shimatani, T., Hosotani, N., Ohnishi, M., Kumagai, K., and Saji, I. (2006) *J. Antibiot.*, **59**, 29.
[2] Urata, H., Kinoshita, A., Misono, K.S., Bumpus, F.M., and Husain, A. (1990) *J. Biol. Chem.*, **265**, 22348.
[3] Mizutani, H., Schechter, N., Lazarus, G., Black, R.A., and Kupper, T.S. (1991) *J. Exp. Med.*, **174**, 821.
[4] Wang, Z., Walter, M., Selwood, T., Rubin, H., and Schechter N.M. (1998) *Biol. Chem.*, **379**, 167.
[5] Takai, S., Shiota, N., Jin, D., and Miyazaki, M. (1998) *Eur. J. Pharmacol.*, **358**, 229.
[6] He, S., Gaca, M.D., McEuen, A.R., and Walls, A.F. (1999) *J. Pharmacol. Exp. Ther.*, **291**, 517.
[7] Sukenaga, Y., Kamoshita, K., Takai, S., and Miyazaki, M. (2002) *Jpn. J. Pharmacol.*, **90**, 218.

Sphingosine

Systematic name: (2*R*,3*R*,6*R*,7*Z*)-2-Aminooctadec-7-en-1,3,6-triol

Physical data: $C_{18}H_{37}O_3N$, light yellow oil, $[\alpha]_D^{22}$ +10° (MeOH, *c* 0.1)

Spiculoic acid A

Structure:

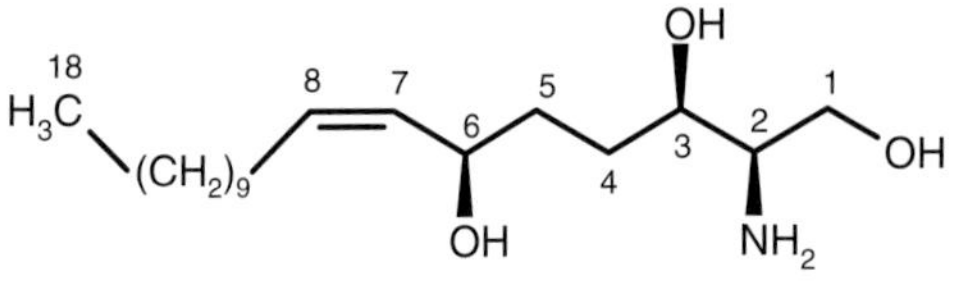

Sphingosine

Compound class: Sphingolipid

Source: *Haliclona* (*Reniera*) sp. (marine sponge)

Pharmaceutical potential: *Cytotoxic*
The terpene hydroperoxide exhibited moderate *in vitro* cytotoxic activity against five human cancer cell lines such as A549 (nonsmall cell lung adenocarcinoma), SK-OV3 (ovarian), SK-MEL-2 (skin melanoma), XF498 (CNS), and HCT15 (colon) with ED_{50} values of 4.33, 4.82, 5.20, 5.75, and 5.40 µg/ml, respectively (doxorubicin was used as a positive control).

Reference

Mansoor, T.A., Park, T., Luo, X., Hong, J., Lee, C.-O., and Jung, J.H. (2007) *Nat. Prod. Sci.*, **13**, 247.

Spiculoic acid A

Physical data: $C_{27}H_{36}O_3$, oil, $[\alpha]_D$ +110° (CH_2Cl_2, *c* 0.1)

Structure:

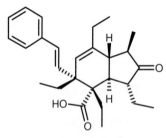

Spiculoic acid A

Compound class: Polyketide

Source: *Plakortis angulospiculatus* (marine sponge)

Pharmaceutical potential: *Cytotoxic*
The isolate displayed significant *in vitro* cytotoxic activity against the human breast cancer MCF-7 cell line with an IC_{50} value of 8.0 µg/ml.

Reference

Huang, X.-H., van Soest, R., Roberge, M., and Andersen, R.J. (2004) *Org. Lett.*, **6**, 75.

Stachsterol

Systematic name: (20S)-20,25-dihydroxy-4-cholesten-3-one

Physical data: $C_{27}H_{44}O_3$, white amorphous powder, $[\alpha]_D^{25.4}$ +31.48° ($CHCl_3$, c 0.667)

Structure:

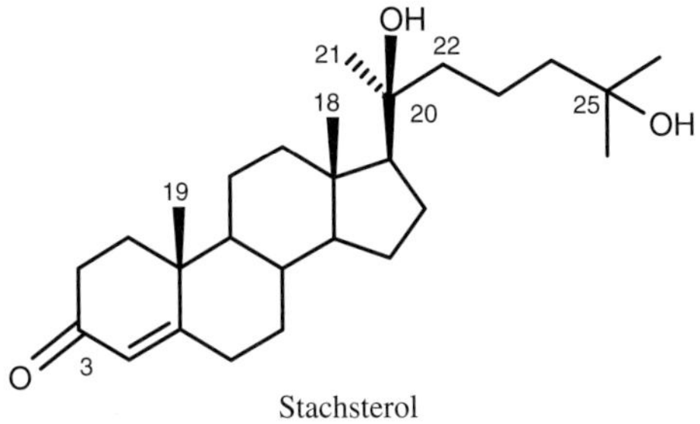

Stachsterol

Compound class: Steroid

Source: *Stachyurus himalaicus* var. *himalaicus* (family: Stachyuraceae)

Pharmaceutical potential: *Cytotoxic*
The steroidal compound was found to possess *in vitro* cytotoxic activity against human HeLa cell lines with an IC_{50} value of 2.5 µg/ml.

Reference

Wang, Y.-S., Yang, J.-H., Luo, S.-D., Zhang, H.-B., and Li, L. (2007) *Molecules*, **12**, 536.

Stevastelins B and B3

Physical data:
- Stevastelin B: $C_{34}H_{61}N_3O_9$, colorless powder
- Stevastelin B3: $C_{34}H_{61}N_3O_9$, colorless powder

Structures:

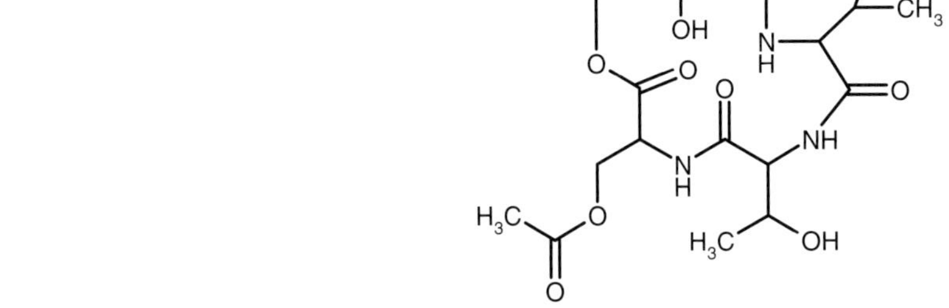

Stevastelin B

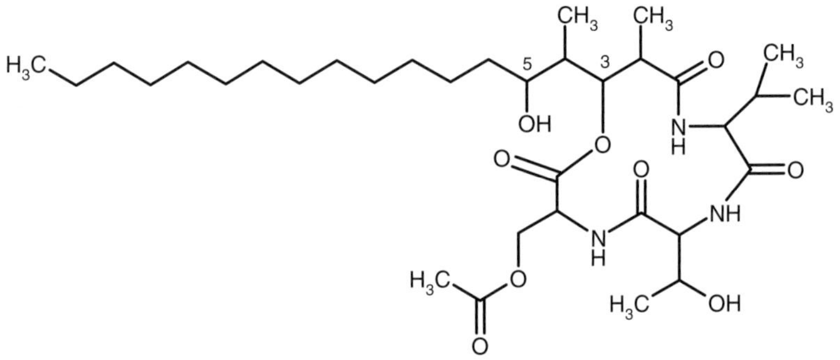

Stevastelin B3

Compound class: Depsipeptides

Source: *Penicillium* sp. (strain NK374186) (culture broth)

Pharmaceutical potential: *Immunosuppressive*
Both the isolates were evaluated to possess significant *in vitro* immunosuppressive effect on T-cell activation; stevastelins B and B3 inhibited the proliferation of human T cells induced by OKT-3 antibody with IC_{50} values of 1.8 and 0.42 µg/ml, respectively. The effects of these compounds on the blastogenesis of mouse spleen lymphocytes were also examined; stevastelins B and B3 inhibited the activation of T cells (induced by concanavalin A) with respective IC_{50} values of 11.0 and 3.8 µg/ml and that of B cells (induced by lipopolysaccharide, LPS) with respective IC_{50} values of 1.5 and 1.2 µg/ml. Thus, stevastelins B and B3 exhibited potent immunosuppressive activity against blastogenesis.

Reference

Morino, T., Masuda, A., Yamada, M., Nishimoto, M., Nishikiori, T., Saito, S., and Shimada, N. (1994) *J. Antibiot.*, **47**, 1341.

Streptokordin

Systematic name: 4-Acetyl-6-methyl-1*H*-pyridine-2-one

Physical data: $C_8H_9NO_2$, colorless amorphous powder

Structure:

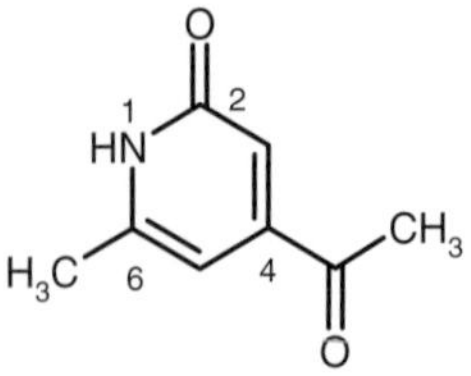

Streptokordin

Compound class: Antibiotic (pyridone derivative)

Source: *Streptomyces* sp. KORDI-3238 (fermentation broth)

Pharmaceutical potential: *Cytotoxic*
The isolate was found to possess moderate *in vitro* cytotoxicity against a number of human cancer cell lines such as MDA-MB-231 (human breast cancer), HCT 15 (human colon cancer), PC-3 (human prostate cancer), NCI-H 23 (human lung cancer), ACHN (human renal cancer), LOXIMVI (human skin cancer), and K-562 (human leukemia) with IC_{50} values of 7.5, 7.8, 3.2, 3.5, 4.7, 7.4, and 8.6 µg/ml, respectively.

Reference

Jeong, S.-Y., Shin, H.J., Kim, T.S., Lee, H.-S., Park, S.-K., and Kim, H.M. (2006) *J. Antibiot.*, **59**, 234.

Streptoverticillin

Physical data: $C_{18}H_{21}NO_3$, yellowish amorphous powder, $[\alpha]_D^{20}$ +18.4° (MeOH, *c* 0.179)

Structure:

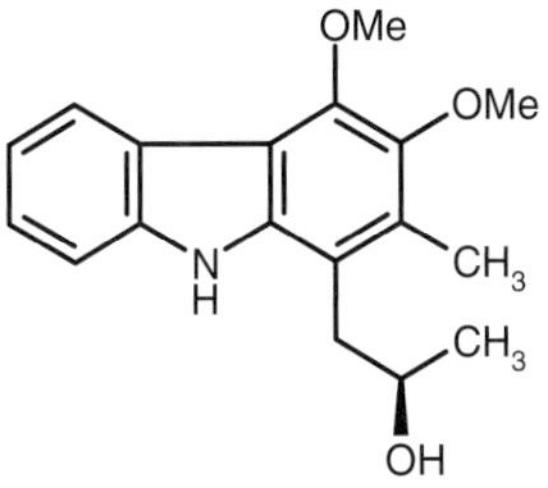

Streptoverticillin

Compound class: Alkaloid

Source: *Streptoverticillium morookaense* strain SC1169 (actinomycete; fermentation broth)

Pharmaceutical potential: *Antifungal*
Streptoverticillin was found to inhibit growth of *Peronophythora litchii* with an MIC of 250 µg/ml equivalent to the potency of carbendazim, a commercial antifungal agent used as a positive control.

Reference

Feng, N., Ye, W., Wu, P., Huang, Y., Xie, H., and Wei, X. (2007) *J. Antibiot.*, **60**, 179.

Strongylin A

Physical data: $C_{22}H_{32}O_3$, $[\alpha]_D^{20}$ +72° (CH_2Cl_2, *c* 0.023)

Strongylophorine-26

Structure:

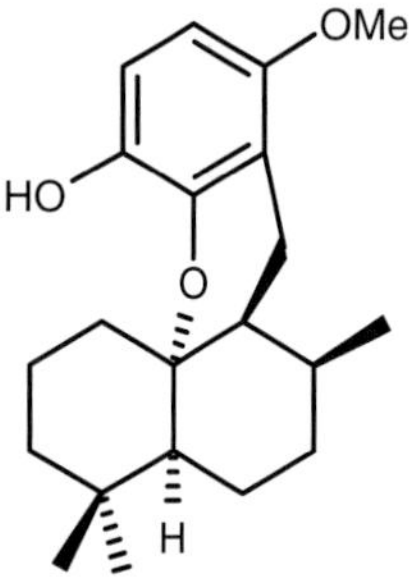

Strongylin A

Compound class: Sesquiterpene hydroquinone derivative

Source: *Strongylophora hartmani* van Soest (marine sponge; Haplosclerida)

Pharmaceutical potentials: *Antiviral; cytotoxic*
The isolate was found to be active *in vitro* against influenza virus (strain PR-8) with an IC$_{50}$ value of 6.5 µg/ml; the compound also displayed cytotoxicity against P-388 murine leukemia cell line with an IC$_{50}$ value of 13.0 µg/ml.

Reference

Wright, A.E., Rueth, S.A., and Cross, S.S. (1991) *J. Nat. Prod.*, **54**, 1108.

Strongylophorine-26

Physical data: $C_{27}H_{36}O_6$, orange amorphous solid, $[\alpha]_D^{28}$ +10.9° (MeCN, *c* 0.35)

Structure:

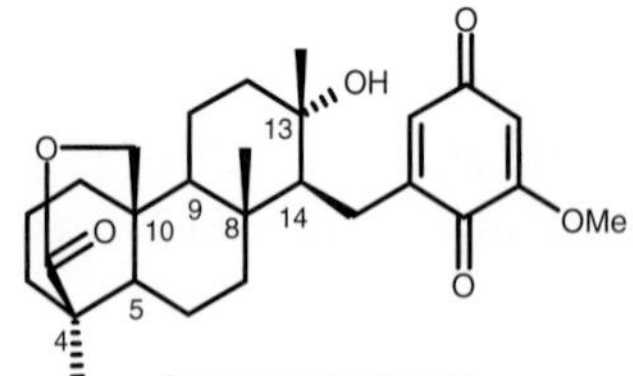

Strongylophorine-26
(Absolute configurations: 4*S*, 5*R*, 8*R*, 9*S*, 10*S*, 13*S*, 14*S*)

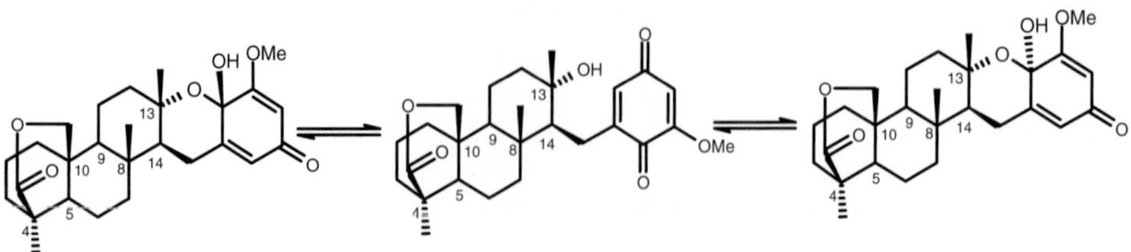

(Tautomeric equilibria observed for strongylophorine-26) [1]

Compound class: Diterpene (meroditerpenoid)

Source: Lipophilic extract of the marine sponge *Petrosia* (*Strongylophora*) *corticata* [1]

Pharmaceutical potential: *Anticancerous (inhibitor of cancer cell invasion)*
Warabi et al. [1] evaluated the anticancerous efficacy of strongylophorine-26 in the cell-based anti-invasion assay with MDA-MB-231 breast cancer cells; the diterpene showed an IC_{50} value of ~1 μg/ml and displayed maximal activity at 2.5 μg/ml. The huge decrease in invasion inhibition observed at concentrations of 2.5 μg/ml and above was due to cell death. From a detailed study on structure–activity relationships, the same group of investigators [2] came to a conclusion that the A ring lactone and the methoxy-substituted quinone moieties in strongylophorine-26 are both indispensable components of its anti-invasion pharmacophore. In a detailed investigation on the biological mechanism of the test compound showing its anti-invasive potential, McHardy et al. [3] suggested that the isolate induces nonpolarized lamellipodial extensions, resulting in inhibition of cell motility, and that its anti-invasive activity depends on the small GTPase Rho – interestingly, the molecule decreases actin stress fibers and increases focal adhesions while causing a dense meshwork of actin filaments to form at the cell periphery, indicating a distinct mechanism of action [3, 4].

References

[1] Warabi, K., McHardy, L.M., Matainaho, L., Van Soest, R., Roskelley, C.D., Roberge, M., and Andersen, R.J. (2004) *J. Nat. Prod.*, **67**, 1387.
[2] Warabi, K., Patrick, B.O., Austin, P., Roskelley, C.D., Roberge, M., and Andersen, R.J. (2007) *J. Nat. Prod.*, **70**, 736–740.
[3] McHardy, L.M., Warabi, K., Andersen, R.J., Roskelley, C.D., and Roberge, M. (2005) *Mol. Cancer Ther.*, **4**, 772.
[4] McHardy, L.M. (2007) A study of the mechanism of action of novel inhibitors of tumor cell invasion, Ph.D. thesis, University of British Columbia.

Strychnogucines A and B

Physical data:
- Strychnogucine A: $C_{42}H_{42}N_4O_3$, white amorphous powder
- Strychnogucine B: $C_{42}H_{42}N_4O_4$, white amorphous powder

Suaveolindole

Structures:

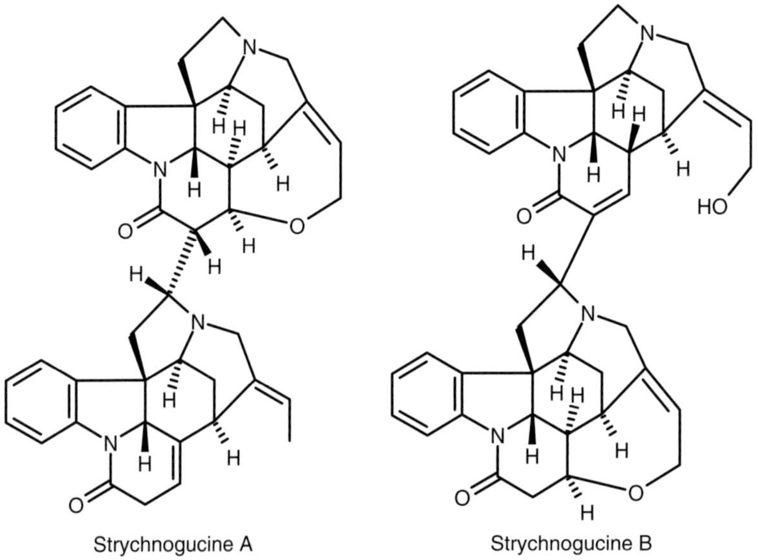

Strychnogucine A   Strychnogucine B

Compound class: Alkaloids (bisindole type)

Source: *Strychnos icaja* Baillon (roots; family: Loganiaceae)

Pharmaceutical potential: *Antimalarial*

Strychnogucines A and B were found to possess potent *in vitro* antiplasmodial activity as evaluated against four different strains of *Plasmodium falciparum* such as FCA 20 Ghana, W2 Indochina, F 32 Tanzania, and PFB Brazil. The respective IC_{50} and IC_{90} values for the alkaloidal isolates were determined to be IC_{50} (µM): 2.31 ± 0.304, nd, 4.813 ± 2.612, and 3.199 ± 0.144; IC_{90} (µM): 6.98, nd, 9.543, and 9.403 (strychnogucine A); IC_{50} (µM): 0.617 ± 0.067, 0.085 ± 0.01, 0.510 ± 0.26, and 0.02; IC_{90} (µM): 3.785, 0.358, 3.228, and 1.15 (strychnogucine B).

In addition, strychnogucine B showed a selective antiplasmodial activity with 25–180 times greater toxicity toward *P. falciparum*, relative to cultured human cancer cells (KB; $IC_{50} \geq 15$ µM) or human fibroblasts (WI38; $IC_{50} = 15.5$ µM).

Reference

Frederich, M., De Pauw, M.-C., Prosperi, C., Tits, M., Brandt, V., Penelle, J., Hayette, M.-P., DeMol, P., and Angenot, L. (2001) *J. Nat. Prod.*, **64**, 12.

Suaveolindole

Physical data: $C_{23}H_{31}NO_2$

Structure:

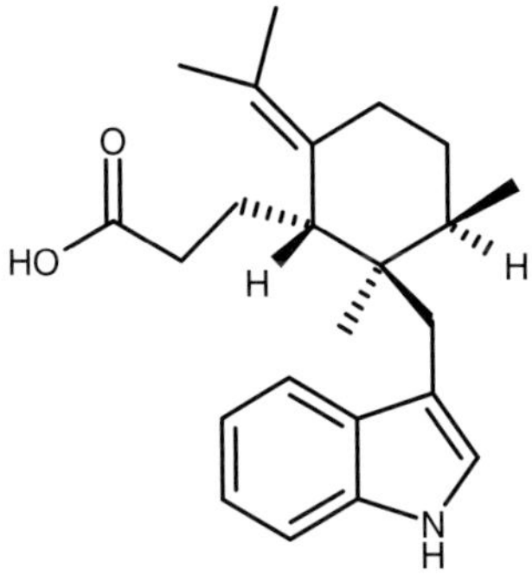

Suaveolindole

Compound class: Indolosesquiterpene

Source: *Greenwayodendron suaveolens* Verdc. (fruits; family: Annonaceae)

Pharmaceutical potential: *Antibacterial*
The investigators tested suaveolindole against a panel of bacteria including both Gram-positive and Gram-negative bacterial strains, but the compound was found to possess significant *in vitro* antibacterial activity only against the Gram-positive bacteria *Bacillus subtilis* (ATCC 43223), *Staphylococcus aureus* (ATTC 6538P), and methicillin-resistant *Staphylococcus aureus* (ATTC 33591) with MIC values of 4, 8, and 8 µg/ml, respectively.

Reference

Yoo, H.-D., Cremin, P.A., Zeng, L., Garo, E., Williams, C.T., Lee, C.M., Goering, M.G., O'Neil-Johnson, M., Eldridge, G.R., and Hu, J.-F. (2005) *J. Nat. Prod.*, **68**, 122.

Suberosol

Systematic name: 24-Methylenelanost-7,9(11)-diene-3β,15α-diol

Physical data: $C_{31}H_{50}O_2$, colorless needles (benzene), mp 179–182 °C, $[\alpha]_D^{20}$ +107° (CHCl$_3$, c 0.19)

Structure:

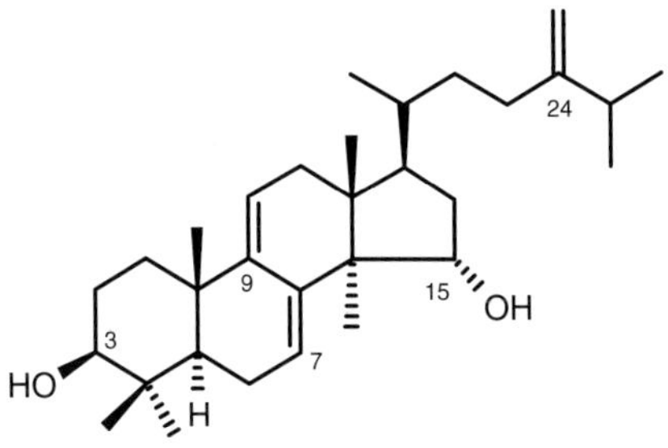

Suberosol

Compound class: Lanostane-type triterpenoid

Source: *Polyalthia suberosa* (stems and leaves; family: Annonaceae)

Pharmaceutical potential: *Anti-HIV*
Suberosol was reported to exhibit inhibition toward HIV replication in H9 lymphocyte cells with an EC_{50} value of 3.0 μg/ml.

Reference

Li, H.Y., Sun, N.J., Kashiwada, Y., Sun, L., Snider, J.V., Cosentino, M., and Lee, K.-H. (1993) *J. Nat. Prod.*, **56**, 1130.

(+)-Subersic acid

Physical data: $C_{25}H_{38}O_3$, $[\alpha]_D^{25}$ +39.3° (MeOH, *c* 3.26)

Structure:

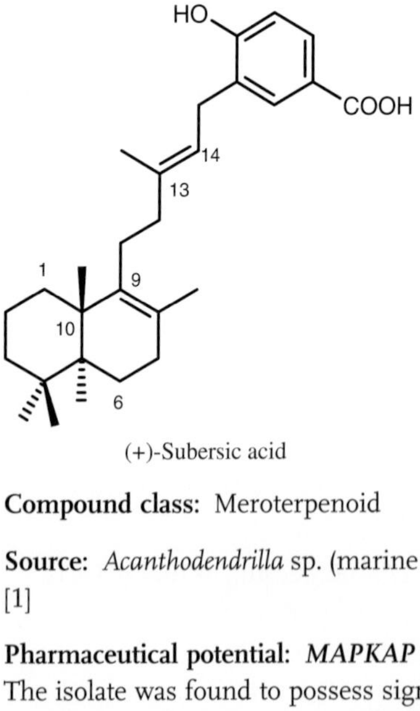

(+)-Subersic acid

Compound class: Meroterpenoid

Source: *Acanthodendrilla* sp. (marine sponge; order: Dendroceratida; family: Dictyodendrillidae) [1]

Pharmaceutical potential: *MAPKAP kinase 2 (MK2) inhibitor*
The isolate was found to possess significant *in vitro* inhibitory activity against MAPKAP kinase 2 (MK2) enzyme with an IC_{50} value of 9.6 μM [1]. TNF-α plays a key role in developing various inflammatory diseases such as rheumatoid arthritis, and MK2 is reported to have a critical role in the regulation of TNF-α production; hence, MK2 kinase inhibitors represent potential therapeutic agents to treat inflammatory diseases [2].

References

[1] Williams, D.E., Telliez, J.-B., Liu, J., Tahir, A., van Soest, R., and Andersen, R.J. (2004) *J. Nat. Prod.*, **67**, 2127.

[2] Kotlyarov, A., Neininger, A., Schubert, C., Eckert, R., Birchmeier, C., Volk, H.-D., and Gaestel, M. (1999) *Nat. Cell Biol.*, **1**, 94.

Subulatin

Physical data: $C_{32}H_{34}O_{17}$, pale yellow amorphous powder, $[\alpha]_D$ −16.8° (MeOH, *c* 0.19)

Structure:

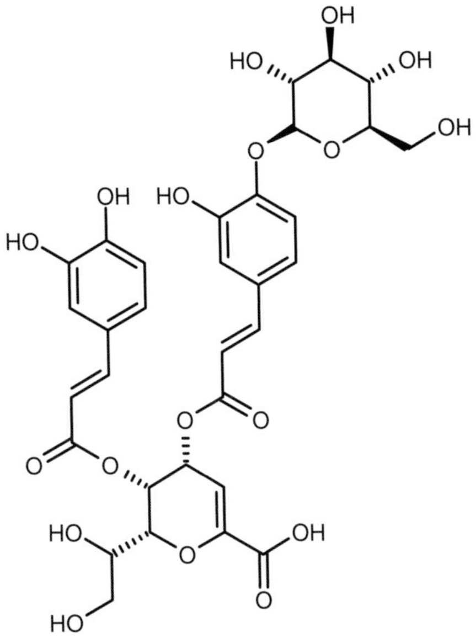

Subulatin

Compound class: Caffeic acid derivative

Sources: *In vitro* cultured liverworts, *Jungermannia subulata*, *Lophocolea heterophylla*, and *Scapania parvitexta*

Pharmaceutical potential: *Antioxidant*
The isolate showed *in vitro* antioxidant activity evaluated by the erythrocyte membrane ghost system; at a dose of 10 mM, the activity of the compound (55% inhibition to lipid peroxidation) was found to be higher than that of α-tocopherol (39% inhibition to lipid peroxidation).

Reference

Tazaki, H., Ito, M., Miyoshi, M., Kawabata, J., Fukushi, E., Fujita, T., Motouri, M., Furuki, T., and Nabeta, K. (2002) *Biosci. Biotechnol. Biochem.*, **66**, 255.

Surugapyrroles A and B

Physical data:
- Surugapyrrole A: $C_9H_{12}N_2O_4$, colorless oil
- Surugapyrrole B: $C_8H_{10}N_2O_4$, colorless oil

Structures:

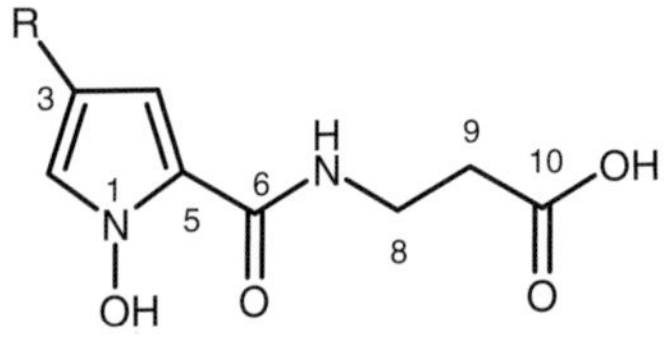

Surugapyrrole A: R = CH_3
Surugapyrrole B: R = H

Compound class: N-Hydroxypyrroles

Source: *Streptomyces* sp. USF-6280 strain (culture broth)

Pharmaceutical potential: *Antioxidants (radical scavengers)*
Surugapyrroles A and B displayed DPPH radical scavenging activity with ED_{50} values of 50.3 and 75.8 µM, respectively (BHT was used as a positive control; ED_{50} = 33.8 µM). It was assumed that the N-hydroxy moiety in the test molecules plays an important role in their radical scavenging activity.

Reference

Sugiyama, Y., Watanabe, K., and Hirota, A. (2009) *Biosci. Biotechnol. Biochem.*, **73**, 230.

Swerchirin (methylbellidifolin)

Systematic name: 1,8-Dihydroxy-3,5-dimethoxyxanthone

Physical data: $C_{15}H_{12}O_6$, mp 185–187 °C

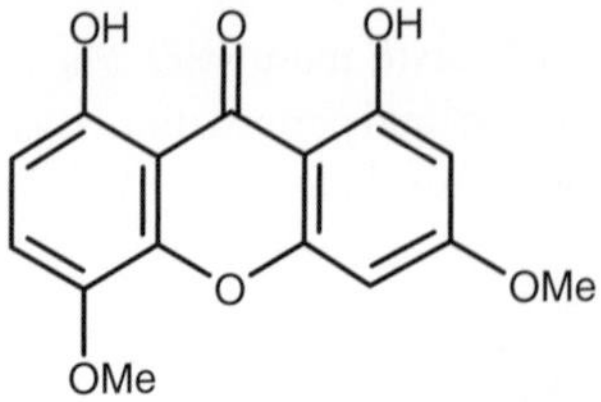

Swerchirin

Compound class: Xanthone (simple tetraoxygenated)

Sources: *Blackstonia perfoliata* (family: Gentianaceae) [1]; *Canscora lucidissima* (family: Gentianaceae) [2]; *Centaurium cachanlahuen* (family: Gentianaceae) [3], *C. erythraea* [4, 5], *C. littorale* [1, 5], *C. pulchellum* [6–10]; *Comastoma pulmonarium* (family: Gentianaceae) [11]; *Frasera albicaulis* (family: Gentianaceae) [12], *F. albomarginata* [13], *F. caroliniensis* [14]; *Gentiana algida* (family: Gentianaceae) [15], *G. bellidifolia* [16], *G. karelinii* [17], *G. lacteal* [18]; *Saponaria vaccaria* (family: Caryophyllaceae) [19]; *Swertia bimaculata* (family: Gentianaceae) [20], *S. chirata* [21–23], *S. chirayita* [24, 25], *S. dilatata* [26], *S. franchetiana* [27], *S. gracilescens* [26], *S. japonica* [28], *S. longifolia* [29], *S. milensis* [30, 31], *S. mussottii* [32], *S. nervosa* [26], *S. paniculata* [33, 34], *S. patens* [35], *S. petiolata* [36], *S. racemosa* [26], *S. speciosa* [37], *S. tetrapetala* [38]

Pharmaceutical potentials: *Hypoglycemic; antimalarial; antihematopoietic; hepatoprotective; chemopreventive against photosensitized DNA damage; mutagenic; vascular activity*
Swerchirin, isolated from the hexane extract of *Swertia chirayita*, was found to have promising antidiabetic activity in fasted, fed, glucose-loaded, and tolbutamide-pretreated albino rat models [39]; the effective dose (ED_{50} value) for lowering blood sugar by 40% in Charles Foster (CF) strain male albino rats (body weight 140–165 g) was determined to be 23.1 mg/kg (oral) [24]. A similar study on the hypoglycemic effect of the molecule in healthy as well as streptozotocin (STZ)-induced diabetic CF albino rats revealed a very significant drop in blood glucose level 7 h after single drug administration (50 mg/kg, p.o., suspension in gum acacia fed through cannula) in both the groups [40]. The work of Saxena et al. [41] offered an insight into the mechanism of blood sugar lowering by crude/impure swerchirin, isolated from the hexane extract of *S. chirayita*, single oral administration of which (50 mg/kg, body weight) to fed CF rats resulted in 60% fall in blood glucose 7 h after treatment. This was found to be associated with marked depression of aldehyde fuchsin strain beta-granules and immunostrained insulin in the pancreatic islets. *In vitro* glucose uptake and glycogen synthesis by muscles (diaphragm) were significantly enhanced by the serum of swerchirin-treated rat. It was observed that at 100, 10, and 1 µg final concentration, swerchirin greatly enhanced glucose (16.7 mM)-stimulated insulin release from isolated islets. On the basis of these findings, the investigators suggested that swerchirin lowers blood glucose level by stimulating insulin release from islets of Langerhans. The same group carried out a comparative study on the antidiabetic efficacy of two chemical compounds, tolbutamide (TB) and centpiperalone (CP), with swerchirin-rich fraction (SW1) of *S. chirayita* in experimental rat models. After a single oral administration of TB, CP, and SW1 to groups of normal and STZ-induced mild and severe diabetic rats, the blood sugar-lowering effect (in terms of ED_{50} values) was determined. Plasma immunoreactive insulin (IRI) levels and the degree of islet beta-cell degranulation were assayed using RIA and histochemical staining, respectively, in normal rats treated with the agents. The percent blood sugar lowering, increase in IRI level, and beta-cell degranulation were highest in CP-treated normal rats (69, 124, and 75%, respectively). In addition, CP was the only agent found active in STZ-induced severely diabetic rats ($P < 0.01$). In STZ-induced mild diabetic rats, however, TB was more effective than CP and SW1. By analysis of data using the ANOVA method, it is concluded that CP is more effective than SW1 ($P < 0.01$) and TB. However, SW1, an impure natural product, showed better blood sugar lowering effect than tolbutamide, which is a drug in use [42].

Swerchirin also exhibited antimalarial activity in a rodent test system infected with *Plasmodium berghei* as reported by Goyal et al. [43]; the compound is effective even at 20% of the standard dose of primaquine administered via either oral or subcutaneous routes. The drug is effective via both routes at 1.6 mg/kg and 320 µg/kg, respectively, by reaching null parasitemia in infected rats [43].

Swerchirin (isolated from the whole herb of *S. calycina* Franch) was investigated for its protective effect on hematopoiesis in mice [44]. A significant increase in colony formation in the spleen (CFU-S) of mice irradiated with 550 rad ^{60}Co γ-rays and an enhancement of proliferative response of bone marrow cells (BMCs) to rmGM-CSF (recombinant murine granulocyte macrophage colony-stimulating factor) treated with swerchirin (10 mg/kg, three times/week, i.p.) were observed. After introduction of swerchirin (single dose of 10 mg/kg, i.p.), a significant increase in the number of peripheral blood leukocytes and a rise in the serum of CSF were also confirmed. The stimulating factors were of the M-CSF type, together with other hematopoietic growth factors, as confirmed by means of McAb of IL-3 and GM-CSF and PcAb of M-CSF. These beneficial effects of swerchirin on hematopoiesis may be related to its activity of inducing CSF-S and other hematopoietic growth factors and demand further evaluation of its usefulness [44].

Hajimehdipoor et al. [45] accessed hepatoprotective potential of aerial parts of *Swertia longifolia* Boiss. and swerchirin, the major component of the plant, against paracetamol (acetaminophen)-induced hepatotoxicity in Swiss mice; pretreatment with total plant extract and swerchirin significantly reduced the elevation of biochemical parameters, aspartate aminotransferase (AST), alanine aminotransferase (ALT), and alkaline phosphatase (ALP), the enzymes that are increased by liver damage ($P < 0.001$). The experimental observations indicated the total plant extract and swerchirin as hepatoprotective in the range of 6–50 mg/kg orally [45]. In addition, swerchirin shows chemopreventive activity, though less significantly, against photosensitized DNA damage that is exhibited in solar UV carcinogenesis, photogenotoxicity, and phototoxicity in human beings [46], along with the exhibition of mutagenic activity when studied against *Salmonella typhimurium* TA100 (S9 mix.) [47].

Swerchirin was also reported to protect mycocardial ischemia reperfusion, possibly due to the effect of the compound on the reduction of mycocardial lipid peroxidation and on the increase in superoxide dismutase (SOD) activity [48].

References

[1] van der Sluis, W.G. (1985) *Plant Syst. Evol.*, **149**, 253.
[2] He, Q., He, L., Xu, S., and Deng, Q. (2000) *Zhong Yao Cai*, **23**, 399.
[3] Versluys, C., Cortes, M., Lopez, J.T., Sierra, J.R., and Razmilie, I. (1982) *Experientia*, **38**, 771.
[4] Kaouadji, M., Vaillant, I., and Mariotte, A.M. (1986) *J. Nat. Prod.*, **49**, 359.
[5] Beerhues, L. and Berger, U. (1994) *Phytochemistry*, **35**, 227.
[6] Yin, Z.-Q., Wang, Y., Ye, W.-C., and Zhaou, S.-X. (2004) *Biochem. Syst. Ecol.*, **32**, 521.
[7] Bibi, H., Ali, I., Choudhury, M.I., and Miana, G.A. (2004) *J. Chem. Soc. Pak.*, **26**, 400.
[8] Valentao, P., Andrade, P.B., Silva, E., Vicente, A., Santos, H., Bastos, M.L., and Seabra, R.M. (2002) *J. Agric. Food Chem.*, **50**, 460.
[9] Jankovic, T., Krstic, D., Savikin-Fodulovic, K., Menkovic, N., and Grubisic, D. (2002) *Planta Med.*, **68**, 944.
[10] Miana, G.A. and Hassan Al-Hazmi, M.G. (1984) *Phytochemistry*, **23**, 1637.
[11] Shufen, F., Beling, H., Jineye, D., and Hougfa, S. (1988) *Zhiwu Xuebao*, **30**, 303.
[12] Stout, G.H., Christensen, E.N., Balkenhol, W.J., and Stevens, K.L. (1969) *Tetrahedron*, **25**, 1961.
[13] Dreyer, D.L. and Bourell, J.H. (1981) *Phytochemistry*, **20**, 493.
[14] Stout, G.H. and Balkenhol, W.J. (1969) *Tetrahedron*, **25**, 1947.
[15] Butayarov, A.V., Batirov, E.K., Tadzhibaev, M.M., and Malikov, V.M. (1993) *J. Nat. Compd.*, **3**, 469.
[16] Markham, K.R. (1964) *Tetrahedron*, **20**, 991.

[17] Tadzhibaev, M.M., Butayarov, A.V., Batirov, E.K., and Malikov, V.M. (1992) *Chem. Nat. Compd.*, **2**, 244.
[18] Schufelberger, D. and Hoistettmann, K. (1988) *Planta Med.*, **54**, 219.
[19] Kazmi, S.N.U.H., Ahmed, Z., and Malik, A. (1989) *Heterocycles*, **29**, 1923.
[20] Ghosal, S., Sharma, P.V., and Choudhuri, R.K. (1975) *Phytochemistry*, **14**, 1393.
[21] Ghosal, S., Sharma, P.V., Choudhuri, R.K., and Bhattacharya, S.K. (1973) *J. Pharm. Sci.*, **62**, 926.
[22] Dalal, S.R. and Shah, R.C. (1956) *Chem. Ind.*, **7**, 664.
[23] Shi, G.-F., Lu, R.-H., Yang, Y.-S., Li, C.-L., Yang, A.M., and Cai, L.-X. (2004) *Acta Cryst. E Struct. Rep. Online*, **60**, o878.
[24] Bajpai, M.B., Asthana, R.K., Sharma, N.K., Chatterjee, S.K., and Mukherjee, S.K. (1991) *Planta Med.*, **57**, 102.
[25] Asthana, R.K., Sharma, N.K., Kulshreshtha, D.K., and Chatterjee, S.K. (1991) *Phytochemistry*, **30**, 1037.
[26] Tomimori, T., Yoshizaki, M., and Namba, T. (1974) *Yakugaku Zasshi*, **94**, 647.
[27] Yang, H., Duan, Y., Hu, F., and Liu, J. (2004) *Biochem. Syst. Ecol.*, **32**, 861.
[28] Ishimaru, K., Sudo, H., Satake, M., Matsunaga, Y., Hasengawa, Y., Takemoto, S., and Shimomuru, K. (1990) *Phytochemistry*, **29**, 1563.
[29] Hajimehdipour, H., Amanzadeh, Y., Sadat Ebrahimi, S.E., and Mozaffariam, V. (2003) *Pharm. Biol.*, **41**, 497.
[30] He, R., Feng, S., and Nie, R. (1984) *Yunnan Zhiwu Yanjin*, **6**, 341.
[31] Liu, J. and Huang, M. (1982) *Yakugaku Zasshi*, **13**, 433; *Chem. Abstr.*, 1983, **98**, 140530.
[32] Sun, H. and Jingye, D. (1981) *Zhiwu Xuebao*, **23**, 464; *Chem. Abstr.*, 1982, **96**, 119014.
[33] Anand, S.M., Basumatary, P.C., Ghosal, S., and Handa, S.S. (1982) *Planta Med.*, **45**, 61.
[34] Menkovic, N., Fodulovic, K.S., Bulatovic, V., and Milosavljevic, S. (2002) *Phytochemistry*, **61**, 415.
[35] He, R., Shuji, F., and Nie, R. (1982) *Yunnan Zhiwu Yanjiu*, **4**, 68.
[36] Khetwal, K.S., Joshi, B., and Bisht, R.S. (1990) *Phytochemistry*, **29**, 1265.
[37] Massias, M., Carbonnier, J., and Molho, D. (1977) *Bull. Mus. Natl. Hist. Nat. Paris*, **13**, 55.
[38] Agato, I., Sekizaki, H., Sakushima, A., Nishibe, S., Hisaba, S., and Kimura, K. (1981) *Yakugaku Zasshi*, **101**, 1067.
[39] Brahmachri, G., Mondal, S., Gangopadhyay, A., Gorai, D., Mukhopadhyay, B., Saha, S., and Brahmachari, A.K. (2004) *Chem. Biodivers.*, **1** (11), 1627; Brahmachari, G. (2005) Progress in the research of naturally occurring xanthones: tetraoxygenated constituents, in Recent Progress in Medicinal Plants, vol. 14 (eds J.N. Govil, V.K. Singh, and K. Ahmed), Studium Press LLC, Texas, USA, pp. 99–194.
[40] Saxena, A.M., Bajpai, M.B., and Mukherjee, S.K. (1991) *Ind. J. Exp. Biol.*, **29**, 674.
[41] Saxena, A.M., Bajpai, M.B., Murthy, P.S., and Mukherjee, S.K. (1993) *Ind. J. Exp. Biol.*, **31**, 178.
[42] Saxena, A.M., Murthy, P.S., and Mukherjee, S.K. (1996) *Ind. J. Exp. Biol.*, **34**, 351.
[43] Goyal, H., Sukumar, S., and Purushottaman, K.K. (1981) *J. Res. Ay. Sid.*, **2**, 286.
[44] Ya, B.Q., Nian, L.C., Li, C., and Gen, X.P. (1999) *Phytomedicine*, **6**, 85.
[45] Hajimehdipoor, H., Sadeghi, Z., Elmi, S., Elmi, A., Ghazi-Khansari, M., Amanzadeh, Y., and Sadat-Ebrahimi, S.E. (2006) *J. Pharm. Pharmacol.*, **58**, 277.
[46] Hirakawa, K., Yoshida, M., Nagatsu, A., Mizukami, H., Rana, V., Rawat, M.S.M., Oikawa, S., and Kawanishi, S. (2005) *Photochem. Photobiol.*, **81**, 314.

[47] Nozaka, T., Morimoto, I., Watanabe, F., and Okitsu, T. (1984) *Shoyakugaku Zasshi*, **38**, 96.
[48] He, Q., Xu, S., and Deng, Q. (2000) *Zhongguo Yaolixue Tongbao*, **16**, 642; He, Q., Xu, S., and Peng, B. (1998) *Zhongguo Zhongyao Zazhi*, **23**, 556.

Swertianolin

See Bellidifolin

Swertifrancheside

Systematic name: 1,5,8-Trihydroxy-3-methoxy-7-(5′,7′,3″,4″-tetrahydroxy-6′-C-β-D-glucopyranosyl-4′-oxy-8′-flavyl)xanthone

Physical data: $C_{35}H_{29}O_{17}$, yellow powder, mp $>320\,°C$, $[\alpha]_D$ $-12.3°$ (MeOH, c 0.17)

Structure:

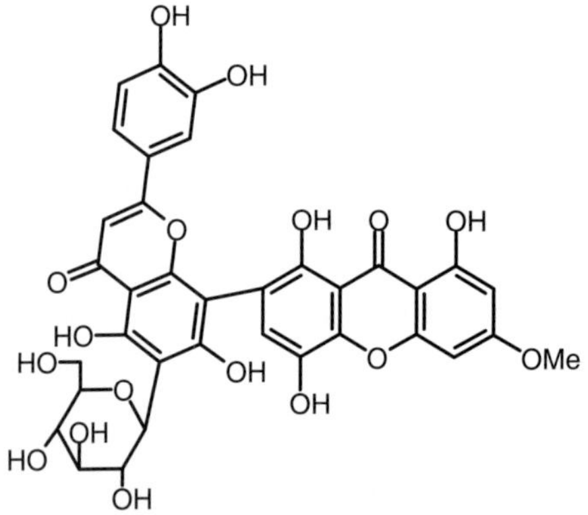

Swertifrancheside

Compound class: Flavone–xanthone dimer (having C-glucosidic linkage)

Source: *Swertia franchetiana* (family: Gentianaceae) [1]

Pharmaceutical potentials: *Anti-HIV; human DNA ligase I (hLI) inhibitor*
Swertifrancheside was reported to inhibit human immunodeficiency virus-1 reverse transcriptase (HIV-1 RTase) activity by 99.8% at 200 μg/ml ($ED_{50} = 30.9$ μg/ml), without being cytotoxic toward cultured mammalian cells [1]. The mode of action was found to be related to its binding property with DNA and may explain why the drug is also an inhibitor of several other polymerases including the DNA polymerase activity of HIV-1 reverse transcriptase; thus, the compound is not a selective HIV-1 RTase inhibitor [1–5]. Mechanistic studies revealed that swertifrancheside binds to DNA and is a competitive inhibitor with respect to the template primer, but a mixed-type competitive inhibitor with respect to TTP; the test compound inhibits the enzyme activity by binding to the template primer [6].

Swertifrancheside was reported to possess human DNA ligase I (hLI) inhibitory activity with an IC_{50} value of 8 μg/ml (i.e., 11 μM); studies with such inhibitors are important for the assignment of specific cellular functions to these enzymes as well as for their development into clinically useful antitumor agents [7].

References

[1] Wang, J.-N., Hou, C.-Y., and Liu, Y.-L. (1994) *J. Nat. Prod.*, **57**, 211.
[2] Tan, P., Hou, C.Y., Liu, Y.L., Lin, L.J., and Cordel, G.A. (1992) *Phytochemistry*, **31**, 4313.
[3] Tan, G.T., Pezzuto, J.M., and Kinghorn, A.D. (1992) in *Natural Products as Antiviral Agent* (eds C.K. Chu and H. Cutler), Plenum Press, New York, p. 195.
[4] Pengsuparp, T., Cai, L., Constant, H., Fong, H.S., Lin, L.-Z., Kinghorn, A.D., Pezzuto, J.M., Cordell, G.A., Ingolfsdottir, K., and Wagner, H. (1995) *J. Nat. Prod.*, **58**, 1024.
[5] Matthee, G., Wright, A.D., and Konig, G.M. (1999) *Planta Med.*, **65**, 493.
[6] Pengsuparp, T. (1995) *Diss. Abstr. Int. [B]*, **56**, 2582.
[7] Tan, G.T., Lee, S., Lee, I.-S., Chen, J., Leitner, P., Besterman, J.M., Kinghorn, A.D., and Pezzuto, J.M. (1996) *Biochem. J.*, **314**, 993.

Swertipunicoside

Systematic name: 1,5,8-Trihydroxy-3-methoxy-7-(1′,3′,6′,7′-tetrahydroxy-9′-oxo-4′-xanthyl)xanthone 2′-C-β-D-glucopyranoside

Physical data: $C_{33}H_{26}O_{17}$, yellow powder (methanol), mp $>360\,°C$, $R_f = 0.15$ (MeOH–H_2O (9:1)), $R_f = 0.42$ (CHCl$_3$–MeOH (3:1))

Structure:

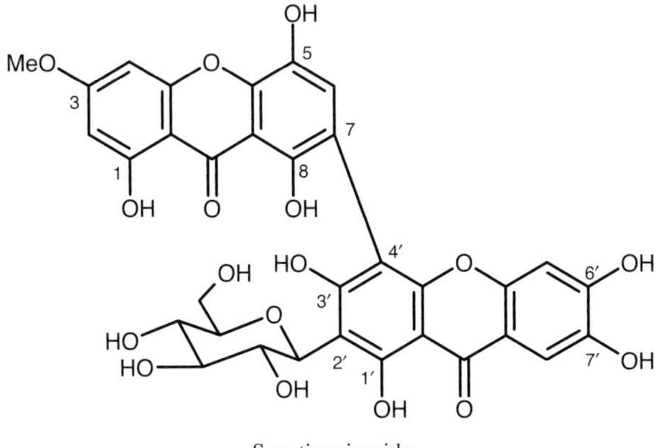

Swertipunicoside

Compound class: Bisxanthone C-glycoside

Source: *Swertia punicea* Hemal. (whole plant; family: Gentianaceae) [1]

Pharmaceutical potential: *Anti-HIV*

Cordell et al. [2] evaluated swertipunicoside as a potent anti-HIV agent; the natural bisxanthone C-glycoside was found to possess anti-HIV reverse transcriptase activity with an IC$_{50}$ value of 3.0 μg/ml.

References

[1] Tan, P., Hou, C.-Y., Liu, Y.-L., Lin, L.-J., and Cordell, G.A. (1991) *J. Org. Chem.*, **56**, 7130.
[2] Cordell, G.A., Angerhofer, C.K., and Pezzuto, J.M. (1994) *Pure Appl. Chem.*, **66**, 2283.

Syncarpamide

Physical data: $C_{28}H_{27}NO_5$, light brown solid, $[\alpha]_D^{25}$ +12.5° (CHCl$_3$, *c* 0.08)

Structure:

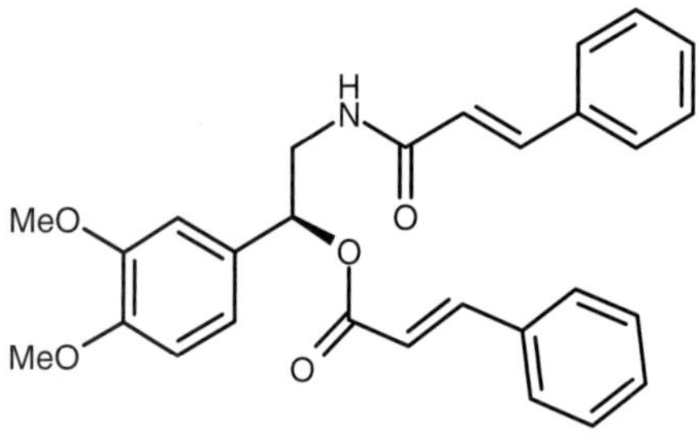

Syncarpamide

Compound class: Alkaloid ((+)-norepinephrine derivative)

Source: *Zanthoxylum integrifolium* (Merr.) Merr. (barks; family: Rutaceae)

Pharmaceutical potential: *Antimalarial*

The isolate was found to possess potent *in vitro* antiplasmodial activity as evaluated against the two different stains of *Plasmodium falciparum* such as D6 clone and W2 clone; the IC$_{50}$ values were determined to be 2.04 and 3.06 µM, respectively.

Reference

Ross, S.A., Sultana, G.N.N., Burandt, C.L., Elsohly, M.A., Marais, J.P.J., and Ferreira, D. (2004) *J. Nat. Prod.*, **67**, 88.

T

Tabucapsanolide A

See Withanolides

Taiwanhomoflavones A and B

Physical data:
- Taiwanhomoflavone A: $C_{33}H_{24}O_{10}$, pale yellow crystals, mp 245–248 °C
- Taiwanhomoflavone B: $C_{32}H_{24}O_{10}$, pale yellow amorphous powder

Structures:

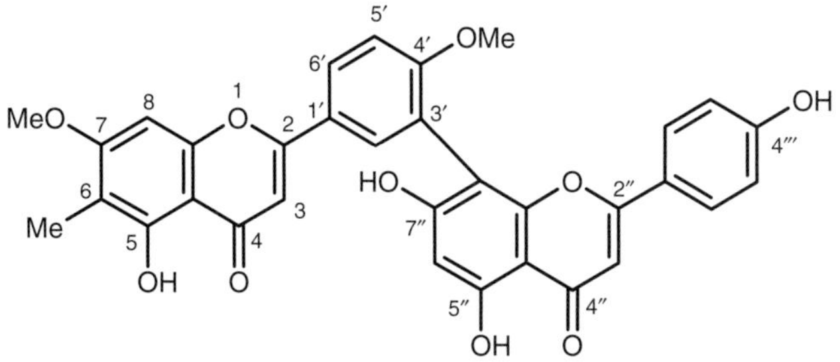

Taiwanhomoflavone A

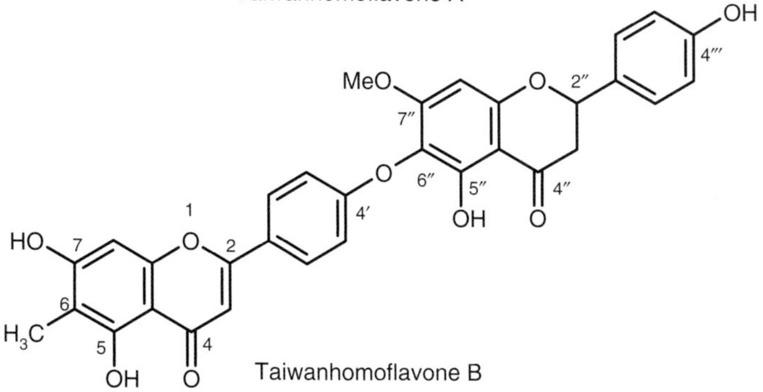

Taiwanhomoflavone B

Compound class: Flavonoids (biflavonoids)

Source: *Cephalotaxus wilsoniana* Hayata (stems and twigs; family: Cephalotaxaceae) [1, 2]

Handbook of Pharmaceutical Natural Products, Volume 2.
Goutam Brahmachari

ISBN: 978-3-527-32148-3

Pharmaceutical potential: *Cytotoxic*

Taiwanhomoflavone A was evaluated to possess significant cytotoxic activities against four cancer cell lines such as KB, COLO-205, Hepa-3B, and HeLa with ED_{50} values of 3.45, 1.06, 2.03, and 2.53 µg/ml, respectively [1]. Another biflavonoid was also found to exhibit promising cytotoxic activity against KB oral epidermoid carcinoma and Hepa-3B hepatoma cells with respective ED_{50} values of 3.8 and 3.5 µg/ml [2].

References

[1] Kuo, Y.-H., Lin, C.-H., Hwang, S.-Y., Shen, Y.-C., Lee, Y.-L., and Li, S.-Y. (2000) *Chem. Pharm. Bull.*, **48**, 440.

[2] Kuo, Y.-H., Hwang, S.-Y., Kuo, L.-M.Y., Lee, Y.-L., Li, S.-Y., and Shen, Y.-C. (2002) *Chem. Pharm. Bull.*, **50**, 1607.

Taiwankadsurin B

Physical data: $C_{31}H_{30}O_{13}$, $[\alpha]_D^{26}$ +62° (CH_2Cl_2, *c* 0.4)

Structure:

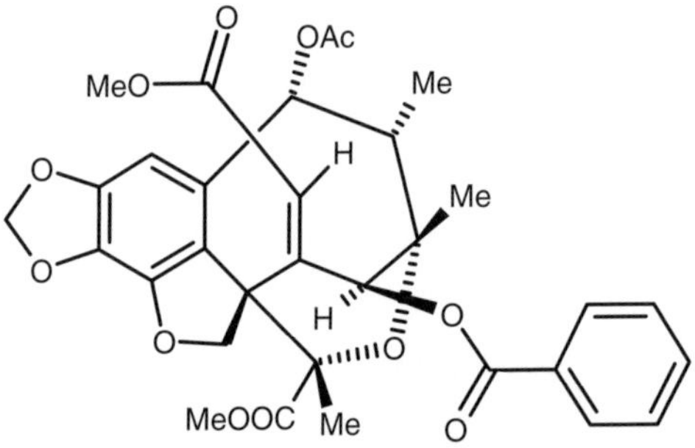

Taiwankadsurin B

Compound class: C-19 homolignan

Source: *Kadsura philippinensis* Elmer (leaves and stems; family: Schisandraceae)

Pharmaceutical potential: *Antitumor*

The isolate exhibited moderate cytotoxicity against human KB and HeLa tumor cells with IC_{50} values of 13.0 and 12.7 µg/ml, respectively.

Reference

Shen, Y.-C., Lin, Y.-C., Cheng, Y.-B., Kuo, Y.-H., and Liam, C.-C. (2005) *Org. Lett.*, **7**, 5297.

Taiwanschirin D

Physical data: $C_{28}H_{34}O_{10}$, light yellow powder, mp 47–50 °C

Structure:

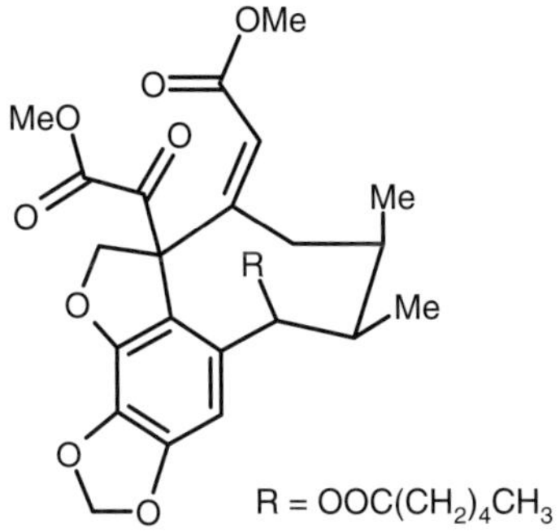

Taiwanschirin D

R = OOC(CH$_2$)$_4$CH$_3$

Compound class: Homolignan

Source: *Kadsura matsudai* Hayata (stems; family: Schisandraceae)

Pharmaceutical potential: *Anti-HBeAg*
The isolate was evaluated to show marginal inhibition toward human type B hepatitis virus in an anti-HBeAg test at a concentration of 94.3 μM (50 μg/ml).

Reference

Li, S.-Y., Wu, M.-D., Wang, C.-W., Kuo, Y.-H., Huang, R.-L., and Lee, K.-H. (2000) *Chem. Pharm. Bull.*, **48**, 1992.

Talosins A and B

Systematic names:
- Talosin A: genistein-7-α-L-6-deoxy-talopyranoside
- Talosin B: genistein-4′,7-di-α-L-6-deoxy-talopyranoside

Physical data:
- Talosin A: C$_{21}$H$_{20}$O$_9$, pale yellow powder, $[\alpha]_D^{25}$ −16.8° (MeOH, *c* 0.1)
- Talosin B: C$_{27}$H$_{30}$O$_{13}$, pale yellow powder, $[\alpha]_D^{25}$ −41.0° (MeOH, *c* 0.1)

Structures:

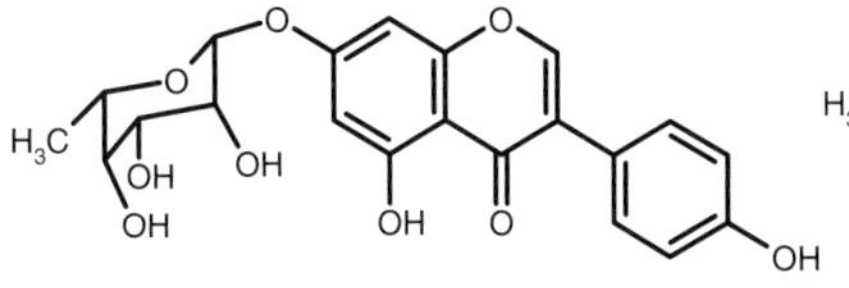

Talosin A 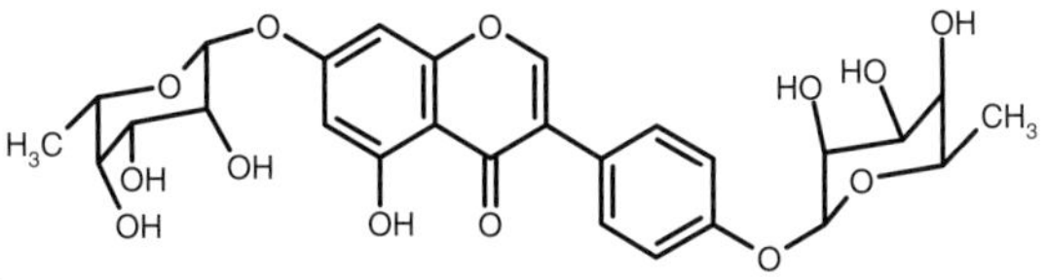   Talosin B

Compound class: Flavonoids (isoflavonol glycosides)

Source: *Kitasatospora kifunensis* MJM341 (culture broth) [1, 2]

Pharmaceutical potential: ***Antifungal***

Talosins A and B were found to exhibit strong antifungal activity against *Candida albicans* (B 02630), *Cryptococcus neoformans* (B 42419), and *Aspergillus niger* (ATCC 6275) with respective MIC values of 15, 7, and 6 μg/ml (talosin A); 7, 3, and 3 μg/ml (talosin B) [1]. However, the talosins did not inhibit *Bacillus subtilis, Serratia marcescens,* and also the dermatomycosis pathogen *Trichophyton mentagrophytes.*

References

[1] Yoon, T.M., Kim, J.W., Kim, J.G., Kim, W.G., and Suh, J.W. (2006) *J. Antibiot.*, **59**, 633.
[2] Kim, J.W., Yoon, T.M., Kwon, H.J., and Suh, J.W. (2006) *J. Antibiot.*, **59**, 640.

Tanariflavanones A–D

Systematic names:

- Tanariflavanone C: (2*S*)-5,7,3′,4′-tetrahydroxy-6-(2-hydroxy-3-methylbut-3-enyl)-2′-(geranyl) flavanone
- Tanariflavanone D: (2*S*)-5,7,3′,4′-tetrahydroxy-6-(6-hydroxy-3,7-dimethyocta-2′,7′-dienyl) flavanone

Physical data:

- Tanariflavanone A: $C_{30}H_{36}O_7$, greenish oil, $[\alpha]_D^{24.6}$ +26.8° ($CHCl_3$, *c* 0.6)
- Tanariflavanone B: $C_{30}H_{34}O_6$, brownish oil, $[\alpha]_D^{24.6}$ +28.2° ($CHCl_3$, *c* 0.5)
- Tanariflavanone C: $C_{30}H_{36}O_7$, yellow amorphous solid, $[\alpha]_D$ −3.1° (MeOH, *c* 0.27)
- Tanariflavanone D: $C_{25}H_{28}O_7$, pale yellow amorphous solid, $[\alpha]_D$ −12.7° (MeOH, *c* 0.24)

Structures:

Tanariflavanone A

Tanariflavanone B

Tanariflavanone C

Tanariflavanone D

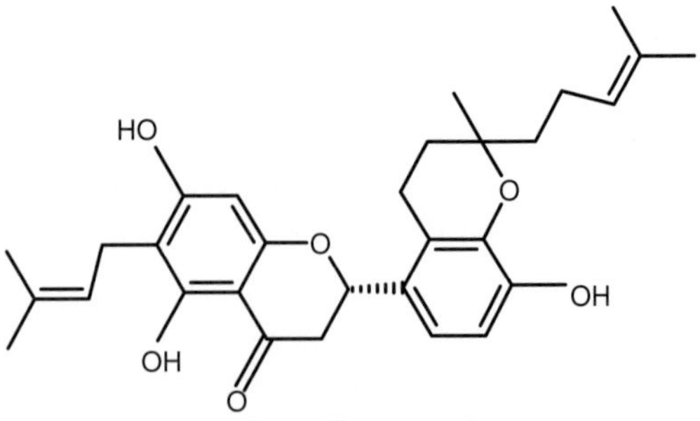

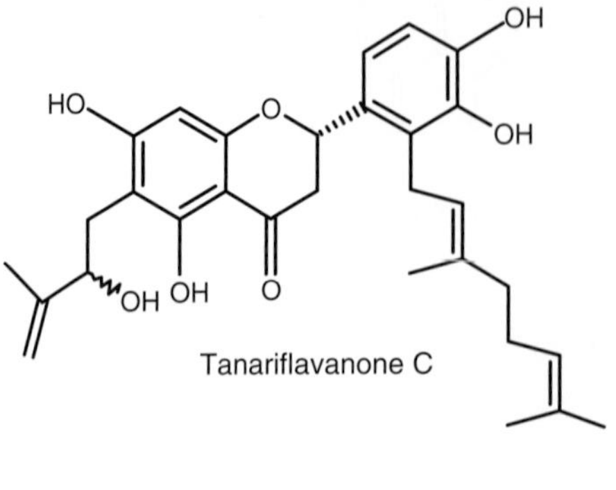

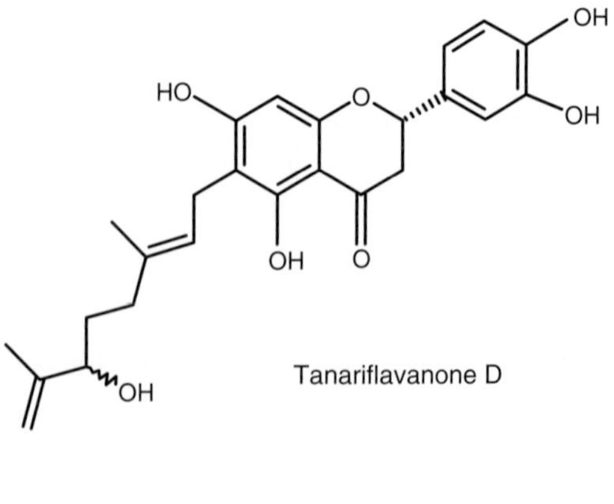

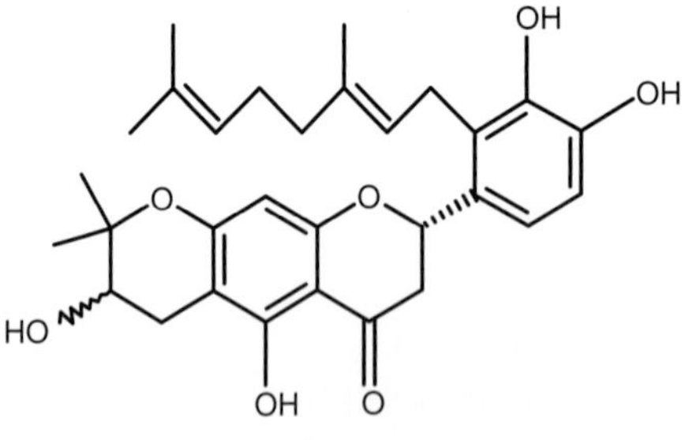

Compound class: Flavonoids (prenylated flavanones)

Source: *Macaranga tanarius* (Linn.) Muell. Arg. (syn *M. tomentosa* Bl. leaves; family: Euphorbiaceae) [1, 2]

Pharmaceutical potentials: *Phytotoxic; cytotoxic; antioxidant*
Tseng et al. [1] evaluated phytotoxic activity of tanariflavanones A and B by determining their effect on the radicle elongation of germinating lettuce; at 200 ppm, tanariflavanone B inhibited radicle length of lettuce up to 30% compared to the distilled H_2O control, while tanariflavanone A exhibited 11% growth inhibition. Hence, tanariflavanones A and B play an allelopathic role in *M. tanarius*.

Tanariflavanone D showed cytotoxicity against human breast cancer (BC) and Vero cell lines with IC_{50} values of 6.5 and 19.5 µg/ml, respectively [2]. Tanariflavanone D also exhibited radical scavenging properties with an IC_{50} value of 20 ± 1 µM and was found to be stronger than tanariflavanone B and 2,6-di (*tert*-butyl)-4-methylphenol (BHT) that showed IC_{50} values of 33 ± 1 and 30 ± 1 µM, respectively [2].

References

[1] Tseng, M.-H., Chou, C.-H., Chen, Y.-M., and Kuo, Y.-H. (2001) *J. Nat. Prod.*, **64**, 827.
[2] Phommart, S., Sutthivaiyakit, P., Chimnoi, N., Ruchirawat, S., and Sutthivaiyakit, S. (2005) *J. Nat. Prod.*, **68**, 927.

Tanghinin and deacetyltanghinin

Systematic names:
- Tanghinin: 3β-*O*-(2′-*O*-acetyl-α-L-thevetosyl)-14β-hydroxy-7,8-epoxy-5β-card-20(22)-enolide
- Deacetyltanghinin: 3β-*O*-α-L-thevetosyl-14β-hydroxy-7,8-epoxy-5β-card-20(22)-enolide

Physical data:
- Tanghinin: $C_{32}H_{46}O_{10}$
- Deacetyltanghinin: $C_{30}H_{44}O_9$, prisms, mp 213–216 °C, $[\alpha]_D^{26}$ −53.5° (MeOH, *c* 0.43) [1]; mp 217 °C [4], $[\alpha]_D$ −59° [5]

Structures:

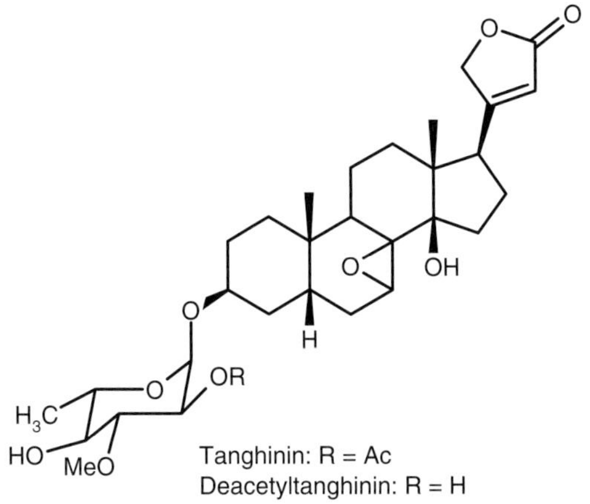

Tanghinin: R = Ac
Deacetyltanghinin: R = H

Compound class: Cardenolide glycosides

Source: *Cerbera manghas* L (seeds, barks, leaves; family: Apocynaceae) [1, 3]

Pharmaceutical potential: ***Cytotoxic***
Tanghinin and deacetyltanghinin were found to display significant *in vitro* cytotoxicity against oral human epidermoid carcinoma (KB), human breast cancer cell (BC), and human small cells lung cancer (NCI-H187); the respective ED_{50} values for tanghinin were determined to be 1.29, 0.77, and 2.3 μg/ml, while deacetyltanghinin showed ED_{50} values of 0.5, 1.48, and 0.1 μg/ml, respectively, against the above cancer cells tested [2].

References

[1] Abe, F. and Yamauchi, T. (1977) *Chem. Pharm. Bull.*, **25**, 2744.
[2] Cheenpracha, S., Karalai, C., Rat-A-Pa, Y., Ponglimanont, C., and Chantrapromma, K. (2004) *Chem. Pharm. Bull.*, **52**, 1023.
[3] Flury, E., Weiss, E., and Reichstein, T. (1965) *Helv. Chim. Acta*, **48**, 1113.
[4] Helfenberger, H. and Reichstein, T. (1965) *Helv. Chim. Acta*, **35**, 1503.
[5] Sigg, H.P., Tamm, C., and Reichstein, T. (1955) *Helv. Chim. Acta*, **38**, 166.

Tasumatrols E, F, I, J, and K

Physical data:
- Tasumatrol E: $C_{33}H_{44}O_{13}$, colorless powder, $[\alpha]_D^{25}$ +36° (MeOH, *c* 0.1)
- Tasumatrol F: $C_{33}H_{44}O_{12}$, colorless powder, $[\alpha]_D^{25}$ +28° (MeOH, *c* 0.1)
- Tasumatrol I: $C_{29}H_{36}O_{10}$, colorless powder, $[\alpha]_D^{25}$ −25.6° (MeOH, *c* 0.1)
- Tasumatrol J: $C_{29}H_{34}O_9$, colorless powder, $[\alpha]_D^{25}$ −21.8° (MeOH, *c* 0.1)
- Tasumatrol K: $C_{29}H_{44}O_8$, colorless powder, $[\alpha]_D^{25}$ +19.3° (MeOH, *c* 0.1)

Structures:

Tasumatrol E: R = OH
Tasumatrol F: R = H

Tasumatrol I

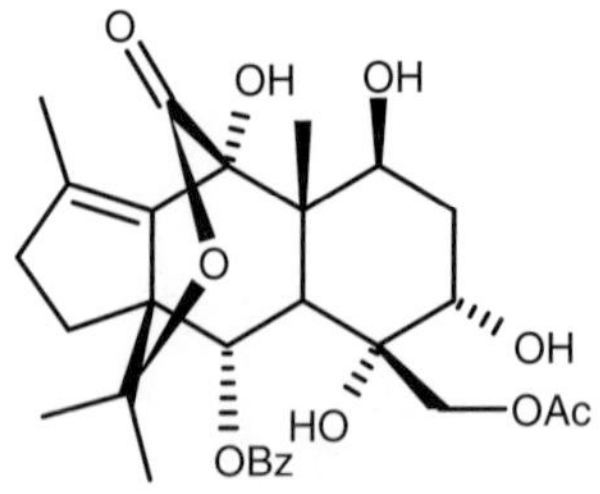

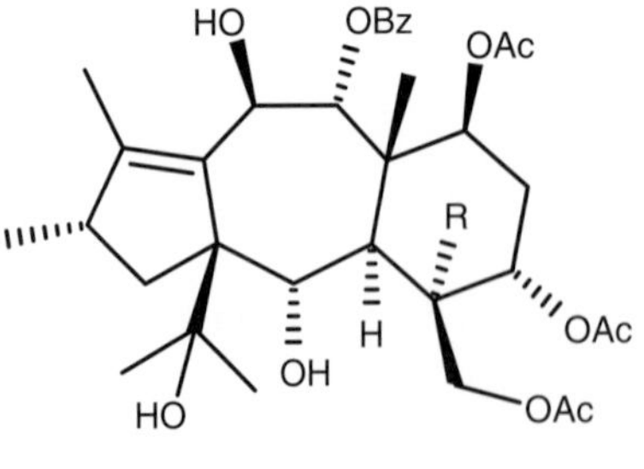

Tasumatrol J

Tasumatrol K

Compound class: Taxane diterpenoids (taxoids of the taxchin type)

Source: *Taxus sumatrana* (leaves and twigs; family: Taxaceae) [1, 2]

Pharmaceutical potential: *Anticancerous (antitumor)*
Shen et al. [1] evaluated the *in vitro* cytotoxic efficacies of tasumatrols E and F against a panel of human cancer cell lines such as A498 (renal cancer), NCI-H226 (non-small cell lung cancer), A549 (non-small cell lung cancer), and PC-3 (prostate cancer) tumor cells using taxol as a standard compound, and both the taxoid molecules were found to exhibit potent cytotoxic activities against the test materials. Tasumatrol E displayed more promising activity than taxol against these four tumor cell lines, while tasumatrol F was more active toward A498 and PC-3 tumor cells. The respective percent (%) inhibition values against A498, NCI-H226, A549, and PC-3 tumor cells by the test compounds and the standard at a concentration of 30 µg/ml were determined to be 100, 84.8, 91.3, and 94.7 (tasumatrol E); 83.0, 78.5, 72.6, and 95.0 (tasumatrol F); and 98.2, 71.2, 79.7, and 91.7 (taxol) [1].

In another communication, the same group of workers [2] isolated tasumatrols I, J, and K from the same plant and evaluated their anticancerous activities; tasumatrols I and K showed significant activity against human liver carcinoma (Hepa59T/VGH), human large cell carcinoma of the lungs (NCI), human cervical epitheloid carcinoma (HeLa), human colon adenocarcinoma (DLD-1), and human medulloblastoma (Med) cell lines. The respective ED_{50} values were determined to be 6.26, 2.45, 2.73, 3.40, and 5.12 µg/ml (tasumatrol I); 16.93, 13.72, 12.69, 11.26, and 10.80 µg/ml (tasumatrol J); and 8.57, 7.28, 7.88, 6.57, and 8.03 µg/ml (tasumatrol K). ED_{50} values for mitomycin, used as a standard, were found to be 0.18 µg/ml (Hepa59T/VGH), 0.19 µg/ml (NCI), 0.18 µg/ml (HeLa), 0.11 µg/ml (DLD-1), and 0.24 µg/ml (Med).

References

[1] Shen, Y.-C., Cheng, K.-C., Lin, Y.-C., Cheng, Y.-B., Khalil, A.T., Guh, J.-H., Chien, C.-T., Teng, C.-M., and Chang, Y.-T. (2005) *J. Nat. Prod.*, **68**, 90.
[2] Shen, Y.-C., Lin, Y.-S., Cheng, Y.-B., Cheng, K.-C., Khalil, A.T., Kuo, Y.-H., Chin, C.-T., and Lin, Y.-C. (2005) *Tetrahedron*, **61**, 1345.

(+)-Tephrorins A and B

Systematic names:
- (+)-Tephrorin A: (2S)-8-[(2S,3R,4S)-4-(acetyloxy)tetrahydro-2-hydroxy-5,5-dimethyl-3-furanyl]-2,3-dihydro-7-methoxy-2-phenyl-4H-1-benzopyran-4-one
- (+)-Tephrorin B: (3S,4S)-4-[(2S)-3,4-dihydro-7-hydroxy-4-oxo-2-phenyl-2H-1-benzopyran-8-yl]tetrahydro-2,2-dimethyl-3-furanyl(2E)-3-phenyl-2-propenoate

Physical data:
- (+)-Tephrorin A: $C_{24}H_{26}O_7$, yellowish oil, $[\alpha]_D^{20}$ +6° ($CHCl_3$, c 0.5)
- (+)-Tephrorin B: $C_{30}H_{28}O_6$, yellowish oil, $[\alpha]_D^{20}$ +8° ($CHCl_3$, c 0.6)

Structures:

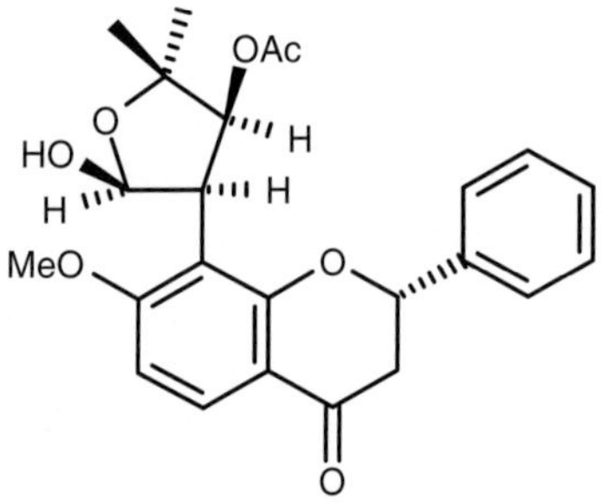

(+)-Tephrorin A 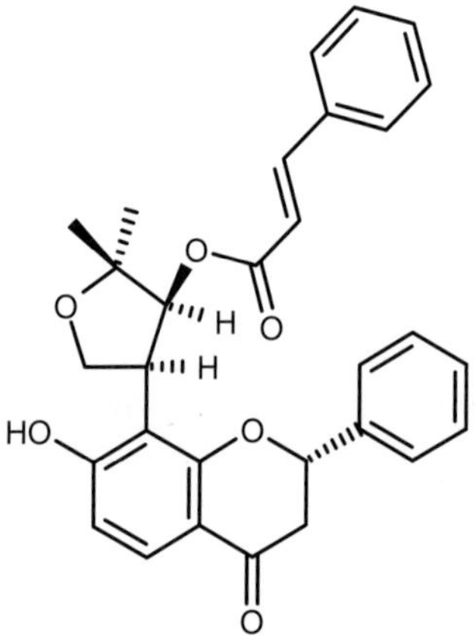   (+)-Tephrorin B

Compound class: Flavonoids (flavanone derivatives)

Source: *Tephrosia purpurea* (family: Leguminosae)

Pharmaceutical potential: *Anticancerous*
The isolates were evaluated for their cancer chemopreventive potentials in the quinone reductase (QR) assay using cultured mouse Hepa 1c1c7 cells; the test compounds showed significant induced QR activity with respective CD (concentration to double enzyme induction) values of 4.0 and 5.9 µM. The CI (i.e., IC_{50}/CD) values of these compounds were determined to be 11.8 and 5.5, respectively. The presence of the bulky cinnamic acid group at C-4″ in (+)-tephrorin B is supposed to affect its biological activity.

Reference

Chang, L.C., Chávez, D., Song, L.L., Farnsworth, N.R., Pezzuto, J.M., and Kinghorn, A.D. (2000) *Org. Lett.*, 2, 515.

(+)-Tephrosone

Systematic name: (+)-(2″R,3″S,4″S)-[2″,3″-b]-Dihydrofurano-5″,5″-dimethyl[4′,5′-h]-6′-hydroxy-4″-tetrahydrofuranohydroxychalcone

Physical data: $C_{21}H_{20}O_5$, yellow needles, $[\alpha]_D^{20}$ +26° (CHCl$_3$, c 0.23)

Structure:

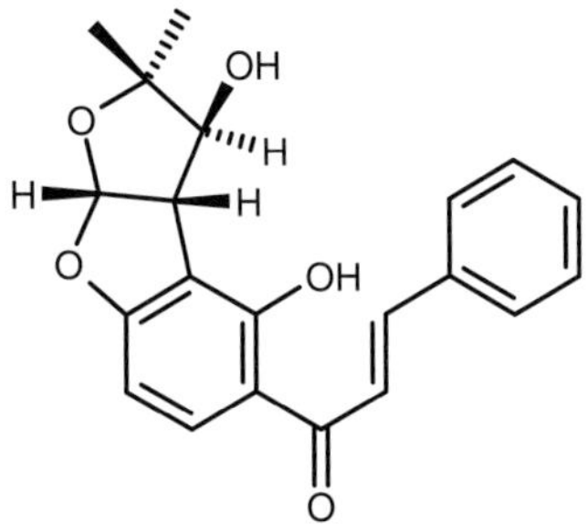

(+)-Tephrosone

Compound class: Flavonoids (chalcone)

Source: *Tephrosia purpurea* (family: Leguminosae)

Pharmaceutical potential: *Anticancerous*
The isolate was evaluated for its cancer chemopreventive potential in the quinone reductase (QR) assay using cultured mouse Hepa 1c1c7 cells; the test compound showed significant induced QR activity with the CD (concentration to double enzyme induction) value of 3.1 µM; the CI (i.e., IC$_{50}$/CD) value of this compound was determined to be 6.1.

Reference

Chang, L.C., Chávez, D., Song, L.L., Farnsworth, N.R., Pezzuto, J.M., and Kinghorn, A.D. (2000) *Org. Lett.*, **2**, 515.

Termilignan B

Physical data: $C_{19}H_{18}O_3$, amorphous solid

Structure:

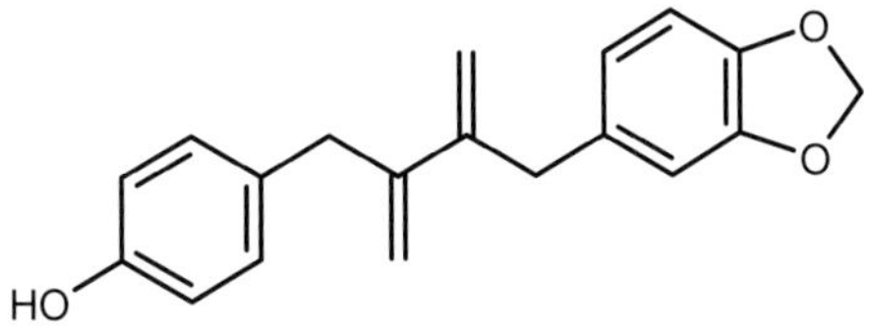

Termilignan B

Compound class: Lignoid

Source: *Terminalia sericea* Burch. ex. DC (roots; family: Combretaceae)

Pharmaceutical potentials: *Antimicrobial; anti-inflammatory*

The isolated compound showed antibacterial activity against two Gram-positive (*Bacillus subtilis* ATCC 6051 and *Staphylococcus aureus* ATCC 12600) and two Gram-negative (*Escherichia coli* ATCC 11775 and *Klebsiella pneumoniae* ATTCC 13883) bacterial strains with respective MIC values of 1.9, 3.9, 15.6, and 7.8 µg/ml (neomycin was used as a reference standard with respective MIC values of 0.1, 1.5, 3.0, and 0.8 µg/ml). The compound was also found to display inhibitory activity against cyclooxygenase-1 and -2 (COX-1 and COX-2) with IC_{50} values of 78 and 156 µM, respectively. Hence, termilignan B may find useful applications as a lead compound against bacterial infections and inflammatory-related diseases.

Reference

Eldeen, I.M.S., Van Heerden, F.R., and Van Staden, J. (2008) *Planta Med.*, **74**, 411.

Ternstrosides A–F

Systematic names:
- Ternstroside A: 2-(3,4-dihydroxyphenyl)ethyl 2-*O*-(3,4-dihydroxyphenylethanoyl)-β-D-glucopyranoside
- Ternstroside B: 2-(3,4-dihydroxyphenyl)ethyl 2-*O*-(4-hydroxyphenylethanoyl)-β-D-glucopyranoside
- Ternstroside C: 2-(3,4-dihydroxyphenyl)ethyl 3-*O*-(3,4-dihydroxyphenylethanoyl)-β-D-glucopyranoside
- Ternstroside D: 2-(3,4-dihydroxyphenyl)ethyl 6-*O*-(3,4-dihydroxyphenylethanoyl)-β-D-glucopyranoside
- Ternstroside E: 2-(4-hydroxyphenyl)ethyl 2-*O*-(3,4-dihydroxyphenylethanoyl)-β-D-glucopyranoside
- Ternstroside F: 2-(2,4,5-trihydroxyphenyl)ethyl 2-*O*-(3,4-dihydroxyphenylethanoyl)-β-D-glucopyranoside

Physical data:
- Ternstroside A: $C_{22}H_{26}O_{11}$, yellow amorphous solid, $[\alpha]_D^{24}$ −33.5° (MeOH, *c* 2.3)
- Ternstroside B: $C_{22}H_{26}O_{10}$, yellow amorphous solid, $[\alpha]_D^{24}$ −19.8° (MeOH, *c* 0.12)
- Ternstroside C: $C_{22}H_{26}O_{11}$, yellow amorphous solid, $[\alpha]_D^{24}$ −31.6° (MeOH, *c* 0.32)
- Ternstroside D: $C_{22}H_{26}O_{11}$, yellow amorphous solid, $[\alpha]_D^{24}$ −28.2° (MeOH, *c* 0.29)
- Ternstroside E: $C_{22}H_{26}O_{10}$, yellow amorphous solid, $[\alpha]_D^{24}$ −36.4° (MeOH, *c* 1.04)
- Ternstroside F: $C_{22}H_{26}O_{12}$, yellow amorphous solid, $[\alpha]_D^{24}$ −21.4° (MeOH, *c* 0.10)

Structures:

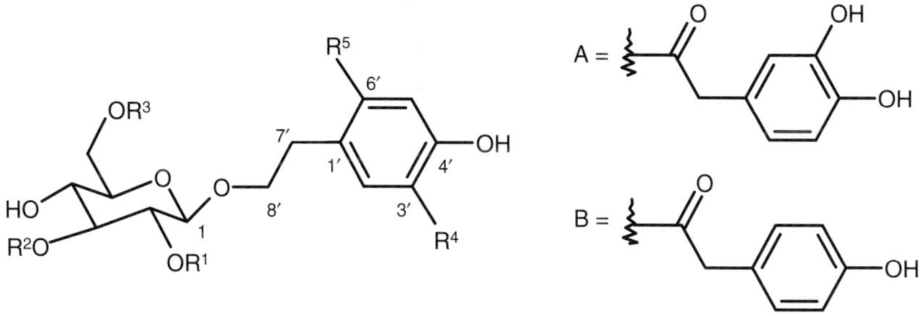

Ternstroside A: R¹ = A; R² = R³ = R⁵ = H; R⁴ = OH
Ternstroside B: R¹ = B; R² = R³ = R⁵ = H; R⁴ = OH
Ternstroside C: R¹ = R³ = R⁵ = H; R² = A; R⁴ = OH
Ternstroside D: R¹ = R² = R⁵ = H; R³ = A; R⁴ = OH
Ternstroside E: R¹ = A; R² = R³ = R⁴ = R⁵ = H
Ternstroside F: R¹ = A; R² = R³ = H; R⁴ = R⁵ = OH

Compound class: Phenylethanoid glucosides

Source: *Ternstroemia japonica* Thunb. (fresh leaves; family: Theaceae)

Pharmaceutical potential: *Antioxidants*

Ternstrosides A–F displayed potent antioxidative activity as evaluated in three different tests such as hydroxyl radical ($^{\cdot}$OH) inhibitory activity test, total ROS (reactive oxygen species) inhibitory activity test, and peroxynitrite (ONOO$^-$) scavenging activity test. The respective IC$_{50}$ values for the compounds (ternstrosides A–F) were determined to be 3.41 ± 0.09, 4.66 ± 0.04, 3.26 ± 0.01, 6.5 ± 0.09, 4.72 ± 0.04, and 4.21 ± 0.02 μM in the hydroxyl radical inhibitory assay; 39.26 ± 1.02, 62.52 ± 0.51, 33.29 ± 0.04, 66.87 ± 3.01, 54.89 ± 2.85, and 45.36 ± 0.62 μM in the total ROS inhibitory assay; and 1.69 ± 0.02, 2.36 ± 0.03, 1.14 ± 0.01, 3.85 ± 0.02, 2.56 ± 0.05, and 1.54 ± 0.01 μM in the peroxynitrite scavenging assay.

Reference

Jo, Y., Kim, M., Shin, M.H., Chung, H.Y., Jung, J.H., and Im, K.S. (2006) *J. Nat. Prod.*, **69**, 1399.

Terpendoles J, K, and L

Physical data:
- Terpendole J: $C_{32}H_{43}NO_5$, white powder, mp 248–250 °C, $[\alpha]_D^{28}$ $-30.3°$ (MeOH, *c* 1.0)
- Terpendole K: $C_{32}H_{39}NO_5$, white powder, $[\alpha]_D^{28}$ $+21.8°$ (MeOH, *c* 1.0)
- Terpendole L: $C_{37}H_{49}NO_5$, white powder, mp 148–150 °C, $[\alpha]_D^{28}$ $+16.5°$ (MeOH, *c* 1.0)

Structures:

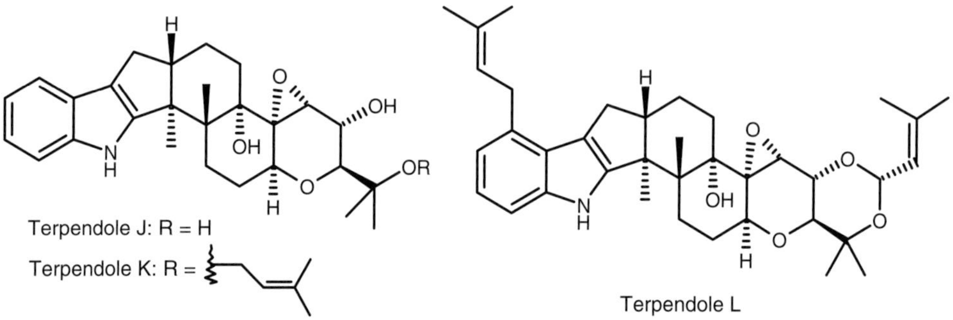

Terpendole J: R = H
Terpendole K: R = (3-methylbut-2-enyl)

Terpendole L

Compound class: Antibiotics

Source: *Albophoma yamanashiensis* (culture broth)

Pharmaceutical potential: *Acyl-CoA:cholesterol acyltransferase (ACAT) inhibitors*
Terpendoles J, K, and L were found to have moderate acyl-CoA:cholesterol acyltransferase (ACAT) inhibitory activity in rat liver microsomes with IC$_{50}$ values of 38.8, 38.0, and 32.4 µM, respectively.

Reference

Tomoda, H., Tabata, N., Yang, D.-J., Takayanagi, H., and Omura, S. (1995) *J. Antibiot.*, **48**, 793.

5,7,3′,4′-Tetrahydroxy-2′-(3,3-dimethylallyl)isoflavone

Physical data: $C_{20}H_{18}O_6$, greenish-white mass

Structure:

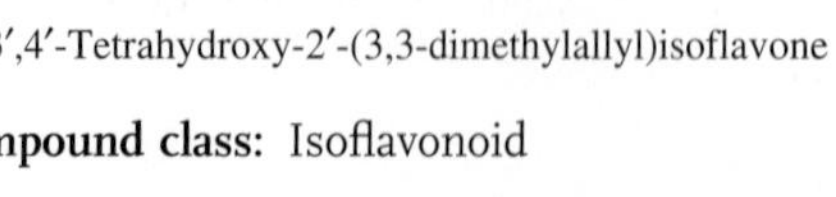

5,7,3′,4′-Tetrahydroxy-2′-(3,3-dimethylallyl)isoflavone

Compound class: Isoflavonoid

Source: *Psorothamnus arborescens* (roots; family: Fabaceae)

Pharmaceutical potential: *Antiprotozoal (leishmanicidal and trypanocidal)*
The isoflavonoid exhibited significant leishmanicidal activity with an IC$_{50}$ value of 13.0 ± 0.8 µM against *Leishmania donovani* axenic amastigotes and also trypanocidal activity against *Trypanosoma brucei brucei* (variant 221) with an IC$_{50}$ value of 12.1 ± 1.9 µM.

Reference

Salem, M.M. and Werbovetz, K.A. (2006) *J. Nat. Prod.*, **69**, 43.

1,3,5,6-Tetrahydroxy-4,7,8-tri(3-methyl-2-butenyl)xanthone

Physical data: $C_{28}H_{32}O_6$, yellow powder, mp 190–192 °C

Structure:

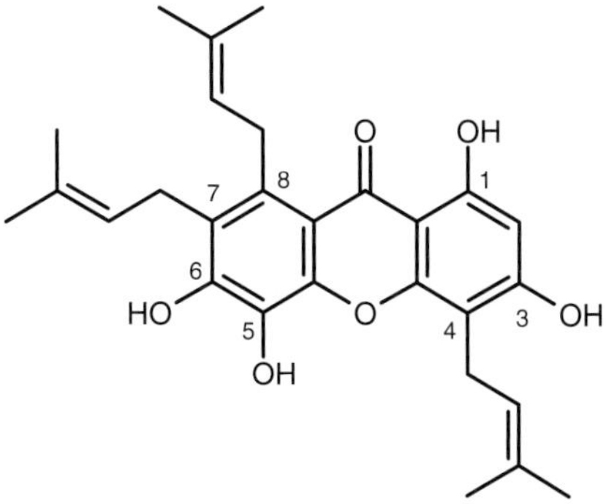

1,3,5,6-Tetrahydroxy-4,7,8-tri(3-methyl-2-butenyl)xanthone

Compound class: Xanthone

Source: *Garcinia xanthochymus* (woods; family: Guttiferae)

Pharmaceutical potential: *Nerve growth factor (NGF) stimulator*
The xanthone was evaluated to have the potentiality in stimulating the effect of nerve growth factor (NGF) as assessed in NGF-mediated neurite outgrowth in PC12D cells; the compound at the concentration range of 1–3 μM did not induce neurite outgrowth from PC12D cells in the absence of NGF, but it (at 3 μM) significantly enhanced the NGF-induced (2 ng/ml) proportion of neurite-bearing cells by 8.3%. However, the test compound exhibited cytotoxicity toward PC12D cells at a concentration of 10 μM.

Reference

Chanmahasathien, W., Li, Y., Satake, M., Oshima, Y., Ishibashi, M., Ruangrungsi, N., and Ohizumi, Y. (2003) *Chem. Pharm. Bull.*, **51**, 1332.

Tetrahydroswertianolin

See Bellidifolin

(2S)-5,7,3′,5′-Tetrahydroxyflavanone-7-O-β-D-glucopyranoside

Physical data: $C_{21}H_{22}O_{11}$, pale yellow amorphous powder, mp 219–221 °C, $[\alpha]_D^{20}$ −2.74° (MeOH, c 1.15)

Structure:

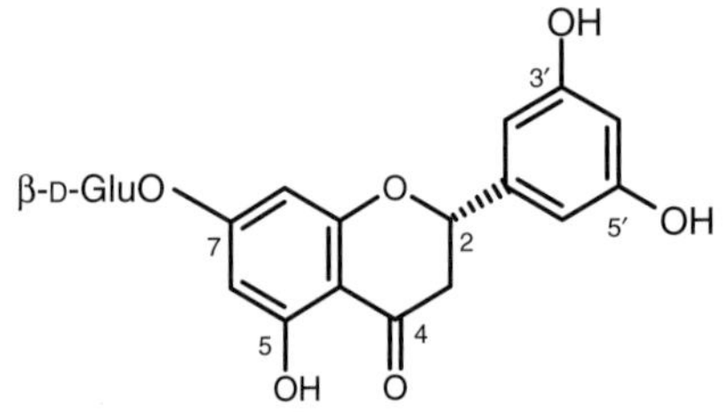

(2S)-5,7,3′,5′-Tetrahydroxyflavanone-7-O-β-D-glucopyranoside

Compound class: Flavonoid (flavanone glycoside)

Source: *Jasminum lanceolarium* (stems and leaves; family: Oleaceae)

Pharmaceutical potential: *Antioxidant (radical scavenger)*
The flavanone glycoside was found to exhibit potent *in vitro* antioxidant activity through scavenging radicals in the DPPH (1,1-diphenyl-2-picrylhydrazyl) assay with an EC_{50} value of 2.0 μg/ml; the activity was observed to be more potent than the positive control, ascorbic acid (EC_{50} = 3.32 μg/ml), under the same conditions.

Reference

Sun, J.-M., Yang, J.-S., and Zhan, H. (2007) *Chem. Pharm. Bull.*, **55**, 474.

5,6,7,4′-Tetrahydroxyflavonol-3-O-rutinoside

Physical data: $C_{27}H_{30}O_{16}$, amorphous powder, $[\alpha]_D^{20}$ −27.4° (MeOH, c 0.1)

Structure:

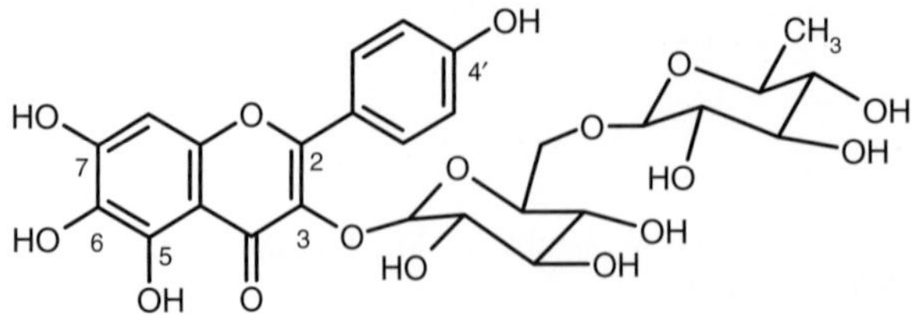

5,6,7,4′-Tetrahydroxyflavonol-3-O-rutinoside

Compound class: Flavonoid glycoside (flavonol glycoside)

Source: *Daphniphyllum calycinum* Benth. (leaves; family: Euphorbiaceae)

Pharmaceutical potential: *Antioxidant*
The flavonol glycoside showed moderate *in vitro* antioxidant activity through scavenging free radical in the DPPH assay; the IC_{50} value was determined to be 43.2 μg/ml (ascorbic acid used as a positive control; IC_{50} = 22 μg/ml).

Reference

Gamez, E.J.C., Luyengi, L., Lee, S.K., Zhu, L.-F., Zhu, L.-F., Zhou, B.-N., Song, H.H.S., Pezzuto, J. M., and Kinghorn, A.D. (1998) *J. Nat. Prod.*, **61**, 706.

Tetrahydroxysqualene

Physical data: $C_{30}H_{50}O_4$, off-white amorphous powder, $[\alpha]_D^{30}$ +75.9° ($CHCl_3$, *c* 1.40)

Structure:

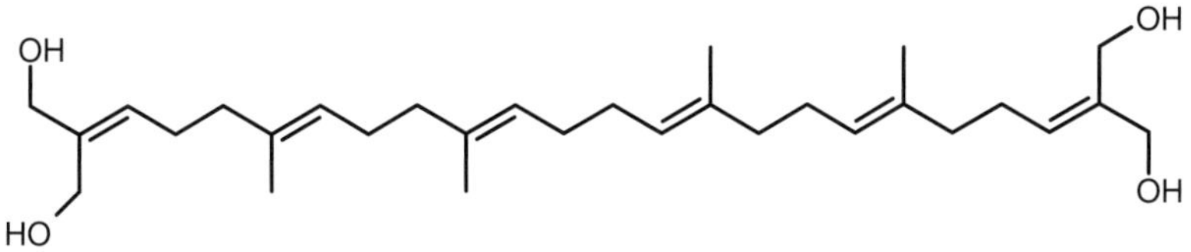

Tetrahydroxysqualene

Compound class: Terpenoid

Source: *Rhus taitensis* (leaves and twigs; family: Anacardiaceae)

Pharmaceutical potential: ***Antimycobacterial***
Tetrahydroxysqualene exhibited *in vitro* antituberculosis activity as studied against *Mycobacterium tuberculosis* H37Ra with an MIC value of 10.0 μg/ml, while showing only modest cytotoxicity (EC_{50} = 27.5 ± 0.8 μg/ml) toward human T cells.

Reference

Noro, J.C., Barrows, L.R., Gideon, O.G., Ireland, C.M., Koch, M., Matainaho, T., Piskaut, P., Pond, C.D., and Bugni, T.S. (2008) *J. Nat. Prod.*, **71**, 1623.

13,14,15,16-Tetranorlabdane-8α,12,14-triol

Physical data: $C_{16}H_{30}O_3$, yellowish-brown powder, $[\alpha]_D^{25}$ +25° ($CHCl_3$, *c* 0.47)

Structure:

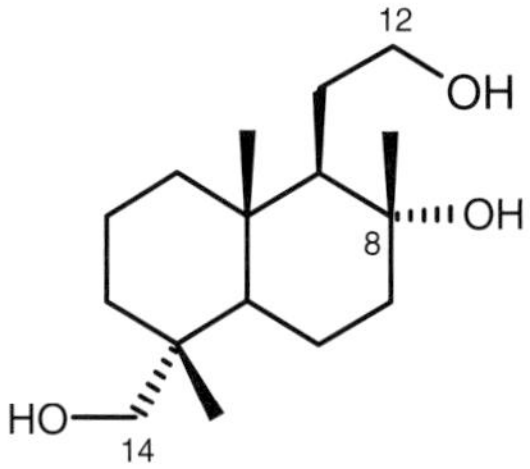

13,14,15,16-Tetranorlabdane-8α,12,14-triol

Compound class: Diterpenoid (labdane-type tetranorditerpene)

Source: *Crassocephalum mannii* Hook (whole plants; family: Asteraceae)

Pharmaceutical potential: *Cyclooxygenase (COX) inhibitor*
At a concentration of 100 μM, the test compound showed inhibitory effect of 44% against inducible COX-1, but it did not display such inhibitory activity against inducible COX-2.

Reference

Hegazy, M.F., Ohta, S., Abdel-Latif, F.F., Albadry, H.A., Ohta, E., Paré, P.W., and Hirata, T. (2008) *J. Nat. Prod.*, **71**, 1070.

Theograndin II

Systematic name: 5,7,3′,4′-Tetrahydroxyflavone-8-*O*-β-D-glucoronopyranoside 3″-*O*-sulfate (or hypolaetin 8-*O*-β-D-glucoronopyranoside 3″-*O*-sulfate)

Physical data: $C_{21}H_{18}O_{16}S$, yellow powder, mp 157.0–158.6 °C, $[\alpha]_D^{20}$ +110° (MeOH, *c* 0.0008)

Structure:

Theograndin II

Compound class: Flavonoid (flavone glycoside)

Source: *Theobroma grandiflorum* (seeds; family: Sterculiaceae)

Pharmaceutical potentials: *Antioxidant; cytotoxic*
Theograndin II exhibited antioxidant activity by scavenging free radicals in the DPPH assay with an IC_{50} value of 120.2 μM. It also showed weak cytotoxicity against the HCT-116 and SW-480 human colon cancer cells with IC_{50} values of 143 and 125 μM, respectively.

It has been observed that the antioxidant activities of sulfated flavonoid glycosides are significantly less than those of their corresponding flavonoid glycosides; the reduced activity may be attributed to the inter- as well as intramolecular hydrogen bonding formed between the oxygen of the sulfated group and the hydrogen of the hydroxyl function connected to a flavonoid skeleton. Formation of such hydrogen bonding is supposed to reduce the available hydrogen donors, thereby decreasing the antioxidant potentiality. However, sulfated flavonoid glycosides are highly soluble in water and are also easily hydrolyzable to their corresponding flavonoid glycosides; hence, sulfated flavonoid glycosides might increase bioavailability and demand for detailed *in vivo* as well as clinical studies to evaluate the significance of such sulfated compounds to human health.

Thiazomycin

Reference

Yang, H., Protiva, P., Cui, B., Ma, C., Baggett, S., Hequet, V., Mori, S., Bernard Weintein, I., and Kennelly, E.J. (2003) *J. Nat. Prod.*, **66**, 1501.

Thiazomycin

Physical data: $C_{61}H_{60}N_{14}O_{18}S_5$

Structure:

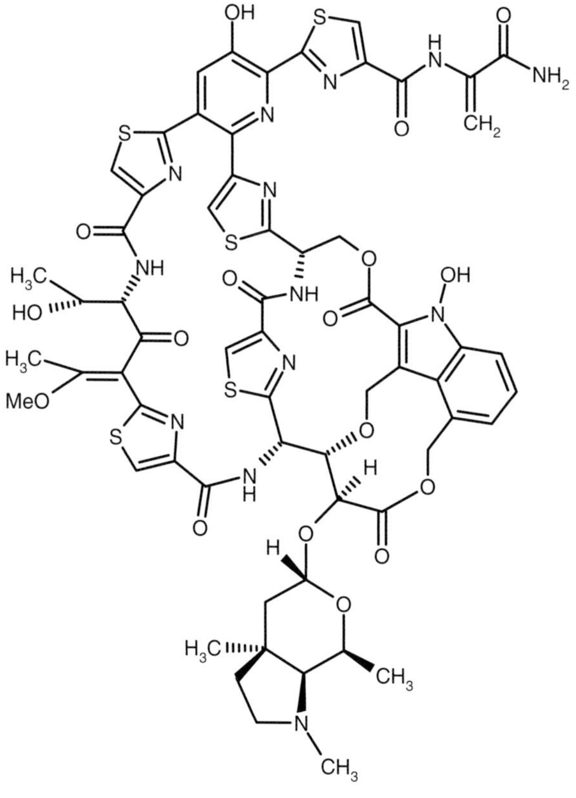

Thiazomycin

Compound class: Antibiotic (thiazolyl peptide)

Source: *Amycolatopsis fastidiosa* (culture broth) [1, 2]

Pharmaceutical potential: *Antibacterial*

Thiazomycin appeared to be an extremely potent antibiotic against Gram-positive bacteria both *in vitro* and *in vivo* and did not show cross-resistance to clinically relevant antibiotic classes such as β-lactams, vancomycin, oxazolidinone, and quinolones [2]. The isolate was evaluated to possess potent bactericidal activity against Gram-positive pathogens with an MIC range of

0.002–0.064 μg/ml. Singh et al. [2] determined MIC values for the compound against various bacterial strains such as *Streptococcus pneumoniae* (penicillin sensitive; CL8002), *S. pneumoniae* (penicillin resistant; CL5771), *S. pyogenes* (CL10440), *Enterococcus faecalis* (vancomycin sensitive; CL8516), *E. faecalis* (vancomycin resistant; CL5246), *E. faecium* (vancomycin and linezolid resistance; CL579), *Staphylococcus aureus* (MSSA; MB2865), *S. aureus* (MRSA; MB5393), *S. aureus* (vancomycin intermediate; CL5706), and *S. epidermidis* (CL8040), respectively, as 0.004, 0.002, 0.002, 0.064, 0.64, 0.008, 0.016, 0.032, <0.03, and 0.064 μg/ml. Thiazomycin was not found to inhibit growth of Gram-negative bacteria such as *Escherichia coli* and *Pseudomonas aeruginosa*, and also the growth of *Candida albicans* at 32 μg/ml, thereby displaying preferential selectivity for Gram-positive bacteria over Gram-negative bacteria and eukaryotic microorganisms. The same group of investigators showed the efficacy of the drug also under *in vivo* conditions; it exhibited good potentiality against *Staphylococcus aureus* infection in mice with an ED_{99} value of 0.15 mg/kg on subcutaneous administration.

The investigators demonstrated that the test compound inhibits bacterial growth by selective inhibition of protein synthesis, and it was thought to interact with L11 protein and 23S rRNA of the 50S ribosome; this mechanism of inhibition of thiazomycin was found to be distinct and different from that of the clinically used drugs. That is why thiazomycin did not exhibit cross-resistance to the drug-resistant organisms that were tested including *Staphylococcus aureus* that was resistant to protein synthesis inhibitors such as linezolid, macrolides, chloramphenicol, aminoglycosides, and tetracycline [2].

References

[1] Jayasuriya, H., Herath, K., Ondeyka, J.G., Zhang, C., Zink, D.L., Brower, M., Gailliot, F.P., Greene, J., Birdsall, G., Venugopal, J., Ushio, M., Burgess, B., Russotti, G., Walker, A., Hesse, M., Seeley, A., Junker, B., Connors, N., Salazar, O., Genilloud, O., Liu, K., Masurekar, P., Barrett, J.F., and Singh, S.B. (2007) *J. Antibiot.*, **60**, 554.

[2] Singh, S.B., Occi, J., Jayasuriya, H., Herath, K., Motyl, M., Dorso, K., Gill, C., Hicky, E., Overbye, K.M., Barrett, J.F., and Masurekar, P. (2007) *J. Antibiot.*, **60**, 565.

Thiomarinols A–G

Physical data:
- Thiomarinol A: $C_{30}H_{44}N_2O_9S_2$, orange crystals, mp 106–110 °C
- Thiomarinol B: $C_{30}H_{44}N_2O_{11}S_2$, yellow powder
- Thiomarinol C: $C_{30}H_{44}N_2O_8S_2$, yellow powder
- Thiomarinol D: $C_{31}H_{46}N_2O_9S_2$, yellow powder
- Thiomarinol E: $C_{32}H_{48}N_2O_9S_2$, yellow powder
- Thiomarinol F: $C_{30}H_{42}N_2O_9S_2$, yellow powder
- Thiomarinol G: $C_{30}H_{44}N_2O_8S_2$, yellow powder

Structures:

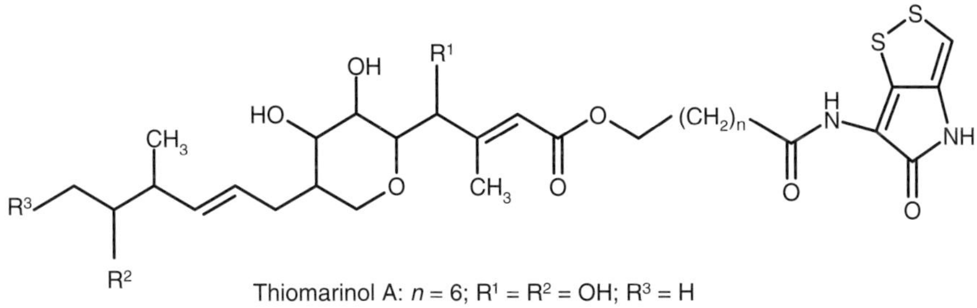

Thiomarinol A: $n = 6$; $R^1 = R^2 = OH$; $R^3 = H$
Thiomarinol C: $n = 6$; $R^1 = R^3 = H$; $R^2 = OH$
Thiomarinol D: $n = 6$; $R^1 = R^2 = OH$; $R^3 = CH_3$
Thiomarinol E: $n = 8$; $R^1 = R^2 = OH$; $R^3 = H$
Thiomarinol F: $n = 6$; $R^1 = OH$; $R^2 = O(oxo)$; $R^3 = H$

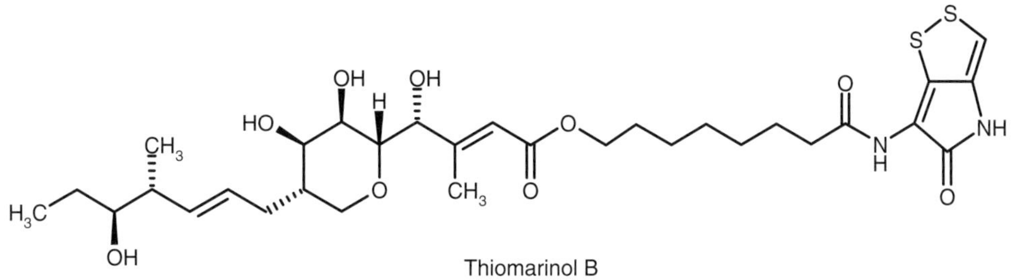

Thiomarinol B

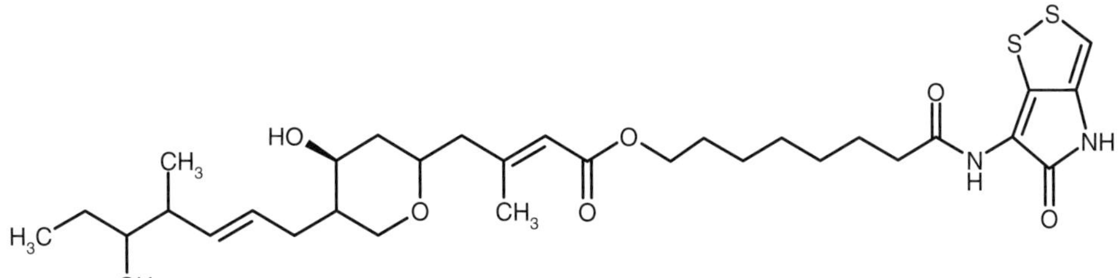

Thiomarinol G

Compound class: Antibiotics

Source: *Alteromonas rava* sp. nov. SANK 73390 [1–3]

Pharmaceutical potential: *Antimicrobial*

Thiomarinols A–G were evaluated to possess significant antimicrobial potentials [1–3]; thiomarinol A displayed potent such efficacy against the test organisms *Staphylococcus aureus* 209P JC-1, *S. aureus* 507 (MRSA), *Enterococcus faecalis* NCTC775, *Escherichia coli* NIHJ JC-2, *Salmonella enteritidis* G, *Klebsiella pneumoniae* IID685, *Enterobacter cloacae* 963, *Serratia marcescens* IAM1184, *Proteus vulgaris* IID874, *Morganella morganii* IFO3848, and *Pseudomonas aeruginosa* PA01 with respective MIC values of <0.01, <0.01, <0.01, 3.13, 3.13, 0.78, 6.25, 25.0, 0.39, 12.5, and 0.39 µg/ml [1].

The same group of investigators [2, 3] evaluated antimicrobial potentials of the remaining thiomarinols also against a variety of microbes such as *Staphylococcus aureus* 209P JC-1, *S. aureus*

56R, *S. aureus* 535 (MRSA), *Enterococcus faecalis* 681, *Escherichia coli* NIHJ JC-2, *E. coli* 609, *Salmonella enteritidis, Klebsiella pneumoniae* 806, *K. pneumoniae* 846 (R), *Enterobacter cloacae* 963, *Serratia marcescens* IAM1184, *Proteus vulgaris* 1420, *Morganella morganii* 1510, *Pseudomonas aeruginosa* 1001, *P. aeruginosa* No. 7, and *P. aeruginosa* PA01 and determined the respective MIC values as ≤0.01, ≤0.01, ≤0.01, 0.05, 0.8, 0.8, 0.4, 0.8, 0.2, 0.8, 3.1, 0.05, 6.2, 0.2, 0.4, and 0.8 µg/ml for thiomarinol B; ≤0.01, ≤0.01, ≤0.01, 0.8, 3.1, 1.5, 1.5, 1.5, 0.8, 3.1, 6.2, 0.2, 12.5, 0.8, 0.8, and 0.4 µg/ml for thiomarinol C; ≤0.01, ≤0.01, ≤0.01, 0.2, 1.5, 1.5, 0.8, 0.8, 0.4, 1.5, 6.2, 0.1, 12.5, 0.4, 1.5, and 0.4 µg/ml for thiomarinol D; ≤0.01, ≤0.01, ≤0.01, 0.05, 1.5, 0.8, 0.8, 0.8, 0.4, 1.5, 6.2, 0.2, 12.5, 0.4, 0.4, and 0.4 µg/ml for thiomarinol E; ≤0.01, ≤0.01, ≤0.01, 0.2, 6.2, 3.1, 3.1, 3.1, 1.5, 12.5, 25.0, 1.5, 100, 3.1, 3.1, and 6.2 µg/ml for thiomarinol F; and ≤0.01, ≤0.01, ≤0.01, 3.1, >200, 50.0, 25.0, 12.5, 6.2, >200, >200, 3.1, >200, 12.5, 12.5, and 12.5 µg/ml for thiomarinol G. Hence, the antibiotics may serve as promising leads in developing antimicrobial drugs.

References

[1] Shiozawa, H., Kagasaki, T., Kinoshita, T., Haruyama, H., Domon, H., Utui, Y., Kodama, K., and Takahashi, S. (1993) *J. Antibiot.*, **46**, 1834.
[2] Shiozawa, H., Kagasaki, T., Torikata, A., Tanaka, N., Fujimoto, K., Hata, T., Furukawa, Y., and Takahashi, S. (1995) *J. Antibiot.*, **48**, 907.
[3] Shiozawa, H., Shimada, A., and Takahashi, S. (1997) *J. Antibiot.*, **50**, 449.

Thonningianins A and B

Physical data:
- Thonningianin A: $C_{42}H_{34}O_{21}$, pale yellow solid, $[\alpha]_D$ −44° (MeOH, *c* 0.08)
- Thonningianin B: $C_{35}H_{30}O_{17}$, pale yellow solid, $[\alpha]_D$ −141° (MeOH, *c* 0.1)

Structures:

Compound class: Ellagitannins

Source: *Thonningia sanguinea* Vahl (family: Balanophoraceae)

Pharmaceutical potential: *Antioxidant*
Thonningianins A and B, the chemical constituents of the African medicinal herb *Thonningia sanguinea*, showed strong free radical scavenging activity against 1,1-diphenyl-2-picrylhydrazyl (DPPH) as shown with the help of ESR study by Ohtani *et al*. The DPPH radical was scavenged completely (100%) by a 25 µM solution of thonningianin A and scavenged 90% by a 34.5 µM solution of thonningianin B. In a separate experiment, IC_{50} values of the compounds were evaluated to be 8 and 21 µM, respectively. The authors assumed that the presence of an additional galloyl group in thonningianin A might be responsible for its more potent efficacy in comparison to thonningianin B.

Reference
Ohtani, I.I., Gotoh, N., Tanaka, J., Higa, T., Gyamfi, M.A., and Aniya, Y. (2000) *J. Nat. Prod.*, **63**, 676.

threo-(7S,8R)-1-(4-Hydroxyphenyl)-2-[4-(E)-propenylphenoxy]-propan-1-ol

Physical data: $C_{18}H_{21}O_3$, yellow oil, $[\alpha]_D^{20}$ +24.99° (CHCl$_3$, *c* 0.5)

Structure:

threo-(7S,8R)-1-(4-Hydroxyphenyl)-2-[4-(E)-propenylphenoxy]-propan-1-ol

Compound class: Neolignoid

Source: *Ribes fasciculatum* var. *chinense* Max. (stems and twigs; family: Saxifragaceae)

Pharmaceutical potential: *NFAT transcription factor inhibitor*
Nuclear factor of activated T cells (NFAT), a cytoplasmic protein, is activated by stimulation of cell surface receptors coupled to Ca^{2+} mobilization, and modulation of NFAT transcription factor is considered useful in the therapy of immune diseases. The test compound exhibited inhibitory activity against the NFAT transcription factor with an IC_{50} value of 15.6 ± 0.35 µM (cyclosporin A was used as a positive control).

Reference
Dat, N.T., Cai, X.F., Shen, Q., Lee, I., and Kim, Y.H. (2005) *Chem. Pharm. Bull.*, **53**, 114.

β-Thujaplicin

See Hinokitiol

Tianshic acid and its methyl ester

Systematic names:
- Tianshic acid: (9E)-8,11,12-trihydroxyoctadecenoic acid
- Tianshic acid methyl ester: (9E)-8,11,12-trihydroxyoctadecenoic acid methyl ester

Physical data:
- Tianshic acid A: $C_{18}H_{34}O_2$
- Tianshic acid methyl ester: $C_{19}H_{36}O_5$, amorphous powder

Structures:

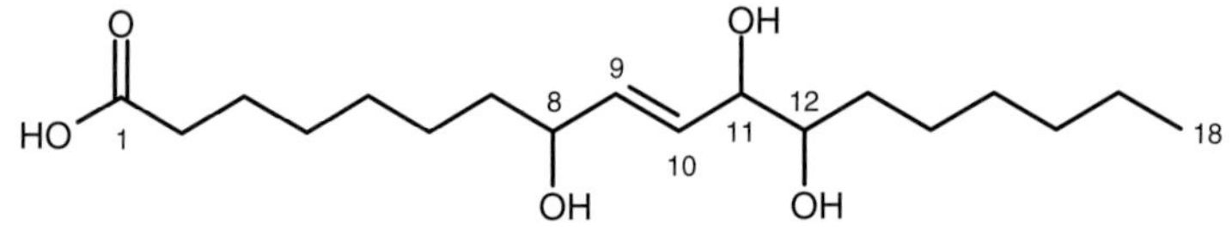

Tianshic acid

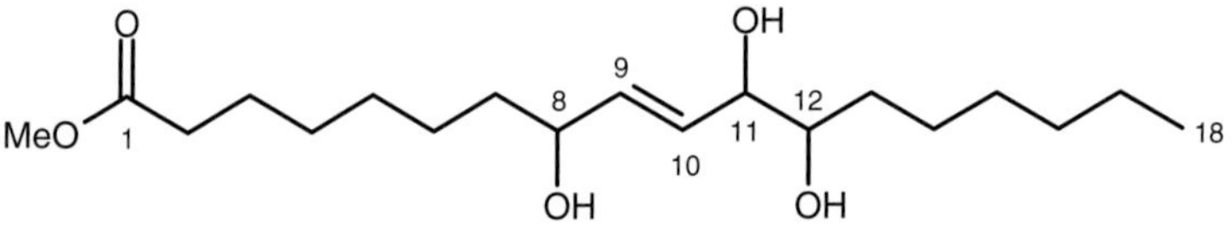

Tianshic acid methyl ester

Compound class: Fatty acid and its ester

Sources: Tianshic acid: *Aesculus wilsonii* L. (family: Sapindaceae/Hippocastanaceae) [1], *Allium fistulosum* L. (seeds; family: Liliaceae) [2]; *Ophiopogon japonicus* (tubers; family: Liliaceae) [3]; *Sambucus williamsii* Hance (stems; family: Caprifoliaceae) [4]; methyl ester: *Sambucus williamsii* Hance (stems) [4].

Pharmaceutical potentials: *Alkaline phosphatase activity inducer; antifungal*
Both the fatty acid and its methyl ester showed stimulating effects on alkaline phosphatase activity of osteoblastic UMR106 cells. They were found to induce about 1.5-fold enzyme activity at a concentration of 30.0 μM; however, the compounds did not exhibit stimulating effects on UMR106 cell proliferation in the predetermined concentration range of 3.0 nM to 30.0 μM [4]. The inducing effects on this enzyme activity by these compounds may indicate positive effects on osteoblast differentiation. In addition, tianshic acid possesses antifungal activity; it was found to inhibit the growth of *Phytophthora capsici* on V8 media [2].

References

[1] Chen, X.S., Chen, D.H., Si, J.Y., Tu, G.A.Z., and Ma, L.B. (2000) *Acta Pharm. Sin.*, **35**, 198.

[2] Sang, S., Lao, A., Wang, Y., Chin, C.-K., Rosen, R.T., and Ho, C.-T. (2002) *J. Agric. Food Chem.*, **50**, 6318.
[3] Cheng, Z.-H., Wu, T., and Yu, B.-Y. (2005) *Nat. Prod. Res. Dev.*, **17**, 1.
[4] Yang, X., Wong, M., Wang, N., Chan, A.S.-C., and Yao, X. (2006) *Chem. Pharm. Bull.*, **54**, 676.

TMC-135A and B

Physical data:
- TMC-135A: $C_{39}H_{49}N_3O_8S$, pale yellow powder, mp 166–168 °C
- TMC-135B: $C_{39}H_{49}N_3O_8S$, pale yellow powder, mp 190–192 °C

Structures:

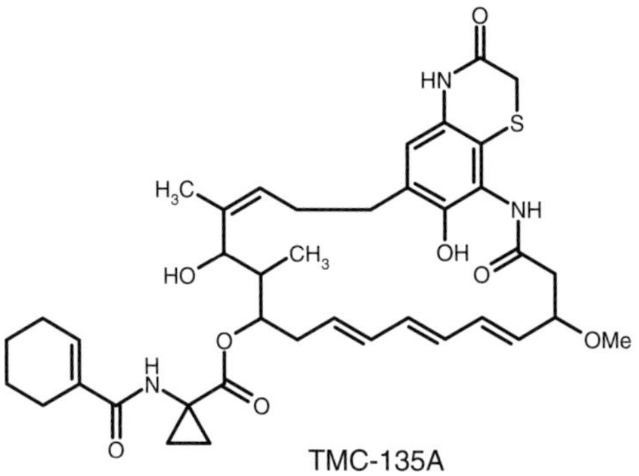

Compound class: Antibiotics (anamycin group)

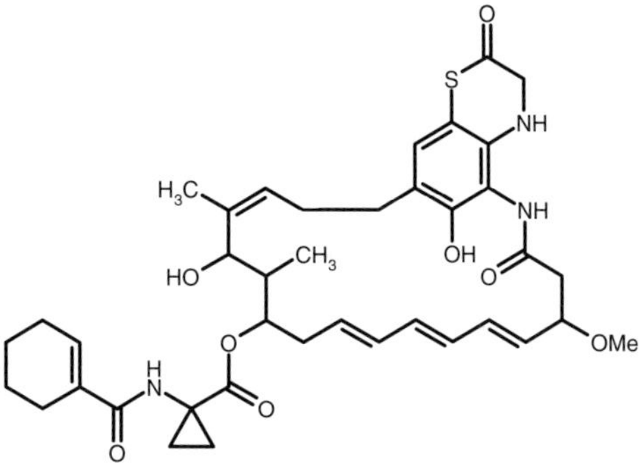

Source: *Streptomyces* sp. TC-1190 (fermentation broth)

Pharmaceutical potential: *Antitumor*
TMC-135A and B were reported to possess significant growth inhibitory effects on a series of human tumor cell lines such HCT-116 human colon carcinoma, SK-BR-3 human breast adenocarcinoma, HeLa S3 human epitheloid carcinoma, U937 human histiocytic lymphoma, WiDr human colon adenocarcinoma, HT29 urinary bladder carcinoma, HL-60 human premyelocytic leukemia, THP1 human monocytic leukemia, Raji Burkitt's lymphoma, Jurkat human lymphoma, P388D1 murine lymphoid neoplasm, and B-16 murine melanoma; the respective IC_{50} (μM) values for the isolates were determined to be 0.07, 0.88, 0.11, 0.11, 0.20, 0.15, 0.05, 0.06, 0.05, 0.13, 0.56, and 0.07 (TMC-135A); 0.86, 6.5, 1.1, 0.6, 1.6, 1.3, 0.30, 0.32, 0.34, 0.47, 2.5, and 0.67 (TMC-135B).

Reference

Nishio, M., Kohno, J., Sakurai, M., Suzuki, S.-I., Okada, N., Kawano, K., and Komatsubara, S. (2000) *J. Antibiot.*, **53**, 724.

TMC-1A, B, C, and D

Physical data:
- TMC-1A: $C_{28}H_{36}N_2O_7$, yellow powder, mp >95 °C, $[\alpha]_D^{24}$ $-55 \pm 3°$ (CHCl$_3$, c 0.1)
- TMC-1B: $C_{28}H_{36}N_2O_7$, yellow powder, mp >75 °C, $[\alpha]_D^{24}$ $\sim 0°$ (CHCl$_3$, c 0.1)
- TMC-1C: $C_{30}H_{38}N_2O_7$, yellow powder, mp >106 °C, $[\alpha]_D^{24}$ $+116 \pm 1°$ (CHCl$_3$, c 0.1)
- TMC-1D: $C_{30}H_{40}N_2O_7$, yellow powder, mp >89 °C, $[\alpha]_D^{24}$ $+16 \pm 2°$ (CHCl$_3$, c 0.1)

Structures:

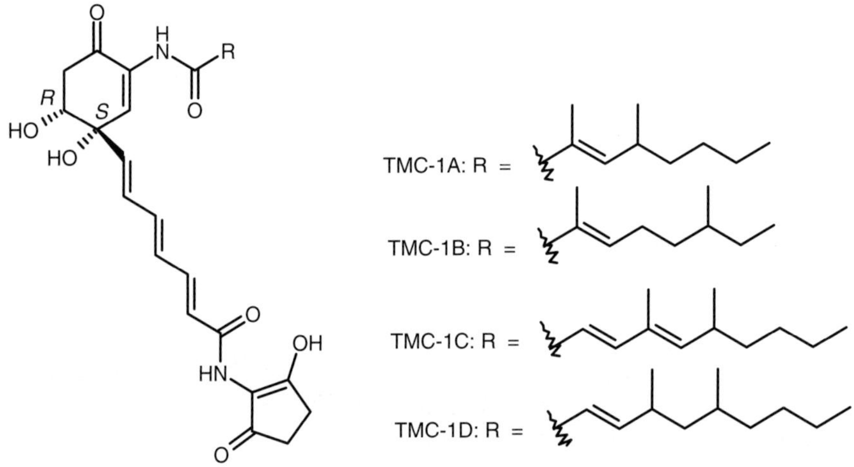

Compound class: Antibiotics (manumycin group)

Source: *Streptomyces* sp. A-230 (fermentation broth)

Pharmaceutical potential: *Antitumor*
TMC-1A–D were reported to possess significant growth inhibitory effects on a series of human tumor cell lines such HCT-116 human colon carcinoma, SW480 human colon adenocarcinoma, Saos-2 human osteogenic sarcoma, WiDr human colon adenocarcinoma, OVCAR-3 human ovarian adenocarcinoma, HL-60 human promyelocytic leukemia, HeLa S3 human epitheloid carcinoma, and P388D1 murine lymphoid neoplasm; the respective IC$_{50}$ (µg/ml) values for the isolates were determined to be 28.0, 47.7, 37.6, 28.6, 28.6, 29.8, 24.2, and 15.5 (TMC-1A); 16.9, 17.1, 23.1, 13.1, 13.5, 13.1, 10.5, and 8.0 (TMC-1B); 3.4, 9.0, 4.6, 10.0, 5.8, 6.3, 6.7, and 3.3 (TMC-1C); and 6.8, 11.0, 5.9, 9.3, 6.0, 12.0, 8.3, and 3.2 (TMC-1D).

Reference

Kohno, J., Nishio, M., Kawano, K., Nakanishi, N., Suzuki, S.-I., Uchida, T., and Komatsubara, S. (1996) *J. Antibiot.*, **49**, 1212.

TMC-52A, B, C, and D

Physical data:
- TMC-52A: $C_{20}H_{30}N_4O_6$, white powder, $[\alpha]_D^{24}$ +22° (H_2O, c 0.4)
- TMC-52B: $C_{20}H_{30}N_4O_6$, white powder, $[\alpha]_D^{24}$ +22° (H_2O, c 0.4)
- TMC-52C: $C_{20}H_{30}N_4O_5$, white powder, $[\alpha]_D^{24}$ +17° (H_2O, c 0.5)
- TMC-52D: $C_{20}H_{30}N_4O_5$, white powder, $[\alpha]_D^{24}$ +11° ((H_2O, c 0.5)

Structures:

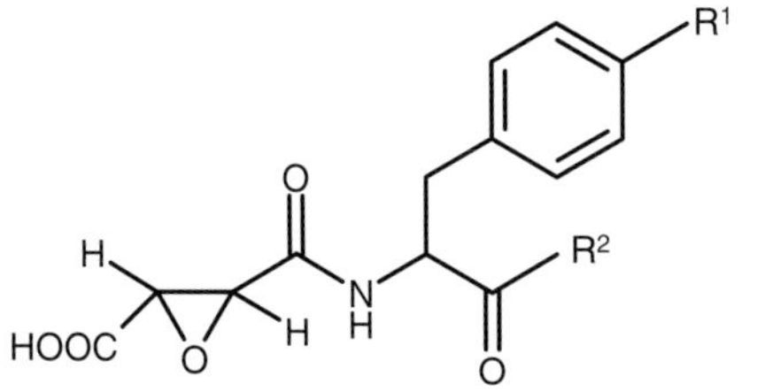

TMC-52A: R^1 = OH; R^2 = —NH(CH$_2$)$_3$NH(CH$_2$)$_4$NH$_2$
TMC-52B: R^1 = OH; R^2 = —NH(CH$_2$)$_4$NH(CH$_2$)$_3$NH$_2$
TMC-52C: R^1 = H; R^2 = —NH(CH$_2$)$_3$NH(CH$_2$)$_4$NH$_2$
TMC-52D: R^1 = H; R^2 = —NH(CH$_2$)$_4$NH(CH$_2$)$_3$NH$_2$

Compound class: Antibiotics (epoxysuccinyl peptides)

Source: *Gliocladium* sp. F-2665 (fungal strain; fermentation broth) [1]

Pharmaceutical potential: *Cysteine proteinase inhibitors*

Cysteine proteinases are reported to be involved in the pathogenesis of bone disorders [2–4] and cancer metastasis [5]; the present investigators [1] evaluated TMC-52A–D as potent inhibitors of cysteine proteinases, particularly cathepsin L, with respective IC$_{50}$ values of 13, 10, 10, and 6 nM.

References

[1] Isshiki, K., Nishio, M., Sakurai, N., Uchida, T., Okuda, T., and Komatsubara, S. (1998) *J. Antibiot.*, **51**, 629.
[2] Kakegawa, H., Nikawa, T., Tagami, K., Kamioka, H., Sumitani, K., Kawata, T., Drobnic-Kosorok, M., Lenarcic, B., and Katsunuma, N. (1993) *FEBS Lett.*, **321**, 247.
[3] Delaisse, J.-M., Eeckhout, Y., and Vaes, G. (1984) *Biochem. Biophys. Res. Commun.*, **125**, 441.
[4] Barret, A. and Kirschke, H. (1985) *Methods Enzymol.*, **80**, 535.
[5] Yagel, S., Warner, A.H., Nellans, H.N., Lala, P.K., Waghorne, C., and Denhardt, D.T. (1989) *Cancer Res.*, **49**, 3553.

TMC-86A, B, and TMC-96

Physical data:
- TMC-86A: $C_{16}H_{26}N_2O_6$, colorless oil, $[\alpha]_D^{20}$ +10° (H_2O, c 0.27)
- TMC-86B: $C_{20}H_{34}N_2O_7$, white powder, $[\alpha]_D^{20}$ +30° (H_2O, c 0.41)
- TMC-96: $C_{18}H_{32}N_2O_6$, colorless sticky solid, $[\alpha]_D^{20}$ +25° (MeOH, c 0.45)

Structures:

TMC-86A

TMC-86B

TMC-96

Compound class: Antibiotics (epoxy-β-aminoketone derivatives)

Sources: TMC-86A and B: *Streptomyces* sp. TC 1084 (fermentation broth) [1, 2]; TMC-96: *Saccharothrix* sp. TC 1094 (fermentation broth) [1, 2]

Pharmaceutical potentials: *Proteasome inhibitors; cytotoxic (antitumor)*
Proteasome is considered to be responsible for activation of TNF-α and plays a key role in developing inflammatory diseases, autoimmune diseases, and pathological muscle wasting [3–5]. The catalytic core of proteasome is 20S proteasome. TMC-86A, B, and TMC-96 were found to display the chymotrypsin-like and peptidylglutamyl-peptide-hydrolyzing inhibitory activities of 20S proteasome with IC_{50} (μM) values of 5.1 and 3.7 (TMC-86A); 1.1 and 31 (TMC-86B); and 2.9 and 3.5 (TMC-96), respectively (respective IC_{50} values for eponemycin used as a reference standard were 1.1 and 5.4 μM) [1]. TMC-86A, B, and TMC-96 exhibited weak inhibitory activity against trypsin-like activity of 20S proteasome with IC_{50} values of 51, 250, and 36 μM, respectively. However, the isolates did not inhibit *m*-calpain, cathepsin L, and trypsin at 100 μM, thereby suggesting their high specificity for proteasome [1].

The present investigators [1] also evaluated their *in vitro* cytotoxic activities against a number of tumor cell lines such as HCT-116 (human colon carcinoma), HeLa S3 (human epitheloid carcinoma), SK-BR-3 (human breast adenocarcinoma), WiDr (human colon adenocarcinoma), HL-60 (human promyelocytic leukemia), B-16 (murine melanoma), and P388D 1 (murine lymphoid neoplasm) – the respective IC_{50} (μM) values were determined to be 0.22, 0.23, 0.27, 0.25, 0.20, 0.22, and 0.22 (TMC-86A); 0.21, 0.75, 0.65, 0.53, 0.43, 0.20, and 0.21 (TMC-86B); and 0.22, 0.21, 0.32, 0.27, 0.24, 0.20, and 0.22 (TMC-96) [1].

References

[1] Koguchi, Y., Kohno, J., Suzuki, S.-I., Nishio, M., Takahashi, K., Ohnuki, T., and Komatsubara, S. (1999) *J. Antibiot.*, **52**, 1069.

[2] Koguchi, Y., Kohno, J., Suzuki, S.-I., Nishio, M., Takahashi, K., Ohnuki, T., and Komatsubara, S. (2000) *J. Antibiot.*, **53**, 63.
[3] Palombella, V.I., Rando, O.J., Goldberg, A.L., and Maniatis, T. (1994) *Cell*, **78**, 773.
[4] Solomon, V., Baracos, V., Sarraf, P., and Goldberg, A.L. (1998) *Proc. Natl. Acad. Sci. U. S. A.*, **95**, 12602.
[5] Rock, K.L., Gramm, C., Rothstein, L., Clark, K., Stein, R., Dick, L., Hwang, D., and Goldberg, A.L. (1994) *Cell*, **78**, 761.

TMC-89A and B

Physical data:
- TMC-89A: $C_{21}H_{36}N_4O_9$, white powder, mp 99 °C, $[\alpha]_D^{20}$ −7.7° (H_2O, c 0.3)
- TMC-89B: $C_{21}H_{36}N_4O_9$, white powder, mp 97 °C, $[\alpha]_D^{20}$ −6.8° (H_2O, c 0.3)

Structures:

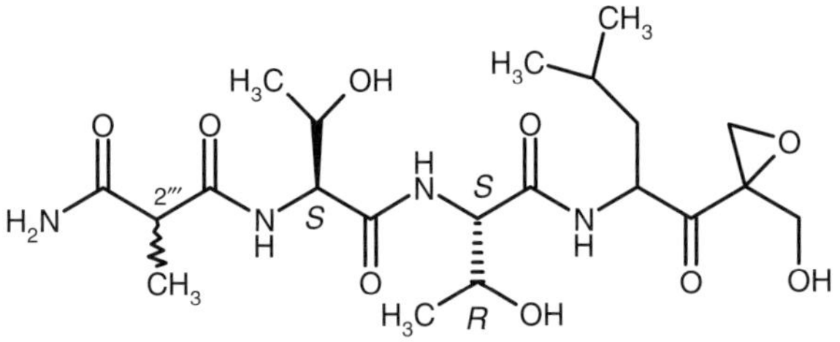

TMC-89A: *R* (C-2′″); TMC-89B: *S* (C-2′″)

Compound class: Antibiotics

Source: *Streptomyces* sp. TC 1087 (fermentation broth) [1]

Pharmaceutical potential: *Proteasome inhibitors*
20S proteasome is regarded as a therapeutic target for inflammatory and autoimmune diseases [2, 3]. TMC-89A and B were found to display chymotrypsin-like, trypsin-like, and peptidylglutamyl-peptide-hydrolyzing inhibitory activities of 20S proteasome with IC_{50} (μM) values of 1.1, 0.39, and 7.2 (TMC-89A); 1.1, 0.51, and 7.1 (TMC-89B), respectively [1]. TMC-89A did not inhibit *m*-calpain, cathepsin L, and trypsin at 100 μM, suggesting its high selectivity for proteasome.

References

[1] Koguchi, Y., Nishio, M., Suzuki, S.-I., Takahashi, K., Ohnuki, T., and Komatsubara, S. (2000) *J. Antibiot.*, **53**, 967.
[2] Rock, K.L., Gramm, C., Rothstein, L., Clark, K., Stein, R., Dick, L., Hwang, D. and Goldberg, A. L. (1994) *Cell*, **78**, 761.
[3] Meyer, S., Kohler, N.G., and Joly, A. (1997) *FEBS Lett.*, **413**, 354.

TMC-95A, B, C, and D

Physical data:
- TMC-95A: $C_{33}H_{38}N_6O_{10}$, colorless powder, $[\alpha]_D^{23}$ +102° (MeOH, c 0.54)
- TMC-95B: $C_{33}H_{38}N_6O_{10}$, colorless powder, $[\alpha]_D^{23}$ +74° (MeOH, c 0.47)
- TMC-95C: $C_{33}H_{38}N_6O_{10}$, colorless powder, $[\alpha]_D^{23}$ −18° (MeOH, c 0.23)
- TMC-95D: $C_{33}H_{38}N_6O_{10}$, colorless powder, $[\alpha]_D^{23}$ −36° (MeOH, c 0.10)

Structures:

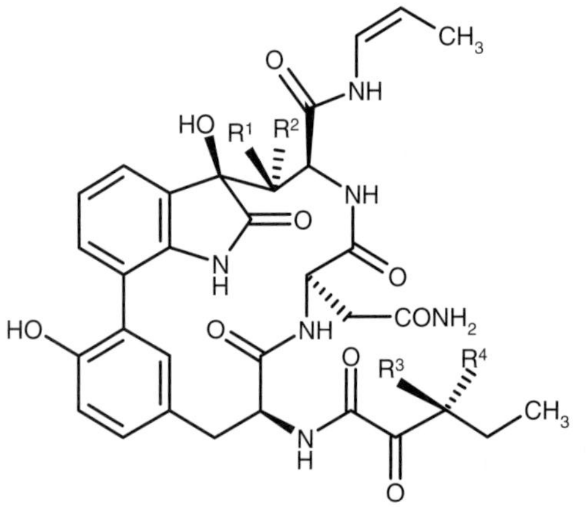

TMC-95A: $R^1 = R^4 = H$; $R^2 = OH$; $R^3 = CH_3$
TMC-95B: $R^1 = R^3 = H$; $R^2 = OH$; $R^4 = CH_3$
TMC-95C: $R^2 = R^4 = H$; $R^1 = OH$; $R^3 = CH_3$
TMC-95D: $R^2 = R^3 = H$; $R^1 = OH$; $R^4 = CH_3$

Compound class: Antibiotics

Source: *Apiospora montagnei* Sacc. TC 1093 (fermentation broth) [1, 2]

Pharmaceutical potentials: *Proteasome inhibitors; cytotoxic*
20S proteasome is regarded as a therapeutic target for inflammatory and autoimmune diseases [3, 4]. TMC-95A–D were found to display chymotrypsin-like, trypsin-like, and peptidylglutamyl-peptide-hydrolyzing inhibitory activities of 20S proteasome with respective IC_{50} (µM) values of 0.0054, 0.20, and 0.060 (TMC-95A); 0.0087, 0.49, and 0.060 (TMC-95B); 0.36, 14, and 8.7 (TMC-95C); and 0.27, 9.3, and 3.3 (TMC-95D) [2]. TMC-95A did not inhibit *m*-calpain, cathepsin L, and trypsin at 30 µM, suggesting its high selectivity for proteasome; furthermore, this isolate exhibited cytotoxic activities against HCT-116 human colon carcinoma cells and HL-60 human promyelocytic leukemia cells with IC_{50} values of 4.4 and 9.8 µM, respectively [2].

References

[1] Kohno, J., Koguchi, Y., Nishio, M., Nakao, K., Kuroda, M., Shimizu, R., Ohnuki, T., and Komatsubara, S. (2000) *J. Org. Chem.*, **65**, 990.
[2] Koguchi, Y., Kohno, J., Nishio, M., Takahashi, K., Okuda, T., Ohnuki, T., and Komatsubara, S. (2000) *J. Antibiot.*, **53**, 105.
[3] Rock, K.L., Gramm, C., Rothstein, L., Clark, K., Stein, R., Dick, L., Hwang, D., and Goldberg, A.L. (1994) *Cell*, **78**, 761.
[4] Meyer, S., Kohler, N.G., and Joly, A. (1997) *FEBS Lett.*, **413**, 354.

Tolybyssidins A and B

Physical data:
- Tolybyssidin A: $C_{71}H_{116}N_{16}O_{17}$, white amorphous solid, $[\alpha]_D^{20}$ $-13°$ (MeOH, c 0.2)
- Tolybyssidin B: $C_{72}H_{115}N_{16}O_{16}S$, yellow–white waxy solid, $[\alpha]_D^{20}$ $-18°$ (MeOH, c 0.3)

Structures:

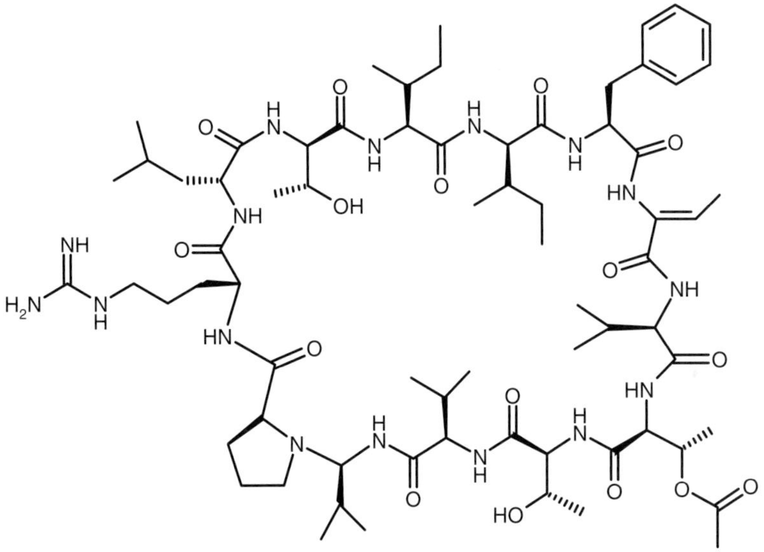

Tolybyssidin A

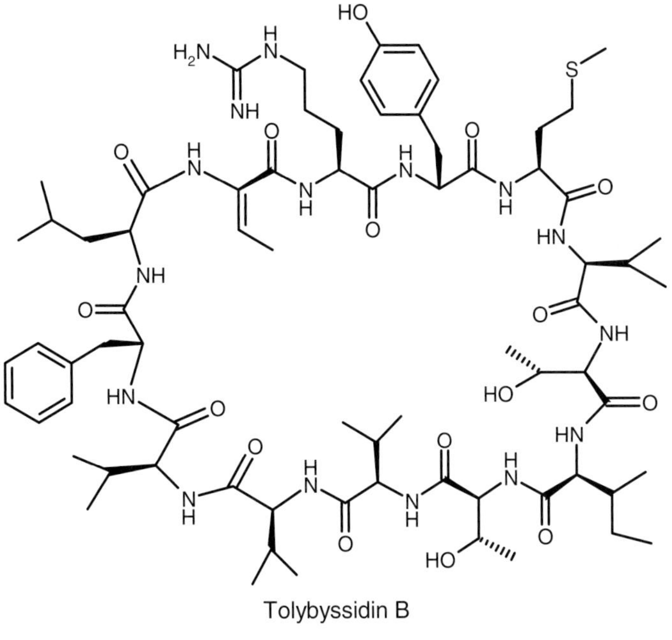

Tolybyssidin B

Compound class: Cyclic peptides

Source: *Tolypothrix byssoidea* EAWAG 195 (cyanobacterium; culture broth)

Pharmaceutical potential: *Antifungal*
Tolybyssidins A and B were found to inhibit *in vitro* growth of *Candida albicans* with MIC values of 32 and 64 μg/ml, respectively (miconazole was used as a reference standard; MIC = 8 μg/ml).

Reference

Jaki, B., Zerbe, O., Heilmann, J., and Sticher, O. (2001) *J. Nat. Prod.*, **64**, 154.

Tolypodiol

Physical data: $C_{28}H_{40}O_5$, white amorphous solid, $[\alpha]_D^{25}$ −2.9° (MeOH, *c* 0.002)

Structure:

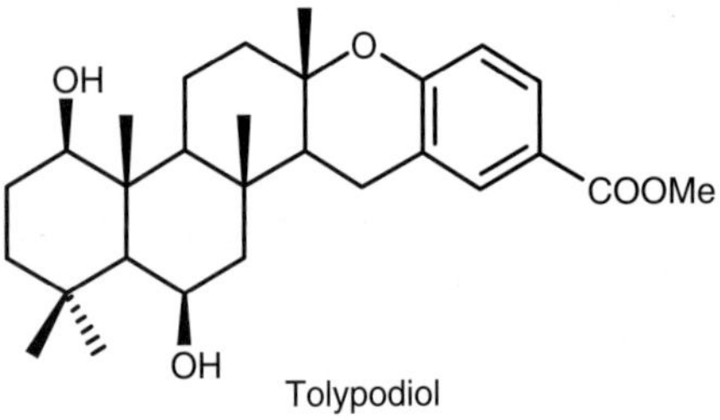

Tolypodiol

Compound class: Diterpenoid

Source: *Tolypothrix nodosa* (HT-58-2; terrestrial cyanobacterium)

Pharmaceutical potential: *Anti-inflammatory*
Tolypodiol was reported to possess potent anti-inflammatory activity in the mouse ear edema assay, with an ED_{50} value of 30 μg/ear – the value is comparable with those obtained for the standards, hydrocortisone (20 μg/ear) and manoalide (100 μg/ear) in the same assay.

Reference

Prinsep, M.R., Thomson, R.A., West, M.L., and Wylie, B.L. (1996) *J. Nat. Prod.*, **59**, 786.

Topopyrones A–D

Physical data:
- Topopyrone A: $C_{18}H_9O_7Cl$, orange powder, >280 °C
- Topopyrone B: $C_{18}H_9O_7Cl$, yellow powder, >280 °C
- Topopyrone C: $C_{18}H_{10}O_7$, orange powder, >280 °C
- Topopyrone D: $C_{18}H_{10}O_7$, orange powder, >280 °C

Structures:

Topopyrone A: R = Cl
Topopyrone C: R = H

Topopyrone B: R = Cl
Topopyrone D: R = H

Compound class: Antibiotics (pyranoanthraquinone derivatives)

Sources: Topopyrones A–D from *Phoma* sp. BAUA2861 (fungal strain); topopyrones C and D also from *Penicillium* sp. BAUA4206 (fungal strain) [1, 2]

Pharmaceutical potentials: *Topoisomerase I inhibitors (antiviral); antitumor; antimicrobial*
Topopyrones A–D were evaluated as promising human topoisomerase I inhibitors [1]; they were found to inhibit selectively recombinant yeast growth in human topoisomerase I inductive conditions with IC_{50} values of 1.22, 0.15, 4.88, and 19.63 ng/ml, respectively. The activity and selectivity (>66 000) of topopyrone B were comparable to those of camptothecin ($IC_{50} = 0.10$ ng/ml; selectivity $\geq$ 100 000), a specific inhibitor of topoisomerase I. The relaxation of supercoiled pBR322 DNA by human DNA topoisomerase I was inhibited by these compounds; however, they did not inhibit human DNA topoisomerase II. Topopyrone B exerted potent inhibitory activity against herpes virus, especially varicella zoster virus (VZV); it inhibited VZV growth with an EC_{50} value of 0.038 μg/ml, which is 24-fold stronger than that of acyclovir (0.9 μg/ml).

All the isolates showed *in vitro* antitumor activity against a series of tumor cell lines, but the activities were found to be weaker than that of camptothecin. The IC_{50} (μg/ml) values of the test compounds and reference standards, respectively, against HeLa, B16, Colon 26, 3LL, P388, L1210, Vero, and HEL 11-21 were determined to be 0.56, 1.5, 0.57, 0.29, 0.52, 0.5, 0.56, and 0.57 μg/ml (topopyrone A); 3.5, 4.1, 2.4, 2.2, 2.5, 2.0, 3.7, and 4.6 μg/ml (topopyrone B); 6.9, 7.5, 6.1, 5.0, 4.4, 4.0, 7.8, and 8.2 μg/ml (topopyrone C); >5, >5, >5, 2.5, 0.5, 0.47, >5, and >5 μg/ml (topopyrone D); 0.01, 0.18, 0.031, 0.031, 0.013, 0.016, 0.05, and 0.073 μg/ml (camptothecin); 0.96, 0.16, 0.031, 0.14, 1.96, 1.25, >2.5, and 0.59 μg/ml (adriamycin) [1].

Topopyrones A, B, and C were also found to possess strong inhibitory activity against Gram-positive bacteria including quinolone-resistant MRSA, while topopyrone D exhibited only weak inhibitory activity against Gram-positive bacteria. However, topopyrones showed no antimicrobial activity against Gram-negative bacteria, yeast, or fungi.

References

[1] Kanai, Y., Ishiyama, D., Senda, H., Iwatani, W., Takahashi, H., Konno, H., Tokumasu, S., and Kanazawa, S. (2000) *J. Antibiot.*, **53**, 863.
[2] Ishiyama, D., Kanai, Y., Senda, H., Iwatani, W., Takahashi, H., Konno, H., and Kanazawa, S. (2000) *J. Antibiot.*, **53**, 873.

Topostatin

Physical data: $C_{36}H_{58}N_4O_{11}S$, yellowish white powder, mp 179–189 °C (decomposition), $[\alpha]_D^{28}$ +18.3° (MeOH, c 0.1)

Structure:

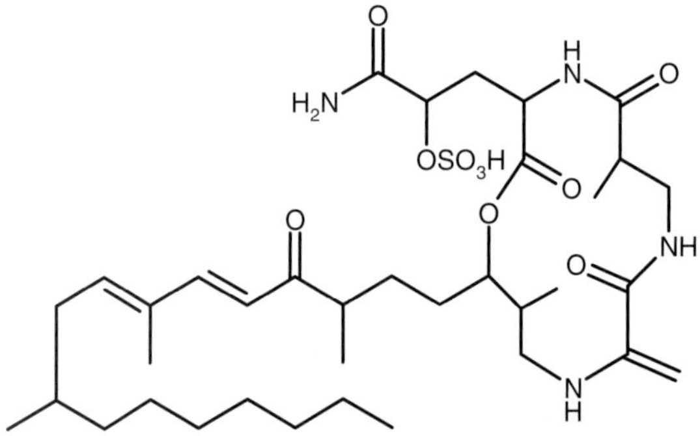

Topostatin

Compound class: Antibiotic

Source: *Thermomonospora alba* strain no. 1520 (culture broth) [1, 2]

Pharmaceutical potentials: *Topoisomerase I and II inhibitor; antitumor*
The isolate was evaluated to possess inhibitory efficacy against the activities of topoisomerase I and II enzymes with IC_{50} values of 13.0 and 3.0 ng/µl, respectively; in addition, the compound showed potent antitumor activity against SNB-75 and SNB-78 (tumor cells of central nervous system) with IC_{50} values of 0.4 and 7.0 µM, respectively [1].

References

[1] Suzuki, K., Nagao, K., Monnai, Y., Yagi, A., and Uyeda, M. (1998) *J. Antibiot.*, **51**, 991.
[2] Suzuki, K., Yahara, S., Kido, Y., Nagao, K., Hatano, Y., and Uyeda, M. (1998) *J. Antibiot.*, **51**, 999.

Torososide B

Systematic name: Physcion 8-O-β-D-glucopyranosyl-(1→6)-β-D-glucopyranosyl-(1→3)-β-D-glucopyranosyl-(1→6)-β-D-glucopyranoside

Physical data: $C_{40}H_{52}O_{25}$, yellow powder, mp 208–209 °C, $[\alpha]_D^{14}$ −71.3° (pyridine, c 0.3)

Structure:

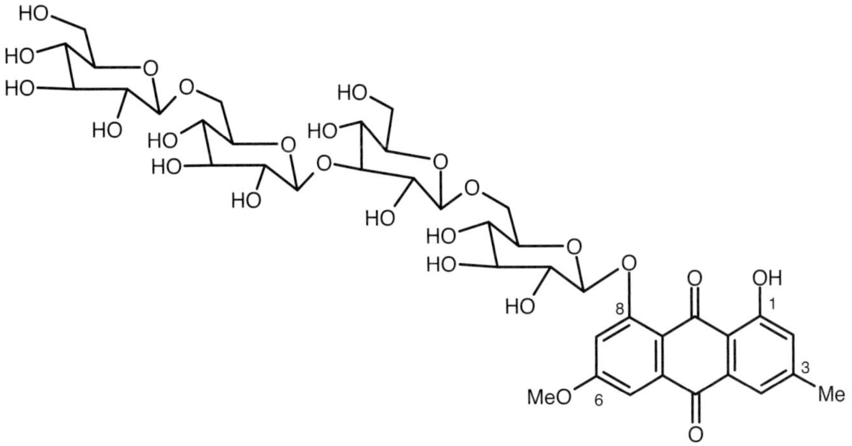

Torososide B

Compound class: Anthraquinone glycoside

Source: *Cassia torosa* Cav. (seeds; family: Leguminosae)

Pharmaceutical potential: *Inhibitor to leukotriene (LT) release (antiallergic)*
Torososide B was evaluated to possess significant antiallergic potential; it inhibited the release of leukotrienes such as LTB_4, LTC_4, LTD_4, and LTE_4 from rat peritoneal mast cells, stimulated by calcium ionophore A23187, by 46.9, 39.7, 41.2, and 43.3%, respectively, at a concentration of 10^{-4} M.

Reference

Kanno, M., Shibano, T., Takido, M., and Kitanaka, S. (1999) *Chem. Pharm. Bull.*, **47**, 915.

Torvanol A

Physical data: $C_{20}H_{20}O_{10}S$, amorphous powder, mp >300 °C, $[\alpha]_D^{29}$ −50° (H_2O, *c* 0.280)

Structure:

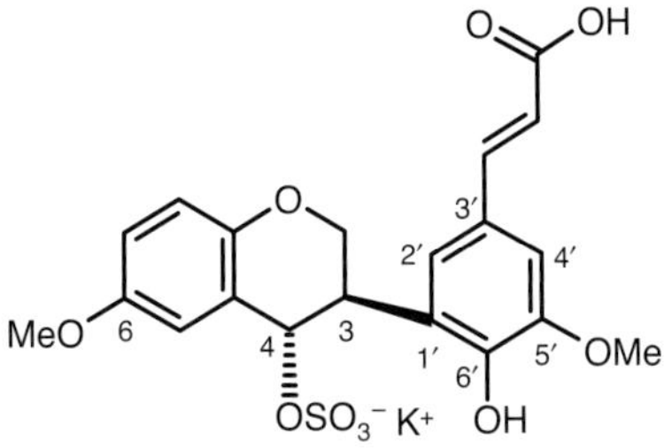

Torvanol A

Compound class: Isoflavonoid (sulfated)

Source: *Solanum torvum* (fruits; family: Solanaceae)

Pharmaceutical potential: *Antiviral*
The sulfated isoflavonoid was found to exhibit antiviral activity against herpes simplest virus type-1 (HSV-1) with an IC_{50} value of 9.6 µg/ml; however, the compound showed no cytotoxicity against BC, KB, and Vero cells up to a dose of 50 µg/ml.

Reference

Arthan, D., Svasti, J., Kittakoop, P., Pittayakhachonwut, D., Tanticharoen, M., and Thebtaranonth, Y. (2002) *Phytochemistry*, **59**, 459.

Torvoside H

Systematic name: (25*S*)-26-*O*-(β-D-Glucopyranosyl)-6α,26-dihydroxy-5α-spirostan-3-one 6-*O*-[α-L-rhamnopyranosyl-(1→3)-β-D-quinovopyranoside]

Physical data: $C_{45}H_{74}O_{18}$, amorphous powder, mp 170–172 °C, $[\alpha]_D^{29}$ −58.15° (MeOH, *c* 0.114)

Structure:

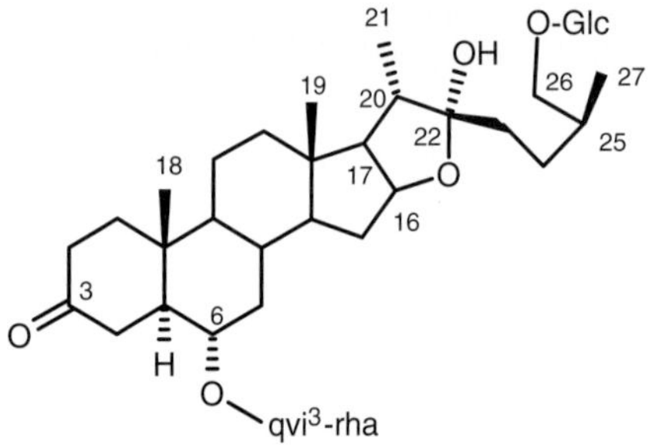

Torvoside H

Compound class: Steroidal glycoside

Source: *Solanum torvum* (fruits; family: Solanaceae)

Pharmaceutical potential: *Antiviral*
The isolate was found to exhibit antiviral activity against herpes simplex virus type-1 (HSV-1) with an IC_{50} value of 23.2 µg/ml; however, the compound showed no cytotoxicity against BC, KB, and Vero cells up to a dose of 50 µg/ml.

Reference

Arthan, D., Svasti, J., Kittakoop, P., Pittayakhachonwut, D., Tanticharoen, M., and Thebtaranonth, Y. (2002) *Phytochemistry*, **59**, 459.

Trachyspic acid

Physical data: $C_{20}H_{28}O_9$, white powder (hygroscopic), $[\alpha]_D^{25}$ +3.1° (MeOH, c 1.0)

Structure:

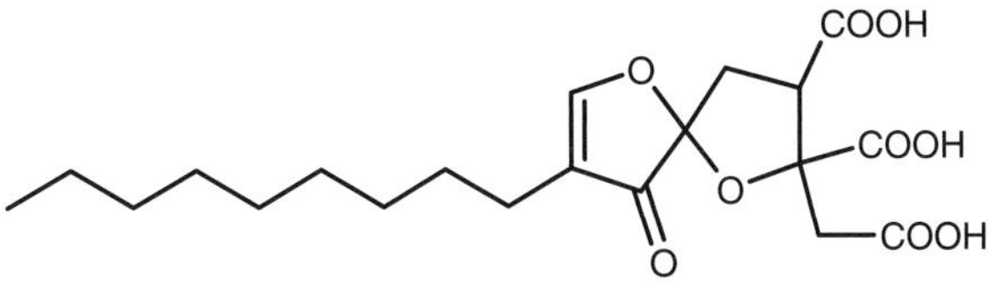

Trachyspic acid

Compound class: Antibiotic (tricarboxylic acid derivative)

Source: *Talaromyces trachyspermus* SANK12191 (culture broth)

Pharmaceutical potential: *Heparanase inhibitor*
The isolate was evaluated to possess inhibitory activity against heparanase with an IC_{50} value of 36 μM.

Reference

Shiozawa, H., Takahashi, M., Takatsu, T., Kinoshita, T., Tanzawa, K., Hosoya, T., Furuya, K., and Takahashi, S. (1995) *J. Antibiot.*, **48**, S 357

Trichodimerol (BMS-182123)

Physical data: $C_{28}H_{32}O_8$, pale yellow crystalline solid, $[\alpha]_D$ −376° (MeOH, c 0.26)

Structure:

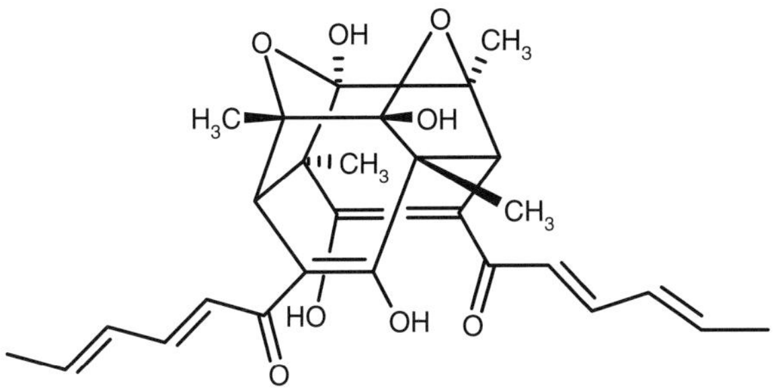

Trichodimerol (BMS-182123)

Compound class: Antibiotic

Sources: *Trichoderma longibrachiatum* Rifai aggr. (fungus) [1]; *Penicillium chrysogenum* strain V39673 (culture broth) [2]

Pharmaceutical potential: *Inhibitor to TNF-α production*
Trichodimerol (BMS-182123), the fungal metabolite, was found to inhibit bacterial endotoxin-induced production of tumor necrosis factor (TNF-α) in murine macrophages and human peripheral blood monocytes (*in vitro*); Warr et al. [2] determined IC_{50} values for the inhibition of lipopolysaccharide-induced TNF-α production in murine macrophages and human monocytes as 600 ng/ml and 4.0 µg/ml, respectively. The isolate suppressed the lipopolysaccharide-induced TNF-α promoter activity without affecting the stability of posttranscriptional mRNA [2].

References

[1] Andrade, R., Ayer, W.A., and Mebe, P.P. (1992) *Can. J. Chem.*, **70**, 2526.
[2] Warr, G.A., Veitch, J.A., Walsh, A.W., Hesler, G.A., Pirnik, D.M., Leet, J.E., Lin, P.-F.M., Medina, I.A., McBrien, K.D., Forenza, S., Clark, J.M., and Lam, K.S. (1996) *J. Antibiot.*, **49**, 234.

Trierixin

Physical data: $C_{378}H_{52}N_2O_8S$, pale pink powder, mp 121–122 °C, $[\alpha]_D^{20}$ +306.2° (CHCl$_3$, *c* 0.2)

Structure:

Trierixin

Compound class: Antibiotic (triene-ansamycin group)

Source: *Streptomyces* sp. AC 654 (fermentation broth) [1, 2]

Pharmaceutical potential: *Inhibitor of ER stress-induced XBP1 activation (antitumor)*
It has been reported that activated X-box-binding protein-1 (XBP1) induces the upregulation of molecular chaperons such as GRP78, PDI, ERdj4, or EDEM to diminish the accumulation of unfolded proteins [3–6]. The activation of XBP1 is considered to be greatly correlated with tumor growth, and an inhibitor of XBP1 activation would be a new class of antitumor drugs [3, 7, 8]. Tashiro et al. [1] evaluated the triene-ansamycin group isolate as a promising inhibitor of thapsigargin-induced XBP1-luciferase activation in HeLa/XBP1-luc cells and also in endogenous XBP1 splicing within HeLa cells with IC_{50} values of 14.0 and 19.0 ng/ml, respectively.

References

[1] Tashiro, E., Hironiwa, N., Kitagawa, M., Futamura, Y., Suzuki, S.-I., Nishio, M., and Imoto, M. (2007) *J. Antibiot.*, **60**, 547.
[2] Futamura, Y., Tashiro, E., Hironiwa, N., Kohno, J., Nishio, M., Nishio, M., and Imoto, M. (2007) *J. Antibiot.*, **60**, 582.
[3] Romero-Ramirez, L., Cao, H., Nelson, D., Hammond, E., Lee, A.H., Yoshida, H., Mori, K., Glimcher, L.H., Denko, N.C., Giaccia, A.J., Le, Q.T., and Koong, A.C. (2004) *Cancer Res.*, **64**, 5943.
[4] Kanemoto, S., Kondo, S., Ogata, M., Murakami, T., Urano, F., and Imaizumi, K. (2005) *Biochem. Biophys. Res. Commun.*, **331**, 1146.
[5] Yoshida, H., Matsui, T., Hosokawa, N., Kaufman, R.J., Nagata, K., and Mori, K. (2003) *Dev. Cell*, **4**, 265.
[6] Lee, A.H., Iwakoshi, N.N., and Glimcher, L.H. (2003) *Mol. Cell. Biol.*, **23**, 7448.
[7] Fujimoto, T., Onda, M., Nagai, H., Nagahata, T., Ogawa, K., and Emi, M. (2003) *Breast Cancer*, **10**, 301.
[8] Shuda, M., Kondoh, N., Imazeki, N., Tanaka, K., Okada, T., Mori, K., Hada, A., Arai, M., Wakatsuki, T., Matsubara, O., Yamamoto, N., and Yamamoto, M. (2003) *J. Hepatol.*, **38**, 605.

5,7,4′-Trihydroxy-6,8-diprenylisoflavone

Physical data: $C_{25}H_{26}O_5$, pale yellow amorphous powder

Structure:

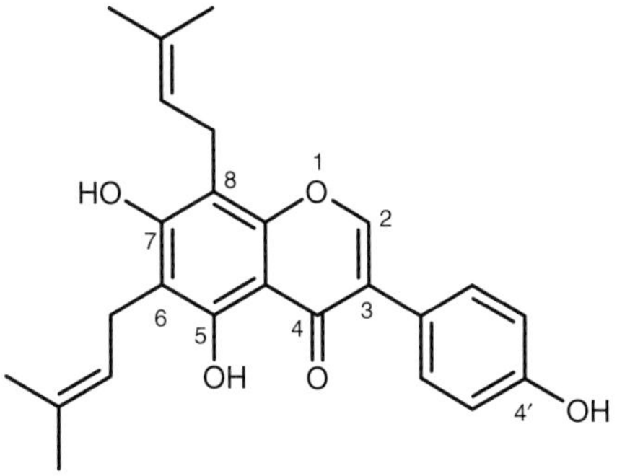

5,7,4′-Trihydroxy-6,8-diprenylisoflavone

Compound class: Flavonoid

Sources: *Derris scandens* (stems; family: Leguminosae) [1, 2]; *Glycyrrhiza uralensis* Fisch. ex DC. (roots; family: Fabaceae) [3]

Pharmaceutical potential: *Antibacterial*
The prenylated isoflavone was reported to have a potent growth inhibitory activity against the oral Gram-positive bacterium *Streptococcus mutans* with an MIC value of 2 µg/ml [3].

References

[1] Rao, M.N., Krupadanam, G.L.D., and Srimannarayana, G. (1994) *Phytochemistry*, **37**, 267.
[2] Sekine, T., Inagaki, M., Ikegami, F., Fujii, Y., and Ruangrungsi, N. (1999) *Phytochemistry*, **52**, 87.
[3] He, J., Chen, L., Heber, D., Shi, W., and Lu, Q.-Y. (2006) *J. Nat. Prod.*, **69**, 121.

5,7,3'-Trihydroxy-4'-methoxy-8,2'-di(3-methyl-2-butenyl)-(2S)-flavanone

Physical data: $C_{26}H_{30}O_6$, pale yellow amorphous powder, mp 162–164 °C, $[\alpha]_D^{27.5}$ −50.8° (CHCl$_3$, c 0.78)

Structure:

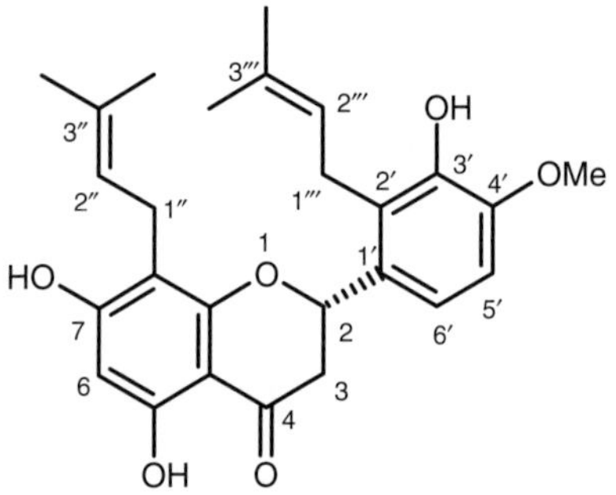

5,7,3'-Trihydroxy-4'-methoxy-8,2'-di(3-methyl-2-butenyl)-(2S)-flavanone

Compound class: Flavonoid (flavanone derivative)

Source: *Dendrolobium lanceolatum* (Dunn) Schindl. (roots; family: Leguminosae-Papilonoideae)

Pharmaceutical potentials: *Antimalarial; antimycobacterial; cytotoxic*
The flavanone derivative displayed significant antimalarial activity by inhibiting the growth of *Plasmodium falciparum* (K1, multidrug-resistant strain) with an IC$_{50}$ value of 3.3 µg/ml; it also showed moderate *in vitro* antimycobacterial activity against *Mycobacterium tuberculosis* H37Ra with an MIC value of 12.5 µg/ml. In addition, the test compound showed cytotoxicity against the NCI-H187 cancer cell line with an IC$_{50}$ value of 8.1 µg/ml.

Reference

Kanokmedhakul, S., Kanokmedhakul, K., Nambuddee, K., and Kongsaeree, P. (2004) *J. Nat. Prod.*, **67**, 968.

2,2',4'-Trihydroxy-6'-methoxy-3',5'-dimethylchalcone

Physical data: $C_{18}H_{18}O_5$, orange powder, mp 155–160 °C

Structure:

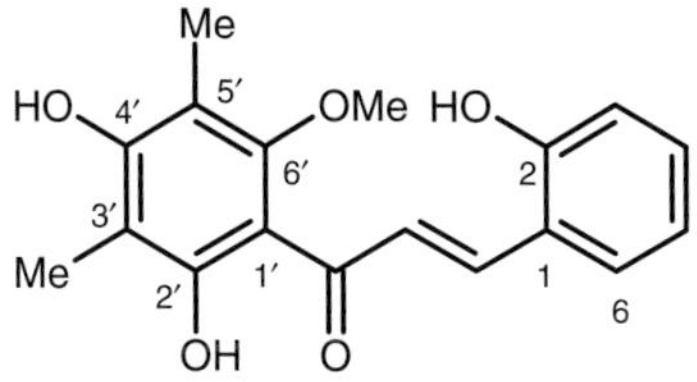

2,2',4'-Trihydroxy-6'-methoxy-3',5'-dimethylchalcone

Compound class: Flavonoid (chalcone derivative)

Source: *Psorothamnus polydenius* (S. Watson) Rydb. (twigs, leaves, and flowers; family: Fabaceae)

Pharmaceutical potential: *Antiprotozoal*
The chalcone isolate was evaluated for its *in vitro* antiprotozoal activity against *Leishmania donovani* and *Trypanosoma brucei*; the compound exhibited leishmanicidal as well as trypanocidal activity with IC$_{50}$ values of 7.5 and 6.8 μg/ml, respectively. Furthermore, treatment of *L. mexicana*-preinfected macrophages with the test compound (at 12.5 μg/ml) reduced the number of infected macrophages by at least 96% while posing no toxicity to the host cell.

Reference

Salem, M.M. and Werbovetz, K.A. (2005) *J. Nat. Prod.*, **68**, 108.

5,6,3'-Trihydroxy-7,8,4'-trimethoxyflavone

Physical data: $C_{18}H_{16}O_8$, yellow prisms, mp 217–219 °C

Structure:

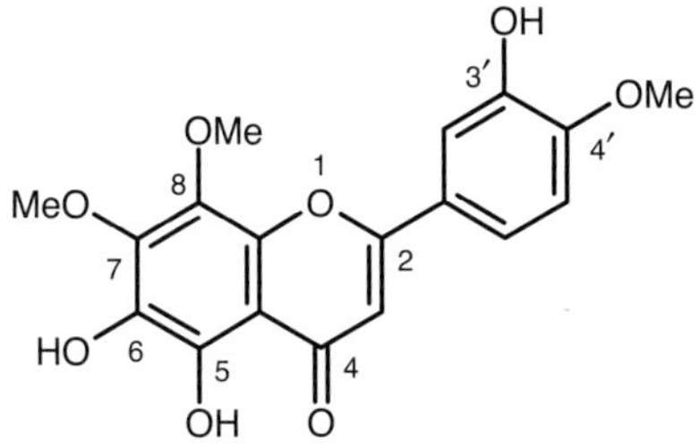

5,6,3'-Trihydroxy-7,8,4'-trimethoxyflavone

Compound class: Flavonoid

Source: *Crinum latifolium* L. (leaves; family: Amaryllidaceae)

Pharmaceutical potentials: *Antiangiogenic; cytotoxic*
The flavonoid was found to exhibit moderate inhibitory activity against the *in vitro* tube-like formation of human umbilical venous endothelial cells (HUVECs); the inhibition percentage (IP) was measured to be 39.2 ± 3.8 at a dose of 1 μg/ml (suramin was used as a positive control;

IP $= 75.3 \pm 4.6$ at 30 µg/ml). It was also found to display significant cytotoxicity against two tumor cell lines such as B16F10 (murine melanoma) and HCT116 (human colon carcinoma) with IC$_{50}$ values of 4.23 and 4.78 µg/ml, respectively (adriamycin was used as a positive control with IC$_{50}$ values of 0.09 and 0.11 µg/ml, respectively).

Reference

Nam, N.-H., Kim, Y., You, Y.-J., Hong, D.-H., Kim, H.-M., and Ahn, B.-Z. (2004) *Nat. Prod. Res.*, **18**, 485.

5,7,3′-Trihydroxy-4′,5′-(2′′′′′,2′′′′′-dimethylpyran)-8,2′-di(3-methyl-2-butenyl)-(2S)-flavanone

Physical data: $C_{30}H_{34}O_6$, pale yellow amorphous powder, mp 188–190 °C, $[\alpha]_D^{26.8}$ −67.7° (CHCl$_3$, c 0.68)

Structure:

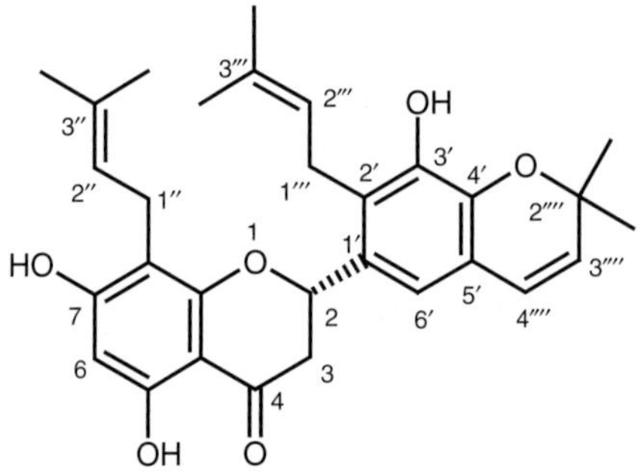

5,7,3′-Trihydroxy-4′,5′-(2′′′′′,2′′′′′-dimethylpyran)-8,2′-di(3-methyl-2-butenyl)-(2S)-flavanone

Compound class: Flavonoid (flavanone derivative)

Source: *Dendrolobium lanceolatum* (Dunn) Schindl. (roots; family: Leguminosae-Papilionoideae)

Pharmaceutical potentials: *Antimalarial; antimycobacterial; cytotoxic*
The flavanone derivative displayed significant antimalarial activity by inhibiting the growth of *Plasmodium falciparum* (K1, multidrug-resistant strain) with an IC$_{50}$ value of 2.6 µg/ml; it also showed moderate *in vitro* antimycobacterial activity against *Mycobacterium tuberculosis* H37Ra with an MIC value of 6.3 µg/ml. In addition, the test compound showed strong cytotoxicity against cancer cell lines KB, BC, and NCI-H187 with IC$_{50}$ values of 1.2, 1.7, and 0.6 µg/ml, respectively.

Reference

Kanokmedhakul, S., Kanokmedhakul, K., Nambuddee, K., and Kongsaeree, P. (2004) *J. Nat. Prod.*, **67**, 968.

1,2,5-Trihydroxyxanthone

Physical data: $C_{13}H_8O_5$, orange needles, mp 248–250 °C

Structure:

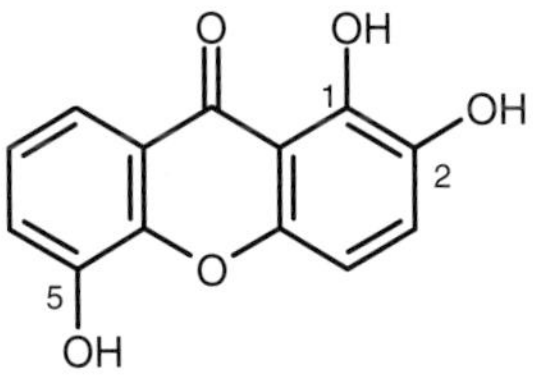

1,2,5-Trihydroxyxanthone

Compound class: Xanthone

Source: *Garcinia subelliptica* (woods; family: Guttiferae)

Pharmaceutical potential: *Antioxidant*
The trihydroxyxanthone displayed potent *in vitro* antioxidant potential as evaluated in lipid peroxidation as well as DPPH and superoxide anion scavenging assays. At a concentration of 10 μg/ml, the test compound inhibited lipid peroxidation by 40.5%, while it was found to scavenge DPPH radicals and superoxide anions (O_2^-) by 94.2 and 67.3%, respectively, at the same concentration (10 μg/ml).

Reference

Minami, H., Kinoshita, M., Fukuyama, Y., Kodama, M., Yoshizawa, T., Sugiura, M., Nakagawa, K., and Tago, H. (1994) *Phytochemistry*, **36**, 501.

Trikendiol

Physical data: $C_{38}H_{46}N_2O_4$, bright red crystal (acetone), mp 160–162 °C, $[\alpha]_D$ +102° (CHCl$_3$, *c* 0.02)

Structure:

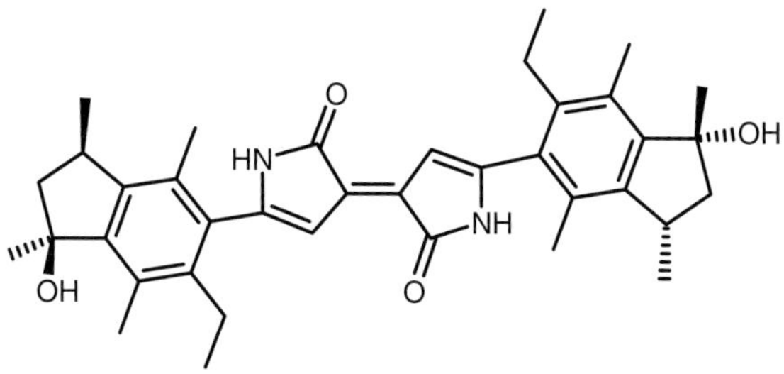

Trikendiol

Compound class: Alkaloid

Source: *Trikentrion loeve* Carter (sponge; Axinellidae; family: Euryponidae)

Pharmaceutical potential: *Anti-HIV*

Loukaci and Guyot established trikendiol as an active chemotype in a CEM-4 HIV-1 infection assay with an IC_{50} value of 2.0 µg/ml as measured by inhibition of cytopathogenic effect of the virus, but the isolate was found to cause no inhibition of HIV-1 aspartyl protease at a concentration of 10^{-5} M.

Reference

Loukaci, A. and Guyot, M. (1994) *Tetrahedron Lett.*, **35**, 6869.

Trilobacin

Physical data: $C_{37}H_{67}O_7$, colorless waxy gum

Structure:

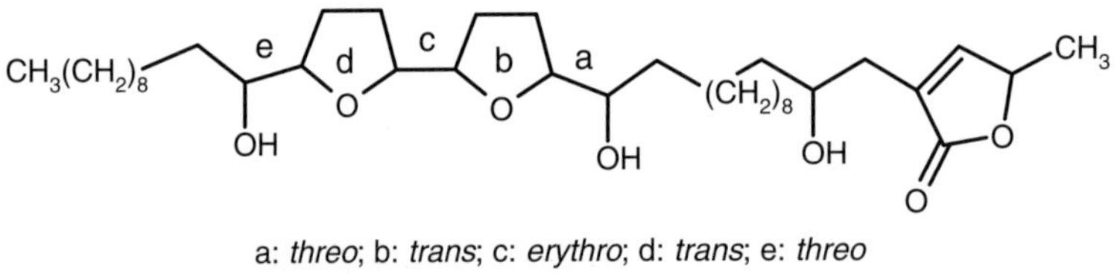

[Structure of Trilobacin: CH₃(CH₂)₈—CH(OH)—[e]—CH—[d: O ring]—CH—[c]—CH—[O ring]—CH—[b]—CH(OH)—[a]—CH₂—(CH₂)₈—CH(OH)—CH₂—[butenolide ring with =O and CH₃]]

a: *threo*; b: *trans*; c: *erythro*; d: *trans*; e: *threo*

Trilobacin

Compound class: Acetogenin derivative

Source: *Asimina triloba* (L.) Dunal (barks; family: Annonaceae)

Pharmaceutical potential: *Cytotoxic*

The compound displayed significant *in vitro* cytotoxicities against the human cancer cell lines such as A-549 (human lung carcinoma), MCF-7 (human breast carcinoma), and HT-29 (human colon carcinoma) with ED_{50} values of 8.02×10^{-4}, 0.329, and $<10 \times 10^{-15}$ µg/ml, respectively. In addition, the test compound was evaluated for its anticancerous potential against a panel of NCI human cancer cell lines, and it was found to exert potent activity in most of the cases.

Reference

Zhao, G., Hui, Y., Rupprecht, J.K., McLaughlin, J.L., and Wood, K.V. (1992) *J. Nat. Prod.*, **55**, 347.

Triphyophyllum naphthylisoquinoline alkaloids

Physical data:
- Dioncophylline A: $C_{24}H_{37}NO_3$, mp 214 °C, $[\alpha]_D^{20}$ −14.9° (CHCl₃, *c* 0.45)
- 7-*epi*-Dioncophylline A: $C_{24}H_{37}NO_3$
- Dioncophylline B: $C_{23}H_{25}NO_3$, needles, $[\alpha]_D^{20}$ −37.6° (CHCl₃, *c* 0.37)
- Dioncophylline C: $C_{23}H_{25}NO_3$, amorphous solid, mp 246 °C, $[\alpha]_D^{20}$ +19.2° (CHCl₃, *c* 0.52)
- Dioncopeltine A: $C_{23}H_{35}NO_4$, crystals, mp 233–234 °C, $[\alpha]_D$ −13.1° (CHCl₃, *c* 0.528)

Structures:

Dioncophylline A: *p*-isomer
7-*epi*-Dioncophylline A: *m*-isomer

Dioncopeltine A

Dioncophylline B

Dioncophylline C

Compound class: Naphthylisoquinoline alkaloids

Source: *Triphyophyllum peltatum* (roots and stem barks; family: Dioncophyllaceae) [1–9]

Pharmaceutical potential: *Antimalarial*

The naphthylisoquinoline alkaloids, isolated from *Triphyophyllum peltatum*, have been found to be quite promising antimalarial agents; the antiplasmodium potentials (IC_{50}) of the isolates against asexual blood stages of *Plasmodium falciparum* (NF 54, clone A1A9) were determined to be 0.014 µg/ml (dioncophylline C), 0.021 µg/ml (dioncopeltine A), 0.190 µg/ml (7-epidioncophylline A), and 0.224 µg/ml (dioncophylline B); the IC_{50} values obtained for these compounds are much lower than those observed for most other plant-derived compounds and compare well with the IC_{50} values for antiplasmodial drugs that are currently in use (chloroquine, IC_{50} = 0.005 µg/ml). Activities against those of *P. berghei* (Anka) were also noted to be effective. Apart from testing the activity of these alkaloids on the asexual erythrocytic *P. falciparum* and *P. berghei in vitro*, François et al. [12] also investigated the activity of a series of naphthylisoquinoline alkaloids on exoerythrocytic malaria parasites; they used *P. berghei*-infected human hepatoma cells (hHepG2) that were incubated with a culture medium containing 10 µg/ml of the test alkaloid. The most active agents were found to be dioncophylline A and dioncophyllacine A.

Owing to the promising *in vitro* antiplasmodium activity exhibited by dioncophylline C and dioncopeltine A, François et al. [10, 11] became motivated for their *in vivo* test against *P. berghei* in mice. Dioncopeltine A suppressed parasitemia almost totally, while dioncophylline C cured infected mice completely after oral treatment with 50 mg/kg body weight/day for 4 days without noticeable toxic effects. Analysis of the dose–response relationship of dioncophylline C revealed a 50% effective dosage (ED_{50}) of 10.71 mg/kg body weight/day under these conditions. The investigators observed that although four daily treatments with 50 mg/kg/day are needed to achieve radical cure, one oral dose is sufficient to kill 99.6% of the parasites. Intravenous application of dioncophylline C was found to be even more effective, with an ED_{50} of 1.90 mg/kg body weight/day without any noticeable toxic effects. Both dioncopeltine A and dioncophylline C are active against the chloroquine-resistant *P. berghei* Anka CRS parasites. Sustained release of these compounds at 20 mg/kg body weight/day by implanted miniosmotic pumps exhibited curative effects.

The naphthylisoquinoline alkaloids can thus be regarded as lead compounds for novel drugs acting against both erythrocytic and exoerythrocytic stages of *Plasmodium* [12]. An *in vivo* experiment pointed out that dioncophylline C, in particular, because of its pharmacokinetic properties as well as bioavailability in the bloodstream and low toxicity has to be considered a promising candidate for further development in the preclinical and clinical phases. The compound is also regarded as a promising lead for studies of structure–activity relationships. The free N- and 8-OH functions were shown to be prerequisites for the outstanding activity of such types of molecules against *P. falciparum*, with the presence of at least one free phenolic hydroxyl function appearing to be essential [11].

Very recently, Schwedhelm *et al.* [13] have presented a structural model of the complex formed between the novel antimalarial compound, dioncophylline C, and its presumed target ferriprotoporphyrin IX heme (FPIX) in heme solution; the conformation of dioncophylline C is sterically stabilized by a water molecule coordinated to iron in FPIX. The derived structural model shows high similarity to complexes formed by FPIX and antimalarials of the quinoline family (e.g., chloroquine, quinine, quinidine, and amodiaquine); hence, this structural feature may provide an important hint at possibilities for a further optimization of novel naphthylisoquinoline alkaloid antimalarial drugs [13].

References

[1] Bringmann, G., Rübenacker, M., Weirich, R., and Aké Assi, L. (1992) *Phytochemistry*, **31**, 4019.

[2] Frosch, T., Schmitt, M., Schenzel, K., Faber, J.H., Bringmann, G., Kiefer, W., and Popp, J. (2006) *Biopolymers*, **82**, 295.

[3] Bringmann, G., Rübenacker, M., Jansen, J.R., and Scheutzow, D. (1990) *Tetrahedron Lett.*, **31**, 639.

[4] Bringmann, G., Rübenacker, M., Geuder, T., and Aké Assi, L. (1991) *Phytochemistry*, **30**, 3845.

[5] Bringmann, G., Rübenacker, M., Vogt, P., Busse, H., Aké Assi, L., Peters, K., and von Schnering, H.G. (1991) *Phytochemistry*, **30**, 1691.

[6] François, G., Bringmann, G., Dochez, C., Schneider, C., Timperman, G., and Aké Assi, L. (1995) *J. Ethnopharmacol.*, **46**, 115.

[7] François, G., Bringmann, G., Phillipson, J.D., Aké Assi, L., Dochez, C., Rübenacker, M., Schneider, C., Wery, M., Warhurst, D.C., and Kirby, G.C. (1984) *Phytochemistry*, **35**, 1461.

[8] Bringmann, G., Saeb, W., God, R., Schaffer, M., François, G., Peters, K., Peters, E.-M., Proksch, P., Hostettmann, K., and Aké Assi, L. (1998) *Phytochemistry*, **49**, 1667.

[9] Bringmann, G., Günther, C., Saeb, W., Mies, J., Brun, R., and Aké Assi, L. (2000) *Phytochemistry*, **54**, 337.
[10] François, G., Timperman, G., Eling, W., Aké Assi, L., Holenz, J., and Bringmann, G. (1997) *Antimicrob. Agents Chemother.*, **41**, 2533–2539.
[11] François, G., Timperman, G., Holenz, J., Assi, L.A., Geuder, T., Maes, L., Dubois, J., Hanocq, M., and Bringmann, G. (1996) *Ann. Trop. Med. Parasitol.*, **90**, 115–123.
[12] François, G., Timperman, G., Steenackers, T., Aké Assi, L., Holenz, J., and Bringmann, G. (1997) *Parasitol. Res.*, **83**, 673.
[13] Schwedhelm, K.F., Horstmann, M., Faber, J.H., Reichert, Y., Bringmann, G., and Faber, C. (2007) *ChemMedChem*, **2**, 541.

Tripterifordin

Physical data: $C_{20}H_{30}O_3$, white needles (acetone), mp 255–256 °C, $[\alpha]_D$ −46.6° (CHCl$_3$, c 0.93)

Structure:

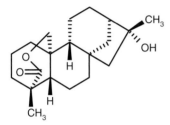

Tripterifordin

Compound class: Diterpenoid (kaurane-type diterpene lactone)

Source: *Tripterygium wilfordii* Hook. (roots; family: Celastraceae)

Pharmaceutical potential: *Anti-HIV*
Tripterifordin was found to exhibit moderate inhibitory activity against HIV replication in H9 lymphocyte cells with an EC$_{50}$ value of 1.0 µg/ml (6.0 µM); it did not inhibit uninfected H9 cell growth at 15 µM.

Reference

Chen, K., Shi, Q., Fujioka, T., Zhang, D.-C., Hu, C.-Q., Jin, J.-Q., Kilkuskie, R.E., and Lee, K.-H. (1992) *J. Nat. Prod.*, **55**, 88.

Triptonines A and B

Physical data:
- Triptonine A: $C_{45}H_{55}NO_{21}$, colorless needles, mp 284.0–285.5 °C, $[\alpha]_D^{25}$ −24.1° (MeOH, c 1.0)
- Triptonine B: $C_{45}H_{55}NO_{22}$, amorphous powder, $[\alpha]_D^{25}$ +15.5° (MeOH, c 0.6)

Structures:

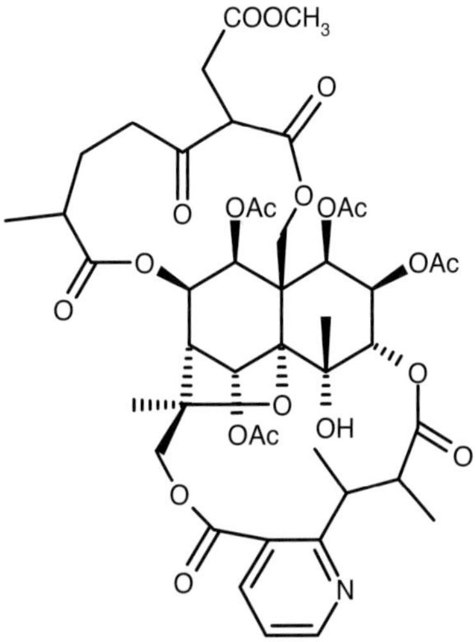

Triptonine A

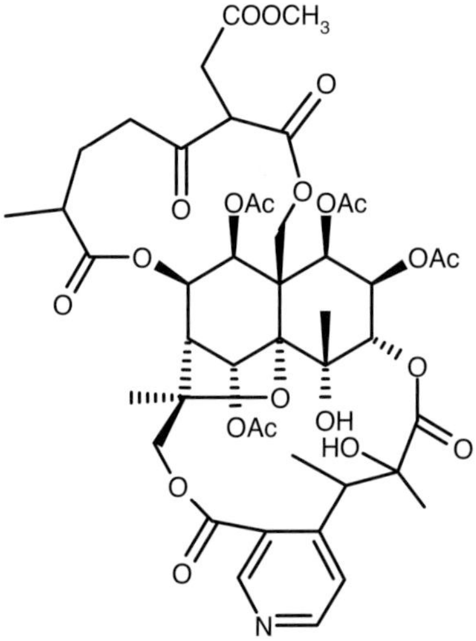

Triptonine B

Compound class: Sesquiterpene pyridine alkaloids

Source: *Tripterygium hypoglaucum* (Levl.) Hutch. (root barks; family: Celastraceae)

Pharmaceutical potential: *Anti-HIV*

Both the sesquiterpene alkaloids were found to exhibit potent anti-HIV activity by inhibiting HIV replication in H9 lymphocytes; triptonine B demonstrated more potent anti-HIV activity with an EC_{50} value of <0.10 µg/ml and an *in vitro* therapeutic index (TI) of >1000. An EC_{50} value of 2.54 µg/ml (with a TI of 39.4) was determined for triptonine A.

References

[1] Duan, H., Takaishi, Y., Imakura, Y., Jia, Y., Li, D., Cosentino, L.M., and Lee, K.-H. (2000) *J. Nat. Prod.*, **63**, 357.
[2] Duan, H., Takaishi, Y., Bando, M., Kido, M., Imakura, Y., and Lee, K.H. (1999) *Tetrahedron Lett.*, **40**, 2969.

Tsugarioside C

Systematic name: 3α-Acetoxy-(Z)-24-methyl-5α-lanosta-8,23,25-trien-21-oic acid ester β-D-xyloside

Physical data: $C_{38}H_{57}O_8$, colorless needles ($CHCl_3$), mp 181–183 °C, $[\alpha]_D^{27}$ +10° ($CHCl_3$, c 0.1)

Structure:

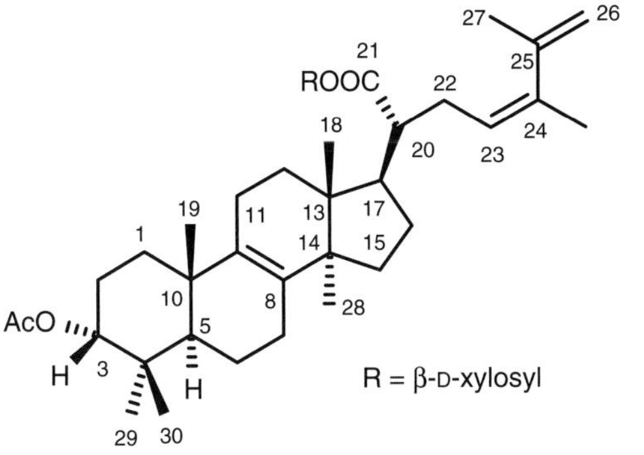

Tsugarioside C

Compound class: Triterpene (lanostanoid)

Source: *Ganoderma tsugae* Murr. (fruiting bodies; family: Polyporaceae)

Pharmaceutical potential: *Cytotoxic*

Tsugarioside C exhibited *in vitro* cytotoxic activity against human hepatoma (PLC/PRF/5 and T-24 cells) and human cervical carcinoma (HT-3 and SiHa cells) with ED_{50} values of 6.5, 8.6, 7.2, and 9.5 μg/ml, respectively.

Reference

Su, H.-J., Fann, Y.-F., Chung, M.-I., Won, S.-J., and Lin, C.-N. (2000) *J. Nat. Prod.*, **63**, 514.

Tubelactomicin A

Physical data: $C_{29}H_{42}O_6$, white powder, $[\alpha]_D^{25}$ +130° (MeOH, *c* 0.64)

Structure:

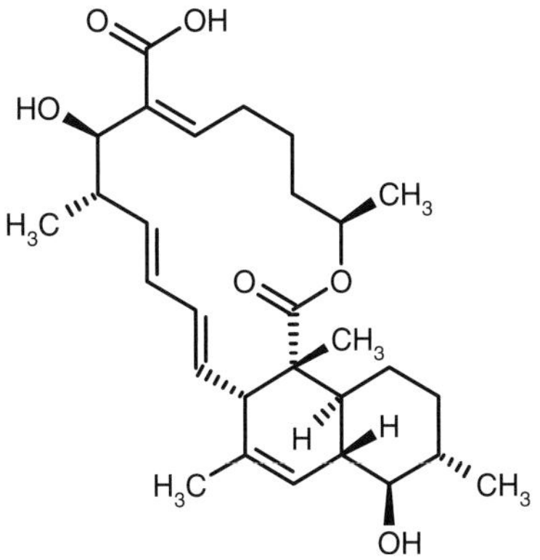

Tubelactomicin A

Compound class: Antibiotic (macrolide)

Source: *Nocardia* sp. MK703-102F1 (culture broth) [1, 2]

Pharmaceutical potential: *Antitubercular*
Tubelactomicin A exhibited strong and specific antimicrobial activities against rapid growers *Mycobacterium* including drug-resistant strains except for a strain that was viomycin resistant (with an MIC value of 50 µg/ml against *M. smegmatis* ATCC 607 VM-R) [1]. The MIC values against *M. smegmatis* ATCC607, *M. smegmatis* ATCC 607 PM-R, *M. smegmatis* ATCC 607 KM-R, *M. smegmatis* ATCC 607 SM-R, *M. smegmatis* ATCC 607 RFP-R, *M. phlei*, *M. vaccae*, and *M. fortuitum* were determined to be 0.10, 0.10, 0.10, 0.10, 0.10, 0.20, 0.10, and 0.78 µg/ml, respectively [2].

References

[1] Igarashi, M., Hayashi, C., Homma, Y., Hattori, S., Kinoshita, N., Hamada, M., and Takeuchi, T. (2000) *J. Antibiot.*, **53**, 1096.
[2] Igarashi, M., Nakamura, H., Naganawa, H., and Takeuchi, T. (2000) *J. Antibiot.*, **53**, 1102.

Tubocapsenolide A

See Withanolides

Tungtungmadic acid

Systematic name: 3-Caffeoyl-4-dihydrocaffeoyl quinic acid

Physical data: $C_{25}H_{26}O_{12}$

Structure:

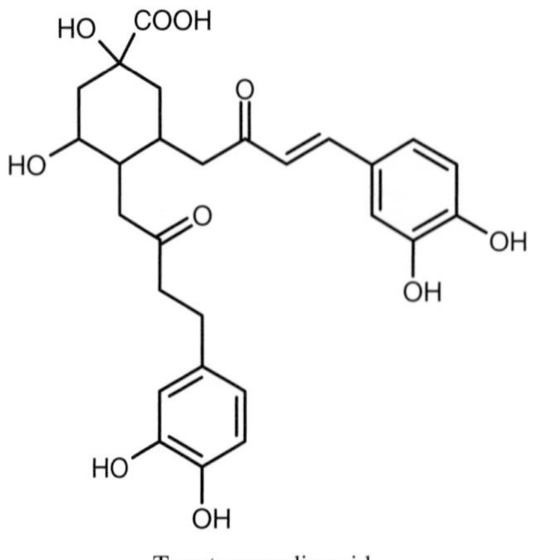

Tungtungmadic acid

Compound class: Chlorogenic acid derivative

Source: *Salicornia herbacea* (whole plants; family: Chenopodiaceae)

Pharmaceutical potential: *Antioxidant*

Tungtungmadic acid was found to exhibit strong free radical scavenging activity and to inhibit lipid peroxidation as evaluated by using various antioxidant assays. In the DPPH assay, the compound showed potent free radical scavenging activity with an IC_{50} value of 5.1 ± 0.37 µM (α-tocopherol was used as a positive control; $IC_{50} = 49.7 \pm 0.27$ µM). It inhibited iron-induced liver microsomal lipid peroxidation (MDA assay) with an IC_{50} value of 9.3 ± 0.43 µM (α-tocopherol was used as a positive control; $IC_{50} = 41.2 \pm 0.37$ µM); in addition, the isolate was found to be effective in protecting the plasmid DNA against strand breakage induced by hydroxyl radicals.

Reference

Chung, Y.C., Chun, H.K., Yang, J.Y., Kim, J.Y., Han, E.H., Kho, Y.H., and Jeong, H.G. (2005) *Arch. Pharm. Res.*, **28**, 1122.

U

UCS1025 A

Physical data: $C_{20}H_{25}NO_5$, colorless needle-shaped crystal, mp 135–137 °C, $[\alpha]_D^{28}$ + 30.1° (MeOH, *c* 0.1)

Structure:

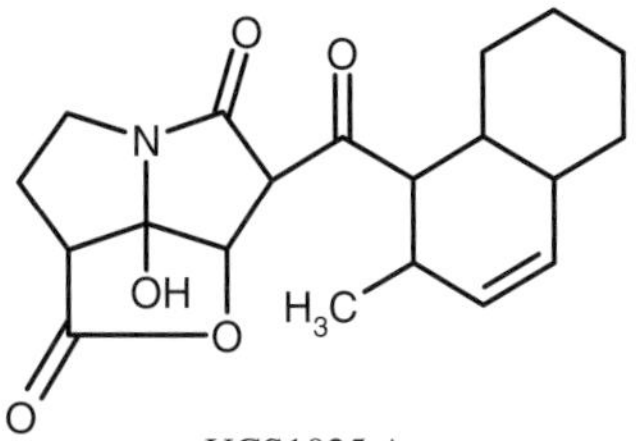

UCS1025 A

Compound class: Antibiotic

Source: *Acremonium* sp. KY4917 (culture broth)

Pharmaceutical potentials: ***Antibacterial; antitumor***
The isolate was found to exhibit significant antimicrobial activity against Gram-positive bacteria, *Staphylococcus aureus, Bacillus subtilis,* and *Enterococcus hirae,* and Gram-negative bacterium, *Proteus vulgaris,* with MIC values of 1.3, 1.3, 1.3, and 5.2 μg/ml, respectively. The antibiotic also showed moderate antiproliferative activity against human tumor cell lines (such as ACHN, A431, MCF-7, and T24) with IC_{50} values ranging from 21 to 58 μM.

Reference

Nakai, R., Ogawa, H., Asai, A., Ando, K., Agatsuma1, T., Matsumiya, S., Akinaga, S., Yamashita, Y., and Mizukami, T. (2000) *J. Antibiot.*, **53**, 294.

UK-2A and B

Physical data:

- UK-2A: $C_{26}H_{31}N_2O_9$, colorless needles, mp 207–209 °C, $[\alpha]_D^{23}$ + 89.11° ($CHCl_3$, *c* 0.8)
- UK-2B: $C_{27}H_{31}N_2O_9$, colorless needles, mp 867–890 °C, $[\alpha]_D^{17}$ + 87.5° ($CHCl_3$, *c* 0.3)

Handbook of Pharmaceutical Natural Products, Volume 2.
Goutam Brahmachari

ISBN: 978-3-527-32148-3

UK-3A

Structures:

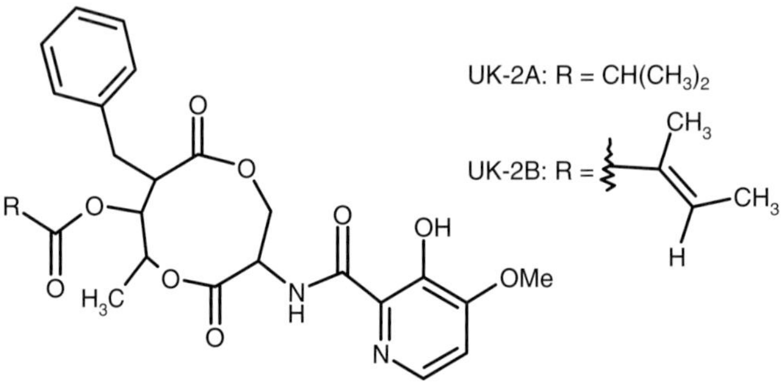

UK-2A: R = CH(CH$_3$)$_2$

UK-2B: R = [2-methyl-2-butenyl group]

Compound class: Antibiotics

Source: *Streptomyces* sp. 517-02 (mycelial cake) [1, 2]

Pharmaceutical potential: *Antifungal*

UK-2A and B were found to display potent *in vitro* antifungal activity against a broad spectrum of test organisms (antimycin A was used as a reference standard) [1]; the respective MIC values against the microbial strains, namely, *Saccharomyces cerevisiae* IFO 0203, *Candida albicans* IFO 1061, *Rhodotorula rubra* IFO 0001, *Schizosaccharomyces pombe* IFO 0342, *Hansenula anomala* IFO 0136, *Torulaspora delbrueckii* DSM 70504, *Torulaspora delbrueckii* DSM 70504, *Aspergillus niger* ATCC6275, *Mucor mucedo* IFO 7684, *Neurospora sitophila* DSM1130, *Penicillium chrysogenum* IFO 4626, *Phycomyces nitens* IFO 5694, *Rhizopus formosaensis* IFO 4732, *Sclerotinia sclerotiorum* IFO 5292, *Thamnidium elegans* IFO 6152, and *Trichophyton mentagrophytes* IFO 6124 were determined to be 0.05, 0.39, 0.78, 0.1, 3.13, 1.56, 0.0125, 0.39, 0.025, 0.1, 0.39, 0.025, 0.0125, 0.05, 0.00156, and 0.1 µg/ml (UK-2A); 0.1, 0.78, 0.78, 0.1, 0.78, 3.13, 0.39, 1.56, 0.0125, 0.1, 0.39, 0.1, 0.0125, 0.1, 0.00156, and 0.1 µg/ml (UK-2B); 0.025, 0.1, 1.56, 0.025, 1.56, 0.05, 25.0, 0.39, 0.00625, 0.2, 0.39, 0.1, >100, 0.05, 0.00313, and 0.1 µg/ml (antimycin A) [1].

References

[1] Ueki, M., Abe, K., Kusumoto, A., Hanafi, M., Shibata, K., Tanaka, T., and Taniguchi, M. (1996) *J. Antibiot.*, **49**, 639.
[2] Hanafi, M., Shibata, K., Ueki, M., and Taniguchi, M. (1996) *J. Antibiot.*, **49**, 1226.

UK-3A

Physical data: C$_{25}$H$_{28}$N$_2$O$_8$, colorless needles

Structure:

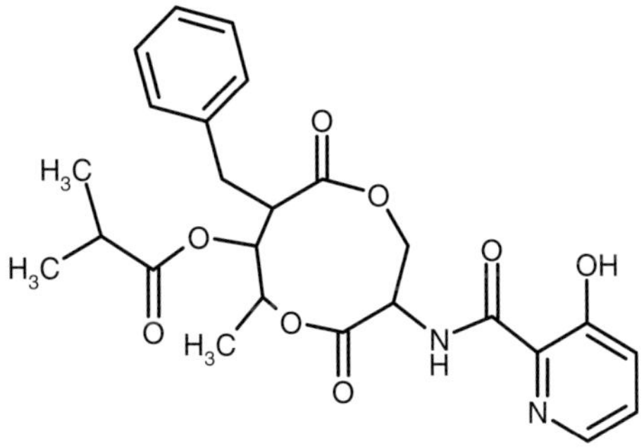

UK-3A

Compound class: Antibiotic

Source: *Streptomyces* sp. 517-02 (mycelial cake)

Pharmaceutical potential: *Antifungal*
Antifungal activity of UK-3A was found to be broad against a variety of test organisms; the antibiotic showed significant *in vitro* inhibitory activity (antimycin A was used as a reference standard) against *Saccharomyces cerevisiae* IFO 0203, *Candida albicans* IFO 1061, *Rhodotorula rubra* IFO 0001, *Schizosaccharomyces pombe* IFO 0342, *Hansenula anomala* IFO 0136, *Aspergillus niger* ATCC6275, *Mucor mucedo* IFO 7684, *Neurospora sitophila* DSM1130, *Penicillium chrysogenum* IFO 4626, *Phycomyces nitens* IFO 5694, and *Rhizopus formosaensis* IFO 4732 with respective MIC values of 1.56, 3.13, 3.13, 1.56, 6.25, 0.78, 0.78, 0.39, 1.56, 0.39, and 0.39 µg/ml, while the respective MIC values for antimycin A were determined to be 0.025, 0.1, 1.56, 0.025, 1.56, 0.39, 0.00625, 0.2, 0.39, 0.1, and >100 µg/ml.

Reference

Ueki, M., Kusumoto, A., Hanafi, M., Shibata, K., Tanaka, T., and Taniguchi, M. (1997) *J. Antibiot.*, **50**, 551.

Ungeremine

Physical data: $C_{16}H_{11}NO_3$

Structure:

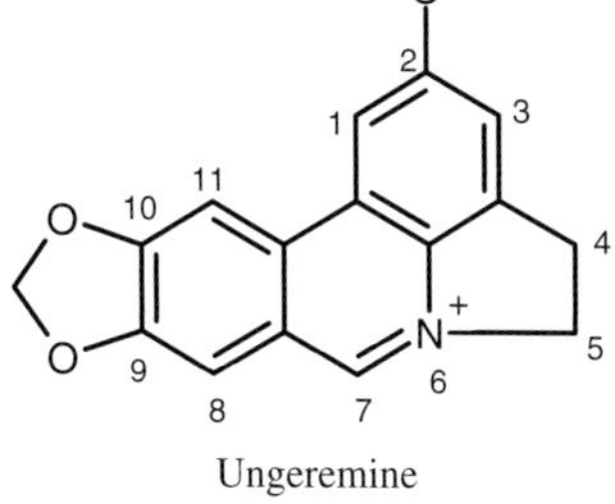

Ungeremine

Compound class: Alkaloid

Sources: *Ungernia minor* (family: Amaryllidaceae) [1], *U. spiralis* [2]; *Crinum asiaticum* (family: Amaryllidaceae) [3], *C. augustum* [4]; *Pancratium maritimum* (family: Amaryllidaceae) [5]; *Hippeastrum solandriflorum* (family: Amaryllidaceae) [6]; *Zephyranthes flava* (family: Amaryllidaceae) [7]; *Nerine bowdenii* Will. Watson (family: Amaryllidaceae) [8]

Pharmaceutical potentials: *Acetylcholinesterase (AChE) inhibitor; cytotoxic; hypotensive*
The isolate exhibited strong *in vitro* inhibitory activity against acetylcholinesterase (AChE) enzyme with an IC$_{50}$ value of 0.35 µM; this alkaloid was found to be about 6–10 times stronger than galantamine (positive control; IC$_{50}$ = 2.2 µM) [1]. Inhibition of acetylcholinesterase (AChE) is a major treatment for Alzheimer's disease. In addition, the test compound was reported to have growth inhibitory and cytotoxic effect [9] and hypotensive properties [10]. The structure–activity relationships concerning its anticancer activity were also studied [11].

References

[1] Normatov, M., Abduazimov, K.A., and Yunusov, S.Y. (1965) *Uzb. Khim. Zh.*, **9**, 25; *Chem. Abstr.*, 1965, **63**, 7061f.
[2] Allayarov, K.H.B., Abdusamatov, A., and Yunusov, S.Y. (1970) *Khim. Prir. Soedin.*, **6**, 143; *Chem. Abstr.*, 1970, **73**, 69771.
[3] Ghosal, S., Kumar, Y., Singh, S.K., and Kumar, A. (1986), *J. Chem. Res. (S)*, **1**, 112.
[4] Ramadan, M.A. (1998) *Bull. Pharm. Sci. Assiut Univ.*, **21**, 97.
[5] Abou-Donia, A.H., Abib, A.A., El Din, A.S., Evidente, A., Gaber, M., and Scopa, A. (1992) *Phytochemistry*, **31**, 2139.
[6] Bastida, J., Codina, C., Porras, C.L., and Paiz, L. (1996) *Planta Med.*, **62**, 74.
[7] Ghosal, S., Singh, S.K., and Srivastava, R.S. (1978) *Phytochemistry*, **25**, 1975.
[8] Rhee, I.K., Appels, N., Hofte, B., Karabatak, B., Erkelens, C., Stark, L.M., Flippin, L.A., and Verpoorte, R. (2004) *Biol. Pharm. Bull.*, **27**, 1804.
[9] Ghosal, S., Singh, S.K., Kumar, Y., Unnikrishnan, S., and Chattopadhyay, S. (1988) *Planta Med.*, **54**, 114.
[10] Zakirov, U.B. (1967) *Med. Zh. Uzb.*, **9**, 42; *Chem. Abstr.*, 1967, **68**, 113264.
[11] He, H.M. and Weng, Z.Y. (1989) *Yaoxue Xuebao*, **24**, 302; *Chem. Abstr.*, 1989, **111**, 108492.

(22E,24R)-3α-Ureido-ergosta-4,6,8(14),22-tetraene

Physical data: $C_{29}H_{44}N_2O$, amorphous powder, $[\alpha]_D^{25}$ +364.5° (MeOH, *c* 2.0)

Structure:

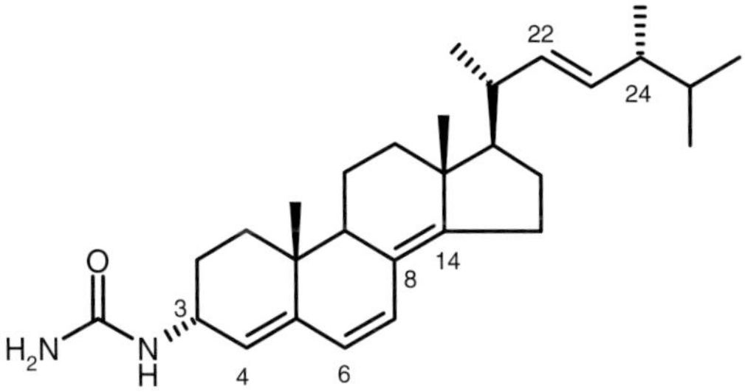

(22E,24R)-3α-Ureido-ergosta-4,6,8(14),22-tetraene

Compound class: Steroid

Source: *Chlorophyllum molybdites* (fruits; family: Agaricaceae)

Pharmaceutical potential: *Cytotoxic*
The steroidal derivative showed moderate cytotoxic activity against human stomach cancer cells Kato-III with an IC_{50} value of 7.60 µg/ml (hinokitiol was used as a positive control with an IC_{50} value of 0.6 µg/ml).

Reference

Yoshikawa, K., Ikuta, M., Arihara, S., Matsumura, E., and Katayama, S. (2001) *Chem. Pharm. Bull.*, **49**, 1030.

Urceolatin

Systematic name: 6-Bromo-1-(3-bromo-4,5-dihydroxybenzyl)phenanthro[4,5-*bcd*]furan-2,3,5-triol

Physical data: $C_{21}H_{12}Br_2O_6$, red amorphous powder

Structure:

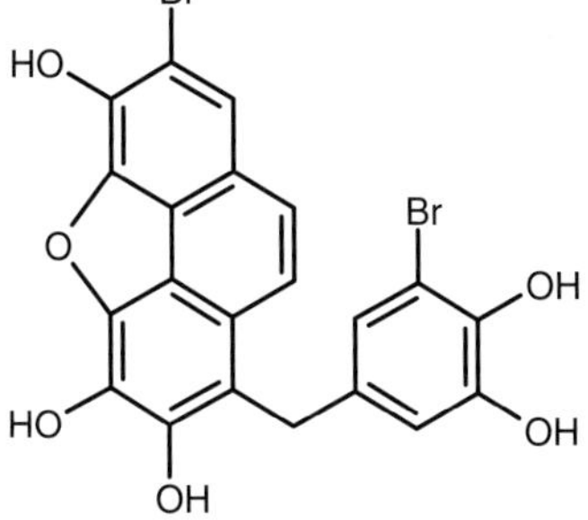

Urceolatin

Compound class: Bromophenol derivative

Source: *Polysiphonia urceolata* (marine red alga; family: Rhodomelaceae)

Pharmaceutical potential: *Antioxidant (free radical scavenging)*
The isolate was evaluated to possess potent ability to scavenge DPPH free radicals with an IC_{50} value of 7.9 µM; the potency was found to be about 10-fold more than that of the positive control, butylated hydroxytoluene (BHT; IC_{50} = 83.8 µM).

Reference

Li, K., Li, X.-M., Ji, N.-Y., Gloer, J.B., and Wang, B.-G. (2008) *Org. Lett.*, **10**, 1429.

Uvaretin

Systematic name: 2′,4′-Dihydroxy-3′-[(2-hydroxyphenyl)methyl]-6′-methoxy-7,8-dihydrochalcone; alternatively, l-[2,4-dihydroxy-3-(2-hydroxybenzyl)-6-methoxyphenyl]-3-phenyl-lpropanone

Physical data: $C_{23}H_{22}O_5$, colorless crystals ($CHCl_3$), mp 167–169 °C

Structure:

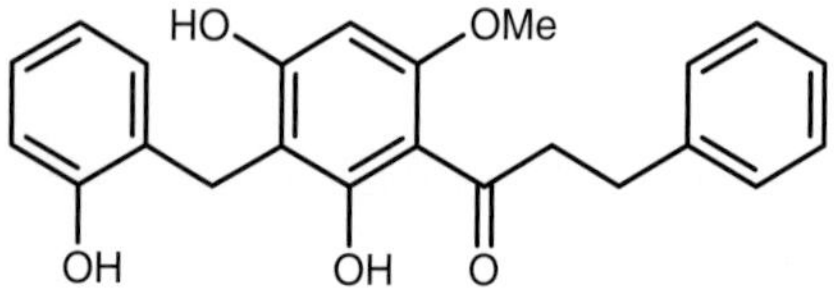

Uvaretin

Compound class: *C*-Benzylated dihydrochalcone

Sources: *Uvaria chamae* Oliv. (roots; family: Annonaceae) [1, 2]; *U. leptocladon* (root barks) [3]; *U. acuminata* (roots; family: Annonaceae) [4]

Pharmaceutical potential: *Cytotoxic (antitumor)*
The compound displayed potent *in vitro* cytotoxicity against human promyelocytic leukemia HL-60 cells with an IC_{50} value of 9.3 µM [4]. It was also reported to demonstrate inhibitory efficacy toward the P-388 (3PS) lymphocytic leukemia test system with a value of 133% test/control (T/C) at 10 mg/kg [1].

References

[1] Cole, J.R., Torrance, S.J., and Wiedhopf, R.M. (1976) *J. Org. Chem.*, **41**, 1852.
[2] Lasswell, W.L., Jr. and Hufford, C.D. (1977) *J. Org. Chem.*, **42**, 1295.
[3] Nkunya, M.H.H., Weenen, H., Renner, C., Waibel, R., and Achenbach, H. (1993) *Phytochemistry*, **32**, 1297.
[4] Ichimaru, M., Nakatani, N., Takahashi, T., Nishiyama, Y., Moriyasu, M., Kato, A., Mathenge, S.G., Juma, F.D., and Nganga, J.N. (2004) *Chem. Pharm. Bull.*, **52**, 138.

V

Vaccihein A

Systematic name: Methyl 2-(3,5-dimethoxy-4-hydroxybenzoyloxy)-4,6-dihydroxyphenyl acetate

Physical data: $C_{18}H_{18}O_9$, white powder

Structure:

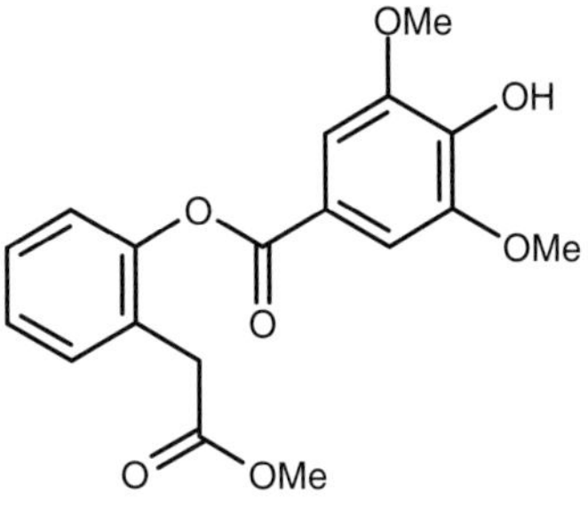

Vaccihein A

Compound class: *ortho*-Benzoyloxyphenyl acetic acid ester

Source: *Vaccinium ashei* (Rabbiteye blueberry, fruits; family: Ericaceae)

Pharmaceutical potential: ***Antioxidant***
The isolate showed moderate antioxidant activity in comparison to α-tocopherol and BHA as evaluated using linoleic acid as the substrate in the ferric thiocyanate method; the compound also displayed free radical scavenging activity against DPPH free radicals more potently than L-cysteine at a concentration of 0.20 mM.

Reference

[1] Ono, M., Matsuoka, C., Koto, M., Tateishi, M., Komatsu, H., Kobayashi, H., Igoshi, K., Ito, Y., Okawa, M., and Nohara, T. (2002) *Chem. Pharm. Bull.*, **50**, 1416.

Valinomycin

Physical data: $C_{54}H_{90}N_6O_{18}$, colorless plates, mp 186–188 °C, $[\alpha]_D$ + 30.8° (MeOH, *c* 0.5)

Handbook of Pharmaceutical Natural Products, Volume 2.
Goutam Brahmachari

ISBN: 978-3-527-32148-3

Valinomycin

Structure:

Valinomycin

Compound class: Cyclodepsipeptide

Source: *Streptomyces padanus* strain TH-04 (an actinomycete)

Pharmaceutical potentials: *Cytotoxic; antitumor; antiviral; antifungal*

Ryoo et al. [4] reported that the compound showed selective cytotoxic activity against HT-29 human colon carcinoma cells via downregulation of glucose-regulated protein 78 (GRP78); such GRP inhibitory action of valinomycin resulting in selective cell death of stressed cancer cells may thus be an excellent target for the use of cancer chemotherapy in the treatment of solid tumors. Antitumor and anticancerous activity of the antibiotic by means of inducing apoptosis is well known [1–3, 6]. Valinomycin is a potassium ionophore and is well known to cause collapse of the mitochondrial membrane potential by inducing uncoupling of respiration and depolarization of isolated mitochondria [7]. Inai et al. [8] also reported that depolarization of intact mitochondria in AH-130 rat ascite hepatoma cells was induced by valinomycin, thereby resulting in apoptosis through degradation of the mitochondrial membrane potential; their detailed observations suggested that there may be a mechanism that transmits the signal from mitochondrial depolarization to subsequent apoptosis execution steps [8].

The antibiotic showed antiviral activity [5] and also displayed potent antifungal efficacy against *Phytophthora capsici* with an IC_{50} value of 15.9 µg/ml [1].

References

[1] Lim, T.H., Oh, H., Kwon, S.Y., Kim, J.-H., Seo, H.-W., Lee, J.-H., Min, B.-S., Kim, J.-C., Lim, C.-H., and Cha, B. (2007) *Nat. Prod. Sci.*, **13**, 144.
[2] Abdalah, R., Wei, L., Francis, K., and Yu, S.P. (2006) *Neurosci. Lett.*, **11**, 68.
[3] Pettit, G.R., Tan, R., Melody, N., Kielty, J.M., Pettit, R.K., Herald, D.L., Tucker, B.E., Mallavia, L.P., Doubek, D.L., and Schmidt, J.M. (1999) *Bioorg. Med. Chem. Lett.*, **7**, 895.
[4] Ryoo, I.J., Park, H.R., Choo, S.J., Hwang, J.H., Park, Y.M., Bac, K., Shin-Ya, K., and Yoo, I.D. (2006) *Biol. Pharm. Bull.*, **29**, 817.

[5] De Clercq, E. (2006) *Expert Rev. Anti-Infect. Ther.*, **4**, 291.
[6] Daniele, R.P. and Holian, S.K. (1976) *Proc. Natl. Acad. Sci. USA*, **10**, 3599.
[7] Furlong, I.J., Mediavilla, C., Ascaso, R., Lopez, R.A., and Collinus, M.K. (1998) *Cell Death Differ.*, **5**, 214.
[8] Inai, Y., Yabuki, M., Kanno, T., Akiyama, J., Yasuda, T., and Utsumi, K. (1997) *Cell Struct. Funct.*, **22**, 555.

3-*O*-Vanillylceanothic acid

Systematic name: 3-*O*-(4-Hydroxy-3-methoxybenzoyl)ceanothic acid

Physical data: $C_{38}H_{52}O_8$, colorless solid, mp 183–185 °C, $[\alpha]_D^{27} - 22°$ (MeOH, *c* 0.229)

Structure:

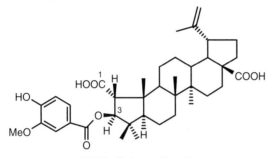

3-*O*-Vanillylceanothic acid

Compound class: Ceanothane-type triterpene

Source: *Ziziphus cambodiana* (root barks; family: Rhamnaceae)

Pharmaceutical potentials: *Antiplasmodial (antimalarial); antimycobacterium*
The isolate was found to exhibit *in vitro* antimalarial and antimycobacterium activity. The triterpene inhibited the growth of *Plasmodium falciparum* with an IC_{50} value of 3.7 µg/ml (IC_{50} value of artemisinin (positive control) = 1 ng/ml), while it showed only moderate antimycobacterium activity with an MIC value of 25 µg/ml (MIC values for the reference standards, rifampicin, isoniazid, and kanamycin sulfate, were determined to be 0.004, 0.06, and 2.5 µg/ml, respectively).

Reference

Suksamararn, S., Panseeta, P., Kunchanawatta, S., Distaporn, T., Ruktasing, S., and Suksamrarn, A. (2006) *Chem. Pharm. Bull.*, **54**, 535.

Variecolin

Physical data: $C_{25}H_{36}O_2$, colorless needles (MeOH), mp 158–159 °C, $[\alpha]_D$ (in varying concentrations as reported in the literature) $= -11.5°$ (MeCN, *c* 0.50) [1], $-72°$ (MeCN, *c* 1.0) [2], $-82°$ (MeCN, *c* 1.16) [4]

Variecolin

Structure:

(2S,3S,6R,10S,11R,14S,15R,16S)
Variecolin

Compound class: Sesquiterpenoid

Sources: *Aspergillus variecolor* MF 138 (fungus) [1]; *Phoma* sp. [2]; *Emericella aurantio-brunnea* (fungus) [3, 4]

Pharmaceutical potentials: *Angiotensin II receptor binding inhibitor; immunosuppressive; anti-HIV*
Hensens et al. [1] evaluated variecolin as a potent angiotensin II receptor binding inhibitor with an IC_{50} value of $3 \pm 1\,\mu M$. The sesquiterpenoid showed promising immunosuppressive activity as determined against concavalin A-induced (T cells) and lipopolysaccharide (LPS)-induced (B cells) proliferation of mouse splenic lymphocytes with respective IC_{50} values of 0.4 and 0.1 µg/ml (respective IC_{50} values of reference standards: 2.7 µg/ml each for azathioprine; 0.04 and 0.07 µg/ml for cyclosporin) [3].

Variecolin also stands as an important lead for anti-HIV drugs; the test compound was found to compete with human macrophage inflammatory protein (MIP)-1α (used as a reference standard in the experiment) for binding to human CCR5 in a scintillation proximity assay (SPA) (respective IC_{50} values: 2.7 nM and 9 µM) [4]. It has been reported that human immunodeficiency virus type-1 makes uses of chemokine receptors (primarily CCR5 and CXCR4) as coreceptors with CD4 for entering into target cells [5, 6], thereby retarding the viral growth. Hence, molecules having potentiality to bind with the CCR5 receptors may be of use in the discovery of anti-HIV drugs [7–10].

From the structure–activity relationships, it is anticipated that the coexistence of a ketonic function at C-5 and a conjugated aldehydic moiety at C-20 in ring B of variecolin might be important for its immunosuppressive as well as anti-HIV activity [3, 4].

References

[1] Hensens O.D., Zink, D., Williamson, J.M., Lotti, V.J., Chang, R.S.L., and Goetz, M.A. (1991) *J. Org. Chem.*, **56**, 3399.

[2] Tezuka, Y., Takahashi, A., Maruyama, M., Tamamura, T., Kutsuma, S., Naganawa, H., and Takeuchi, T. (1998) Jpn. Patent JP 10045662.

[3] Fujimoto, H., Nakamura, E., Okuyama, E., and Ishibashi, M. (2000) *Chem. Pharm. Bull.*, **48**, 1436.

[4] Yoganathan, K., Rossant, C., Glover, R.P., Cao, S., Vittal, J.J., Ng, S., Huang, Y., Buss, A.D., and Butler, M.S. (2004) *J. Nat. Prod.*, **67**, 1681.

[5] Kazmierski, W.M., Boone, L., Lawrence, W., Watson, C., and Kenakin, T. (2002) *Curr. Drug Targets Infect. Disord.*, **2**, 265.

[6] Kedzierska, K., Crowe, S.M., Turville, S., and Cunningham, A.L. (2003) *Rev. Med. Virol.*, **13**, 39 and references therein.
[7] Kazmierski, W., Bifulco, N., Yang, H., Boone, L., DeAnda, F., Watson, C., and Kenakin, T. (2003) *Bioorg. Med. Chem.*, **11**, 2663.
[8] Yoganathan, K., Rossant, C., Ng, S., Huang, Y., Butler, M.S., and Buss, A.D. (2003) *J. Nat. Prod.*, **66**, 1116.
[9] Cao, S., Rossant, C., Ng, S., Buss, A.D., and Butler, M.S. (2003) *Phytochemistry*, **64**, 987.
[10] Yoganathan, K., Yang L.-K., Rossant, C., Huang, Y., Ng, S., Butler, M.S., and Buss, A.D. (2004) *J. Antibiot.*, **57**, 59.

Variecolorquinones A and B

Systematic names:
- Variecolorquinone A: (2S)-2,3-dihydroxypropyl-1,6-dihydroxy-8-methoxy-3-methyl-9,10-dioxoanthracene- 2-carboxylate
- Variecolorquinone B: methyl 2-hydroxy-6-[(5-methoxy-3,6-dioxocyclohexa-1,4-dienyl)methyl]-4-methylbenzoate

Physical data:
- Variecolorquinone A: $C_{20}H_{17}O_9$, yellow amorphous powder, $[\alpha]_D - 18°$ (MeOH, *c* 0.03)
- Variecolorquinone B: $C_{17}H_{15}O_6$, yellow amorphous powder

Structures:

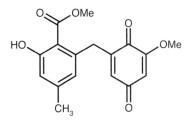

Variecolorquinone A   Variecolorquinone B

Compound class: Quinone derivatives (antibiotics)

Source: *Aspergillus variecolor* B-17 (culture broth)

Pharmaceutical potential: *Cytotoxic (antitumor)*
Both the fungal metabolites were found to possess cytotoxic efficacy against tumor cells; variecolorquinone A displayed selective cytotoxicity against A-549 cells with an IC_{50} value of 3.0 µM, whereas variecolorquinone B showed such efficacy against HL60 and P388 cells with IC_{50} values of 1.3 and 3.7 µM, respectively.

Reference

Wang, W., Zhu, T., Tao, H., Lu, Z., Fang, Y., Gu, Q., and Zhu, W. (2007) *J. Antibiot.*, **60**, 603.

Vedelianin

Systematic name: 2α,3α-Dihydroxy-7(6'-isoprenyl-5',7'-dihydroxystyryl)-1,1-dimethyl-2,3,4,4a,9,9a-hexahydro-1H-xanthene

Physical data: $C_{29}H_{36}O_6$, brownish powder, $[\alpha]_D^{22} +37°$ (MeOH, c 2.88)

Structure:

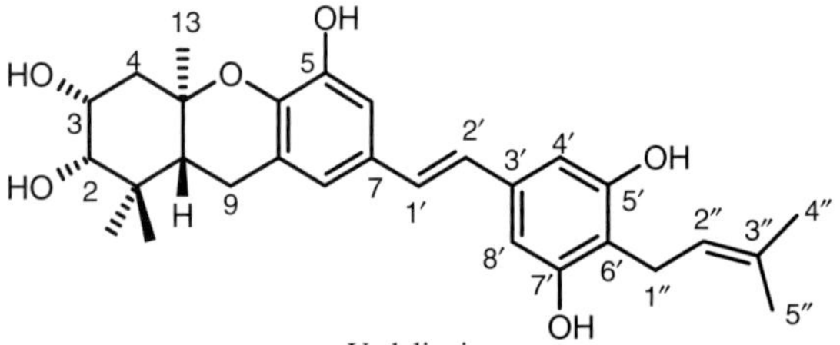

Vedelianin

Compound class: Substituted cyclized geranylstilbene (hexahydroxanthene derivative)

Sources: *Macaranga alnifolia* Muell.-Arg. (leaves; family: Euphorbiaceae) [1]; *Macaranga alnifolia* Baker (fruits) [2]

Pharmaceutical potential: *Anticancerous*
Yoder et al. [2] evaluated vedelianin as a strong antiproliferative agent against the A2780 human ovarian cancer cell line with an IC_{50} value of 0.13 µM. From their study, the investigator also assumed that the presence of free hydroxyls at C-3 and C-5 positions is quite important as per structure–activity relationship; however, the compound did not exhibit any hypotensive activity in the anesthetized rat [1].

References

[1] Thoison, O., Hnawia, E., Gueritte-Voegelein, F., and Sevenet, T. (1992) *Phytochemistry*, **31**, 1439.
[2] Yoder, B.J., Cao, S., Norris, A., Miller, J.S., Ratovoson, F., Razafitsalama, J., Andriantsiferana, R., Rasamison, V.E., and Kingston, D.G. (2007) *J. Nat. Prod.*, **70**, 342.

Venturamides A and B

Physical data:
- Venturamide A: $C_{21}H_{24}N_4O_4S_2$, colorless glass, $[\alpha]_D^{25} +53.4°$ (MeOH, c 0.001)
- Venturamide B: $C_{22}H_{26}N_4O_5S_2$, colorless glass, $[\alpha]_D^{25} +53.6°$ (MeOH, c 0.0004)

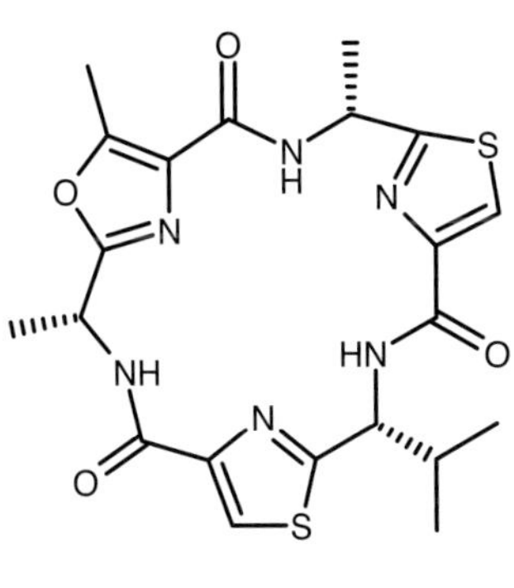

Venturamide A 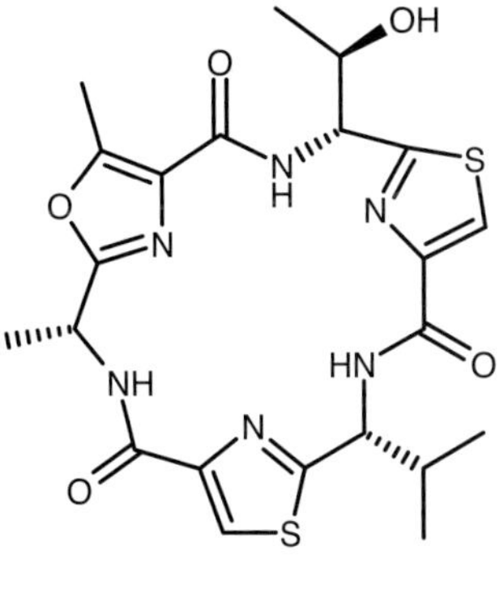   Venturamide B

Compound class: Cyclic hexapeptides (dendroamide-type)

Source: *Oscillatoria* sp. (Panamanian marine cyanobacterium)

Pharmaceutical potential: *Antimalarial*

Venturamides A and B were evaluated for their *in vitro* antimalarial activity against the W2 chloroquine-resistant strain of the malaria parasite, *Plasmodium falciparum*. The investigators showed that venturamide A exhibits strong *in vitro* activity against the parasite (IC_{50} value 8.2 µM), with only mild cytotoxicity to mammalian Vero cells ($IC_{50} = 86$ µM), thus giving an order of magnitude difference in activity to the parasite over host cells. Venturamide B also displayed potent antimalarial activity against *P. falciparum* ($IC_{50} = 5.6$ µM) and mild cytotoxicity to mammalian Vero cells ($IC_{50} = 56$ µM). Hence, the selective antimalarial activity of both the cyclic peptides makes them significant lead structures for further possible development.

Reference

Linington, R.G., González, J., Urena, L.-D., Romero, L.I., Ortega-Barria, E., and Gerwick, W.H. (2007) *J. Nat. Prod.*, **70**, 397.

Verongamine

Systematic name: 3-Bromo-α-(hydroxyimino)-*N*-[2-(1-imidazole-4-yl)ethyl]-4-methoxybenzene propanamide

Physical data: $C_{15}H_{17}N_4O_3Br$, yellow semisolid/oil; $[\alpha]_D^{25}$ +2.6° ($CHCl_3$, *c* 0.5)

Verrucoside

Structure:

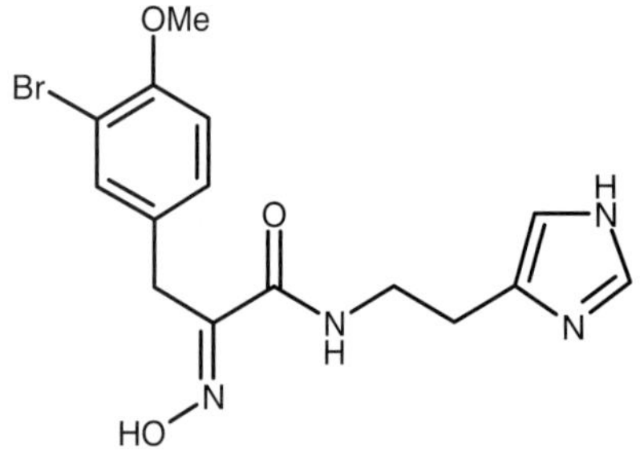

Verongamine

Compound class: Bromotyrosine derivative

Source: *Verongula gigantea* (marine sponge)

Pharmaceutical potential: *Histamine H_3 antagonist*
Verongamine behaved as a specific histamine H_3 antagonist at concentrations as low as 1.0 μg/ml; it was found to bind with an IC_{50} value of 0.5 μM to the H_3 receptor isolated from guinea pig brain membranes. It also displayed H_3-antagonist activity in the electrical field stimulated (EFS) contracted guinea pig ileum.

Reference

Mierzwa, R., King, A., Conover, M.A., Tozzi, S., Puar, M.S., Patel, M., Coval, S.J., and Pomponi, S.A. (1994) *J. Nat. Prod.*, **57**, 175.

Verrucoside

Systematic name: 4β-O-[2-O-Acetyl-α-L-digitalopyranosyl]-5β-pregn-20-en-3β-ol

Physical data: $C_{30}H_{48}O_7$, amorphous powder, $[\alpha]_D$ −30° (CHCl$_3$, *c* 2.0)

Structure:

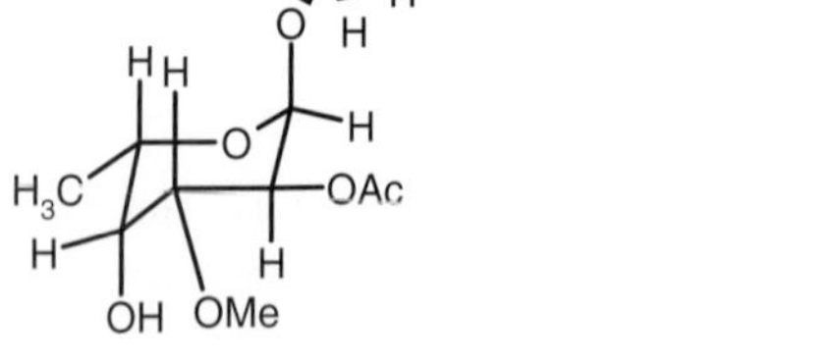

Verrucoside

Compound class: Terpenoid (pregnane glycoside)

Source: *Eunicella verrucosa* Verrill (a gorgonian; family: Gorgonacea)

Pharmaceutical potential: *Cytotoxic*
The isolate displayed significant cytotoxic activity *in vitro* against human lung carcinoma (A-549), murine leukemia (P-388), and human colon carcinoma (HT-29) cells with IC$_{50}$ values of 7.2, 5.9, and 6.3 µg/ml, respectively.

Reference

Kashman, Y., Green, D., Garcia, C., and Arevalos, D.G. (1991) *J. Nat. Prod.*, **54**, 1651.

Verticillin G

Physical data: $C_{30}H_{28}N_6O_7S_4$, white powder, $[\alpha]_D^{25}$ +467.6° (MeOH, *c* 0.2)

Structure:

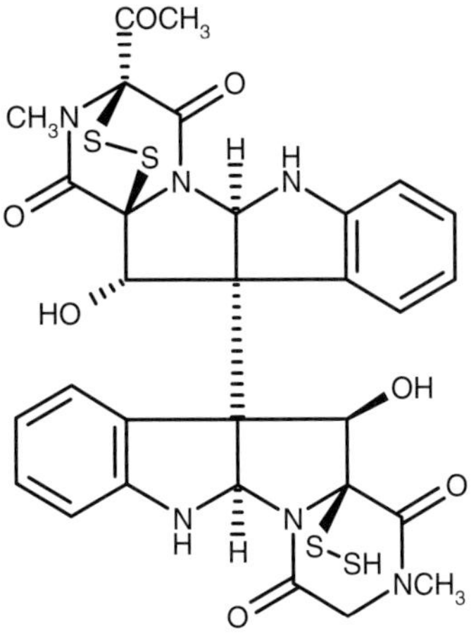

Verticillin G

Compound class: Alkaloid (epidithiodioxopiperazine derivative)

Source: *Bionectra byssicola* F120 (fungal strain; fermentation broth)

Pharmaceutical potential: *Antibacterial*
The fungal metabolite was evaluated to inhibit growth of *Staphylococcus aureus* including methicillin-resistant and quinolone-resistant *S. aureus* with an MIC value of 3–10 µg/ml.

Reference

Zheng, C.-J., Park, S.-H., Koshino, H., Kim, Y.-H., and Kim, W.-G. (2007) *J. Antibiot.*, **60**, 61.

Vilmorrianone

Physical data: $C_{23}H_{26}NO_5$

Structure:

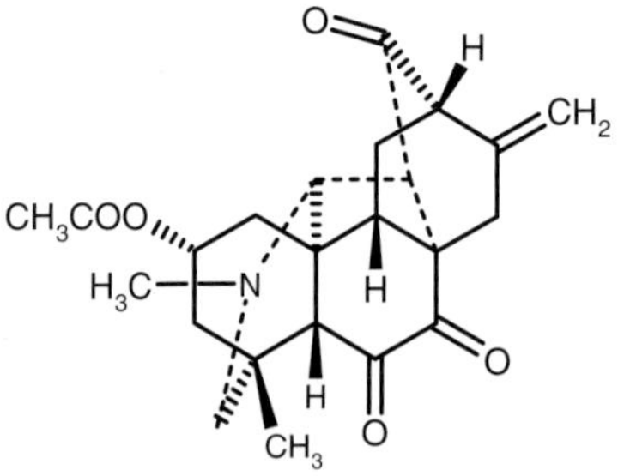

Vilmorrianone

Compound class: Alkaloid (atisine-type diterpenoid alkaloid)

Sources: *Aconitum vilmorrianum* Kom. (roots; family: Ranunculaceae) [1, 2]; *Delphinium denudatum* (roots; family: Ranunculaceae) [2]

Pharmaceutical potential: *Antifungal*
The alkaloid showed antifungal activities against some pathogenic fungi such as *Alleschería boydii*, *Aspergillus niger*, *Epidermophyton floccosum*, and *Pleurotus ostreatus* with respective MIC values of 150, 100, 225, and 175 µg/ml (using the agar diffusion tube method; nystatin and griseofulvin were used as positive controls) [3].

References

[1] Ding, L., Chen, Y., and Wu, F. (1991) *Planta Med.*, **57**, 275.
[2] Ding, L., Wu, F., and Chen, Y. (1994) *Acta Chim. Sin.*, **52**, 932.
[3] Atta-ur-Rahman, Nasreen, A., Akhtar, F., Shekhani, M.S., Clardy, J., Parvez, M., and Choudhary, M.I. (1997) *J. Nat. Prod.*, **60**, 472.

(+)-α-Viniferin

Physical data: $C_{42}H_{30}O_9$, colorless plates (*n*-hexane–acetone), mp 231–233 °C, $[\alpha]_D^{24} + 50.7°$ (EtOH, *c* 1.02)

Structure:

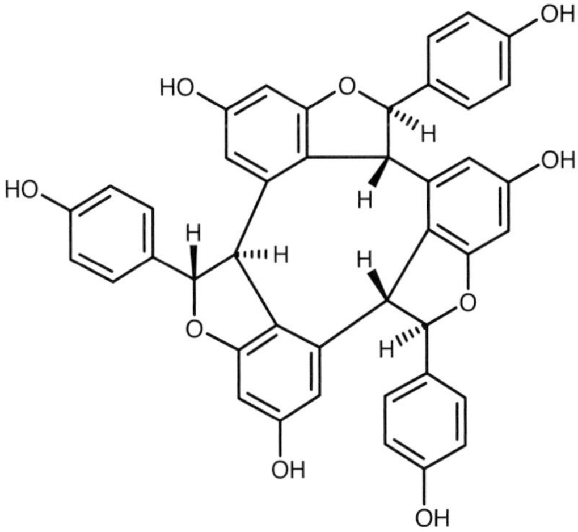

(+)-α-Viniferin

Compound class: Stilbene trimer

Source: *Caragana chamlague* Lamarck (roots; family: Leguminosae) [1, 2]

Pharmaceutical potentials: *Acetylcholinesterase (AChE); inhibitor anti-inflammatory*
Sung et al. [2] evaluated the isolate to possess significant inhibitory effect against acetylcholinesterase activity in a dose-dependent manner with an IC_{50} value of 2.0 µM; at a dose of 10 µM, it showed 96.4% inhibitory activity. The enzyme inhibitory activity was found to be specific, reversible, and noncompetitive. Inhibitors of AChE are reported to find useful applications in the treatment of Alzheimer's disease. The stilbene trimer was also reported to exert weak anti-inflammatory potential (28% inhibition against edema formation at 10 mg/kg body weight) as evaluated using carrageenin-induced hind paw edema in mice [1].

References

[1] Kitanaka, S., Ikezawa, T., Yasukawa, K., Yamanauchi, S., Takido, M., Sung, H.K., and Kim, I.M. (1990) *Chem. Pharm. Bull.*, **38**, 432.
[2] Sung, S.H., Kang, S.Y., Lee, K.Y., Park, M.J., Kim, J.H., Park, J.H., Kim, Y.C., Kim, J., and Kim, Y.C. (2002) *Biol. Pharm. Bull.*, **25**, 125.

Vinylamycin

Physical data: $C_{26}H_{43}N_3O_6$, colorless microcrystalline, mp 182–184 °C, $[\alpha]_D^{25}$ +173° (DMSO, *c* 0.85)

Structure:

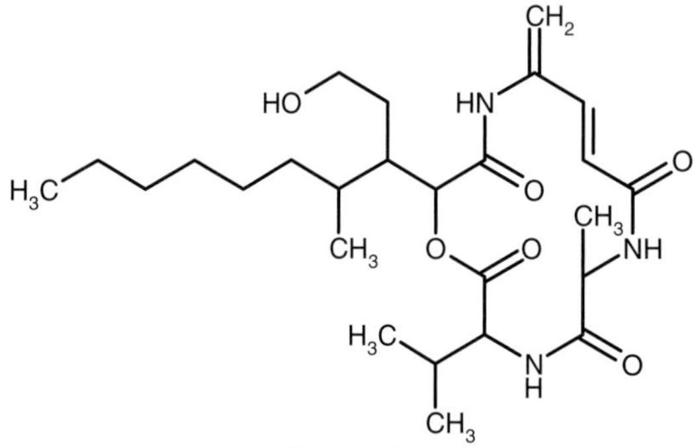

Vinylamycin

Compound class: Antibiotic (depsipeptide)

Source: *Streptomyces* sp. M1982-63F1

Pharmaceutical potential: Antibacterial
Vinylamycin was evaluated to display broad and moderate antimicrobial activities against Gram-positive bacteria including *Staphylococcus aureus* FDA209P, *S. aureus* Smith, *S. aureus* MS9610, *S. aureus* MS16526 (MRSA), *S. aureus* TY-04282 (MRSA), *Micrococcus luteus* IFO3333, *M. luteus* PCI1001, *Bacillus subtilis* NRRL B-558, *B. cereus* ATCC10702, and *Corynebacterium bovis* 1810 with MIC values of 1.56, 3.13, 1.56, 3.13, 6.25, 1.56, 3.13, 12.5, 50, and 3.13 mg/ml, respectively. The test compound did not show acute toxicity in mice at a dose of 100 mg/kg when administered intraperitoneally.

Reference

Igarashi, M., Shida, T., Sasaki, Y., Kinoshita, N., Naganawa, H., Hamada, M., and Takeuchi, T. (1999) *J. Antibiot.*, **52**, 873.

Viridamide A

Physical data: $C_{46}H_{79}N_5O_{10}$, colorless glassy oil, $[\alpha]_D$ −107.4° (CDCl$_3$, *c* 0.05)

Structure:

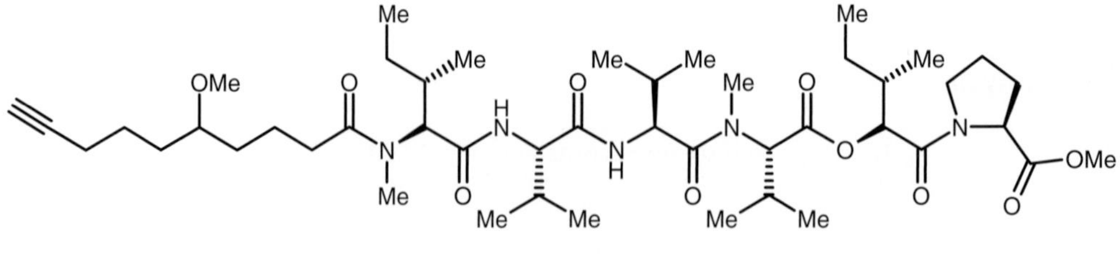

Viridamide A

Compound class: Lipodepsipeptide

Source: *Oscillatoria nigroviridis* (marine cyanobacterium)

Pharmaceutical potential: *Antiparasitic (leishmanicidal; trypanocidal; antiplasmodial)*
The benzoic acid derivative exhibited significant *in vitro* antiparasitic activities against *Leishmania mexicana*, *Trypanosoma cruzi*, and *Plasmodium falciparum* with IC_{50} values of 1.5 ± 0.15, 1.1 ± 0.1, and $5.8 \pm 0.6\,\mu M$, respectively.

Reference

Simmons, T.L., Engene, N., Urena, L.D., Romero, L.I., Ortega-Barria, E., Gerwick, L., and Gerwick, W.H. (2008) *J. Nat. Prod.*, **71**, 1544.

Vitex norditerpenoids 1 and 2

Physical data:
- **1**: $C_{19}H_{30}O_2$, colorless oil, $[\alpha]_D$ $-3.4°$ (MeOH, c 0.64)
- **2**: $C_{21}H_{32}O_4$, colorless oil, $[\alpha]_D$ $-1.4°$ (MeOH, c 0.5)

Structures:

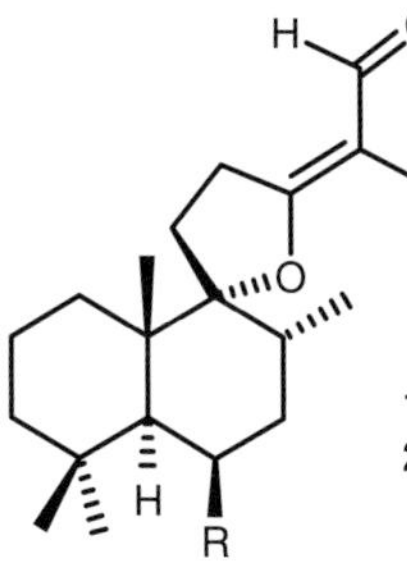

1: R = H
2: R = OAc

Compound class: Norditerpenoids

Source: *Vitex trifolia* L. (dried fruits; family: Verbenaceae)

Pharmaceutical potential: *Trypanocidal*
The norditerpenoid aldehydes **1** and **2** were reported to exhibit *in vitro* trypanocidal activity against epimastigotes of *Trypanosoma cruzi*; the test compounds **1** and **2** showed minimum lethal concentration (MLC) values of 11 and 36 μM, respectively (gentian violet was used as a positive control; MLC = 6.3 μM).

Reference

Kiuchi, F., Matsuo, K., Ito, M., Qui, T.K., and Honda, G. (2004) *Chem. Pharm. Bull.*, **52**, 1492.

W

Weigelic acid

Systematic name: 1β,2α,3α,23-Tetrahydroxy-urs-12-en-28-oic acid

Physical data: $C_{30}H_{48}O_6$, white amorphous powder, mp 244–246 °C, $[\alpha]_D^{20}$ +8.8° (MeOH, *c* 0.25)

Structure:

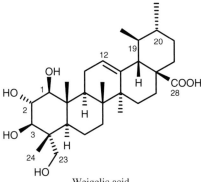

Weigelic acid

Compound class: Triterpenoid

Source: *Weigela subsessilis* L.H. Bailley (leaves and stems; family: Caprifoliaceae)

Pharmaceutical potential: ***Anticomplementary***
Weigelic acid was evaluated to have a moderate anticomplement activity against complement-induced hemolysis via the classical pathway with an IC_{50} value of 152 ± 10.3 μM (rosmarinic acid was used as a positive control; IC_{50} = 182 ± 27.7 μM). From detailed structure–activity relationships of such type of compounds, Thuong *et al.* suggested that the carboxylic group (at C-28) of ursane-type triterpenoids seems to play an important role in inhibiting the hemolytic activity of human serum against erythrocytes.

Reference

Thuong, P.T., Min, B.-S., Jin, W.Y., Na, M.K., Lee, J.P., Seong, R.S., Lee, Y.-M., Song, K.S., Seong, Y. H., Lee, H.-K., Bae, K.H., and Kang, S.S. (2006) *Biol. Pharm. Bull.*, **29**, 830.

WF14865A and B

Systematic names:

- WF14865A: 4-[3-[*N*-[[(2*S*,3*S*)-3-*trans*-carboxyoxiran-2-yl]carbonyl]-L-isoleucyl]aminopropanyl]-1*H*-imidazol-2-ylamine

Handbook of Pharmaceutical Natural Products, Volume 2.
Goutam Brahmachari

ISBN: 978-3-527-32148-3

- WF14865B: 4-[3-[N-[[(2S,3S)-3-trans-carboxyoxiran-2-yl]carbonyl]-L-leucyl]aminopropanyl]-1H-imidazol-2-ylamine

Physical data:
- WF14865A: $C_{16}H_{25}N_5O_5$, white powder
- WF14865B: $C_{16}H_{25}N_5O_5$, white powder

Structures:

WF14865A

WF14865B

Compound class: Antibiotics

Source: *Aphanoascus fulvescens* no. 14865 (fungal strain) [1]

Pharmaceutical potential: *Cathepsin B and L inhibitors*

Cathepsins B and L are reported to be involved in various pathophysiological processes including osteoclastic bone resorption [2–5], ischemic neuronal cell death [6], invasion and metastasis of carcinoma [7], and inflammatory myopathy [8]. Hence, these lysosomal cysteine proteases are useful targets in the treatment of associated diseases. WF14865A and B, the two fungal metabolites, were evaluated to possess significant inhibitory activity against human cathepsins B and L [1] – the respective IC_{50} values against human liver cathepsins B and L were determined to be 8.4 and 66.0 nM for WF14865A and 13.0 and 72.0 nM for WF14865B. WF14865A and B or their derivatives thus appeared to be promising candidates, particularly for the treatment of rheumatic arthritis, osteoarthritis, and inflammatory myopathy.

References

[1] Otsuka, T., Muramatsu, Y., Nakanishi, T., Hatanaka, H., Okamoto, M., Hino, M., and Hashimoto, S. (2000) *J. Antibiot.*, **53**, 449.
[2] Maciewicz, R.A. and Wotton, S.F. (1991) *Biomed. Biochim. Acta*, **50**, 561.
[3] Flipo, R.-M., Huet, G., Maury, F., Zerimech, F., Fontaine, C., Duquesnoy, B., Degand, P., and Delcambre, B. (1993) *Arthritis Rheum.*, **36**, S252.
[4] Page, A.E., Warburton, M.J., Chambers, T.J., Pringle, J.A., and Hayman, A.R. (1992) *Biochim. Biophys. Acta*, **1116**, 57.
[5] Katunuma, N., Matsunaga, Y., Matsui, A., Kakegawa, H., Endo, K., Inubishi, T., Saibara, T., Ohba, Y., and Kakiuchi, T. (1998) *Adv. Enzyme Regul.*, **38**, 235.
[6] Yamashita, T., Kohda, Y., Tsuchiya, K., Ueno, T., Yamashita, J., Yoshioka, T., and Kominami, E. (1998) *Eur. J. Neurosci.*, **10**, 1723.
[7] Navab, R., Mort, J.S., and Brodt, P. (1997) *Clin. Exp. Metastasis*, **15**, 121.
[8] Kumamoto, T., Ueyama, H., Sugihara, R., Kominami, E., Goll, D.E., and Tsuda, T. (1997) *Eur. Neurol.*, **37**, 176.

Withanolide D and its hydroxy derivative

See Withanolides

Withanolides

Systematic names:
- Tubocapsenolide A: 4β,16α-dihydroxy-5β,6β-epoxy-1-oxowitha-2,13,24-trienolide
- Tubocapsanolide A: 5β,6 β:16α,17α-diepoxy-4β-hydroxy-1-oxo-witha-2,24-dienolide
- 20-Hydroxytubocapsanolide A: 5β,6 β:16α,17α-diepoxy-4β,20-dihydroxy-1-oxo-witha-2,24-dienolide
- 23-Hydroxytubocapsanolide A: 5β,6 β:16α,17α-diepoxy-4β,23-dihydroxy-1-oxo-witha-2,24-dienolide
- Tubocapsanolide F: 5β,6β-epoxy-4β,17α-dihydroxy-1-oxo-witha-2,24-dienolide
- Withanolide D: 4β,20α-dihydroxy-1-oxo-5β,6β-epoxy-20S,22R-witha-2,24-dienolide
- 17α-Hydroxywithanolide D: 4β,17α,20α-trihydroxy-1-oxo-5β,6β-epoxy-20S,22R-witha-2,24-dienolide

Physical data:
- Anomanolide B: $C_{28}H_{40}O_8$, white powder, mp 168–170 °C, $[\alpha]_D^{24.4}$ +26.5° (MeOH, c 0.1)
- Tubocapsenolide A: $C_{28}H_{36}O_6$, colorless prisms (MeOH), mp 223–225 °C, $[\alpha]_D^{24.3}$ −0.57° (MeOH, c 0.1)
- Tubocapsanolide A: $C_{28}H_{36}O_6$, white powder, mp 233–235 °C, $[\alpha]_D^{24.4}$ +22.3° (MeOH, c 0.1)
- 20-Hydroxytubocapsanolide A: $C_{28}H_{36}O_7$, white powder, mp 245–247 °C, $[\alpha]_D^{26.5}$ +14.4° (MeOH, c 0.12)
- 23-Hydroxytubocapsanolide A: $C_{28}H_{36}O_7$, white powder, mp 223–225 °C, $[\alpha]_D^{25.2}$ −34.0° (MeOH, c 0.1)
- Tubocapsanolide F: $C_{28}H_{38}O_6$, white powder, mp 200–202 °C, $[\alpha]_D^{25}$ +75.7° (MeOH, c 0.07)
- Withanolide D: $C_{28}H_{38}O_6$
- 17α-Hydroxywithanolide D: $C_{28}H_{38}O_7$

Withanolides

Structures:

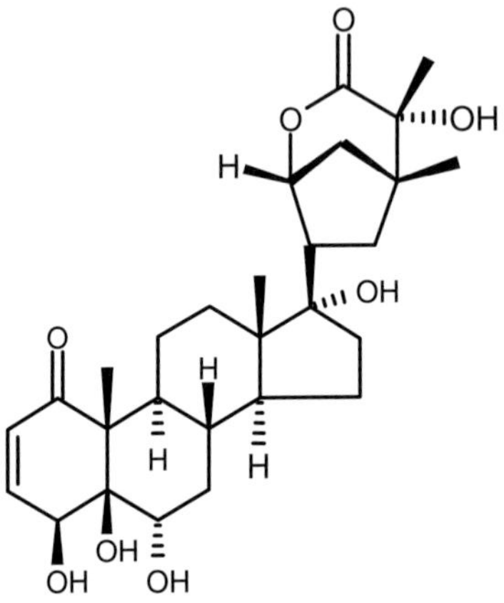

Anomanolide B

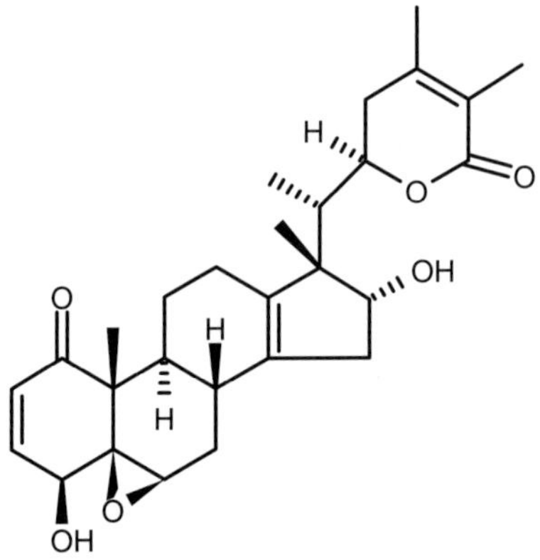

Tubocapsenolide A

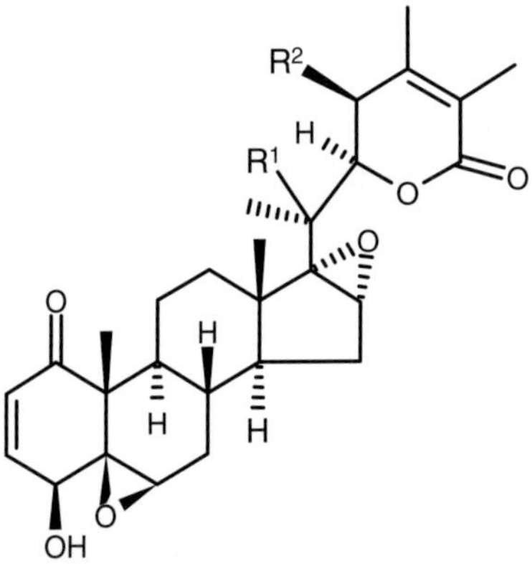

Tubocapsanolide A : $R^1 = R^2 = H$
20-Hydroxytubocapsanolide A : $R^1 = OH$; $R^2 = H$
23-Hydroxytubocapsanolide A : $R^1 = H$; $R^2 = OH$

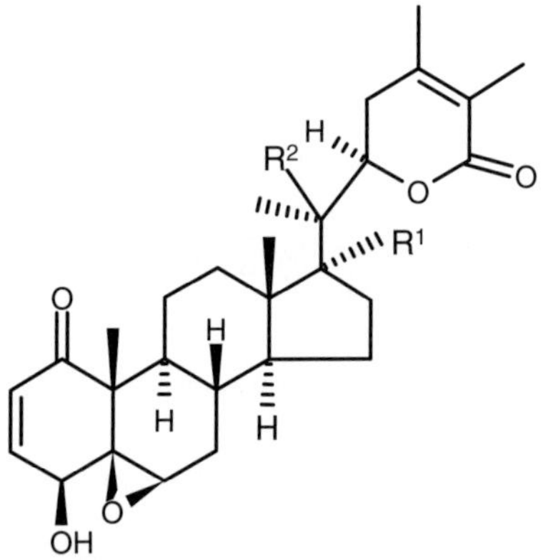

Tubocapsanolide F : $R^1 = OH$; $R^2 = H$
Withanolide D : $R^1 = H$; $R^2 = OH$
17α-Hydroxywithanolide D : $R^1 = OH$; $R^2 = OH$

Compound class: Withanolides (steroidal lactones)

Sources: All the compounds from *Tubocapsicum anomalum* (family: Solanaceae) [1]; withanolide D and 17α-hydroxywithanolide D also from *Withania somnifera* chemitype II (leaves; family; Solanaceae) [2, 3]

Pharmaceutical potential: *Cytotoxic (anticancerous)*
Hsieh et al. [1] evaluated cytotoxic activity of all the isolated compounds against five human cancer cell lines such as hepatocellular carcinoma HepG2 and Hep3B, breast carcinoma MCF-7 and MDA-MB-231, and lung carcinoma A-549 and against the embryonic lung cell line MRC-5 using withaferin A and doxorubicin as positive controls; all of them were found to be promising cytotoxic agents. The respective IC_{50} values (μg/ml) of the tested compounds and controls are as follows: 0.97, 3.17, 4.84, 0.70, 1.49, and 0.20 (anomanolide B); 0.44, 0.26, 0.97, 0.13, 0.15, and 0.20

(tubocapsenolide A); 0.86, 0.42, 1.47, 0.22, 0.47, and 0.73 (tubocapsanolide A); 0.73, 0.99, 1.77, 0.99, 1.42, and 1.36 (20-hydroxytubocapsanolide A); 0.44, 0.49, 2.05, 1.19, 0.79, and 0.90 (23-hydroxytubocapsanolide A); 0.64, 0.80, 1.98, 0.99, 0.88, and 0.81 (tubocapsanolide F); 0.21, 0.47, 0.37, 0.28, 0.33, and 0.80 (withanolide D); 0.49, 0.85, 1.45, 0.71, 0.72, and 1.89 (17α-hydroxywithanolide D); 0.06, 0.06, 0.05, 0.02, 0.02, and 0.07 (withaferin A (control)); and 0.46, 0.45, 0.34, 0.19, 0.42, and 0.77 (doxorubicin (control)).

References

[1] Hsieh, P.-W., Huang, Z.-Y., Chen, J.-H., Chang, F.-R., Wu, C.-C., Yang, Y.-L., Chiang, M.Y., Yen, M.-H., Chen, S.-L., Yen, H.-F., Lubken, T., Hung, W.-C., and Wu, Y.-C. (2007) *J. Nat. Prod.*, **70**, 747.
[2] Lavie, D., Kirson, I., and Glotter, E. (1968) *Isr. J. Chem.*, **6**, 671.
[3] Abraham, A., Kirson, I., Lavie, D., and Glotter, E. (1975) *Phytochemistry*, **14**, 189.

Wrightiamine A

Physical data: $C_{21}H_{34}N_2$, colorless amorphous solid, $[\alpha]_D^{25}$ −14° (MeOH, *c* 0.2)

Structure:

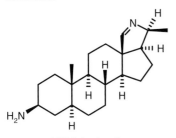

Wrightiamine A

Compound class: Pregnane alkaloid

Source: *Wrightia javanica* DC. (leaves; family: Apocynaceae)

Pharmaceutical potential: *Cytotoxic*
The present investigators studied cytotoxic activity of the isolate against vincristine-resistant murine leukemia P388 cells; the test compound exhibited cytotoxicity against the targeted cell line with an IC_{50} value of 2.0 µg/ml in the presence of vincristine (12.5 ng/ml) and 3.1 µg/ml in the absence of vincristine.

Reference

Kawamoto, S., Koyano, T., Kowithayakorn, T., Fujimoto, H., Okuyama, E., Hayashi, M., Komiyama, K., and Ishibashi, M. (2003) *Chem. Pharm. Bull.*, **51**, 737.

Xanthohumols

Systematic names:
- Xanthohumol: 1-[2,4-dihydroxy-3-(3-methylbut-2-enyl)-6-methoxyphenyl]-3-(4-hydroxyphenyl)propenone
- Dihydroxanthohumol: 1-[2,4-dihydroxy-3-(3-methylbut-2-enyl)-6-methoxyphenyl]-3-(4-hydroxyphenyl)propanone
- Oxygenated xanthohumol: 1-[2,4-dihydroxy-3-(2-hydroxy-3-methoxy-3-methylbutyl)-6-methoxyphenyl]-3-(4-hydroxyphenyl)propenone
- Xanthohumol B: 6-[3,4-dihydro-3,5-dihydroxy-7-methoxy,2,2-dimethyl-2H-benzo[b]pyrano]-3-(4-hydroxyphynyl)-2-propen-1-one
- Xanthohumol D: 1-[2,4-dihydroxy-3-(2-hydroxy-3-methylbut-3-enyl)-6-methoxyphenyl]-3-(4 hydroxyphenyl)propenone

Physical data:
- Xanthohumol: $C_{21}H_{22}O_5$, yellow–orange crystals
- Dihydroxanthohumol: $C_{21}H_{24}O_5$
- Oxygenated xanthohumol: $C_{22}H_{26}O_7$, yellow powder
- Xanthohumol B: $C_{21}H_{22}O_6$
- Xanthohumol D: $C_{21}H_{22}O_6$

Structures:

Compound class: Prenylated chalcones (prenylated flavonoids)

Source: *Humulus lupulus* L. (family: Cannabinaceae) [1–6]

Pharmaceutical potentials: *Nitric oxide (NO) production inhibitory; antitumor; diacylglycerol acyltransferase (DGAT) inhibitor*

Zhao et al. [1] studied the inhibitory activity of the prenylated chalcones on the production of nitric oxide (NO) in macrophage RAW 264.7 cells; macrophages are considered to play crucial roles in inflammation and host defense mechanisms against bacterial and viral infections [7]. Excessive and prolonged generation of NO that influences the production of vascular epidermal growth factor (VEGF) [8–10] during both acute and chronic inflammation usually causes severe injury to host cells and tissues [11], particularly leading to epithelial carcinogenesis [12, 13].

All the isolated chalcones were reported to exhibit significant inhibitory activity against NO production without showing cytotoxicity at concentrations lower than 10 µM (cell viability >95%) [1] – the respective IC_{50} values were recorded as 8.3, 23, 6.5, 5.6, and 9.4 µM. Dihydroxanthohumol, lacking the double bond at the α,β-position, did not show cytotoxicity even at the concentration of 20 µM, but it is a much weaker inhibitory agent in comparison to the other four chalcone derivatives in the series, thereby suggesting that the double bond plays a role in the inhibitory activity of chalcones, whereas the prenyl chain may not be necessary for the activity since these four chalcone derivatives having the same backbone structure but differing in the prenyl side chain showed almost the same NO production inhibitory activities (IC_{50} values: 8.3, 6.5, 5.6, and 9.4 µM, respectively).

Xanthohumol has been suggested to have potential cancer chemopreventive activities. Most of the cancer chemopreventive agents show antiangiogenic properties *in vitro* and *in vivo*, a concept termed as "angioprevention"; Albini et al. [13] reported that xanthohumol can inhibit growth of a vascular tumor *in vivo*. The compound suppressed both the NF-ϰB and Akt pathways in endothelial cells, indicating that components of these pathways are major targets in the molecular mechanism of the test compound. Moreover, using *in vitro* analyses, the investigators found that it interferes with several points in the angiogenic process, including inhibition of endothelial cell invasion and migration, growth, and formation of a network of tubular-like structures; hence, xanthohumol can be added to the expanding list of antiangiogenic chemopreventive drugs whose potential in cancer prevention and therapy should be evaluated.

Xanthohumol and xanthohumol B inhibited DGAT (diacylglycerol acyltransferase) activity with IC_{50} values of 50.3 and 194 µM in rat liver microsomes, respectively; both of them showed preferential inhibition of triacylglycerol formation in intact Raji cells, indicating that they inhibit DGAT activity preferentially in living cells [6].

References

[1] Zhao, F., Watanabe, Y., Nozawa, H., Daikonnya, A., Kondo, K., and Kitanaka, S. (2005) *J. Nat. Prod.*, **68**, 43.
[2] Hansel, R. and Schulz, J. (1988) *Arch. Pharm. (Weinheim)*, **321**, 37.
[3] Stevens, J.F., Taylor, A.W., Nickerson, G.B., Ivancic, M., Henning, J., Haunold, A., and Deinzer, M.L. (2000) *Phytochemistry*, **53**, 759.
[4] Stevens, J.F., Ivancic, M., Hsu, V.L., and Deinzer, M.L. (1997) *Phytochemistry*, **44**, 1575.
[5] Verzele, M. and Keukeleire, D. (1991) Chemistry and Analysis of Hops and Beer Bitter Acids, Elsevier, New York, p. 204.
[6] Tabata, N., Ito, M., Tomoda, H., and Omura, S. (1997) *Phytochemistry*, **46**, 683.
[7] Adam, D.O. and Hamilton, T.A. (1984) *Annu. Rev. Immunol.*, **2**, 283.
[8] Zhao, F., Nozawa, H., Daikonnya, A., Kondo, K., and Kitanaka, S. (2003) *Biol. Pharm. Bull.*, **26**, 61.

[9] Xiong, M., Elson, G., Legarda, D., and Leibovich, S.J. (1998) *Am. J. Pathol.*, **153**, 587.
[10] Larcher, F., Murillas, R., Bolontrade, M., Conti, C.J., and Jorcano, J.L. (1998) *Oncogene*, **17**, 303.
[11] Knowles, R.G. and Moncade, S. (1994) *J. Biochem. (Tokyo)*, **298**, 249.
[12] Xie, K., Huang, S., Dong, Z., Juang, S.H., Wang, Y., and Fidler, I.J. (1997) *J. Natl. Cancer Inst.*, **89**, 421.
[13] Tsuji, S., Kawano, S., Tsuji, M., Takei, Y., Tanaka, M., Sawaoka, H., Nagano, K., Fusamoto, H., and Kamada, T. (1996) *Cancer Lett.*, **108**, 195.
[14] Albini, A., Dell'Eva, R., Vene, R., Ferrari, N., Buhler, D.R., Noonan, D.M., and Fassina, G. (2006) *FASEB J.*, **20**, 527.

Xenovulene A

Physical data: $C_{22}H_{30}O_4$, white powder, $[\alpha]_D +520°$ (MeOH, c 0.08)

Structure:

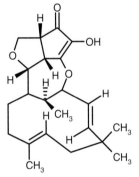

Xenovulene A

Compound class: Sesquiterpenoid

Source: *Acremonium strictum* (fungus; fermentation broth)

Pharmaceutical potential: *Inhibitor of flunitrazepam binding to the GABA-benzodiazepine receptor*
The sesquiterpenoid was found to inhibit binding of flunitrazepam to GABA (γ-aminobutyric acid)-benzodiazepine receptor with an IC_{50} value of 40 nM in an *in vitro* assay using bovine synaptosome membrane preparations.

Reference

Ainsworth, M., Chicarelli-Robinson, M.I., Copp, B.R., Fauth, U., Hylands, P.J., Holloway, J.A., Latif, M., O'Berne, G.B., Porter, N., Renno, D.V., Richards, M., and Robinson, N. (1995) *J. Antibiot.*, **48**, 568.

Xylocarpin J

Systematic name: (4R,4aR,6aR,6bS,7S,9S,10aS,11aS,11bR)-4-(Furan-3-yl)-dodecahydro-10a-hydroxy-7-(2-methoxy-2-oxoethyl)-4a,6b,8,8-tetramethyl-2-oxo-2H,4H-10,11a-methano[1]benzofuro[2,3-f][2]benzopyran-9-yl (2E)-2-methylbut-2-enoate

Physical data: $C_{32}H_{42}O_9$, white amorphous powder, $[\alpha]_D^{25} - 42°$ (MeOH, c 0.6)

Structure:

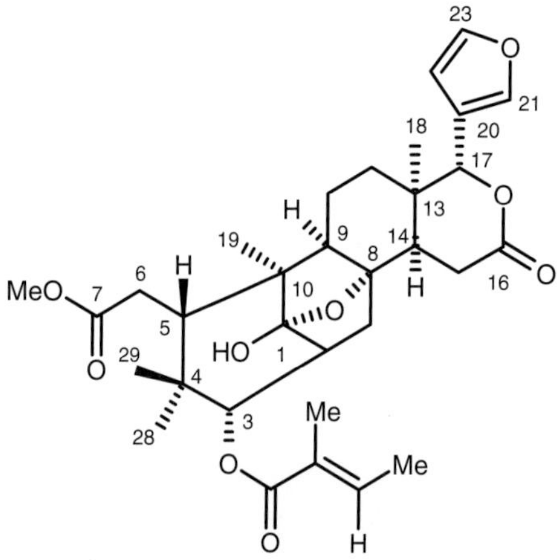

Xylocarpin J

Compound class: Limonoid

Source: *Xylocarpus granatum* Koenig (dried fruit rinds; family: Meliaceae)

Pharmaceutical potential: *Cytotoxic (antitumor)*
The isolate was evaluated for its *in vitro* antitumor activity and was found to exert potent cytotoxicity against the tumor cell lines HCT-8, Bel-7402, BGC-823, A-549, and A-2780 with IC_{50} values of 7.75, 8.22, 8.38, 5.35, and 4.77 μM (taxol was used as a positive control; IC_{50} values were determined to be 0.21, 0.02, 1.37, 2.76, and 0.03 μM).

Reference

Cui, J., Deng, Z., Xu, M., Proksch, P., Li, Q., and Lin, W. (2009) *Helv. Chim. Acta*, **92**, 139.

Yahyaxanthone

Physical data: $C_{22}H_{22}O_8$, yellow crystals, mp 194–196 °C

Structure:

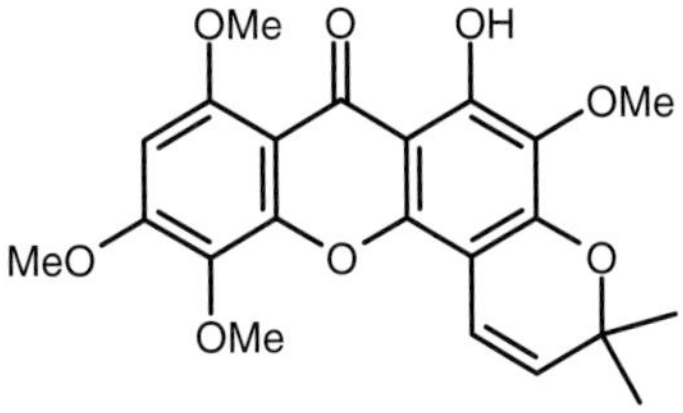

Yahyaxanthone

Compound class: Xanthone

Source: *Garcinia rigida* (leaves; family: Guttiferae/Clusiaceae)

Pharmaceutical potential: *Cytotoxic*
The isolate was found to exhibit *in vitro* cytotoxic activity against L1210 murine leukemia cells with an IC_{50} value of 4.08 μg/ml. The compound also showed toxicity to *Artemia salina* in the brine shrimp lethality test with an LC_{50} value of 3.09 μg/ml.

Reference

Elya, B., He, H.P., Kosela, S., Hanafi, M., and Hao, X.J. (2008) *Fitoterapia*, **79**, 182.

Yinyanghuo A and yinyanghuo B

Physical data:
- Yinyanghu A: $C_{25}H_{24}O_6$, yellow needles, mp 130–132 °C, $[\alpha]_D^{25} - 1°$ (DMSO, *c* 1.0)
- Yinyanghu B: $C_{25}H_{26}O_6$, yellow needles, mp 100–102 °C, $[\alpha]_D^{25} - 2.3°$ (Me$_2$CO, *c* 1.3)

Structures:

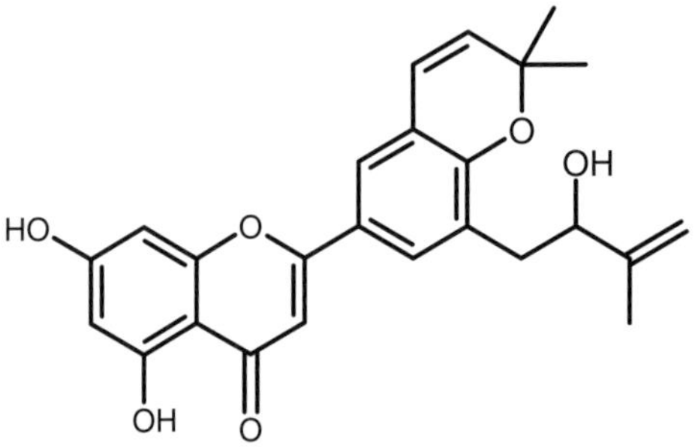

Yinyanghuo A

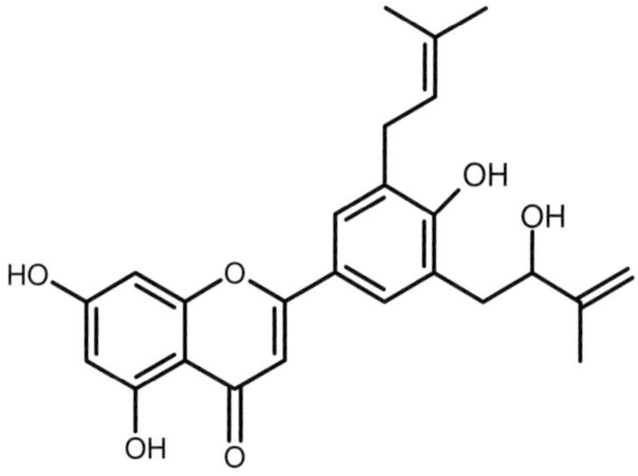

Yinyanghuo B

Compound class: Flavonoids (prenylflavones)

Source: *Epimedium sagittatum* (leaves; family: Berberidaceae)

Pharmaceutical potential: *Inhibitors to platelet aggregation*
Yinyanghu A and yinyanghu B were found to inhibit *in vitro* platelet aggregation induced by arachidonic acid with IC$_{50}$ values of 7.14 and 1.67 µM, respectively.

Reference

Chen, C.-C., Huang, Y.-L., Sun, C.-M., and Shen, C.-C. (1996) *J. Nat. Prod.*, **59**, 412

YM-47141 and YM-47142

Physical data: YM-47141: $C_{46}H_{62}N_8O_{13}$, $[\alpha]_D^{25} - 10.1°$ (MeCN, *c* 0.2)
YM-47142: $C_{43}H_{64}N_8O_{13}$, $[\alpha]_D^{25} - 1.8°$ (MeCN, *c* 0.5)

Structures:

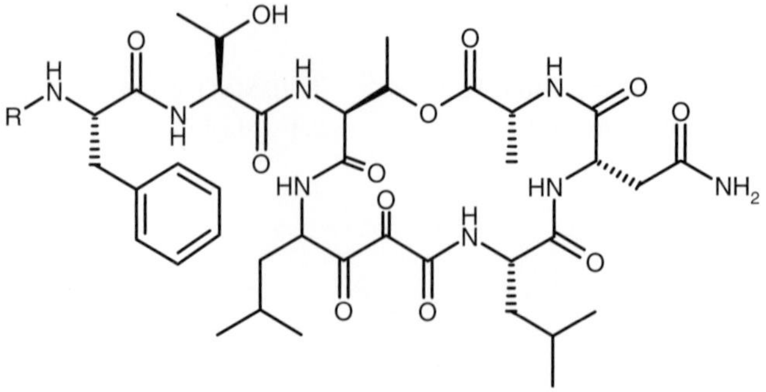

YM-47141: R = COCH$_2$C$_6$H$_5$
YM-47142: R = COCH$_2$CH(CH$_3$)$_2$

Compound class: Antibiotics

Source: *Flexibacter* sp. Q17897 (culture broth) [1, 2]

Pharmaceutical potential: *Human leukocyte elastase inhibitors (anti-inflammatory)*
Human leukocyte elastase (HLE), a neutral serine proteinase, has been implicated in the pathogenesis of a variety of inflammatory diseases such as emphysema, acute respiratory distress syndrome, and rheumatoid arthritis; therefore, inhibitors of HLE are considered as potential therapeutic agents for such inflammatory diseases [3–6]. YM-47141 and YM-47142 were found to possess potent inhibitory activity against human leukocyte elastase with IC_{50} values of 0.15 and 0.30 µM, respectively [1].

References

[1] Yasumuro, K., Suzuki, Y., Shibazaki, M., Teramura, K., Abe, K., and Orita, M. (1995) *J. Antibiot.*, **48**, 1425.
[2] Orita, M., Yasumuro, K., Kokubo, K., Shimizu, M., Abe, K., Tokunaga, T., and Kaniwa, H. (1995) *J. Antibiot.*, **48**, 1430.
[3] Starkey, P.M. and Barrett, A.J. (1976) *Biochem. J.*, **155**, 265.
[4] Janoff, A. (1978) *Neutral Proteases of Human Polymorphonuclear Leukocytes* (eds K. Havermann and A. Janoff), Urban and Schwartzenberg, Baltimore, MD, pp. 390–470.
[5] Merritt, T.A., Cochrane, C.G., Holcomb, K., Bohl, B., Hallman, M., Strayer, D., Edwards, D., and Gluck, L. (1983) *J. Clin. Invest.*, **72**, 656.
[6] Janoff, A. (1985) *Am. Rev. Respir. Dis.*, **132**, 417.

YM-47522

Physical data: $C_{24}H_{33}NO_4$, colorless gum, $[\alpha]_D^{25}$ – 106.1° (MeOH, *c* 0.33)

Structure:

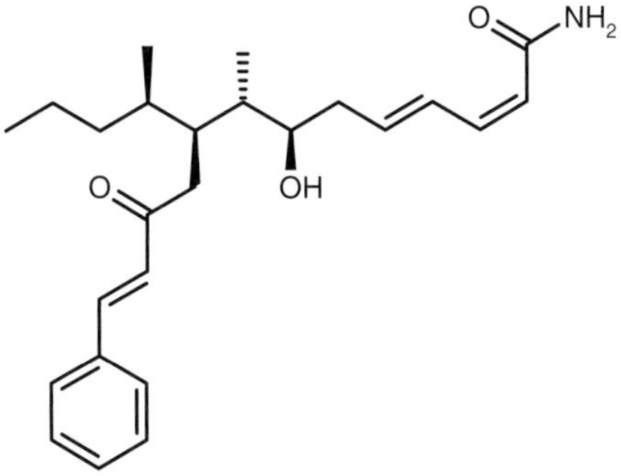

YM-47522

Compound class: Antibiotic

Source: *Bacillus* sp. YL-03709B (fermentation broth) [1, 2]

Pharmaceutical potentials: *Antifungal; cytotoxic*
The metabolite displayed potent *in vitro* antifungal activity, especially against *Rhodotorula acuta* and *Pichia angusta*, with MIC values of 0.05 and 0.75 µg/ml, respectively; it also showed moderate

or weak antifungal activity against *Candida albicans* and *Cryptococcus neoformans* with respective MIC values of 25.0 and 6.25 µg/ml, whereas it was inactive against filamentous fungi and bacteria [1]. The antibiotic was also found to be cytotoxic against lymphoid leukemia L1210 with an IC_{50} value of 0.41 µg/ml [1].

References

[1] Shibazaki, M., Sugawara1, T., Nagai, K., Shimizu, Y., Yamaguchi, H., and Suzuki, K. (1996) *J. Antibiot.*, **49**, 340.
[2] Sugawara, T., Shibazaki, M., Nakahara, H., and Suzuki, K. (1996) *J. Antibiot.*, **49**, 345.

Ys-II and Ys-IV

Systematic names:
- Ys-II: smilagenin 3-*O*-β-D-glucopyranosyl-(1→2)-β-D-galactopyranoside
- Ys-IV: smilagenin 3-*O*-β-D-glucopyranosyl-(1→2)-[β-D-glucopyranosyl-(1→3)]-β-D-galactopyranoside

Physical data:
- Ys-II: $C_{39}H_{64}O_{13}$, colorless needles, mp 252–255 °C, $[\alpha]_D^{26} - 51°$ ($CHCl_3$–MeOH 1:1, *c* 1.0)
- Ys-II: $C_{45}H_{74}O_{18}$, colorless needles, mp 265–267 °C, $[\alpha]_D^{26} - 43°$ (MeOH, *c* 1.0)

Structures:

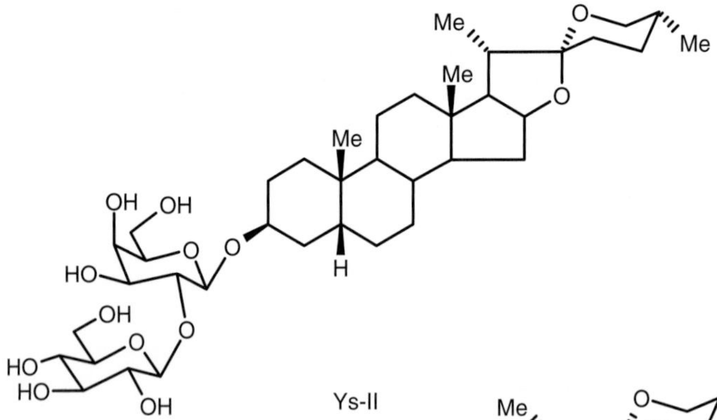

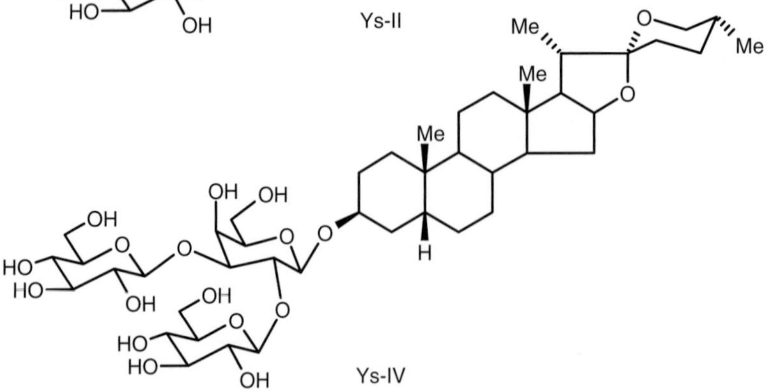

Compound class: Steroidal glycosides (spirostanol glycosides)

Sources: *Yucca gloriosa* (fresh caudex; family: Agavaceae) [1]; *Y. elephantipes* Regel. (stems) [2]

Pharmaceutical potential: *Antifungal*
Both the steroidal saponins, Ys-II and Ys-IV, exhibited moderate growth inhibitory activity against *Candida albicans* with IC_{50} values of 5.0 and 15.0 µg/ml, respectively, and also against *Cryptococcus neoformans* with IC_{50} values of 6.0 and 15.0 µg/ml, respectively [2].

References

[1] Nakano, K., Yamasaki, T., Imamura, Y., Murakami, K., Takaishi, Y., and Tomimatsu, T. (1989) *Phytochemistry*, **28**, 1215.
[2] Zhang, Y., Zhang, Y.-J., Jacob, M.R., Li, X.-C., and Yang, C.-R. (2008) *Phytochemistry*, **69**, 264.

Yuccalan

Systematic name: 3-O-β-D-Glucopyranosyl-(1→6)-β-D-glucopyranosyl-(3β,5α,6α,25 S)spirostan-3,6,27-triol

Physical data: $C_{39}H_{64}O_{15}$, amorphous powder, mp 192–195 °C

Structure:

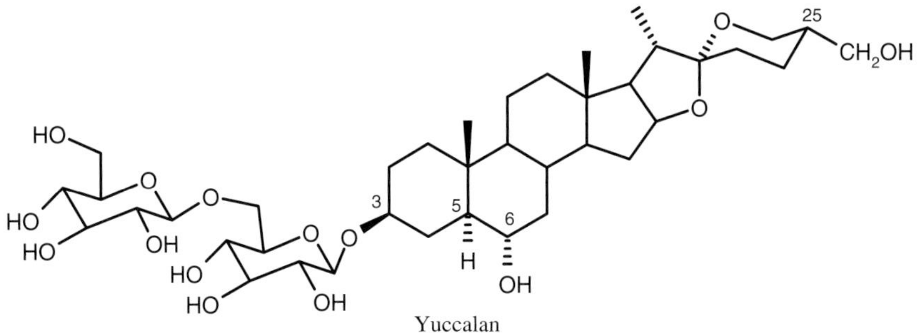

Yuccalan

Compound class: Steroidal saponin

Source: *Yucca smalliana* Fern. (leaves; family: Agavaceae)

Pharmaceutical potential: *Antifungal*
The isolate was found to exhibit antifungal activity against *Rhizoctonia solani* and *Fusarium oxysporum* with MIC values of 65 and 250 ppm, respectively.

Reference

Jin, Y.L., Kuk, J.-H., Oh, K.-T., Kim, Y.-J., Piao, X.-L., and Park, R.-D. (2007) *Arch. Pharm. Res.*, **30**, 543.

Z

Zaluzanin D

Physical data: $C_{17}H_{20}O_4$, crystalline solid, mp 103–104 °C, $[\alpha]_D$ ±0°

Structure:

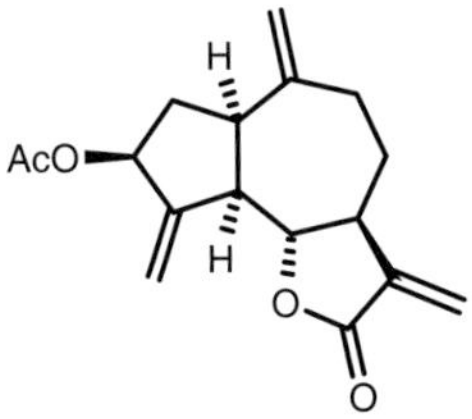

Zaluzanin D

Compound class: Terpenoid (guaianolide derivative)

Sources: *Zaluzania triloba* Lag. (family: Heliantheae) [1]; *Laurus nobilis* L. (leaves; family: Lauraceae) [2]

Pharmaceutical potential: ***Antitrypanosomal***
The guaianolide derivative showed strong trypanocidal activity against epimastigotes of *Trypanosoma cruzi* with a minimum lethal concentration (MLC) of 2.5 μM; it was found to inhibit proliferation of the parasite by 38% at a dose of 1 μg/ml as evaluated in HeLa infection assay [2].

References

[1] Romo de Vivar, A., Cabrera, A., Ortega, A., and Romo J. (1967) *Tetrahedron*, **23**, 3903.
[2] Uchiyama, N., Matsunaga, K., Kiuchi, F., Honda, G., Tsubouchi, A., Nakajima-Shimada, J., and Aoki, T. (2002) *Chem. Pharm. Bull.*, **50**, 1514.

Zarzissine

Physical data: $C_5H_5N_5$, colorless powder, sublimes at 270–271 °C

Structure:

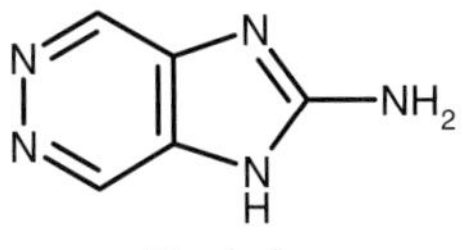

Zarzissine

Compound class: Guanidine alkaloid (4,5-guanidino-pyridaine compound)

Handbook of Pharmaceutical Natural Products, Volume 2.
Goutam Brahmachari

ISBN: 978-3-527-32148-3

Source: *Anchinoe paupertas* (sponge)

Pharmaceutical potential: *Cytotoxic*
The isolate displayed significant cytotoxic activity *in vitro* against human lung carcinoma (A-549), murine leukemia (P-388), and human nasopharyngeal carcinoma (KB) cells with IC_{50} values of 10, 12, and 5 µg/ml, respectively. The compound was also found to have a weak antimicrobial activity.

Reference

Bouaicha, N., Amade, P., Puel, D., and Roussakis, C. (1994) *J. Nat. Prod.*, **57**, 1455

Zerumbone

Systematic name: $2E,6E,9E$-Humulatrien-8-one

Physical data: $C_{15}H_{22}O$

Structure:

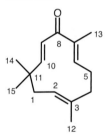

Zerumbone

Compound class: Sesquiterpenoid

Sources: *Zingiber zerumbet* (L.) J.E. Smith (rhizomes; family: Zingiberaceae) [1–3]; *Z. aromaticum* (rhizomes) [4]; *Z. cassumunar* Roxb. (rhizomes) [5, 6]

Pharmaceutical potentials: *NO production inhibitor; anti-inflammatory; antitumor; anticancer; antiproliferative; anti-HIV; fungitoxic*
Zerumbone was found to exert promising inhibition against lipopolysaccharide (LPS)-induced nitric oxide production in murine macrophage RAW 264.7 cells with an IC_{50} value of 5.4 µM (99.6% inhibition of inducible nitric oxide synthase (iNOS) enzyme activity at 20 µg/ml), which is more potent than the positive control, N^ω-monomethyl-L-arginine (L-NAMA), with an IC_{50} value of 21.3 µM (70% inhibition of iNOS enzyme activity at 20 µg/ml). The compound was reported to have anti-HIV activity with an EC_{50} value of 0.04 µM (IC_{50} value was determined to be 0.14 µM) [3]. The sesquiterpenoid also exhibited strong fungitoxic action against *Rhizoctonia solani* with minimum effective dose of 1000 ppm, much lower than those of some commercial fungicides [6].

Sharifah Sakinah *et al.* [8] demonstrated that zerumbone shows significant antiproliferative activity in human hepatoma (HepG2) cells *in vitro* by inducing apoptosis in target cancer cells in a time-dependent manner with an IC_{50} value of 3.45 ± 0.026 µg/ml; the inhibition was caused by decreasing the levels of antiapoptotic protein Bcl-2 and upregulation of proapoptotic Bax without involving p53. Zerumbone was also found to inhibit proliferation of nonmalignant Chang liver and MDBK cells; however, the IC_{50} value obtained was higher (>10 μg/ml) compared to that for HepG2 cells. Hence, zerumbone could be considered as a new alternative chemotherapeutic agent for human hepatoma [8].

Zerumbone was reported to possess effective anticancer potential, possibly by its apoptosis-inducing and antiproliferative influences [7]; it was found to suppress tumor promoter 12-O-tetradecanoylphorbol-13-acetate (TPA)-induced Epstein–Barr virus activation with a significant potential [1]. Murakami et al. [9] also showed that zerumbone suppressed the production of TPA-induced superoxide anion effectively from both NAPDH oxidase in DMSO-differentiated HL-60 human acute promyelocytic leukemia cells and xanthine oxidase in AS52 Chinese hamster ovary cells. Following their experiments, it was revealed that the test compound inhibits proliferation of a number of human colonic adenocarcinoma cell lines in a dose-dependent manner, while the growth of human normal dermal and colon fibroblasts was less affected. Zerumbone was also reported to suppress skin tumor initiation and promotion stages in ICR mice [10]. The same group also pointed out that α-humulene, a structural analogue lacking only the α,β-unsaturated carbonyl group, remained virtually inactive in all experiments they carried out, thereby indicating that α,β-unsaturated carbonyl group in zerumbone might be closely associated with the activities of the molecule [9].

It was established that zerumbone inhibits the activation of NF-ϰB and NF-ϰB-regulated gene expression induced by carcinogens, thereby providing a molecular basis for the prevention and treatment of cancer by it [11]; zerumbone is thus a potent lead compound for the development of anti-inflammatory, antitumor, and anticancerous drugs [9–13].

References

[1] Murakami, A., Takahashi, D., Jiwajinda, S., Koshimizu, K., and Ohigashi, H. (1999) *Biosci. Biotechnol. Biochem.*, **63**, 1811.
[2] Murakami, A., Takahashi, D., Koshimizu, K., and Ohigashi, H. (2003) *Mutat. Res.*, **523–524**, 151.
[3] Jang, D.S., Min, H.-Y., Kim, M.-S., Han, A.-R., Windono, T., Jeohn, G.-H., Kang, S.S., Lee, S.K., and Seo, E.-K. (2005) *Chem. Pharm. Bull.*, **53**, 829.
[4] Dai, J.-R., Cardellina, J.H., II, McMahon, J.B., and Boyd, M.R. (1997) *Nat. Prod. Lett.*, **10**, 115.
[5] Ozaki, Y., Kawahara, N., and Harada, M. (1991) *Chem. Pharm. Bull.*, **39**, 2353.
[6] Kishore, N. and Dwivedi, R.S. (1992) *Mycopathologia*, **120**, 155.
[7] Kirana, C., McIntosh, G.H., Record, I.R., and Jones, G.P. (2003) *Nutr. Cancer*, **45**, 218.
[8] Sharifah Sakinah, S.A., Tri Handayani, S., and Azimahtol Hawariah, L.P. (2007) *Cancer Cell Int.*, **7** (4). doi: 10.1186/1475-2867-7-4.
[9] Murakami, A., Takahashi, D., Kinoshita, T., Koshimizu, K., Kim, H.W., Yoshihira, A., Nakamura, Y., Jiwajinda, S., Terao, J., and Ohigashi, H. (2002) *Carcinogenesis*, **23**, 795.
[10] Murakami, A., Tanaka, T., Lee, J.-Y., Surh, Y.-J., Kim, H.W., Kawabata, K., Nakamura, Y., Jiwajinda, S., and Ohigashi, H. (2004) *Int. J. Cancer*, **110**, 481.
[11] Takada, Y., Murakami, A., and Aggarwal B.B. (2005) *Oncogene*, **24**, 6957.
[12] Huang, G.C., Chien, T.Y., Chen, L.G., and Wang, C.C. (2005) *Planta Med.*, **71**, 219.
[13] Nakamura, Y., Yoshida, C., Murakami, A., Ohigashi, H., Osawa, T., and Uchida, K. (2003) *FEBS Lett.*, **572**, 245.

Zerumboneoxide

Systematic name: (2R,3R)-Epoxy-6E,9E-humulatrien-8-one

Physical data: $C_{15}H_{22}O_2$, colorless crystal, mp 96–97 °C, $[\alpha]_D^{20}$ +0.0° (CHCl$_3$, c 0.24)

Structure:

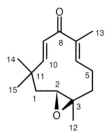

Zerumboneoxide

Compound class: Sesquiterpenoid

Source: *Zingiber zerumbet* (L.) J.E. Smith (rhizomes; family: Zingiberaceae) [1–4]

Pharmaceutical potential: *NO production inhibitor*
The test compound was found to exert significant inhibition against lipopolysaccharide (LPS)-induced nitric oxide production in murine macrophage RAW 264.7 cells with an IC$_{50}$ value of 14.1 µM (99.6% inhibition of inducible nitric oxide synthase (iNOS) enzyme activity at 20 µg/ml), which is almost equipotent to the positive control, N^ω-monomethyl-L-arginine (L-NAMA), with an IC$_{50}$ value of 21.3 µM (70% inhibition of iNOS enzyme activity at 20 µg/ml) [4].

References

[1] Gupta, V.N., Kataky, J.C.S., and Mathur, R.K. (1979) *Indian J. Chem. B*, **18**, 290.
[2] Bhatti, K., Wadia, M.S., Bhatia, I.S., and Kalsi, P.S. (1969) *Chem. Ind.*, 47.
[3] Kitayama, T., Yamamoto, K., Utsumi, R., Takatani, M., Hill, R.K., Kawai, Y., Sawada, S., and Okamoto, T. (2001) *Biosci. Biotechnol. Biochem.*, **65**, 2193.
[4] Jang, D.S., Min, H.-Y., Kim, M.-S., Han, A.-R., Windono, T., Jeohn, G.-H., Kang, S.S., Lee, S.K., and Seo, E.-K. (2005) *Chem. Pharm. Bull.*, **53**, 829.

Zhankuic acids A–C

Systematic names:
- Zhankuic acid A: 4α-methylergosta-8,24(28)-dien-3,7,11-trione-26-oic acid
- Zhankuic acid B: 3α-hydroxy-4α-methylergosta-8,24(28)-dien-7,11-dione-26-oic acid
- Zhankuic acid C: 3α,12α-dihydroxy-4α-methylergosta-8,24(28)-dien-7,11-dione-26-oic acid

Physical data:
- Zhankuic acid A: $C_{29}H_{40}O_5$, pale yellow needles, mp 136–138 °C, $[\alpha]_D^{25}$ +77.6° (CHCl$_3$, c 1.47)
- Zhankuic acid B: $C_{29}H_{42}O_5$, pale yellow needles, mp 188–191 °C, $[\alpha]_D^{25}$ +44.4° (CHCl$_3$, c 0.26)

- Zhankuic acid C: $C_{29}H_{40}O_6$, pale yellow needles, mp 164–168 °C, $[\alpha]_D^{25}$ +118° (CHCl$_3$, c 0.125)

Structures:

Zhankuic acid A

Zhankuic acid B: R^1 = OH; R^2 = H
Zhankuic acid C: R^1 = R^2 = OH

Compound class: Triterpenes (steroidal acids)

Sources: *Antrodia cinnamomea* Chang & Chou, sp. nov. (fruiting bodies; family: Polyporaceae), a fungal parasite of the tree *Cinnamomum micranthum* (Hayata) Hayata [1]; *A. camphorata* (fruiting bodies) [2]

Pharmaceutical potentials: *Cytotoxic; anticholinergic and antiserotonergic; anti-inflammatory*
Zhankuic acids A and C exhibited *in vitro* cytotoxicity against P-388 murine lymphocytic leukemia cells with IC$_{50}$ values of 1.8 and 5.4 µg/ml, respectively [1]. Among them, zhankuic acid B showed anticholinergic and antiserotonergic activities as tested on the guinea pig ileum preparation at a concentration of 10 µg/ml [1]. All these three compounds were evaluated as major immunomodulatory principles in human mononuclear cell proliferation and expression of hepatitis B surface antigen (HBsAg) [3].

Recently, Shen *et al.* [2] reported that zhankuic acids A–C possess anti-inflammatory effects as evaluated by using an acute cellular model in isolated peripheral human neutrophils activated by N-formyl-methionyl-leucyl-phenylalanine (fMLP) and/or phorbol 12-myristate-13-acetate (PMA); reactive oxygen species (ROS) production and firm adhesion by neutrophils are known to display two important responses during inflammation. The present investigators observed that pretreatment with 1–25 µM of zhankuic acid A, B, or C reduced fMLP- or PMA-induced ROS production in a concentration-dependent manner, with IC$_{50}$ value(s) around 5–20 µM. From their detailed studies, it was revealed that the zhankuic acids exhibit leukocyte-modulating activity by inhibiting both ROS production and firm adhesion by neutrophils without significant cytotoxic effects, thereby suggesting their potentiality as promising anti-inflammatory lead candidates.

References

[1] Chen, C., Yang, S.-W., and Shen, Y.-C. (1995) *J. Nat. Prod.*, **58**, 1655.
[2] Shen, Y.-C., Wang, Y.-H., Chou, Y.-C., Chen, C.-F., Lin, L.-C., Chang, T.-T., Tien, J.-H., and Chou, C.-J. (2004), *Planta Med.*, **70**, 310.
[3] Yang, Y.-Y., Lin, H.-C., Hou, M.-C., Chiang, J.-H., Chiou, Y.-Y., Tasy, S.-H., Chang, F.-Y., and Lee, S.-D. (2003), *J. Chin. Med. Assoc.*, **66**, 247.

Zinolol

Systemic name: 2-(*O*-β-D-Glucopyranosyl)-6-(*N*-methylaminomethyl)-1,4-dihydroxybenzene

Physical data: $C_{14}H_{21}NO_8$, amorphous solid, mp 164–166 °C, $[\alpha]_D^{25}$ −375° (MeOH, c 0.04)

Structure:

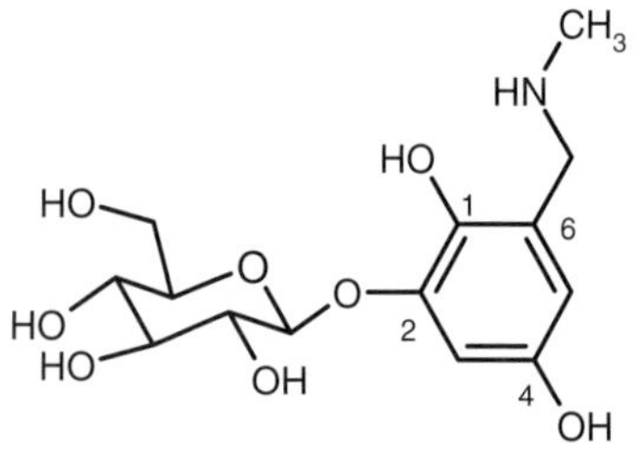

Zinolol

Compound class: Hydroxyhydroquinone glycoside

Source: *Anagallis monelli* ssp. *linifolia* (L.) (aerial parts; family: Primulaceae).

Pharmaceutical potentials: *Antimutagenic; antioxidant*

The compound was found to be nonmutagenic (nongenotoxic) at different concentrations and even at high concentration (50 mg/ml) as assessed by the *Escherichia coli* PQ37 mutagenicity assay; the present investigators also showed that the compound possesses effective antioxidant potential with significant scavenging activity for both $ABTS^+$ and DPPH radicals. It exhibited $ABTS^+$ cation radical scavenging activity of 97.1% at a concentration of 1 mg/ml; at a lower concentration of 0.03 mg/ml, it displayed 70% $ABTS^+$ cation radical inhibition immediately after 5 min of incubation (trolox: 97.9% inhibition at 1 mg/ml and after 5 min of incubation). In the DPPH assay, the hydroxyhydroquinone showed radical scavenging activity in a concentration-dependent manner; percentage of inhibition varies from 25.8 to 97.8 with the varying concentrations from 0.03 to 1 mg/ml after 30 min of incubation (trolox: 99.3% inhibition at 1 mg/ml after 30 min of incubation). Thus, the *in vitro* test of zinolol against toxic radical entities proved the efficiency of the molecule as a promising antioxidant agent. Hence, zinolol appears as a strong agent against the genotoxicity as well as the oxidative stresses caused by free radicals.

Reference

Ammar, S., Mahjoub, M.A., Charfi, N., Skandarani, I., Chekir-Ghedira, L., and Mighri, Z. (2007) *Chem. Pharm. Bull.*, **55**, 385.

Tables

All the compounds described in this book are classified into seven tables (Tables A.1–A.7) on the basis of their structural skeletons, namely, alkaloids, antibiotics; flavonoids; lignoids; terpenoids; xanthonoids, and miscellaneous. Each of the tables enlists compounds as per alphabetic order, and highlights origin as well as documented drug use of each of the respective compounds.

Table A.1 Alkaloids

Compound	Type	Origin	Bioactivity
(−)-Dibromophakellin	Guanidine alkaloid	Marine sponge	$α_{2b}$-Adrenoceptor agonist
(−)-Salutaridine	Morphinane alkaloid	Plant	Antihepatitis B virus
(+)-Aplysinillin	Bromotyrosine-derived alkaloid	Marine sponge	Inhibitor to hyphae formation; cytotoxic (anticancerous)
(+)-N-(Methoxycarbonyl)-N-norboldine	Isoquinoline alkaloid	Plant	Antibacterial
1,2-Dehydrogeissoschizoline	Indole alkaloid	Plant	Antimalarial
1-Methoxycanthinone	β-Carboline	Plant	Anti-HIV; antitumor
2-(Methyldithio)pyridine-N-oxide	Pyridine-n-oxide alkaloid	Plant	Antimicrobial; antiproliferative
5-Bromo-8-methoxy-1-methyl-β-carboline	β-Carboline	Marine bryozoan	Antitumor; antimicrobial
6-Methoxyspirotryprostatin B	Diketopiperazine alkaloid	Fungus	Cytotoxic
6-O-Nicitinoyl-7-O-acetylscutebarbatine G	neo-Clerodane diterpenoid alkaloid	Plant	Cytotoxic
7-epi-Dioncophylline A	Naphthylisoquinoline alkaloid	Plant	Antimalarial
7′-O-Demethylisocephaeline	Benzoquinolizidine alkaloid	Plant	Antileishmanial
7-O-Nicitinoyl scutebarbatine H	neo-Clerodane diterpenoid alkaloid	Plant	Cytotoxic
7-Oxo-3,8,9-trihydroxystaurosporine	Indolo[2,3-a]carbazole alkaloid	Marine ascidian	Cytotoxic
7-Oxo-8,9-dihydroxy-4′-N-demethylstaurosporine	Indolo[2,3-a]carbazole alkaloid	Marine ascidian	Cytotoxic
7-Oxohernangerine	Isoquinoline alkaloid	Plant	Anti-HIV
8-Acetylheterophyllisine	Heteratiosine-type diterpenoid alkaloid	Plant	Antifungal
9-Deacetoxyfumigaclavine C		Fungus	Cytotoxic
9-Hydroxycanthin-6-one		Plant	Cytotoxic (antitumor)
9-Methoxycanthin-6-one		Plant	Cytotoxic (antitumor)
14-Norpseurotin A	Diketopiperazine alkaloid	Fungus	Antimicrobial
18-Oxotryprostatin A	Diketopiperazine alkaloid	Fungus	Cytotoxic
Ageladine A		Marine sponge	Matrix metalloproteinase (mmp) inhibitor (anticancerous)

Ambiguine H and I isonitriles	Indole alkaloids	Cyanobacteria	Antimicrobial (antibacterial and antifungal)
Ancistrotanzanines A and B	Naphthylisoquinoline alkaloids	Plant	Antiplasmodial; antileishmanial; antrypanocidal
Ammomontine	Pyrimidine-β-carboline alkaloid	Plant	Antileishmanial
Ascidiathiazones A and B	Thiazine alkaloids	Ascidian	Anti-inflammatory
Batzelladines A and B	Guanidine alkaloids	Marine sponge	Anti-HIV
Benzastatin C	Dihydroquinoline alkaloid	Bacteria	Antioxidant (free radical scavenger); antiviral
Bidebilines C and D	Bis-dehydroaporphine alkaloids	Plant	Antimalarial
Buchapine and its isomer	Quinoline alkaloids	Plant	Anti-HIV
Cassiarins A and B	Isoquinoline alkaloids	Plant	Antimalarial
Castanospermine	Tetrahydroxyindolizidine alkaloid	Plant	Anti-HIV; enzyme inhibitor; anti-inflammatory; antidiabetic
Caulophine		Plant	Antimyocardial ischemia activity
Ceratamines A and B		Marine sponge	Antimitotic
Chaetominine		Fungus	Cytotoxic (anticancerous)
Chinese bittersweet alkaloid II		Plant	Cytotoxic
Circumdatin H		Fungus	Mitochondrial NAPDH oxyadse inhibitor
Cladoniamide G	Tryptophan-derived alkaloid	Bacteria	Cytotoxic
Corynoline	Isoquinoline alkaloid	Plant	Cytotoxic; anticancerous; acetylcholiesterase (AChE) inhibitor; hepatoprotective
Crambescidins 800 and 826	Guanidine alkaloids	Marine sponge	Anti-HIV; anti-HSV-1
Cystodytins D, E, F, G, and I	Tetracyclic aromatic alkaloids	Marine tunicate	Cytotoxic (antitumor)
Decarine	Isoquinoline alkaloid	Plant	Antimalarial; antibacterial
Dioncopeltine A	Naphthylisoquinoline alkaloid	Plant	Antimalarial
Dioncophyllines A, B, and C	Naphthylisoquinoline alkaloids	Plant	Antimalarials
Drymaritin		Plant	Anti-HIV
Esculeoside A	Steroidal alkaloid glycoside	Plant	Cytotoxic
Hamacanthins A and B	Bis-indole alkaloids	Marine sponge	Antimicrobial
Harman and its semisynthetic derivatives	β-Carboline alkaloids	Plant	Anti-HIV
Hederacines A and B		Plant	Anticancerous
Hernandonine		Plant	Anti-HIV
Herquline B	Tropane alkaloids	Fungus	Platelet aggregation inhibitor

(Continued)

Table A.1 (Continued)

Compound	Type	Origin	Bioactivity
Hookerianamides H and I	Steroidal alkaloids	Plant	Cholinesterase inhibitors (anti-Alzheimer's disease); antiplasmodial
Hookerianamides J and K	5α-Pregnane-type steroidal alkaloids	Plant	Inhibitors of acetylcholinesterase (AChE) and butyrylcholinesterase (BChE) enzymes activity; antileishmanial; antibacterial
Hypoglaunine B	Sesquiterene pyridine alkaloid	Plant	Anti-HIV
Isatinones A and B	Indole alkaloids	Plant	Antifungal
Jerantinines A, B, C, D, E, and F	Indole alkaloids	Plant	Cytotoxic
Jineol	Quinoline alkaloid	Centepede	Cytotoxic (antitumor)
Klugine	Benzoquinolizidine alkaloid	Plant	Antileishmanial; antiplasmodial
Lamellarin α 20-sulfate	Lamellarin-type alkaloid	Ascidian	Anti-HIV
Laurolistine (norboldine)		Plant	Anti-HIV
Lindechunine A		Plant	Anti-HIV
Lissoclibadin 1	Polysulphur alkaloid	Ascidian	Cytotoxic; antitumor; anticancerous; antibacterial;
Lissoclibadin 2	Polysulphur alkaloid	Ascidian	Cytotoxic; antitumor; anticancerous; antibacterial; antifungal
Lissoclibadin 3	Polysulphur alkaloid	Ascidian	Cytotoxic; antitumor; anticancerous
Lycojapodine A		Moss	Acetylcholinestrease inhibitor; anti-HIV-1
Lycoperine A		Moss	Acetylcholinestrease inhibitor
Manzamine A and its hydroxy-derivatives	Complex β-carboline alkaloids	Marine sponge	Antitumor; anti-HSV-II; GSK-3 and CDK-5 inhibitor; antibacterial; antifungal; anti-HIV
Martefragin A	Indole alkaloid	Red alga	Antioxidant
Massadine		Marine sponge	Geranylgeranyltransferase type I (GGTase I) inhibitor
Michellamines A and B	Naphthalene tetrahydroisoquinoline alkaloids	Plant	Anti-HIV
Militarinone A	Pyridone alkaloid	Fungus	Neurotrophic
Nitensidine E	Guanidine alkaloid	Plant	Cytotoxic (anticancerous)
N-Methylnarceimicine	Quaternary alkaloid	Plant	Antihepatitis B virus

Nostocarboline	Quaternary β-carboline alkaloid	Cyanobacteria	Butyrylcholinesterase (BChE)-inhibitor (-Glucosidase inhibitory; antioxidant (radical scavenging)
Oriciacridones C and F	Acridone alkaloids	Plant	Anti-HIV, anticancerous
ortho-Demethylbuchenavianine	Flavonoid alkaloid (piperidine-flavone alkaloid)	Plant	Antifungal
Panicutine	Hetidine-type diterpenoid alkaloid	Plant	Antihepatitis B virus (HBV); anti-HIV
Periglaucines A, B, C, and D	Hasubanane-type alkaloids	Plant	CYP2D6 inhibitory
Pipercyclobutanamide A		Plant	Vasorelaxant
Pordamacrines A and B		Plant	Antitubercular
Pseudopteroxazole and seco-pseudopteroxazole	Diterpenoid alkaloids	Coral	
Saxicolaline A	Quaternary alkaloid	Fungus	Antihepatitis B virus
Scalusamide A	Pyrrolidine alkaloid	Plant	Antifungal; antibacterial
Scutebarbatines C, D, E, and F	neo-Clerodane diterpenoid alkaloids	Plant	Cytotoxic
Scutebarbatines G and H	neo-Clerodane diterpenoid alkaloids	Plant	Cytotoxic
Scutebarbatines I, J, K, and L	neo-Clerodane diterpenoid alkaloids	Plant	Cytotoxic
Siamenol	Carbazole alkaloid	Plant	Anti-HIV
Sieboldine A	Fused-tetracyclic ring system consisting of an aza-cyclononane ring	Moss	Acetylcholinesterase (AChE) inhibitor; cytotoxic
Streptoverticillin	Carbazole alkaloid	Actinomycete	Antifungal
Strychnogucines A and B	Bisindole-type alaloids	Plant	Antimalarial
Syncarpamide	(+)-Norepinephrine derivative	Plant	Antimalarial
Trikendiol		Marine sponge	Anti-HIV
Triphyophyllum naphthylisoquinoline alkaloids		Plant	Antimalarial
Ungeremine	Quaternary alkaloid	Plant	Acetylcholinesterase (AChE) inhibitor; cytotoxic; hypotensive
Verticillin G	Epidithiodioxopiperazine derivative	Fungus	Antibacterial
Vilmorrianone	Atisine-type diterpenoid alkaloid	Plant	Antifungal
Wrightiamine A	Pregnane alkaloid	Plant	Cytotoxic
Zarzissine	Guanidine alkaloid (4,5-guanidino-pyridine compound)	Marine sponge	Cytotoxic

Table A.2 Antibiotics

Compound	Type	Origin	Bioactivity
1-Hydroxycrisamicin	Isochromaquinone derivative	Bacteria	Antibacterial
3-Acetamino-6-isobutyl-2,5-dioxopiperazine	Cyclodipeptide	Fungus	Cytotoxic
4849F		Bacteria	Inhibitor of IL-4 receptor
Adxanthromycins A and B	Dimeric anthrone peroxide galactosides	Bacteria	Inhibitors of ICAM-1/LFA-1 mediated cell adhesion molecule (antitumor)
Agrochelin		Bacteria	Cytotoxic
Antimycin A₉	Alkaloid derivative	Bacteria	Antimicrobial; nematocidal; insecticidal; bovine heart NADH oxidase inhibitor
Apratoxin D	Cyclodepsipeptide	Marine cyanobacteria	Cytotoxic
Apratoxins A and E	Cyclodepsipeptides	Marine cyanobacteria	Cytotoxic(antiproliferative)
Arenamides A and B	Cyclohexadepsipeptides	Marine acetinomycete	Antitumor (NFκB inhibitors); inhibitors to NO production; cytotoxic
Arisugacins A, B, C, and D		Fungus	Acetylcholinesterase (AChE) inhibitors
AS-186a, b, c, d, and g		Fungus	Inhibitors of acyl-CoA:cholesterol acyltransferase (ACAT)
Aspergillides A, B, and C	Macrolides	Fungus	cytotoxic
Azoxybacilin		Bacteria	Antifungal
Bassiatin		Fungus	Platelet aggregation inhibitor
BE-23372M		Fungus	Protein tyrosine kinase inhibitor; antitumor
BE-24566B		Bacteria	Antimicrobial
BE-54238A and BE-54238B		Bacteria	Cytotoxic and antitumor (anticancer)
Brartemicin		Actinomycete	Antitumor
Brasilicardin A	Diterpenoid antibiotic	Bacteria	Immunosuppressive; antitumor
Brasilinolide A	Macrolide	Bacteria	Antifungal; immunosuppressive
Brasilinolide B	Macrolide	Bacteria	Antifunga
Brasiliquinones A, B, and C	Benz[a]anthraquinone antibiotics	Bacteria	Antibacterial; antitumor
Bistramides E, F, G, H, I, and J	Cyclic hexapeptides	Ascidian	Antitumor
Callipeltin A	Cyclic depsidecapeptide	Sponge	Anti-HIV; antifungal
Caminoside A	Glycolipid	Marine sponge	Antimicrobial

Name	Structure/Class	Source	Activity
Carmabin A, Dragomabin, and Dragomamide A	Lipopeptides	Cyanobacteria	Antimalarial
Cathestains A and B		Fungus	Cysteine proteinase inhibitors
Chemomicin A	Angucyclinone-type antibiotic	Bacteria	Antimicrobial
Chloropeptin I	Chlorinated peptide; bicyclic hexapeptide	Bacteria	Anti-HIV; antimicrobial
Chloropeptin II	Chlorinated peptide; bicyclic hexapeptide	Bacteria	Anti-HIV; antimicrobial
Ciliatamides A and B	Lipopeptides	Marine sponge	Antileishmanial; cytotoxic
Clethramycin	Polyene antibiotic	Bacteria	Broad spectrum antifungal
Complestatin	Chlorinated peptide; bicyclic hexapeptide	Bacteria	Anti-HIV; antimicrobial
Complestatins A and B	Peptides (bicyclic hexapeptides)	Bacteria	Anti-HIV
Cordyheptapeptide A	Cycloheptadepsipeptide	Fungus	Antimalarial
CT2108A and B	Azaphilone-type antibiotics	Fungus	Fatty acid synthase (FAS) inhibitors; antifungal
Cyclomontanins A, C, and D	Cyclopeptides	Plant	Anti-inflammatory
Cystothiazole A		Myxobactrium	Antifungal; cytotoxic
Cytotrienin		Bacteria	Antitumor; antibacterial
Daunomycin	Anthrcycline antibiotic	Actinomycete	Antifungal; antibacterial
Decatromicins A and B		Bacteria	Antimicrobial
Diperamycin	Cyclic hexadepsipeptide antibiotic	Bacteria	Antimicrobial; antitumor
Durhamycins A and B	Aureolic acid family	Bacteria	Anti-HIV
EI-1511-3 and EI-1511-5		Bacteria	Interleukin-1β converting enzyme (ICE) inhibitors (anti-inflammatory); antimicrobials
EI-1625-2		Bacteria	interleukin-1β converting enzyme (ICE) inhibitors (anti-inflammatory); antimicrobial
Epichlicin	Cyclic peptide	Fungus	Antifungal
epi-Cochioquinone A		Fungus	Acyl-CoA:cholesterol acyltransferase (ACAT) inhibitor
Espicufolin	Anthrquinone derivative	Bacteria	Neuronal cell protecting
Etnangien	Macrolactone antibiotic	Myxobactrium	Antibacterial
F390, F390B and C	Dihydroxanthone derivatives	Fungus	Antitumor
Fattiviracin A1		Bacteria	Antiviral
FD-211		Fungus	Antitumor
Fleephilone		Fungus	Anti-HIV
Formobactin		Bacteria	Free radical scavenger (antioxidant); neuronal cell protecting
FR177391		Bacteria	Antihypertriglyceridemic

(*Continued*)

Table A.2 (Continued)

Compound	Type	Origin	Bioactivity
FR191512	Polyphenolic compound	Fungus	Antiviral (anti-influenza virus)
FR207944	Triterpne glucoside	Fungus	Antifungal
FR901463, FR901464, and FR901465		Bacteria	Antitumor
FR901469	A 40-membered macrocyclic lipopeptidolactone consisting of 12 amino acids and a 3-hydroxypalmitoyl moiety	Fungus	Antifungal; antipneumonic
Fumimycin	Highly functionalized benzofuranone derivative	Fungus	Peptide deformylase (PDF) inhibitor; antibacterial
Funalenone		Fungus	Collagenase inhibitor; bacterial cell wall synthesis enzymes inhibitors; anti-HIV
Furanocandin		Bacteria	Antifungal
Gilvusmycin		Bacteria	Antitumor
Glomosporin	Cyclic depsipeptide	Fungus	Antifungal
Grassypeptolide	Macrocyclic depsipeptide	Cyanobacteria	Anticancerous
Harziphilone		Fungus	Anti-HIV
Hesseltin A		Fungus	Antiviral (HSV-1)
Hirsutellide A	Cyclohexadepsipeptide	Fungus	Antimycobacterial; antimalarial
Hispidospermidin		Fungus	Phospholipase C (PLC) inhibitor
Hypocrellin D		Ascomycete	Cytotoxic
IC202A, B, and C	Ferrioxamine family	Bacteria	Immunosuppressive
Isocomplestatin	Peptide (a bicyclic hexapeptide)	Bacteria	Anti-HIV
Jaspamides B, C, D, E, F, G, and H. and Jaspamides J, K, L, M, N, O, and P	Depsipeptides (tryptophan modified jaspamide derivatives)	Marine sponge	Cytotoxic (antimicrofilament)
Kalimantacins A, B, and C		Bacteria	Antimicrobials
Karalicin		Bacteria	Antiviral; cytotoxic; antifungal
Kempopeptins A and B	Cyclodepsipeptides	Cyanobacteria	Serine protease inhibitors
Kitastatin -1	Cyclodepsipeptide	Bacteria	Anticancerous

Koshikamide A₂	Peptide	Marine sponge	Cytotoxic
Lactonamycin		Bacteria	Antimicrobial; cytotoxic
Largamides D, E, F, and G	Cyclic peptides	Marine cyanobacteria	Chymotrypsin inhibitors
Longicalycinin A	Cyclic peptide	Plant	Cytotoxic
Lucensimycin E		Bacteria	Antibacterial
Manoalide	Sesquiterpenoid antibiotic	Marine sponge	Anti-inflammatory; antibacterial; cobra venom and bee venom inhibitor
Marinopyrroles A and B	Halogenated bipyrrole derivatives	Bacteria	Antibacterial; cytotoxic
Microspinosamide	A cyclic depsidecapeptide	Marine sponge	Anti-HIV
Migrastatin	A glitarimide antibiotic having a unique 14-membered lactone ring	Bacteria	Antitumor
Moromycins A and B	C-Glycosylangucycline-type antibiotics	Bacteria	Cytotoxic (anticancer)
Multipolides A and B	10-Membered lactones	Fungus	Antifungal
Neopyrrolomycin B		Bacteria	Antibacterial
NG-061	Phenyl acetic acid hydrazide derivative	Fungus	Potentiator of nerve growth factor (NGF)
Nidulalin A (F390)	Dihydroxanthone derivative	Fungus	Antitumor
NK154183A and B		Bacteria	Antitumor; antifungal
NK372135A, B, and C		Fungus	Antifungal
Nothramicin	Anthraquinone derivative	Bacteria	Antimycobacterial
NP-101A	Acetamidobenzamide derivative	Bacteria	Antifungal
Obyanamide	Depsipeptide	Marine cyanobacteria	Cytotoxic
Ochracenomicin A	Benz[α]-anthraquinone derivative	Bacteria	Antimicrobial
Paecilopeptin		Fungus	Cathepsin S enzyme inhibitor
Palau'amide	Cyclopeptide	Cyanobacteria	Cytotoxic (antitumor)
Pelagiomicin A		Bacteria	Antimicrobial; cytotoxic
Penicillide		Fungus	Inhibitor of acyl-CoA:cholesterol acyltransferase (ACAT)
PF 1163A and B		Fungus	Antifungal
Phakellistatin 14	Cyclic peptide	Marine sponge	Cytotoxic
Phomacins A, B, and C	Cytochalasan derivatives	Fungus	Antitumor
Phosphatoquinones A and B	Naphthoquinone derivatives	Fungus	Protein tyrosine phosphatase (PTPase) inhibitors
Piptamine		Mushroom	Antimicrobial

(Continued)

Table A.2 (*Continued*)

Compound	Type	Origin	Bioactivity
PM-94128	Isocoumarin antibiotic	Bacteria	Antitumor
Pochonin G	Macrocyclic antibiotic (resorcylic acid lactone)	Fungus	WNT-5A expression inhibitor (thus acts as hair-growth stimulator)
Polyketomycin		Bacteria	Antibacterial; cytotoxic
Pterocidin	Polyketide δ-lactone	Bacteria	Cytotoxic
Pulchellalactum		Fungus	Tyrosine phosphatase inhibitor
Purpactin A		Fungus	Inhibitor of acyl-CoA:cholesterol acyltransferase (ACAT)
Pyripyropenes I, J, K, and L		Fungus	Acyl-CoA:cholesterol acyltransferase (ACAT) inhibitors
Ravenic acid	Polyene tetramic acid	Fungus	Antibacterial
Respirantin and a related cyclodepsipeptide	Cyclodepsipeptides	Bacteria	Anticancerous
Rhodiocyanoside A	Cyanoglycoside	Plant	Antiallergic
Rostratins A–D	Dithioalkaloids	Fungus	Cytotoxic
Sch 213766		Fungus	Chemokine receptor CCR-5 inhibitor (anti-HIV)
Sch642305		Fungus	Inhibitor to bacterial DNA primase
Scytalidamides A and B	Cyclic peptides	Marine sponge	Cytotoxic (antitumor)
Semicochliodinols A and B	Indole-quinone derivatives	Fungus	HIV-1 protease inhibitor; epidermal growth factor receptor protein tyrosine kinase (EGF-R PTK)
Sg17-1-4	Isocoumarin derivative	Fungus	Cytotoxic
SPF-32629A and B		Fungus	Human chymase inhibitors
Stevastelins B and B3	Depsipeptides	Actinomycete	Immunosuppressive
Streptokordin	Pyridone derivative	Bacteria	Cytotoxic
Terpendoles J, K, and L		Fungus	Acyl-CoA:cholesterol acyltransferase (ACAT) inhibitors
Thiazomycin	Thiazolyl peptide	Bacteria	Antibacterial

Thiomarinols A, B, C, D, E, F, and G		Bacteria	Antimicrobial
TMC-135 A and B	Manumycin group	Bacteria	Antitumor
TMC-1A, B, C, and D	Manumycin group	Fungus	Antitumor
TMC-52A, B, C, and D	Epoxysuccinyl peptides	Bacteria	Cysteine proteinase inhibitors
TMC-86A, B, and TMC-96	Epoxy-β-aminoketone derivatives	Bacteria	Proteasome inhibitors; cytotoxic (antitumor)
TMC-89A and B		Bacteria	Proteasome inhibitors
TMC-95A, B, C, and D		Fungus	Proteasome inhibitors; cytotoxic
Tolybyssidins A and B	Cyclic peptides	Cyanobacteria	Antifungal
Topopyrones A–D	Pyranoanthraquinone derivatives	Fungus	Topoisomerase I inhibitors (antiviral); antitumor; antimicrobial
Topostatin		Actinomycete	Topoisomerase I and II inhibitor, antitumor
Trachyspic acid	Tricarboxylic acid derivative	Fungus	Heparanase inhibitor
Trichodimerol (BMS-182123)		Fungus	Inhibitor to TNF-α production
Trierixin	Triene-ansamycin group	Bacteria	Inhibitor of ER stress-induced XBP1 activation (antitumor)
Tubelactomicin A	Macrolide	Actinomycete	Antitubercular
UCS1025 A		Fungus	Antibacterial; antitumor
UK-2A and B		Bacteria	Antifungal
UK-3A		Bacteria	Antifungal
Valinomycin	Cyclodepsipeptide	Bacteria	Cytotoxic; antitumor; antiviral; antifungal
Venturamides A and B	Cyclic hexapeptides (dendroamide-type)	Cyanobacteria	Antimalarial
Vinylamycin	Depsipeptide	Bacteria	Antibacterial
Viridamide A	Lipodepsipeptide	Cyanobacteria	Antiparasitic (leishmanicidal; trypanocidal; antiplasmodial)
WF14865A and B		Fungus	Cathepsins B and L inhibitors
YM-47141 and YM-47142		Bacteria	Human leucocyte elastase (HLE) inhibitors (anti-inflammatory)
YM-47522		Bacteria	Antifungal; cytotoxic

Table A.3 Flavonoids

Compound	Type	Origin	Bioactivity
(−)-Epicatechin 5-gallate	Flavan	Plant	μ-Calpain inhibitor
(+)-Dihydromyricetin	Flavanone	Plant	Antitumor; anticancerous; antiviral; anti-HIV
(+)-Tephrorins A and B	Flavanone	Plant	Anticancerous
(+)-Tephrosone	Chalcone	Plant	Anticancerous
(2S) (−)-Kurarinone and (2S)-2′-methyl ether of kurarinone	Flavanones	Plant	Cytotoxic
(2S)- and (2R)-Eriodictyol 7-O-β-D-glucopyranosiduronic acids	Flavanone glycosides	Plant	Aldose reductase inhibitors
(2S)-2′-Hydroxy-7,8,3′,4′,5′-pentamethoxyflavan	Flavan	Plant	Cytotoxic
(2S)-5,7,3′,5′-Tetrahydroxy flavanone-7-O-β-D-glucopyranoside	Flavanone glycoside	Plant	Antioxidant
(2S)-5,7-Dimethoxy-8-(2S-hydroxy-3-methyl-3-butenyl)-3′,4′-methylenedioxyflavanone	Flavanone	Plant	Quinone reductase inducer
(2S)-5,7-Dimethoxy-8-formylflavanone	Flavanone	Plant	Quinone reductase inducer
(2S)-5-Hydroxy-7-methoxy-8-[(E)-3-oxo-1-butenyl]flavanone	Flavanone	Plant	Quinone reductase inducer
(2S)-Poncirin	Flavone glycoside	Plant	Anti-inflammatory
2,2′,4′-Trihydroxy-6′-methoxy-3′,5′-dimethylchalcone	Chalcone	Plant	Antiprotozoal
2,3-dihydro-4′,4-di-O-methylamentoflavone	Flavone	Plant	Inhibitor of tyrosinase and melanin activity
2′,4′-Dihydroxy-3′-(2-hydroxybenzyl)-6′-methoxychalcone	Chalcone	Plant	Cytotoxic; antimalarial; antibacterial
2′-Hydroxyneobavaisoflavanone	Isoflavonoid (isoflavanone)	Plant	Anti-HIV
2-Methylchromone glycosides	Chromone glycosides	Plant	Cytotoxic
3′-(3-Methyl-2-butenyl)-4′-O-β-D-glucopyranosyl-4,2′-dihydroxychalcone	Chalcone glycoside	Plant	Antioxidant
3-O-Methylcalopocarpin	Pterocarpan	Plant	anti-HIV
4′,5′-Dihydro-11,5′-dihydroxy-4′-methoxytephrosin	Rotenoid	Plant	Quinone reductase inducer
5,6,3′-Trihydroxy-7,8,4′-trimethoxyflavone	Flavone	Plant	Antiangiogenic; cytotoxic
5,6,7,4′-Tetrahydroxyflavonol-3-O-rutinoside	Flavonol glycoside	Plant	Antioxidant
5,7,3′,4′-Tetrahydroxy-2′(3,3-dimethylallyl)isoflavone	Isoflavonoid	Plant	Antiprotozoal (leishmanicidal and trypanocidal)
5,7,3′-Trihydroxy-4′-methoxy-8,2′-di(3-methyl-2-butenyl)-(2S)-flavanone	Flavanone	Plant	Antimalarial; antimycobacterial; cytotoxic
5,7,3′-Trihydroxy-4′,5′-(2,2-dimethylpyran)-8,2′-di(3-methyl-2-butenyl)-(2S)-flavanone	Flavanone	Plant	Antimalarial; antimycobacterial; cytotoxic

Compound	Type	Source	Activity
5,7,4′-Trihydroxy-6,8-diprenylisoflavone	Prenylated flavone	Plant	Antibacterial
5-Deoxyglyasperin F	Isoflavanone	Plant	Anti-HIV
5-Galloylquercetin-3-O-α-L-arabinofuranoside	Flavonol glycoside	Plant	Antimalarial; cytotoxic
5-Hydroxy-7,2′,4′,5′-tetramethoxtflavone	Flavone	Plant	Antibacterial
6- and 8-(2-Pyrrolidinone-5-yl)-(−)-epicatechin	Flavan-3-ols	Plant	Inhibitors to the formation of advanced glycation end products (AGEs)
6-Hydroxyluteolin 7-O-laminaribioside	Flavone glycoside	Plant	Antioxidant
7,4′-Dihydroxy-3′-methoxyisoflavone	Isoflavone	Plant	Antimalarial
7,8-Dihydroxyflavanone	Flavanone	Plant	Cytotoxic; inhibitor to Jun-Fos-DNA complex formation
7-O-galloyltricetifavan and 7,4′-di-O-galloyltricetifavan	Flavan	Plant	Antiviral
8-Lavandulylkaempferol	Flavonol	Plant	Antioxidant
Abacopterins A, B, C, and D	Flavan-4-ol glycosides	Plant	Cytotoxic
Abyssinones A, C, and D	Chalcones	Plant	Cytotoxic
Acacetin-7-O-β-D-galactopyranoside	Flavone glycoside	Plant	Anti-HIV
Acumitin	C-Benzylated dihydrochalcone	Plant	Cytotoxic
Aloe C-glucosylchromone (8-[C-β-D-[2-O-(*E*)-cinnamoyl]glucopyranosyl]-2-[(*R*)-2-hydroxypropyl]-7-methoxy-5-methylchromone)	C-Glucosylchromone	Plant	Anti-inflammatory
Amentoflavone	Biflavonoid	Plant	Antitumor; anticancerous; inhibitor of human cathepsin B; anti-inflammatory; antiviral; antifungal; human neutrophil elastase (HNE) inhibitor; vasodilator; caffeine-like Ca^{2+} releaser
Ampelopsin [(+)-Dihydromyricetin]	Flavanone	Plant	Antitumor; anticancerous; antiviral; anti-HIV
Artocarpin	Prenylated-flavone	Plant	Cytotoxic
Artochamin C	Pyrano-flavone	Plant	Cytotoxic
Artoindonesianin E1	Prenylated oxepinoflavone	Plant	Cytotoxic (anticancerous)
Artoindonesianin P	Prenylated flavone	Plant	Cytotoxic
Avicularin	Flavonol glycoside	Plant	Antioxidant
Barbadensis chromone	C-Glucosyl chromone	Plant	Anti-inflammatory
BF-4, 5 and 6	Flavanones	Plant	Cytotoxic

(*Continued*)

Table A.3 (Continued)

Compound	Type	Origin	Bioactivity
Bipinnatones A and B	Prenylated dihyro chalcones	Plant	Inhibitors of hemoglobinase II enzyme activation (antimalarial)
C-2′-Decoumaroyl-aloeresin G	Chromone glycoside	Plant	β-Secretase (BACE1) inhibitor
Calabricosides A and B	Flavonol glycosides	Plant	Antioxidant (radical scavenging)
Calycosin	Isoflavonoid	Plant	Antiprotozoal; antiplasmodial; antileishmanial; antigiardial; antioxidant
Camellianoside	Flavonol glycoside	Plant	Antioxidant
Catiguanins A and B	Phenylpropanoid-substituted epicatechins	Plant	Antioxidant
Chamaejasmenins A and D	Biflavanones	Plant	Antimitotic; antifungal
Conferols A and B	4-Hydroxyisoflavones	Plant	Anti-inflammatory
Cube'Resin Flavanones	Prenylated flavanones	Plant	Cytotoxic
Cudranians 1 and 2	Flavonol glycosides	Plant	Antioxidants (radical scavengers)
Cupressuflavone	Biflavonoid	Plant	Human neutrophil elastase (HNE) inhibitor
Daedalin A	Chromene derivative	Fungus	Tyrosinase inhibitor; antioxidant; inhibitor to melanin synthesis
Desacetylpyramidaglain D	Flavagline (cyclopenta[bc]benzopyran type)	Plant	Antituberculosis; antiviral
Diandraflavone	Flavone glycoside	Plant	Antioxidant
Diinsininol and diinsimin	Biflavonoids	Plant	Anti-inflammatory
Diuvaretin	C-Benzylated dihydrochalcone	Plant	Cytotoxic (antitumor)
Dorsilurins F, G, H, I, J, and K	Prenylated flavonols	Plant	Antioxidant
Epigallocatechin-(4β → 8,2β → O-7)-epicatechin	A-type proanthocyanidin dimer	Plant	Anti-HIV
Eriodictyol 7-O-sophoroside	Flavanone glycoside	Plant	Antioxidant
Erybraedin A	Pterocarpan	Plant	Antibacterial; antimycobacterial; antiplasmodial; cytotoxic
Erylatissins A, B, and C	Flavonoids (isoflavones and flavanone)	Plant	Antimicrobial; antioxidant (radical scavenger)
Eupatilin	Flavone	Plant	Anti-inflammatory; antitumor; antiproliferative; anticancerous; antioxidant
Exiguaflavanones A and B	Flavanones	Plant	Antimalarial
Flavokavain B	Chalcone	Plant	Leishmanicidal
Furowanin B	Isoflavonoid	Plant	Antiestrogenic

Name	Type	Source	Activity
Ginkgetin	Biflavone	Plant	Anti-inflammatory; antiarthritic; analgesic; cytotoxic; antiviral
Glycyrrhizols A and B	Pterocarpenes	Plant	Antibacterial
Hinokiflavone	Biflavonoid (a diflavonyl ether)	Plant	Cytotoxic; inhibitor of procoagulant activity of adherent human monocytes; neuroprotective
Isochamaejasmenin B	Biflavanone	Plant	Antimitotic; antifungal
Isochamuvaritin	C-Benzylated dihydrochalcone	Plant	Cytotoxic
Isoliquiritigenin	Neoflavonoid	Plant	Anticancerous; antitumor; anti-inflammatory; antioxidant; antimalarial; vasorelaxant; antihistaminic; antihyperpigmentation
Irilone glycoside	Isoflavonoid glycoside	Plant	Antioxidant
Isoscutellarein glycoside	Flavonoid glycoside	Plant	Antioxidant
Isonymphaeol-B	Prenylated flavanone	Propolis	Antioxidant
Isorhamnetin-3-O-(-L-[6---p-coumaroyl-β-D-glucopyranosyl-(1 → 2)-rhamnopyranoside	Flavonol glycoside	Plant	Antioxidant
Isowightheone hydrate	Isoflavone	Plant	Cytotoxic
Jaceosidin	Flavone	Plant	Anticancerous; antitumor; antiallergic; anti-inflammatory
Kaempferol 3-O-(-L-[6---p-coumaroyl-β-D-glucopyranosyl-(1 → 2)-rhamnopyranoside]-7-O-β-D-glucopyranoside	Flavonol glycoside	Plant	Antioxidant
Kaempferol 3-O-α-L-[6---p-coumaroyl-β-D-glucopyranosyl-(1 → 2)-rhamnopyranoside]	Flavonol glycoside	Plant	Antioxidant
Kaempferol 7-O-(2,3-di-E-p-coumaroyl-α-L-rhamnopyranoside)	Flavonol glycoside	Plant	Cytotoxic
Kaempferol 7-O-(2-E-p-coumaroyl-α-L-rhamnopyranoside)	Flavonol glycoside	Plant	cytotoxic
Leachianone A	Flavanone	Plant	Cytotoxic; antitumor; antimalarial
Lespeflorins A_3, B_{2-4}, C_3, D_1, G_{2-3}, G_5, G_8, G_{10}, H_2	Flavonids	Plant	Melanin synthesis inhibitors
Leufolins A and B	Flavanones	Plant	Butyrylcholinesterase (BChE) inhibitors
Littorachalcone	Dihydrochalcone dimer	Plant	Nerve growth factor (NGF) stimulator
Maackiaflavanones A and B	Prenylated flavanones	Plant	Cytotoxic
Maackiapterocarpan A	Prenylated pterocarpan	Plant	Cytotoxic
Macaflavanone G	Prenylated flavanone	Plant	Cytotoxic

(Continued)

Table A.3 (Continued)

Compound	Type	Origin	Bioactivity
Macaranone A	Prenylated flavonol	Plant	Cytotoxic
Mallotophilippens C, D, and E	Chalcones	Plant	Inhibitors to NO production and iNOS gene expression
Matsudone A	Flavonol glycoside	Plant	Cyclooxygenase inhibitory (anti-inflammatory)
Millewanins G and H	Isoflavonoids	Plant	Antiestrogenic
Miquelianin	Flavonol glycoside	Plant	Antioxidant; antidepressant
Monodictyochromones A and B	Chromanone derivatives	Fungus	Anticancerous
Monotesone A	Flavanone	Plant	Antifungal
Muntingia biflavans 1 and 2	Biflavans	Plant	Cytotoxic
Myrciacitrins I, II, III, IV, and V	Flavanone glycosides	Plant	Aldose reductase inhibitors (antidiabatics)
Myricitrin-5-methyl ether	Flavonol glycoside	Plant	Antioxidant
Neobavaisoflavone	Isoflavone	Plant	Platelet aggression inhibitor
Nymphaeols A, B, and C	Prenylated flavanones	Plant	Cytotoxic; antioxidant; anti-inflammatory
Obochalcolactone	Chalcone	Plant	Cytotoxic
Ophioglonin	Homoflavonoid	Plant	Antihepatitis-B virus (anti-HBV)
ortho-Demethylbuchenavianine	Flavonoid alkaloid (piperidine-flavone alkaloid)	Plant	Anti-HIV, anticancerous
Panduratin A	Cyclohexenyl chalcone derivative	Plant	Antioxidative; anti-inflammatory; cytotoxic; anticancerous; antimutagenic; antiaging; antibacterial
Papyriflavonol A	Prenylated flavonol	Plant	Secretory phospholipase A_2s ($sPLA_2$s) inhibitor (anti-inflammatory); antimicrobial; cytotoxic; tyrosinase activity inhibitor
Perforamone B	Chromone	Plant	Antiplasmodial; antimycobacterial
Pestalotheol C	Chromenone type of metabolite	Fungus	Anti-HIV
Phellamurin	Prenylated dihydroflavanol glycoside	Plant	DNA strand cleaving agent
Propolis neoflavonoids 1 and 2	Neoflavonoids	Propolis	Inhibitors of nitric oxide (NO) production
Quercetin 3-O-(2″,3″-digalloyl)-β-D-galactopyranoside	Flavonol glycoside	Plant	Insulin-like activity
Quercetin 3-O-α-L-[6→*p*-coumaroyl-β-D-glucopyranosyl-(1 → 2)-rhamnopyranoside]	Flavonol glycoside	Plant	Antioxidant
Quercetin 3-O-α-L-[6→*p*-coumaroyl-β-D-glucopyranosyl-(1 → 2)-rhamnopyranoside]-7-O-β-D-glucopyranoside	Flavonol glycoside	Plant	Antioxidant

Compound	Class	Source	Activity
Quercetin 3-O-β-D-glucopyranosyl-(1 → 2)-rhamnopyranoside	Flavonol glycoside	Plant	Antioxidant
Quercetin 4'-O-rhamnopyranosyl-3-O-β-D-allopyranoside	Flavonol glycoside	Plant	Cyclooxygenase (COX) inhibitor
Quercetin-3-O-α-(6''-caffeoylglucosyl-β-1,2-rhamnoside	Flavonol glycoside	Plant	Angiotensin converting enzyme (ACE) inhibitory
Remangiflavanones A and B	Flavanones	Plant	Antibacterial
Robustaflavone and its 7,4',7''-trimethyl ether	Biflavonoid	Plant	Antihepatitis B virus (anti-HBV); human neutrophil elastase (HNE) inhibitor
Rotenone	Flavonoid	Plant	Cytotoxic; antitumor; NO production inhibitor; antiproliferative; insecticidal and piscicidal
Sandwicensin	Pterocarpan (isoflavonoid)	Plant	Anti-HIV; antimycobacterial
Sappanchalcone	Chalcone	Plant	Anticonvulsant; anti-inflammatory; cytoprotective (antioxidative)
Schizolaenones A and B	Prenylated flavanones	Plant	Cytotoxic
Selinone	Flavanone	Plant	Antifungal
Sigmoidins A and B	Flavanones	Plant	Antioxidant; anti-inflammatory; antibacterial
Sophoraflavanone G	Prenylated flavanone	Plant	Cytotoxic; antioxidant; antimicrobial; antimalarial
Sophoraflavanone L	Prenylated flavanone	Plant	Cytotoxic
Taiwanhomoflavones A and B	Biflavonoid	Plant	Cytotoxic
Talosins A and B	Isoflavonol glycosides	Bacteria	Antifungal
Tanariflavanones A, B, C, and D	Prenylated flavanones	Plant	Phytotoxic; cytotoxic; antioxidant
Theograndin II	Flavone glycoside	Plant	Antioxidant; cytotoxic
Torvanol A	Isoflavonoid (sulfated)	Plant	Antiviral
Uvaretin	C-Benzylated dihydrochalcone	Plant	Cytotoxic (antitumor)
Xanthohumols	Prenylated chalcones	Plant	Nitric oxide (NO) production inhibitory; antitumor; diacylglycerol acyltransferase (DGAT) inhibitor
Yinyanghuo A and Yinyanghuo B	Prenylflavones	Plant	Inhibitors to platelet aggregation

Table A.4 Lignoids

Compound	Type	Origin	Bioactivity
(−)-Cubebin and (−)-3,4-dimethoxy-3,4-desmethylenedioxycubebin		Plant	Testosterone 5α-reductase inhibitory activity (antiandrogenic); stimulatory effect on melanogenesis (hair re-growth); anti-inflammatory; analgesic; trypanocidal
(−)-Demethoxylpinoresinol		Plant	iNOS inhibitor
(+)-Lyoniresinol 4,4′-bis-O-β-D-glucopyranoside	Lignan glycoside	Plant	Lipoxygenase enzyme inhibitor
(+)-Lyoniresinol-3-α-O-β-D-glucopyranoside	Lignan glycoside	Plant	Antibacterial; antifungal
(+)-Sesamin		Plant	Cytotoxic
4-Methoxymagnaldehyde B		Plant	Cytotoxic
8′-epi-Cleomiscosin A	Coumarinolignoid	Plant	Tyrosinase inhibitor
Aiphanol	Stilbenolignan	Plant	Cyclooxygenase (COX) inhibitor
Anolignans A and B	Dibenzylbutadiene type	Plant	Anti-HIV
Bidenlignasides A and B	Neolignan glucosides	Plant	Antihistaminic (antiallergic)
Cleistanone	Arylanphthalide lignan	Plant	Cytotoxic
Crassifogenin C	Norlignan	Plant	Antioxidant (radical scavenging)
Dehydrodiconiferyl dibenzoate		Plant	Antimalarial
Dichotomoside D	Neolignan glycoside	Plant	Antiallergic
Futokadsurins A, B, and C	Tetrahydrofuran lignans	Plant	NO production inhibitors
Graminone B		Plant	Vasodilator
Imperatorin and isoimperatorin		Plant	Anti-HIV
Longipedumin A		Plant	HIV-1 protease inhibitor
Manassantins A and B	Dineolignoids	Plant	Inhibitor of transcription factor NF-κB and tumour necrosis factor-α (TNF-α); anti-inflammatory, anticancerous; neuroleptic; antiplasmodial
Negundin B		Plant	Lipoxygenase (LOX) enzyme inhibitor
Rubrisandrin A	Dibenzocyclooctadiene lignan	Plant	Anti-HIV
Taiwankadsurin B	C₁₉-homolignan	Plant	Antitumor
Taiwanschirin D	Homolignan	Plant	Anti-HBeg
Termilignan B		Plant	Antimicrobial; anti-inflammatory
Threo-(7S,8R)-1-(4-Hydroxyphenyl)-2-[4-(E)-propenylphenoxy]-propan-1-ol	Neolignoid	Plant	NFAT transcription factor inhibitor

Table A.5 Terpenoids

Compound	Type	Origin	Bioactivity
(+)-Makassaric acid	Meroterpenoid	Marine sponge	MAPKAP kinase 2 (MK2) inhibitor
(+)-Subersic acid	Meroterpenoid	Marine sponge	MAPKAP kinase 2 (MK2) inhibitor
(1R,4S)-1-Hydroperoxy-p-menth-2-en-8-ol acetate	Monoterpene	Plant	Antitrypanosomal
(20R)-28-Hydroxylupen-30-al-3-one	Triterpenoid	Plant	Antitumor
(22E,24R)-3α-Ureido-ergosta-4,6,8(14),22-tetraene	Steroid	Plant	Cytotoxic
(22E,24R)-5α,8α-epidioxyergosta-6,9,22-triene-3β-ol 3-O-β-D-glucopyranoside	Steroid	Plant	Cytotoxic
(2E,6R)-8-Hydroxy-2,6-dimethyl-2-octanoic acid	Monoterpene	Plant	Antiosteoporotic
1,4-Endoperoxy-bisabola-2,10-dienes	Sesquiterpenes	Plant	Cytotoxic (antitumor)
9-Hydroxy heychenone	Diterpenoid	Plant	Cytotoxic (anticancerous)
12-Deacetoxy-21acetoxyscalarin	Sesterterpenoid	Marine sponge	Cytotoxic
12-Deoxyphorbol 13-(3E,5E-decadienoate)		Plant	Anti-HIV
12-Methyl-5-dehydroacetylhorminone	Abietane-diterpenoid	Plant	Antitubercular
12-Methyl-5-dehydrohorminone	Abietane-diterpenoid	Plant	Antitubercular
12-O-Acetyl-16-O-deacetyl-16-epi-scalarolbutenolide	Sesterterpenoid	Marine sponge	Cytotoxic
12-O-Deacetyltrichilin H	Limnoid	Plant	Cytotoxic
13,14,15,16-Tetranorlabdane-8(,12,14-triol	Diterpenoid (labdane-type tetranorditerpene)	Plant	Cyclooxygenase (COX) inhibitor
15-Acetoxyorbiculin G	Sesquiterpenoid (dihydroagarofuranoid sesquiterpene)	Plant	Antitubercular
15-O-Deacetylnimbolidin	C-seco Limonoid	Plant	Cytotoxic
16,23-Epoxy-5β-cholestane triglycoside	Steroidal glycoside	Plant	Immunosuppressive

(Continued)

Table A.5 (Continued)

Compound	Type	Origin	Bioactivity
16-O-Deacetyl-16-epi-scalarolbutenolide	Sesterterpenoid	Marine sponge	Cytotoxic
16β,17-Dihydroxy-ent-kauran-19-oic acid	Kaurane diterpenoid	Plant	Anti-HIV; anti-inflammatory
18-Beta-L-3′,5′diacetoxy arabinofuranosyl-ent-kaur-16-ene	Diterpenoid (ent-kaurane glycoside)	Plant	Antibacterial
1-Beta-hydroxy arbusculin A	Sesquiterpene lactone	Plant	Melanogenesis inhibitor
1-O-Acetyl-4R,6S-britannilactone	Sesquiterpene lactone	Plant	Inhibitor to nitric oxide production
20(R),22(ξ),24(S)-Dammar-25(26)-ene-3β,6α,12β,20,22,24-hexanol	Triterpenoid	Plant	Cytotoxic
20-epi-Bryonolic acid	Triterpenoid (D:C-Friedooleanane-type)	Plant	Cytotoxic; COX-1 and -2 inhibitor
21β-Hydroxyolean-12-en-3-one	Triterpenoid	Plant	Antigiardial
25-O-Methoxycimigenol 3-O-α-L-arabinopyranoside	Triterpenoid (cycloartane glycoside)	Plant	Cytotoxic
28,29-Dihydroxyfriedelan-3-one	Triterpenoid	Plant	Cytotoxic (anticancerous)
2-alpha,19-alpha-Dihydroxy-3-oxo-12-ursen-28-oic acid	Triterpenoid	Plant	Anti-HIV
2-O-Caffeoyl maslinic acid	Triterpenoid	Plant	Antioxidant (inhibitor of NO production, and radical scavenger)
2-O-E-p-coumaroyl alphitolic acid	Lupane-type triterpene	Plant	Antiplasmodial (antimalarial); antimycobacterium
3-(Z)-Caffeoyllupeol	Triterpenoid	Plant	Antimalarial
3,21-Dioxo-olean-18-en-oic acid	Triterpenoid	Plant	Tie2 kinase inhibitor
3,4-Secoisopimara-4(18),7,15-triene-3-oic acid	Diterpenoid	Plant	Antispasmodic
3-O-trans-p-Coumaroyl actinidic acid	Coumaroyl triterpene	Plant	Pancreatic lipase inhibitor
3-O-Vanillylceanothic acid	Ceanothane-type triterpene	Plant	antiplasmodial (antimalarial); antimycobacterium
3β,12-Dihydroxy-13-methyl-5,8,11,13-podocarpatetraen-7-one	Podocarpane-type diterpenoid	Plant	Cytotoxic
3β,12-Dihydroxy-13-methyl-6,8,11,13-podocarpatetraene	Podocarpane-type diterpenoid	Plant	Cytotoxic
3β-O-(E)-Coumaroyl-D:C-friedooleana-7,9(11)-dien-29-oic acid	Triterpenoid (D:C-friedooleanane-type)	Plant	Cytotoxic

4β,14-Dihydroxy-6α,7β-H-1(10)-cadinene	Sesquiterpenoid (cadinene-type)	Basidiomycete	Anti-HIV; cytotoxic
5-Hydroxyzerumbone	Sesquiterpenoid	Plant	Inhibitor to nitric oxide production
5α,8α-Epidioxycholest-6-en-3-ol	Steroid	Sea urchin	Cytotoxic
5α,8α-Epidioxysterol	Steroid	Soft coral	Cytotoxic
5α-Epoxyalantolactone	Sesquiterpene lactone (eudesmanolide)	Plant	Antiproliferative; antimycobacterial
6-(9'-Purine-6',8'-diolyl)-2-beta-suberosanone	Sesquiterpene alkaloid	Coral	Cytotoxic
6,7-Di-O-nicitinoylscutebarbatine G	neo-Clerodane diterpenoid alkaloid	Plant	Cytotoxic
6-O-Nicitinoyl-7-O-acetylscutebarbatine G	neo-Clerodane diterpenoid alkaloid	Plant	Cytotoxic
6α-Malonyloxymanoyl oxide	Diterpenoid (labdane-type)	Plant	Antibacterial
6β,16β-Diacetoxy-25-hydroxy-3,7-dioxo-29-nordammara-1,17(20)-dien-21-oic acid	Triterpenoid (29-nordammarane triterpenoid)	Fungus	Antimicrobial
7,8-Dehydrocerberin	Cardenolide glycoside	Plant	Cytotoxic
7-Formamido-20-isocyanoisocycloamphilectane	Diterpenoid formamide derivative	Sponge	Antimalarial
7-O-Nicitinoyl scutebarbatine H	neo-Clerodane diterpenoid alkaloid	Plant	Cytotoxic
7-Oxo-10α-cucurbitadienol	Triterpenoid (cucurbitane-type triterpene)	Plant	Anti-inflammatory
7α-Hydroperoxymannol	Sesquiterpene hydroperoxide	Plant	Cytotoxic
8-Acetoxy-4,10-dihydroxy-2,11(13)-guaiadiene-12,6-olide	Guaianolide	Plant	Cytotoxic
8-Acetylheterophyllisine	Diterpenoid alkaloid	Plant	Antifungal
8α,19-Dihydroxylabd-13E-en-15-oic acid	Diterpenoid (labdane-type)	Plant	Cyclooxygenase (COX) inhibitor
9(11)-Dehydroaxinysterol	Steroid (epidioxyergostane-type)	Marine sponge	Anticancerous
Acutoside A	Triterpenoid saponin	Plant	Inhibitor to matrix metalloproteinase (MMP-1)
Ainsliadimer A	Sesquiterpenoid	Plant	Inhibitor to NO production
Albizosides A, B, and C	Triterpenoid saponins	Plant	Cytotoxic (antitumor)

(Continued)

Table A.5 (Continued)

Compound	Type	Origin	Bioactivity
Alisiaquinol	Meroterpenoid	Marine sponge	Antimalarial
Alisiaquinones A, B, and C	Meroterpenoids	Marine sponge	Antimalarial
Alismorientol A	Sesquiterpenoid	Plant	Anti-HBV
Andrographolide	Diterpeneoid lactone	Plant	Hepatoprotective and hepatostimulative; antiviral; anti-HIV; anti-inflammatory; antibiotic; antiplatelet aggregation activity; antiallergic; inhibitor of NO synthesis; antihyperglycemic;. antithrombotic, hypotensive; antiatherosclerotic; anticancerous; cytotoxic
Anomanolide B	Steroidal lactone	Plant	Cytotoxic (anticancerous)
Aplysinoplides A, B, and C	Sesterterpenoids	Marine sponge	Cytotoxic
Ardisia saponin 1	Triterpenoid saponin	Plant	Cytotoxic (antiglioblastoma agent)
Ardistanosides A, B, C, D, and E	Riterpenoid saponins	Plant	Cytotoxic
Asprellic acids A and C	Triterpenoids	Plant	Cytotoxic (antitumor)
Astragaloside IV	Triterpenoid saponin	Plant	Antihepatitis B virus (anti-HBV)
Ballodiolic acid	Diterpenoid (clerodane-type)	Plant	Lipoxygenase inhibitor
Ballotenic acid	Diterpenoid (clerodane-type)	Plant	Lipoxygenase inhibitor
Beta-sesquiphellandrene	Sesquiterpenoid	Plant	Antirhinoviral
Beta-Sitosterol glucoside-3'-O-hexacosannoicate	Steroidal glycoside	Plant	Anticomplement activity
Betunilic acid and its derivatives	Trirerpenoids	Plant	Anti-HIV
Bielschowskysin	Diterpenoid (highly oxygenated hexacyclic diterpene)	Coral	Antimalarial; anticancer
Blumenol A	Sesquiterpenoid	Plant	Cytotoxic
Brasilicardin A	Diterpenoid antibiotic	Bacteria	Immunosuppressive; antitumor
Bromophycolide A	Macrolide (diterpene-benzoate)	Red alga	Antineoplastic; antibacterial; antifungal, anti-HIV
Bruceanols D, E, and F	Quassinoids (modified triterpenes)	Plant	Cytotoxic (antitumor)
Bruceolide and its acetyl derivatives	Quassinoids (modified triterpenes)	Plant	Antimalarial

Brunfelsia saponin	Furostan-type saponin	Plant	Antileishmanial
Caesalpinins C, D, and F	Furanocassane and furanonorcassane-type diterpenoids	Plant	Antimalarial
CAF-603	Sesquiterpenoid (carotene sesquiterpene)	Fungus	Antifungal
Callophycoic acids A–H	Diterpenoids	Red alga	Antibacterial; antifungal; antimalarial; anticancer
Callophycols A and B	Diterpenoids	Red alga	Antibacterial; antifungal; antimalarial; anticancer
Candelalides A, B, and C	Pyrone diterpenoids	Fungus	Voltage-gated potassium channel Kv1.3 blockers
Capilliposide B	Triterpenoid saponin	Plant	Cytotoxic
Capilloide	Diterpenoid (cembranolide)	Soft coral	Cytotoxic
Cardivins A, B, C, and D	Sesquiterpenoids (germacranolide type)	Plant	Cytotoxic (antitumor)
Caribenols A and B	Norditerpenoids	Coral	Antitubercular; antimalarial
Carnosic acid	Diterpenoid	Plant	Anti-HIV
Caseamembrols A and B	Clerodane diterpenoids	Plant	Cytotoxic
Caseanigrescens A, B, C, and D	Clerodane diterpenoids	Plant	Cytotoxic
Caseargrewins A, B, C, and D	Clerodane diterpenoids	Plant	Antimalarial; antimycobaterial; cytotoxic
Caudatin 3-O-β-cymaropyranoside	Steroidal glycoside	Plant	Cytotoxic
Celasdin B	Friedelane-type triterpenoid	Plant	Anti-HIV
Celastrol	A quinone methide triterpene	Plant	Antioxidant; anticancer; antitumor; anti-inflammatory; immunosuppressive; antiasthma; antiangiogenic; neuroprotective; inhibitor of neurodegenerative diseases like Alzheimer disease, amyotrophic lateral sclerosis (ALS), Huntington's disease (HD) and Parkinson's disease (PD); antifungal; inhibitor of adjuvant arthritis, TNF-α and IL-1β activation, induced expression of class II MHC molecules, NF-κB activation, NO production, and lipid peroxidation
Cholestane glycosides 1–10	Steroidal glycosides	Plant	Cytotoxic
Cimifoetisides A and B	Cyclolanostane triterpene diglycosides	Plant	Immunosuppressive
CJ-01	Sesquiterpene furan	Plant	Chitin synthase inhibitor; antifungal
Clathsterol	Steroid (sterol sulfate)	Marine sponge	Anti-HIV

(*Continued*)

Table A.5 (Continued)

Compound	Type	Origin	Bioactivity
Clionastatins A and B	Steroids (tri-and tetrachlorinated androstane derivatives)	Marine sponge	Cytotoxic
Colossolactones V, VII, and VIII	Triterpenoids (lanostane-type)	Mushroom	Anti-HIV
Convallasaponin A	Steroidal saponin (cardenolide glycoside)	Plant	Cytotoxic (antitumor)
Convallatoxin	Cardenolide glycoside	Plant	Cytotoxic (antitumor)
Corymbolone	Sesquiterpenoid	Plant	Antimalarial
Costunolide	Sesquiterpene lactone	Plant	Melanogenesis inhibitor; cytotoxic; anticancerous; antifungal; anti-inflammatory; antitumor; inhibitor to TNF-α and IL-1/IL-6; inhibitor to blood-ethanol elevation
Crassolide	Terpenoid (cembranolide)	Soft coral	Cytotoxic
Crassurolide A	Cembranoid	Soft coral	Anti-inflammatory
Crossosoma cardenolide	Cardenolide	Plant	Cytotoxic
Cuminganosides A, B, C, D, E, and F	Triterpene glycosides	Plant	Cytotoxic (anticancerous)
Debromolaurinterol	Sesquiterpenoid	Sea hare	Na,K-ATPase inhibitor
Dehydrocostus lactone	Terpenoid (guaianolide derivative)	Plant	Antitrypanosomal; inhibitor to cytotoxic T lymphocytes (CTL); anticancerous;COX-2 inhibitor; antimycobacterial
Dianversicosides A, B, C, D, E, F, and G	Triterpenoid saponins	Plant	Cytotoxic (anticancerous)
Dihydroagarofurans 1 and 2	Dihydroagarofuranoid sesquiterpenes	Plant	Antitubercular
Dilopholide	Diterpenoid (xenicane-type diterpene)	Brown alga	Cytotoxic
Dinochrome A	Carotenoid	Red tide	Anticarcinogenic
Dysidine	Sesquiterpene aminoquinone	Marine sponge	Human synovial phospholipase A_2 (PLA_2) inhibitor
Dysoxyhainanin A	Triterpenoid	Plant	Antibacterial
Dzununcanone	Triterpenoid	Plant	Antigiardial

E and Z-Senegasaponins a and b	Triterpenoid saponins	Plant	Hypoglycemic; inhibitors of alcohol absorption
Elaeodendrosides F, G, T, U, V, and W	Cardenolides	Plant	Antiproliferative (anticancerous)
Elephantopus sesquiterpenes 1, 2, and 3	Sesquiterpene lactones	Plant	Cytotoxic
Elephantopus sesquiterpenes 4 and 5	Sesquiterpene lactones	Plant	Cytotoxic; anti-inflammatory
Elongatols A and E	Sesquiterpenoid (nardosinane-type)	Soft coral	Cytotoxic
Emarginatines B and F	Sesquiterpene pyridine alkaloids	Plant	Cytotoxic
Ergosta-7,22-dien-3β-ol	Steroid	Fungus	Antiviral
Ergosterol peroxide	Steroid	Fungus	Anticancerous; antitumor; inhibitor of melanin synthesis
Eucalyptals A, B, and C	Terpenoids (phloroglucinol-coupled terpenoids)	Plant	Cytotoxic (antitumor)
Eunicea sesquiterpenoids 1–5	Sesquiterpenoids	Coral	Antimalarial
Euonymus beta-dihydroagarofuran	Sesquiterpene polyester	Plant	Antitumor
Eupaheliangolide A, heliangin and 3-epi-heliangin	Sesquiterpene lactones	Plant	Cytotoxic
Euphorbia diterpenoids 1–3	Diterpenoid esters (myrsinol-type skeleton)	Plant	Inhibitors of prolyl endopeptidase (PEP) and urease enzymes
Euphornin L	Macrocyclic diterpenoid	Plant	Cytotoxic
Eurycomanone	Quassinoid	Plant	Cytotoxic (antitumor); antimalarial
Eurysterols A and B	Steroidal sulphates	Marine sponge	Cytotoxic; antifungal
Floribundasaponin A	Steroidal saponin	Plant	Anti-inflammatory
Fomitopinic acid A, and Fomitosides E and F	Lanostane triterpenoid, and its glycosides	Mushroom	COX-1 and COX-2 enzyme inhibitory
Forskoditerpenosides C, D, and E	Labdane diterpene glycosides	Plant	Relaxative
FR207944	Triterpne glucoside (antibiotic)	Fungus	Antifungal
Fucosterol	Steroid	Marine alga	Antioxidant; hepatoprotective
Fukanemarins A and B, Fukanefuromarins A–G, and related compounds	Sesquiterpene coumarins	Plant	Inhibitors to NO production

(*Continued*)

Table A.5 (Continued)

Compound	Type	Origin	Bioactivity
Ganoderone A and C	Lanostane-type triterpenes	Fungus	Antiviral
Gedunin	Limonoid (ring D-seco)	Plant	Antimalarial, cytotoxic
Gelomulides K and M	ent-Abietane diterpenes	Plant	Cytotoxic
Gelsemiol 6'-trans-caffeoyl-1-glucoside	Iridoid glycoside	Plant	Nerve growth factor (NGF) stimulator
Geumonoid	Triterpene	Plant	Anti-HIV
Gibberosins K and L	Diterpenoids (xeniaphyllane-derived)	Soft coral	Cytotoxic
Globostellatic acids A, B, C, and D	Isomalabaricane triterpenoids	Marine sponge	Cytotoxic
Glochierriosides A and B	Triterpenoid saponins	Plant	Cytotoxic (anticancerous)
Gnidilatimonoein	Diterpene ester	Plant	Cytotoxic; anticancerous; antimetastastic
Gummiferaosides A, B, and C	Triterpenoid saponins	Plant	Cytotoxic
Gutolactone	Quassinoid	Plant	Antimalarial
Gypsosaponins A, B, and C	Triterpenoid (oleanane-type) saponins	Plant	Pancreatic lipase-inhibitor
Halidrys monoditerpene	Monoditerpenoid	Marine brown alga	Antibacterial
Halistanol sulfates F and G	Steroids	Marine sponge	anti-HIV
Halosterols A and B	Steroids	Plant	Chymotrypsin inhibitory (antihepatitis C virus)
Hedychilactone D	Diterpenoid	Plant	Cytotoxic (anticancerous)
Holistanol sulphates F and G	Steroids (sterol sulphates)	Marine sponge	Anti-HIV
Holothurins A_3 and A_4	Triterpenoid saponin	Sea cucumber	Cytotoxic
Hookerianamides H and I	Steroidal alkaloids	Plant	Cholinesterase inhibitor (anti-Alzheimer's disease); antiplasmodial
Hookerianamides J and K	Steroidal alkaloids	Plant	Inhibitors of acetylcholinesterase and butyrylcholinesterase enzymes activity; antileishmanial; antibacterial
Hyperinols A and B	Taraxastane-type triterpenoids	Plant	Chymotrypsin inhibitory (antihepatitis C virus)
Hypoglaunine B	Sesquiterpene polyalcohol ester (sesquiterene pyridine alkaloid)	Plant	Anti-HIV
Hyrtios seterterpenes 1–3	Scalarane sesterterpenes	Marine sponge	Cytotoxic
Igalan	Sesquiterpene lactone (α-*exo*-methylene-γ-lactone)	Plant	Antiproliferative

Indicanone	Guaiane-type sesquiterpene	Plant	Inhibitor to NO production
Inflexin	Diterpenoid (*ent*-kauane type)	Plant	Cytotoxic; aromatase inhibitor
Integric acid	Sesquiterpenoid (eremophilane-type)	Fungus	Anti-HIV
Isoiguesterin and 20-*epi*-isoiguesterinol	Bisnortriterpene quinone methides	Plant	Cytotoxic, antileishmanial
Iso-secotanapartholide and its 3-O-methyl ether	Sesquiterpenes	Plant	Inhibitors of inducible nitric oxide synthase (iNOS) expression
Isotanshinone IIB	Abietane-diterpenoid	Plant	Platelet aggregation inhibitor
Ixerochinolide	Sesquiterpenoid (guaianolide-type sesquiterpene lactone)	Plant	Cytotoxic
Kadlongilactones A and B	Triterpenoids (triterpene dilactones)	Plant	Antitumor
Kadsuphilactone B	Triterpenoid (triterpene dilactone)	Plant	Antihepatitis B virus (HBV)
Kansuiphorins A and B	Diterpenoids (diterpene esters)	Plant	Antileukaemia (anticancerous)
Kidjoranin glycosides 1 and 2	Steroidal glycosides (pregnane glycosides)	Plant	Cytotoxic
Komaroviquinone	Diterpenoid (icetexane diterpene)	Plant	Trypanocidal
Lancifodilactones F and G	Nortriterpenoids	Plant	Anti-HIV
Lancilactone C	Triterpene lactone	Plant	Anti-HIV
Lanigerol	Icetaxane diterpene	Plant	Antimicrobial
Lansioside A	Triterpene glycoside (secoonocerane-type amino sugar glycoside)	Plant	Inhibitor of leukotriene D_4 induced contraction of ileum
Laurebiphenyl	Dimeric sesquiterpene (cyclolaurane-type)	Marine red alga	Cytotoxic (anticancerous)
Laurinterol	Sesquiterpenoid	Marine plant	Na,K-ATPase inhibitor
Leucisterol	Steroid	Plant	Inhibitor of butyrylcholinesterase enzyme (BChE)
Leucosesterlactone	Sesterterpenoid	Plant	Prolylendopeptidase (PEP) inhibitor
Leucosesterterpenone	Sesterterpenoid	Plant	Prolylendopeptidase (PEP) inhibitor
Leucospiroside A	Steroidal saponin	Plant	Cytotoxic
Liphagal	Meroterpenoid	Marine sponge	PI3 kinase-(inhibitor; cytotoxic (antitumor)
Lobohedleolides	Cembranoid diterpenes	Soft coral	Anti-HIV; antitumor
Lucialdehyde B	Lanostane-type triterpene	Fungus	Cytotoxic, antiviral

(*Continued*)

Table A.5 (Continued)

Compound	Type	Origin	Bioactivity
Lucialdehyde C	Lanostane-type triterpene	Fungus	Cytotoxic
Lucidenic acid N	Triterpene	Fungus	Cytotoxic
Lupulin A	Diterpenoid	Plant	Antibacterial
Lychnostatins 1 and 2	Germacranolides	Plant	Anticancerous
Lyratols C and D	Sesquiterpenoids	Plant	Cytotoxic
Madecassoside	Triterpenoid saponin	Plant	Wound healing; protective against myocardial ischemia-reperfusion injury; anti-inflammatory; antirheumatoid arthritic; skin-care agent; antioxidant
Madhucosides A and B	Triterpenoid glycosides	Plant	Antioxidant
Mannioside A	Steroidal saponin	Plant	Anti-inflammatory
Manoalide	Sesquiterpenoid antibiotic	Marine sponge	Anti-inflammatory; antibacterial; cobra venom and bee venom inhibitor
Maoecrystal Z	Diterpenoid	Plant	Cytotoxic (antitumor)
Maprounic acid derivatives	Pentacyclic triterpenoids	Plant	anti-HIV
Marianine and Marianosides A and B	Lanostane triterpenoid and its glycosides	Plant	Chymotrypsin inhibitory (antihepatitis C virus)
Maslinic acid and *epi*-maslinic acid	Triterpenoids	Plant	Antioxidant; anti-inflammatory; anticarcinogenic; hypolipidemic; growth promoting; anti-HIV, antiproliferative
Maytenfolone-A	Triterpenoid	Plant	Anti-HIV; cytotoxic
Maytenonic acid	Triterpenoid	Plant	Cytotoxic (anticancerous)
Megathyrin A	Diterpenoid	Plant	Cytotoxic
Melemeleone B	Sesquiterpene quinone	Marine sponge	Protein tyrosin kinase (PTK) inhibitor
Metachromins S and T	Sesquiterpenoids	Marine sponge	Cytotoxic
Michaolides A-K	terpenoids (cembranolides)	Soft coral	Cytotoxic
Microclavatin	Diterpenoid	Soft coral	Cytotoxic
Minheryin G	*ent*-Kaurane diterpenoid	Plant	Cytotoxic
Moronic acid	Triterpenoid	Plant	Anti-HIV; anti-HSV; cytotoxic
Multicaulin and 12-demethylmulticaulin	Norditerpenoids	Plant	Antituberculer
Multiorthoquinone and 2-demethylmultiorthoquinone	Norditerpenoids	Plant	Antituberculer
Muqubilone	Norsesterterpene acid	Marine sponge	Antiviral

Compound	Type	Source	Activity
Mustakone	Sesquiterpenoid	Plant	Antimalarial
Myrsine saponin	Triterpenoid saponin	Plant	Cytotoxic
Nalanthalide	Pyrone diterpenoid	Fungus	Voltage-gated potassium channel Kv1.3 blocker (immunosuppressive)
Nardoperoxide and Isonardoperoxide	Sesquiterpenoids (guaiane-type endoperoxides)	Plant	Antimalarial
Nerolidol glycoside	Sesquiterpene glycoside	Plant	Antihyperglycemic
Neurolenins A, B, C, and D	Sesquiterpenoids (germacranolide type)	Plant	Antimalarial; cytotoxic; antidysentric; antifeedant
Nigranoic acid	A ring-secocycloartene triterpenoid	Plant	Anti-HIV; cytotoxic
Norcaesalpinins A, B, D, and E	Furanocassane and furanonorcassane-type diterpenoids	Plant	Antimalarial
Oblonganoside A	Triterpenoid saponin	Plant	Antiviral
Oleanolic acid glycoside	Triterpenoid saponin	Plant	Cytotoxic (antitumor)
Oleoyl danshenxinkun A	Abietane-diterpenoid	Plant	Platelet aggregation inhibitor
Oleoyl neocryptotanshinone	Abietane-diterpenoid	Plant	Platelet aggregation inhibitor
Orbiculins (A, D–I), Celafolin A-1, Ejap-2, and Triptogelin C-1	Sesquiterpenes (β-agarofurans)	Plant	Cytotoxic; antitubercular
Pacificins C and H	Renylbicyclogermacrane type diterpenoids	Soft coral	Cytotoxic
Pacificins K and L	Diterpenoids (prenylbicyclogermacrane-type)	Soft coral	Cytotoxic
Paeonilide	Monoterpenoid	Plant	Anti-PAF activity
Paeonins A and B	Monoterpene galactosides	Plant	Lipoxygenase inhibitors
Platanic acid	Terpenoid	Plant	Anti-HIV
Platycodon saponins	Triterpenoid saponins	Plant	Cytotoxic
Plectranthols A and B	Diterpenoids	Plant	Antioxidants
Polacandrin	Dammarane triterpenoid	Plant	Cytotoxic (antitumor)
Polpunonic acid (maytenonic acid)	Triterpenoid	Plant	Cytotoxic (anticancerous)
Pomolic acid	Triterpenoid	Plant	Anti-HIV; anti-inflammatory; cytotoxic; anticancerous; anticomplementary

(*Continued*)

Table A.5 (Continued)

Compound	Type	Origin	Bioactivity
Propindilactone L	Nortriterpenoid [18(13 → 14)-*abeo*-schiartane skeleton]	Plant	Antihepatitis B virus (anti-HBV)
Protoxylocarpins A–E	Protolimonoids	Plant	Cytotoxic (antitumor)
Przewalskin B	Diterpenoid	Plant	Anti-HIV
Pseudopteroxazole and *seco*-pseudopteroxazole	Diterpenoid alkaloids	Coral	Antitubercular
Pulsatilla saponin D	Triterpenoid saponin	Plant	Antitumor
Quassimarin	Quassinoid (simaroubolide)	Plant	Antitumor
Quillaic acid glycosidic ester	Monodesmosidic triterpene saponin	Plant	Immunomodulatory
Rediocides A, C, E, and F	Daphnane diterpenoids	Plant	Acaricidal (antiallergic; insecticidal
Reynosin	Sesquiterpene lactone	Plant	Melanogenesis inhibitor; cytotoxic; antitumor; inhibitor to TNF-α and CINC-1/IL-8
Rotundifoliosides A, H, I, and J	Triterpenoid saponins (ursane-type saikosaponin analogs)	Plant	Antiproliferatives
Rubriflordilactone B	Bisnortriterpenoid	Plant	Anti-HIV
Rubrisandrin A	Dibenzocyclooctadiene lignan	Plant	Anti-HIV
Salaspermic acid	Triterpenoid (friedelane-type)	Plant	Anti-HIV
Salvileucalin B	Diterpenoid (neoclerodane-type)	Plant	Cytotoxic
Salzmannianosides A and B	Saponins (triterpenoid glycosides)	Plant	Antifungal
Samaderines B, C, E, X, Y, and Z	Quassinoids (modified triterpenes)	Plant	Cytotoxic; antimalarial; anti-inflammatory
Santolina sesquiterpene	Germacrane sesquiterpene	Plant	Cytotoxic
Sapinmusaponins Q and R	Triterpene (tirucallane-type triterpenoids) saponins	Plant	Antiplatelet aggregation
Saponaceol A	Triterpenoid ester	Fungus (mushroom)	Cytotoxic
Sarasinosides J	Norlanostane triterpenoidal saponin	Marine sponge	Antimicrobial
Savins A and B	Triterpenoids	Plant	Butyrylcholinesterase (BChE) inhibitors

Sch 725432	Sesquiterpenoid (caryophyllene-type)	Fungus	Antifungal
Schiprolactone A	Lanostane triterpenoid	Plant	Cytotoxic (anticancerous)
Schisanlactone A	Trterpenoid	Plant	HIV-1 protease inhibitor
Schisanwilsonene A	Sesquiterpenoid (carotene-type)	Plant	Antihepatitis B virus (anti-HBV)
Scisellascilloside E-1	Nortriterpenoid oligoglycoside	Plant	Cytotoxic (antitumor)
Scoparic acid A	Labdane-type diterpenoid acid	Plant	β-Glucuronidase inhibitor
Scutebarbatines C–F	neo-Clerodane diterpenoid alkaloids	Plant	Cytotoxic
Scutebarbatines G and H	neo-Clerodane diterpenoid alkaloids	Plant	Cytotoxic
Scutebarbatines I, J, K, and L	neo-Clerodane diterpenoid alkaloids	Plant	Cytotoxic
Senegins II and III	Triterpenoid saponins	Plant	Hypoglycemic
Sesquiterpene hydroperoxides 1–3	Sesquiterpenoids (sesquiterpene hydroperoxides)	Plant	Trypanocidal
Sesterstatins 1–3	Sesterpenoids	Marine sponge	Antineoplastic; antibacterial
Sibiskoside	Monoterpene glycoside	Plant	Antiobestic
Simalikalactone D	Quassinoid	Plant	Antimalarial; antiviral, antitumor
Sinularia glycoside	Steroidal glycoside	Soft coral	Spermatostatic
Sinularolides B–E	Cembranoids	Soft coral	Antitumor
Stachsterol	Steroid	Plant	Cytotoxic
Strongylin A	Sesquiterpene hydroquinone derivative	Marine sponge	Antiviral; cytotoxic
Stronglyophorine-26	Diterpene (meroditerpenoid)	Marine sponge	Anticancerous (inhibitor of cancer cell invasion)
Suaveolindole	Indolosesquiterpene	Plant	Antibacterial
Suberosol	Lanostane-type triterpenoid	Plant	Anti-HIV
Tabucapsanolide A and its hydroxy derivatives	Steroidal lactones	Plant	Cytotoxic (anticancerous)
Tanghinin and deacetyltanghinin	Cardenolide glycosides	Plant	Cytotoxic
Tasumatrols E, F, I, J, and K	Taxane diterpenoids	Plant	Anticancerous (antitumor)
Tetrahydroxysqualene	Terpenoid	Plant	Antimycobacterial
Tolypodiol	Diterpenoid	Cyanobacteria	Anti-inflammatory
Torvoside H	Steroidal glycoside	Plant	Antiviral

(*Continued*)

Table A.5 (Continued)

Compound	Type	Origin	Bioactivity
Tripteriforden	Diterpenoid (kaurane-type diterpene lactone)	Plant	Anti-HIV
Triptonines A and B	Sesquiterpene pyridine alkaloids	Plant	Anti-HIV
Tsugarioside C	Triterpene (lanostanoid)	Fungus	Cytotoxic
Tubocapsanolide F	Steroidal lactone	Plant	Cytotoxic (anticancerous)
Tubocapsenolide A	Steroidal lactone	Plant	Cytotoxic (anticancerous)
Variecolin	Sesquiterpenoid	Fungus	Angiotensin II receptor binding inhibitor; immunosuppressive; anti-HIV
Verrucoside	Terpenoid (pregnane glycoside)	Gorgonian (sea whip/sea fan)	Cytotoxic
Vitex norditerpenoids 1 and 2	Norditerpenoids	Plant	Trypanocidal
Weigelic acid	Triterpenoid	Plant	Anticomplementary
Withanolide D and its hydroxy derivative	Steroidal lactones	Plant	Cytotoxic (anticancerous)
Withanolides	Steroidal lactones	Plant	Cytotoxic (anticancerous)
Xenovulene A	Sesquiterpenoid	Fungus	Inhibitor of flunitrazepam binding to the GABA-benzodiazepine receptor
Xylocarpin J	Limonoid	Plant	Cytotoxic (antitumor)
Ys-II and Ys-IV	Steroidal glycosides (spirostanol glycosides)	Plant	Antifungal
Yuccalan	Steroidal saponin	Plant	Antifungal
Zaluzanin D	Terpenoid (guaianolide derivative)	Plant	Antitrypanosomal
Zerumbone	Sesquiterpenoid	Plant	NO production inhibitor; anti-inflammatory; antitumor; anticancer; antiproliferative; anti-HIV; fungitoxic
Zerumboneoxide	Sesquiterpenoid	Plant	NO production inhibitor
Zhankuic acids A, B, and C	Triterpenes (steroidal acids)	Fungus	cytotoxic; anticholinergic and antiserotonergic; anti-inflammatory

Table A.6 Xanthonoids

Compound	Type	Origin	Bioactivity
1,2,5-Trihydroxyxanthone	Simple xanthone	Plant	Antioxidant
1,3,5,6-Tetrahydroxy-4,7,8-tri(3-methyl-2-butenyl)xanthone	Prenylated xanthone	Plant	Nerve growth factor (NGF) stimulator
5-O-Demethylpaxanthonin	Prenylated xanthone	Plant	Antioxidant
5-O-Demethylpaxanthonin and its 6-deoxy derivative	Cyclopentanylxanthones	Plant	Free radical scavengers (antioxidants)
6-O-Methyl-2-deprenylrheediaxanthone B	Furanoxanthone	Plant	Antioxidant
7-Methoxydeoxymorrellin	Caged xanthone	Plant	Cytotoxic
Allanxanthone C	Prenylated xanthone	Plant	Antiplasmodial; cytotoxic
Bellidifolin	Simple xanthone	Plant	Hypoglycemic; antioxidant; mutagenic; acetylcholinesterase (AChE) inhibitor
Bellidin (desmethylbellidifolin)	Simple xanthone	Plant	Antioxidant; cardioprotective; acetylcholinesterase (AChE) inhibitor; antitubercular
Blancoxanthone	Pyranoxanthone	Plant	Antiviral
Brasixanthones B, C, and D	Prenylated xanthones	Plant	Anticancerous
Celebixanthone and 5-O-methylcelebiaxanthone	Prenylated xanthones	Plant	Cytotoxic; antimalarial
Chaetoxanthones A, B, and C	Chromanoxanthones	Fungus	Antiprotozoal
Cochinchinone A	Prenylated xanthone	Plant	Cytotoxic; antioxidant
Cochinchinone B	Prenylated xanthone	Plant	Antioxidant
Cochinchinone C	Prenylated xanthone	Plant	Cytotoxic; antimalarial; antioxidant
Cratoxyarborenones A, B, C, D, E, and F	Prenylated xanthones	Plant	Cytotoxic
Cudrania xanthone	Prenylated xanthone	Plant	Cytotoxic
Decussatin	Simple xanthone	Plant	Antioxidant, vascular activator
Desmethylbellidifolin	Simple xanthone	Plant	Antioxidant; cardioprotective; acetylcholinesterase (AChE) inhibitor; antitubercular
Dicerandrols A, B, and C	2,2'-Dimeric tetrahydroxanthones	Fungus	Antibacterial; cytotoxic

(Continued)

Table A.6 (Continued)

Compound	Type	Origin	Bioactivity
Ehretianone	Quinonoid xanthone	Plant	Antisnake venom
Garciniaxanthones F, G, and H		Plant	Antioxidant
Gaudichaudic acids F, G, H, and I	Polyprenylated heptacyclic xanthonoids	Plant	Cytotoxic
Hypericum xanthone	Prenylated xanthone	Plant	Antioxidant
Hyperixanthone A	Prenylated xanthone	Plant	Antibacterial
Hyperxanthones C and E	Prenylated and chromanoxanthones	Plant	Antitumor
Jacarelhyperols A and B	Bisxanthones	Plant	PAF antagonist (vascular activity)
Macluraxanthone B and C	Prenylated tetraoxygenated xanthones	Plant	Anti-HIV, cytotoxic
Mangiferin	C-Glucoxanthone	Plant	Antidiabetic; antihyperlipidemic; antiatherogenic; antitubercular; antioxidant; anti-inflammatory; immunomodulatory; antitumor, anticancerous, analgesic; hepatoprotective; cardioprotective; antiviral
Mangiferin-7-O-β-glucoside	Xanthone glycoside	Plant	Antidiabetic
Methylbellidifolin	Simple xanthone	Plant	Hypoglycaemic; antimalarial; antihematopoietic; hepatoprotective; chemopreventive against photosensitized DNA damage; mutagenic; vascular activity
Monodictysins B and C	Hexahydroxanthones	Fungus	Anticancerous
Morellic acid and 8,8a-epoxymorellic acid	Prenylated xanthones	Plant	Cytotoxic; antibacterial; anti-HIV
Nidulalin A (F390)	Dihydroxanthone derivative	Fungus	Antitumor
Norathyriol	Simple tetraoxygenated (aglycon of magiferin)	Plant	Enzyme inhibitor; anti-inflammatory; antioxidant; analgesic
Norswertianolin	Simple xanthone	Plant	Antioxidant; cardioprotective; acetylcholinesterase (AChE) inhibitor; antitubercular
Pancixanthone B		Plant	Antileishmanial
Phomoxanthones A and B	Bixanthonolignoids	Fungus	Antimalarial; antitubercular; cytotoxic
Pyranojacaeubin	Pyranoxanthone	Plant	Antiviral

Pyranoxanthones 1 and 2	Pyranoxanthone	Plant	Cytotoxic
Rheediaxanthone A	Prenylated xanthone	Plant	Antiviral
Rubraxanthone		Plant	Cytotoxic; anticancerous; platelet-activating factor (PAF) inhibitor; antimicrobial
Sch54445	Polycyclic xanthone	Bacteria	Antifungal
Swerchirin	Simple xanthone	Plant	Hypoglycaemic; antimalarial; antihematopoietic; hepatoprotective; chemopreventive against photosensitized DNA damage; mutagenic; vascular activity
Swertianolin	Simple xanthone	Plant	Hypoglycemic; antioxidant; mutagenic; acetylcholinesterase (AChE) inhibitor
Swertifrancheside	Flavone-xanthone dimer (having C-glucosidic linkage)	Plant	Anti-HIV; human DNA ligase I (hLI) inhibitor
Swertipunicoside	Bisxanthone C-glycoside	Plant	Anti-HIV
Tetrahydroswertianolin	Simple xanthone	Plant	Hypoglycemic; antioxidant; mutagenic; acetylcholinesterase (AChE) inhibitor
Vieillardixanthone	Prenylated xanthone	Plant	Antioxidant
Yahyaxanthone	Chromenoxanthone	Plant	Cytotoxic

Table A.7 Miscellaneous

Compound	Type	Origin	Bioactivity
(−)-Calanolide B	Dipyranocumarin	Plant	Anti-HIV
(+)-Calanolide A	Dipyranocumarin	Plant	Anti-HIV
(+)-α-Viniferin	Stilbene trimer	Plant	Acetylcholinesterase (AChE); inhibitor anti-inflammatory
(10E,12Z,15Z)-9-Hydroxy-10,12,15-octadecatrienoic acid methyl ester	Unsaturated hydroxyl fatty acid ester	Plant	Anti-inflammatory
(3R,4R)-(−)-6-Methoxy-1-oxo-3-n-pentyl-3,4-dihydro-1H-isochromen-4-yl acetate	Dihydroisocoumarin	Plant	Antifungal; aromatase inhibitor; antiproliferative
(3R,4S,1′R)-3-(1′-hydroxyethyl)-4-methyldihydrofuran-2(3H)-one	Furanone derivative	Plant	Endoplasmic reticulum (ER) stress protector
(5R)-5-Hydroxy-7-(4″-hydroxy-3″-methoxyphenyl)-1-phenyl-3-heptanone	Diarylheptanoid	Plant	Cytotoxic
(5R)-7-(4″-hydroxy-3″-methoxyphenyl)-5-methoxy-1-phenyl-3-heptanone	Diarylheptanoid	Plant	Cytotoxic; inhibitor of PG biosynthesis (anti-inflammatory)
(R)-4″-Methoxydalbergione	3,3-Diarylpropene	Plant	Antimalarial
1-(3′,5′-Dihydroxyphenoxy)-7-(2″,4″,6-trihydroxyphenoxy)-2,4,9-trihydroxydibenzo-1,4-dioxin	Phlorotannin	Brown alga	Antioxidant (radical scavenging)
1-(4′-Hydroxy-3′-methoxyphenyl)-7-phenyl-3-heptanone	Diarylheptanoid	Plant	Inhibitor of PG biosynthesis (anti-inflammatory)
1,7-Bis(4-hydroxyphenyl)-1,4,6-heptatrien-3-one	Diarylheptanoid	Plant	Antioxidant
1,7-Bis(4-hydroxyphenyl)-2,4,6-heptatrienone	Diarylheptanoid	Plant	Antioxidant
15-Hydroxy-tetracosa-6,9,12,16,18-pentaenoic acid	Oxylipin	Soft coral	Antitube-forming (antiangiogenic agent; anticancerous)
19-(2-Furyl)nonadeca-5,7-diynoic acid	2-Substituted furan	Plant	Antimalarial; antiviral
1-Hydroxy-5-phenyl-3-pentanone	Hydroxy-pentanone derivative	Mushroom	Endoplasmic reticulum (ER) stress protector
2-(2,4-Dihydroxy-5-prenylphenyl)-5,6-methylenedioxybenzofuran	Arylbenzofuran	Plant	Cytotoxic
2-(3,4-Dihydroxyphenyl)-ethyl-O-β-D-glucopyranoside	Phenylenthanoid glucoside	Plant	Antioxidant (radical scavenger)
2,5-Dihydroxy-3-methanesulfinylbenzyl alcohol	Phenolic derivative	Fungus	Antibacterial
3,4-Dihydroxy-5-prenylcinnamic acid	Cinnamic acid derivative	Propolis	Antioxidant
3,5-Dihydro-2-(1′-oxo-3′-hexadecenyl)-2-cyclohexen-1-one	Polyketide	Plant	Antitumor (antiproliferative)

Compound	Type	Source	Activity
3,5-Di-O-caffeoyl quinic acid	Phenolic compound	Plant; propolis	Hepatoprotective; cytotoxic
3-Geranyl-2,4,6-trihydroxybenzophenone	Benzophenone derivative	Plant	Antileishmanial
3-Hydroxymethyl-4-methylfuran-2(5H)-one	Furanone derivative	Mushroom	Endoplasmic reticulum (ER) stress protector
4-Senecioyloxymethyl-6,7-dimethoxycoumarin	Coumarin	Plant	Antiangiogenic
4,4′-Dihydroxybenzyl sulfone	Phenolic compound	Plant	Inhibitor to platelet aggregation
4-Methylaeruginoic acid	Phenylthiazoline derivative	Bacteria	Cytotoxic (antitumor)
4′-O-Demethylknipholone-4′-O-β-D-glucopyranoside	Phenylanthraquinone glycoside	Plant	Antiplasmodial; antitrypanosomal
4′,5′-Dihydro-11,5′-dihydroxy-4′-methoxytephrosin	Rotenoid	Plant	Quinone reductase inducer
5- and 8-Hydroxy-2-(1′-hydroxyethyl)naphtho-[2,3-b]-furan-4,9-diones	Furanonaphthoquinones	Plant	Antimalarial, antitumor, cytotoxic
5,5′-Dibuthoxy-2,2′-bifuran	Bifuran derivative	Plant	Inhibitor of hACAT-1/hACAT-2
5-Methoxy-8-(2′-hydroxy-3′ buthoxy-3′-methylbutyloxy)-psoralen	Furanocoumarin	Plant	Inhibitor of COX-2 and 5-LOX (anti-inflammatory)
6,6′-Bieckol	Phlorotannin	Marine brown alga	Antioxidant (radical scavenging)
6″,7″-Dihydro-5′,5″-dicapsaicin	Substituted amide (capsaicinoid dimer)	Plant	Antioxidant
6′-Coumaroyl-1′-O-[2-(3,4-dihydroxyphenyl)ethyl]-β-D-glucopyranoside	Phenolic glycosidic derivative	Plant	Antioxidant
6-Desmethyl-N-methylfluvirucin A1 and N-methylfluvirucin A1	Macrolactums	Actinomycete	Anthelmintic
6-n-Pentyl-α-pyrone	(-Pyrone derivative	Fungus	Tyrosinase inhibitory activity; antibiotic
6-O-Galloyl-D-glucose	D-Glucose derivative	Plant	Antihypertensive
7-(4″-Hydroxy-3″-methoxyphenyl)-1-phenyl-4E-hepten-3-one	Diaryheptanoid	Plant	Cytotoxic
7-Butyl-6,8-dihydroxy-3(R)-pent-11-enylisochroman-1-one	Dihydroisocoumarin	Fungus	Antimalarial; antitubercular; antifungal
7-Epiclusianone	Tetraprenylated benzophenone derivative	Plant	Antiallergic; antibacterial; trypanosomicidal; vasodilator
7-Hydroxy-5-hydroxymethyl-2H-benzo[1,4]thiazin-3-one	Phenolic derivative	Fungus	Antibacterial
7-Methoxypraecansone B	Coumarin	Plant	Quinone reductase inducer
8-Acetoxy-4,10-dihydroxy-2,11(13)-guaiadiene-12,6-olide	Guaianolide	Plant	Cytotoxic

(Continued)

Table A.7 (Continued)

Compound	Type	Origin	Bioactivity
8'-*epi*-Cleomiscosin A	Coumarinolignoid	Plant	Tyrosinase inhibitor
Aculeatol E	Dioxadispiroketal-type derivative	Plant	Anticancerous
Aglafolin	Benzofuran	Plant	Cytotoxic (antitumor); paf-antagonist; inhibitor of platelet aggregation; antifungal; insecticidal
Albibrissinoside B	Phenolic glycoside	Plant	Antioxidant
Alterporriols G and H	Alterporriol-type anthraquinoid dimers	Plant	Cytotoxic (antitumor); inhibitor of kinase enzyme activity
Alvaradoins E–N	Anthracenone C-glycosides	Plant	Antitumor; cytotoxic; antileukemic
Amarogentin	*seco*-Iridoid glycoside	Plant	Antiulcerogenic; antileishmanial, mutagenic
Amburoside A	Phenolic glucoside	Plant	Antimalarial; neuroprotective; antioxidant; hepatoprotective
Amomols A and B	Oxaspiroketal-type derivatives	Plant	Anticancerous
Ansaetherone	Tetrapetalone glycoside	Bacteria	Antioxidant (radical scavenger)
Antrocamphin A	Benzonoid	Fungus	Anti-inflammatory
Antroquinonol	Ubiquinone derivative	Fungus	Cytotoxic (anticancerous)
Aplysiallene	Bromoallene	Sea hare	NA,K-ATPase inhibitors
Ardimerin digallate	Dimeric lactone	Plant	Anti-HIV
Ardisiphenols A, B, and C	Alk(en)ylphenols	Plant	Antioxidant; cytotoxic
Asphodelin A and its 4'-O-glucoside	Coumarin derivatives	Plant	Antimicrobial (antibacterial and antifungal)
Aureoquinone	Naphthoquinone	Fungus	Protease inhibitor; antibacterial
Australifungin	Polyketide	Fungus	Antifungal; inhibitor of ceramide synthase (sphinganine *N*-acyltransferase); anti-HIV
Balsaminolate	1,4-Naphthaquinone derivative	Plant	Cyclooxygenase (COX)-2 inhibitor
Bauhinoxepins A and B	Dibenzoxepin derivatives	Plant	Antimycobacterial
Bauhinoxepin F	Dibenzoxepin derivative	Plant	Antimycobacterial; anti-inflammatory; antifungal; cytotoxic

Bauhinoxepin I	Dibenzoxepin derivative	Plant	Antimycobacterial; antimalarial; anti-inflammatory; cytotoxic
Bauhinoxepin J	Dibenzoxepin derivative	Plant	Antimycobacterial; antimalarial; cytotoxic
Benthophoenin		Bacteria	Antioxidant (free radical scavenger)
Beta- and gama-Sanshool	Aliphatic acid amides	Plant	Inhibitors of cholesterol acyltransferase (hACAT) activity
Beta-Lapachone	Quinone (*ortho*-naphphoquinonic compound)	Plant	Anticancer; antitumor; antiproliferative; antiviral; antibacterial; antifungal; antiparasitic; anti-inflammatory
Beta-Thujaplicin	Tropolone derivative	Plant	Antimicrobial; antitumor; anticancerous; iron chelating; inhibitor of TNF-α production; platelet-type 12-lipoxygenase inhibitor
Bis-(4-hydroxybenzyl) sulphide	Sulphur compound	Plant	Histone deacetylase (HDAC) inhibitor; antiproliferative (antitumor)
Bis(4-hydroxybenzyl)ether	Phenolic ether compound	Plant	Inhibitor to platelet aggregation
Biyouyanagin A	Spirolactone derivative	Plant	Anti-HIV; inhibitor for cytokines
Brevipolides A and B	5,6-Dihydro-α-pyrone derivatives	Plant	Cytotoxic
Brunfelsia saponin	Furostan-type saponin	Plant	Antileishmanial
Cadiyenol	Polyacetylene compound	Plant	Antitumor; NO production inhibitor
Caffeoylglycolic acid methyl ester	Caffeoyl acid derivative	Plant	Antioxidant (radical and superoxide anion scavenging)
Calanolides E2 and F	Pyranocoumarins	Plant	Anti-HIV
Callysponginol Sulfate A	Sulfated C_{24} acetylenic fatty acid	Plant	Membrane type 1 matrix metalloproteinase (MT1-MMP inhibitor) [anticancerous]
Calotropone	Pregnanone (lineolon-type) derivative	Plant	Cytotoxic
Cassiaside C_2	Naphthopyrone glycoside	Plant	Antiallergic
Citrafungins A and B	Polyketides	Plant	Geranylgeranyltransferase type I (GGTase I) inhibitor; antifungal

(*Continued*)

Table A.7 (Continued)

Compound	Type	Origin	Bioactivity
Citrifolinoside	Iridoid	Plant	Inhibitor of UV B-induced Activator Protein-1 (AP-1)
Colpol	Dibromo C_6-C_4-C_6 metabolite	Marine red alga	Cytotoxic
Cordyol C	Diphenyl ether	Fungus	Anti-HSV-1; cytotoxic; antimalarial; antimycobacterial
Curtisians A–D	p-Terphenyl derivatives	Fungus	Antioxidant
Cuscuta propenamides 1 and 2	Amides	Plant	α-Glucosidase inhibitors
Cyathusals A, B, and C	Polyketide-types	Mushroom	Antioxidant
Cylindol A	Biphenyl ether derivative	Plant	5-Lipoxygenase inhibitor
Cytonic acids A and B	p-Tridepsides	Fungus	Human cytomegalovirus (hCMV) protease-inhibitor (antiviral)
Cytosporic acid	Polyketide	Fungus	Anti-HIV
Decursinol	Coumarin	Plant	Acetylcholinesterase (AChE) inhibitor
Demethoxycurcumin	Diarylheptanoid	Plant	Cytotoxic; antitumor; antioxidant
Dendrocandins C, D, and E	Bibenzyl derivatives	Plant	Antioxidant
Desacetylpyramidaglain D	Flavagline (cyclopenta[bc]benzopyran type)	Plant	Antituberculosis; antiviral
Dicranin	Acetylenic derivative (unsaturated C_{18}-fatty acid derivative)	Moss	15-Lipoxygenase inhibitor; antimicrobial
Dieckol	Phlorotannin	Marine brown alga	Antioxidant (radical scavenging)
Diplosoma Ylidenes 1 and 2	Ylidenes	Marine sponge; ascidian	Antimicrobial; cytotoxic
Drummondins D, E, and F	Filicinic acid derivatives	Plant	Antimicrobial

Eckol	Phlorotannin	Marine brown alga	Antioxidant (radical scavenging)
Eckstolonol	Phlorotannin	Marine brown alga	Antioxidant (radical scavenger)
Embelin, 5-O-Methylembelin and 5-O-ethylembelin	1,4-Benzoquinonoids	Plant	Cytotoxic; anticancerous; antioxidant; antitumor; antimitotic; inhibition of hepatitis C virus protease; analgesic; antipyretic; anti-inflammatory; antifertile; NF-κB blocker; antimicrobial; wound healer; anthelmintic
Epigallocatechin-(4β → 8,2(→ O-7)-epicatechin	A-type proanthocyanidin dimmer	Plant	Anti-HIV
Eugeniin	Tannin (ellagitannin)	Plant	α-Glucosidae and maltase activity inhibitor
Evolvoid A	Phenyl propanoid	Plant	Antistress
Flavaspidic acids AB and PB	Phloroglucinol derivatives	Plant	Antioxidant; antimicrobial
Fucodiphloroethol G	Phlorotannin	Marine brown alga	Antioxidant (radical scavenging)
Fuzanin D	Pyridine derivative	Actinomycete	WNT-5A expression inhibitor (thus acts as hair-growth stimulator); cytotoxic
Gaboroquinone A	Phenylanthraquinone	Plant	Antiplasmodial; antitrypanosomal
Gelsemiol 6'-trans-caffeoyl-1-glucoside	Iridoid glycoside	Plant	Nerve growth factor (NGF) stimulator
Gigantetronenin	Mono-tetrahydrofuran acetogenin derivative	Plant	Cytotoxic (antitumor)
Gigantrionenin	Mono-tetrahydrofuran acetogenin derivative	Plant	Cytotoxic (antitumor)
Girolline	2-Aminoimidazole derivative	Marine sponge	Antitumor; antimalarial
Glionitrin A	Diketopiperazine disulfide	Bacteria; fungus	Antimicrobial; antitumor
Glioperazine B	Dioxopiperazine	Fungus	Antibacterial
GTRI-02	Tetralone derivative	Actinomycete	Antioxidant (lipid peroxidation inhibitor)

(Continued)

Table A.7 (Continued)

Compound	Type	Origin	Bioactivity
Guangsangons K–N	Aromatics (Diels–Alder type adducts)	Plant	Antioxidant
Guttiferone A	Prenylated benzophenone	Plant	Anti-HIV; antileishmanial; anticholinesterase activity
Guttiferone F	Prenylated benzophenone	Plant	Anti-HIV; antileishmanial; anticholinesterase activity
Hinokitiol	Tropolone derivative	Plant	Antimicrobial; antitumor; anticancerous; iron chelating; inhibitor of TNF-α production; platelet-type 12-lipoxygenase inhibitor
Hydnellin A	Nitrogen-containing terphenyl	Inedible mushroom	Antioxidant
Hyemalosides A and B	Phenolic glycosides	Plant	Anti-HIV
Impatienolate	1,4-Naphthaquinone derivative	Plant	Cyclooxygenase (COX)-2 inhibitor
Indigofera acylphloroglucinol glycosides	Acylphloroglucinol glycosides	Plant	Lipoxygenase enzyme inhibitor
Integracins A, B, and C	Dimeric alkyl aromatics	Fungus	Anti-HIV
Integrastatins A and B	Tetracyclic aromatic heterocyles	Fungus	Anti-HIV
Interiotherins A and B	Furanocoumarins	Plant	Acetylcholinesterase (AChE) inhibitors
Isariotin F	Spirocyclic and bicyclic hemiacetal	Fungus	Antimalarial; antitubercular; antifungal; cytotoxic
Isodrummondin D	Filicinic acid derivative	Plant	Antimicrobial
Isoliquiritigenin	Coumarin (neoflavonoid)	Plant	Anticancerous; antitumor; anti-inflammatory; antioxidant; antimalarial; vasorelaxant; antihistaminic; antihyperpigmentation
Juglarins A and B	Diarylheptanoids	Plant	Cytotoxic
Kavapyrone	Pyrone derivative	Plant	Leishmanicidal

Kweichowenol B	Cyclohexene derivative	Plant	Antitumor
Laccaridiones A and B	ortho-Naphthoquinone derivatives	Mushroom	Protease inhibitors; antiproliferative
Litseaefoloside C	Phenolic glycoside	Plant	Enzyme (α-glucosidase and lipase) inhibitory
LMG-4	GD$_3$-type ganglioside molecular species	Starfish	Neuritogenic activity
Lucidumoside C	Secoiridoid glucoside	Plant	Antioxidant
Locoracemosides B and C	Benzylated glycosides	Plant	α-Chymotrypsin inhibitors
Luzonial A	Iridoid (iridoid aldehyde)	Plant	Cytotoxic
Luzonial B and luzonidial B	Iridoids (iridoid aldehyde	Plant	Cytotoxic
Lychnostatins 1 and 2	Germacranolides	Plant	Anticancerous
Macabarterin	Ellagitannin	Plant	Inhibitor to human neutrophil respiratory burst activity (anti-inflammatory)
Mallotophilippens A and B	Phloroglucinol derivatives	Plant	Inhibitors to NO production; anti-inflammatory; antiallergic
Mallotus benzopyrans 1 and 2	Benzopyrans	Plant	Cytotoxic
Marineosins A and B	Spiroaminals	Bacteria	Cytotoxic
Marsupsin	Phenolic compound	Plant	Antidiabetic
Melophlins P–S	Tetramic acid derivatives	Marine sponge	Cytotoxic
Methyl [2-(2,3-dihydroxy-3-methylbutyl), 4-amino] benzoate	Methylaminobenzoate	Fungus	Cytotoxic
Methyl 3,4-dihydroxy-5-(3′-methyl-2′-butenyl)benzoate	Benzoic acid derivative	Plant	Antiparasitic (leishmanicidal; trypanocidal; antiplasmodial)
Methyl 3,5-di-O-caffeoyl quinate	Phenolic compound	Plant	Hepatoprotective
Methyl brevifolincarboxylate	Benzopyran derivative	Plant	Inhibitor of platelet aggregation; vasorelaxant
Methyl rocaglate	Benzofuran	Plant	Cytotoxic (antitumor); FAF-antagonist; inhibitor of platelet aggregation; antifungal; insecticidal
Mimosifolenone	Aromatic compound (C$_{16}$styrylcycloheptenone derivative)	Plant	Cytotoxic
Motualevic acids A and F	Brominated long-chain acids	Marine sponge	Antibacterial

(Continued)

Table A.7 (Continued)

Compound	Type	Origin	Bioactivity
Multipolides A and B	10-Membered lactones	Fungus	Antifungal
Myriaporones 3 and 4	Polyketide-derived metabolites	Bryozoan	Cytotoxic
Myrothenone A	Cyclopentenone derivative	Fungus	Tyrosinase inhibitory activity
Myrsinoic acids A, B, C, and F	Hydroxybenzoic acid derivatives	Plant	Anti-inflammatory
Nepalensinols D–F	Stilbenoids (resveratrol oligomers)	Plant	Human topoisomerase II-inhibitor (antitumor)
NG-061	Phenyl acetic acid hydrazide derivative	Fungus	Otentiator of nerve growth factor (NGF)
Niruriside	Carbohydrate	Plant	Anti-HIV
Nocardione A	Tricyclic polyketide *ortho*-quinone antibiotic (naphtha[1,2-*b*]furan-4,5-dione derivative)	Bacteria	Cdc25B tyrosine phosphatase inhibitor; antifungal; cytotoxic
Obolactone	α-Pyrone	Plant	Cytotoxic
Obtusafuran	Arylbenzofuran	Plant	Antimalarial
Oceanalin A	Sphingolipid (tetrahydroquinoline-containing dimeric sphingolipid)	Marine sponge	Antifungal
Ochrocarpinones A, B, and C	Benzophenones	Plant	Cytotoxic
Ochrocarpins A–G	Coumarins	Plant	Cytotoxic
Onosmins A and B	Aromatic compounds (N-aryl aminobenzoic acid derivatives)	Plant	Lipoxygenase (LOX) enzyme inhibitor
Oryzafuran	2-Arylbenzofuran	Plant (rice ban)	Antioxidant
Oxypeucedanin hydrate acetonide	Coumarin derivative	Plant	Cytotoxic (antitumor)
Panaxynol and Panaxydol	Polyacetylene derivatives	Plant	Cytotoxic

Compound	Class	Source	Activity
Persenones A and B	Long chain esters	Plant	Antioxidants (inhibitors to production of NO and superoxide)
Petasiformin A	Phenylpropenoyl sulphonic acid	Plant	Antioxidant
Peucedanone	Coumarin	Plant	Acetylcholinesterase (AChE) inhibitor
Phelligridimer A	Pyrano-benzopyranone derivative (dimeric form)	Fungus	Antioxidant
Phelligridin G	Pyrano-benzzzopyranone derivative	Fungus	Antioxidant; cytotoxic
Phoyunbenes A–D	Stilbenes	Plant	Antioxidant (inhibitor to NO production)
Plastoquinones 1 and 2	2-Geranylgeranyl-6-methylbenzoquinone and its hydroquinone	Brown alga	Antioxidants
Pleosporone	Naphthoquinone derivative	Fungus	Antibacterial; cytotoxic
Plumbagin	Naphthoquinone derivative	Plant	Antimalarial; antimicrobial; anticancer; antitumor; cardiotonic; antifertile; anti-atherosclerotic; radio-sensitizing
PM-94128	Isocoumarin antibiotic	Bacteria	Antitumor
Polyhydroxylated cyclic sulphoxide	Polyhydroxylated cyclic sulphoxide	Plant	α-Glucosidase inhibitor
Porrigenic acid	Conjugated ketonic fatty acid	Mushroom	Cytotoxic
Primin	Benzoquinone	Plant	Antimicrobial; antifeedant; antineoplastic
Pterostilbene	Stilbenoid	Plant	Antidiabetic; anticancerous, antioxidant; antihypercholesterolemic, antihypertriglyceridemic; anti-inflammatory
Pulvinatal	Polyketide	Mushroom	Antioxidant
Pyrolaside B	Phenolic glycoside trimer	Plant	Antibacterial
Rhinacanthin C and D	Naphthoquinones	Plant	Antiviral; antiproliferative (antitumor)
Rhodiocyanoside A	Cyanoglycoside	Plant	Antiallergic
Rhuscholide A	Benzofuran lactone	Plant	Anti-HIV1
Rocaglamide	Benzofuran	Plant	Antileukemic; cytotoxic; antitumor; inhibitor of platelet aggregation; insecticidal

(Continued)

Table A.7 (Continued)

Compound	Type	Origin	Bioactivity
Sarcodonin delta	Nitrogen-containing terphenyl	Inedible mushroom	Antioxidant
Sarcophyton polyhydroxysterol	Polyhydroxysterol	Soft coral	Cytotoxic
Schweinfurthin E–H	Prenylated stilbenes	Plant	Antiproliferative
Scleropyric acid	Unsaturated aliphatic carboxylic acid	Plant	Antimycobacterial; antiplasmodial
Scroside D	Phenylenthanoid glycoside	Plant	Antioxidant (radical scavenger)
Scutianthraquinones A–C	Anthraquinones	Plant	Antiproliferative; antimalarial
Secoaggregatalactone A	Secobutanolide derivative	Plant	Cytotoxic
Seselidiol	Polyacetylene derivative	Plant	Cytotoxic (antitumor)
Silvestrol and *epi*-silvestrol	Rocaglate derivatives	Plant	Cytotoxic
Somocystinamide A	Disulphide dimer	Cyanobacteria	Cytotoxic
Sorocenols G and H	Oxygen heterocyclic Diels–Alder type adducts	Plant	Antibacterial; antifungal
Soulattrolide	Dipyranocumarin	Plant	Anti-HIV
Sphingosine	Sphingolipid	Marine sponge	Cytotoxic
Spiculoic acid A	Polyketide	Marine sponge	Cytotoxic
Subulatin	Caffeic acid derivative	Plant	Antioxidant
Surugapyrroles A and B	N-Hydroxypyrroles	Bacteria	Antioxidants (radical scavengers)
Ternstrosides A–F	Phenylethanoid glucosides	Plant	Antioxidants
Thomingianins A and B	Ellagitannins	Plant	Antioxidant
Tianshic acid and its methyl ester	Fatty acid and its ester	Plant	Alkaline phosphatase activity-inducer; antifungal
Torososide B	Anthraquinone glycoside	Plant	Antiallergic (inhibitor to leukotriene release)
Trans-2,3-epoxydeca-4,6,8-triyn-1-ol	Polyacetylenic compound	Fungus	Antifungal
Trans-2-hydroxyisoxypropyl-3-hydroxy-7-isopentene-2,3-dihydrobenzofuran-5-carboxylic acid	Prenylated dihydrobenzofuran derivative	Plant	Cytotoxic (anticancerous)

Trans-N-p-coumaroyl tryamine	Phenolic amide derivative	Plant	Cytotoxic; acetylcholinesterase (AChE) inhibitor
Trilobacin	Acetogenin derivative	Plant	Cytotoxic
Tungtungmadic acid	Chlorogenic acid derivative	Plant	Antioxidant
Urceolatin	Bromophenol derivative	Marine red alga	Antioxidant (free radical scavenging)
Vaccihein A	ortho-Benzoyloxyphenyl acetic acid ester	Plant	Antioxidant
Variecolorquinones A and B	Quinone derivatives (antibiotics)	Fungus	Cytotoxic (antitumor)
Vedelianin	Substituted cyclized geranyistilbene (hexahydroxanthene derivative)	Plant	Anticancerous
Verongamine	Bromotyrosine derivative	Marine sponge	Histamine H_3-antagonist
Zinolol	Hydroxyhydroquinone glycoside	Plant	Antimutagenic; antioxidant